Lecture Notes in Mechanical Engineering

Lecture Notes in Mechanical Engineering (LNME) publishes the latest developments in Mechanical Engineering—quickly, informally and with high quality. Original research reported in proceedings and post-proceedings represents the core of LNME. Volumes published in LNME embrace all aspects, subfields and new challenges of mechanical engineering. Topics in the series include:

- Engineering Design
- Machinery and Machine Elements
- Mechanical Structures and Stress Analysis
- Automotive Engineering
- Engine Technology
- Aerospace Technology and Astronautics
- Nanotechnology and Microengineering
- Control, Robotics, Mechatronics
- MEMS
- Theoretical and Applied Mechanics
- Dynamical Systems, Control
- Fluid Mechanics
- Engineering Thermodynamics, Heat and Mass Transfer
- Manufacturing
- Precision Engineering, Instrumentation, Measurement
- Materials Engineering
- Tribology and Surface Technology

To submit a proposal or request further information, please contact the Springer Editor of your location:

China: Ms. Ella Zhang at ella.zhang@springer.com
India: Priya Vyas at priya.vyas@springer.com
Rest of Asia, Australia, New Zealand: Swati Meherishi at swati.meherishi@springer.com
All other countries: Dr. Leontina Di Cecco at Leontina.dicecco@springer.com

To submit a proposal for a monograph, please check our Springer Tracts in Mechanical Engineering at http://www.springer.com/series/11693 or contact Leontina.dicecco@springer.com

Indexed by SCOPUS. All books published in the series are submitted for consideration in Web of Science.

More information about this series at http://www.springer.com/series/11236

José Machado · Filomena Soares ·
Justyna Trojanowska · Vitalii Ivanov
Editors

Innovations in Industrial Engineering

Springer

Editors
José Machado
Department of Mechanical Engineering
University of Minho
Guimarães, Portugal

Filomena Soares
Department of Industrial Electronics
University of Minho
Guimarães, Portugal

Justyna Trojanowska
Poznan University of Technology
Poznan, Poland

Vitalii Ivanov
Department of Manufacturing Engineering, Machines and Tools
Sumy State University
Sumy, Ukraine

ISSN 2195-4356 ISSN 2195-4364 (electronic)
Lecture Notes in Mechanical Engineering
ISBN 978-3-030-78169-9 ISBN 978-3-030-78170-5 (eBook)
https://doi.org/10.1007/978-3-030-78170-5

This Springer imprint is published by the registered company Springer Nature Switzerland AG
The registered company address is: Gewerbestrasse 11, 6330 Cham, Switzerland

Preface

This volume of Lecture Notes in Mechanical Engineering gathers selected papers presented at the first International Scientific Conference ICIE'2021, held in Guimarães, Portugal, on June 28–30, 2021. The conference was organized by School of Engineering of University of Minho, troughout MEtRICs and Algoritmi Research Centres.

The aim of the conference was to present the latest engineering achievements and innovations, and to provide a chance for exchanging views and opinions concerning the creation of added value for the industry and for the society. The main conference topics include (but are not limited to):

- Innovation
- Industrial Engineering
- Mechanical Engineering
- Mechatronics Engineering
- Systems and Applications
- Societal Challenges
- Industrial Property

The organizers received 213 contributions from 24 countries around the world. After a thorough peer review process, the committee accepted 126 papers written by 412 authors from 18 countries for the conference proceedings (acceptance rate of 59%), which were organised in three volumes of the Springer Lecture Notes in Mechanical Engineering.

This volume, with the title "Innovations in Industrial Engineering", specifically reports on current developments in the field of industrial engineering, with a special focus on research and industrial applications aimed at improving quality of processes and products, and contributing to sustainable economy. It also describes innovative technologies and strategies associated to industry 4.0., discussing new ways to improve industrial production and supply chain management by applying mathematical and computer methods. Last but not least, it analyses important issues relating to sustainability, education and collaborations between industry, universities and interface units, as well as national strategies for digitalisation. This book consists of 41 chapters, prepared by 136 authors from 11 countries.

Extended versions of selected best papers from the conference will be published in the following journals: Sensors, Applied Sciences, Machines, Management and Production Engineering Review, International Journal of Mechatronics and Applied Mechanics, SN Applied Sciences, Dirección y Organización, Smart Science, Business Systems Research and International Journal of E-Services and Mobile Applications.

A special thank to the members of the International Scientific Committee for their hard work during the review process.

We acknowledge all that contributed to the staging of ICIE'2021: authors, committees, and sponsors. Their involvement and hard work were crucial to the success of the ICIE'2021 conference.

June 2021

José Machado
Filomena Soares
Justyna Trojanowska
Vitalii Ivanov

Contents

Long-Term History Data and Integrated Expert Knowledge in Asset Lifecycle Management

Toni Ahonen(✉), Antti Rantala, Helena Kortelainen, and Pasi Valkokari

VTT Technical Research Centre of Finland, Tampere, Finland
toni.ahonen@vtt.fi

Abstract. Modern asset lifecycle management needs to deal with a combination of strategic level information for defining right allocations, timings and resources for investments, tactical level information for allocating maintenance tasks based on their criticality, maintenance windows and e.g. production schemes as well as operational level information for effective implementation of daily maintenance. There is a need for a process that better uses lifecycle data as an integral part of lifecycle decision-making and thus for maintenance and investment planning. In current paper, we focus on the aspects of strategic and tactical asset management where development targets are identified for maintenance actions and investments. Based on previous research and an interview study the paper proposes a process for maintenance and investment planning which utilizes a combination of expert knowledge and maintenance and process event data.

Keywords: Maintenance · Lifecycle · Data · Digitalization

1 Introduction

Data and knowledge driven planning processes are in the core of the asset lifecycle management. Availability of relevant data is crucial in order to make the best asset management decisions, but the long-term availability of asset information is a challenge [1]. The exploitation of gathered data and utilization of information management systems are challenges that require an integrated asset information management strategy [2]. It is generally recognized that various data, such as technical and financial data are needed, and they are usually scattered among separate information systems. Crespo et al. [1] propose that a common database should be developed where relevant data of all assets could be stored altogether to support decision making.

The data needed for asset life cycle management related decision making is currently stored and managed in different systems including the product data management systems (PDM, PLM) applied by the product vendor and computerized maintenance management systems (CMMS) that is utilised by the product user/owner. Manual data collection is typically conducted by maintenance personnel when handling work orders in a CMMS. The manually recorded data in a CMMS is often scare and has poor quality and thus the maintenance event history data seldom supports the hierarchical and systemic approach for the management of the assets [3, 4]. Further sources of failure data like diaries that

J. Machado et al. (Eds.): icieng 2021, LNME, pp. 1–11, 2022.
https://doi.org/10.1007/978-3-030-78170-5_1

operators use to share and transfer acute issues in their everyday work, are often in a format that also hinders the exploitation of the gathered data for instance in maintenance organisation [5]. The role of CMMS in the maintenance management and asset life cycle management could be significantly improved by investing in improving the data quality and quantity, and by integrating data sources. Kortelainen et al. [6] have presented a framework for systemic approach for asset management data refinement where the structure of the information produced is based on a thorough understanding of the asset hierarchy and the asset management decision-making. In addition to the need to have data scientists translating data into knowledge [7], companies need to define their strategic goals for data refinement [8] which should provide guidance on the selection of proper tools and methods.

Human expertise and tacit knowledge [9, 10] offer a valuable contribution to the asset lifecycle related decision-making. If companies can extract and exploit the tacit knowledge they have, it could improve their decision-making processes significantly [11]. Altogether, professional experience, tacit knowledge and recorded descriptions of earlier events - i.e. expert knowledge - are used for maintenance management and the management of the assets over their lifecycle. In today's industrial practices these largely manual processes often form the basis for maintenance planning - rather than effective data-driven processes. The primary challenges are the collection and integration of relevant data into an exploitable format throughout the levels of decision-making - at strategic, tactical and operational levels.

The research activities in the ongoing co-innovation project "Solid value from digitalisation in forest industry (SEED)" project [12] cover the asset lifecycle from engineering to operations and asset management. SEED-project aims at developing methods and tools for business driven maintenance and asset replacement decision making. In this paper, we study the challenges that the fragmented maintenance data create and the possibilities that properly managed maintenance data offer to asset managers. The objective of this paper is to provide a conceptual assessment framework that enables the utilisation of both history data and expert knowledge for replacement investment and maintenance development planning.

2 Asset Lifecycle Decisions

Strategic and tactical levels of asset management address the optimization of investment allocation, maintenance strategy and programme choices, and the timing and scope of the major maintenance activities such as replacements or shutdowns [13]. A further question is the optimization of the maintenance interval in terms of costs related to planned downtime and the risk of increase failure-related downtime. However, there is also a need to combine production and maintenance related knowledge together in order to cross-optimize the production schemes and planned maintenance shutdown and thus fully exploit the resulting maintenance windows. This endeavour draws from the idea of opportunistic replacement policies [14] and of Total Productive Maintenance (TPM) [15]. The need for cross-optimization leads in a complex optimization problem where data needs to be better used.

Decision-makers have to take a multi-objective approach to driving their decisions and have to weigh the preferences of different stakeholders during the decision-making process [1]. For example, maintenance and production organizations have to agree on the timing of the major maintenance activities and replacement investment plans. It is essential that in the asset management decisions making process also considers the impacts that the outcome may have on the value of the asset [16]. Rosqvist [17] states that maintenance has no intrinsic value but maintenance is performed to support the strategic objectives of the plant and the fundamental objectives of the company. Understanding of the value creating mechanisms is thus of particular importance.

In the industry, companies are generally aiming at reliable production systems with higher availability performance. Industrial production lines are composed successive process stages or sub-processes that might also be partly parallel, and storage capacity is placed between stages. Intermediate storages ensures a steady material flow through the whole system and the probability of production stops is reduced even disturbances or maintenance actions take place in some parts of the system. In order to direct the improvement measures effectively, the whole system should be studied systematically and the items that contribute most to the system unavailability, or items that are critical, should be identified As the asset systems are often complex, a reliability or availability performance model of the system is necessary to understand the impacts of such activities [18]. Often it is not feasible or necessary to model whole system on the same level of detail. Functional modelling methods (e.g. Structured Analysis and Design Technique, SADT and Integration Definition for Function Modeling, IDEF) offer a hierarchical top-down approach that is applicable also in the industrial context.

Rosqvist et al. [17] introduce "the definition of the fundamental, strategic and maintenance (means) objectives of the production plant" as a key phase in the determination of a maintenance programme for a plant. They also present the hierarchic relationship between a company's fundamental objectives, the strategic objectives at plant level, and the operational objectives of the maintenance organisation. This phase provides the maintenance organization with a profound reasoning for the selection of the most profitable maintenance strategies for the assets.

Gopalakrishnan et al. [19] state that manufacturing companies carry out criticality assessments but they are found ineffective for making maintenance decisions. The study showed that companies did not reach reliable results with the tools used and thus the right critical assets were not identified in practice. A data-driven approach with clear objectives for maximizing system productivity was proposed. Lopes et al. [20] propose a criticality assessment method with "rules for defining priorities for corrective and preventive maintenance tasks" and a second one to prioritize equipment based on its performance. The first method particularly focuses on asset and task criticality rather than purely addressing criticality of the equipment itself. This focus is based on the observation that criticality assessment of the equipment should be based on understanding the failure behaviour and thus understanding the failure mode criticality. The second method, on the other hand, aims to point out the assets with poor performance. The methods of Lopes et al. [20] thus target at managing the failure-related risks as well as management of the system performance. Overall, there has been a long tradition of applying qualitative criticality assessments for supporting long-term maintenance decision-making and developing methods for asset level assessments (see e.g. [21]).

Applications of RCM (see e.g. [17]) can be utilized to identify the most appropriate and cost-effective tasks of preventive maintenance. Lifecycle cost and profit analysis (LCC/LCP) may be utilized to evaluate the effects of each smaller or larger investment in maintenance. Quite typically, a coarse cost analysis may be applied when making a decision on each of the maintenance task in RCM [22].

Asset life cycle plans are to be created and reviewed annually, both for asset replacement and as required in the event of business developments. The age and condition of assets is to be monitored and taken into account in planning maintenance, replacements and other investments, so that issues of reliability and risk, and disruptions to business plans are minimized. However, the funding of such investments often poses constraints, and methods to allocate the resources based on factors such as condition and criticality are needed [13].

Development of the approaches to meet the challenges described above require a systematic formulation of the problems in terms of asset management strategy, asset criticality, maintenance programmes and the maintenance processes. Sharma et al. [23] state that "data problems still exist and the gap between theory and practice is quite wide". It is also argued that literature on integrating qualitative and quantitative approaches is still rather scarce. We conclude that there is a need for an approach that utilizes data from assets and exploits complementary expert judgment to formulate a thorough understanding of the system criticality from the perspective of failure behaviour and performance.

3 Research Design and Approach

The research applies qualitative research strategy to understand the current state of maintenance data and expert knowledge utilization in process industry when planning for asset replacement investments and maintenance development. The methodology development was based on the work carried out in previous projects (Table 1) and this work was supplemented by the insights collected in the ongoing SEED-project (Table 2).

Table 1. Research data from research projects and assignments

Project type and schedule	Method	Sources
Commercial assignments in different industries (2010–2019)	Asset replacement assessment process	[5]
Research project with industrial partners (MittaMerkki, 2015–2016)	Risk-based investment assessment	[24]
Co-innovation project with research organizations and industrial companies (SEED, 2020 - ongoing)		

The research data was collected by interviews from experts that are responsible for planning major shutdowns and who have also major role in replacement decision making. As companies in process industry have either outsourced their maintenance function or otherwise rely on the offering of external firms, experts from service providers were included in interviews. The interviews were aimed at providing insights on the practical decision making processes, challenges and solutions during the planning, execution and post-processing stage. Examples of specific interview questions include:

1. What are the key problems in asset management?
2. How do you measure your asset managements' success?
3. How much have you outsourced your asset management processes?
4. How do you prioritize your maintenance tasks?

Table 2. Research data collection in SEED-project

Company	Characterization of the interviewed persons
Process industry	Asset Management Manager Development Project Manager Maintenance Development Manager Maintenance Director Maintenance Manager Maintenance Support Manager Reliability Engineer Senior Manager, Reliability
Maintenance service provider	Process Owner, Reliability Reliability Engineer

The interview data was coded using a qualitative data analysis software Nvivo. Data triangulation, i.e. using several types of data sources from different actors, increases the reliability and validity of the research [25].

4 Findings from the Interviews

Below we present the key findings from our interviews from the viewpoint of challenges of data refinement in maintenance and investment planning. Data fragmentation is a major challenge in many organizations. Following is a comment from our interviewee regarding the problems with data:

"Data fragmentation is a challenge. We have many systems and huge amounts of data, but we don't know how to pick the right data to support our decision making."

Data management in overall is one of the main challenges in asset management. Especially availability of history data is a major problem. Crespo et al. [1] noted that long-term availability of asset information is a challenge. Interviews bolstered this argument, as one of the interviewees commented:

> *"Especially old assets' history data management is challenging. It is hard to get access into history data, because systems and employees have changed."*

Crespo et al. [1] proposed that a common database should be developed where relevant data of all assets could be stored altogether to support decision making. One of the interviewed companies had an ongoing project in which the goal is to link different databases into one place and utilize the data to make better asset management decisions. The interviewee commented:

> *"We have an ongoing project in which we try to process maintenance data into valuable information. We are trying to integrate many kinds of data sources into the same cloud platform and construct situation pictures for our workers"*

Also, poor quality of report data is a considerable problem. According to the interviews, maintenance management needs more detailed information on the failure modes and their mechanisms as well as the corrective actions and suggestions for the future. Thus, these expectations should be reflected to how to implement the structure of the maintenance task reporting and how the reporting is instructed. One interviewee said:

> *"We emphasize the importance of quality when we update information about defects into our systems. Instead of writing "pump is broken", there should be more detailed information concerning the problem. Comprehensive reports would help other workers' tasks."*

Sufficient report quality could result in better predictability of asset failures. Predicting failures is one of the most important areas that should be improved according to our interviews. Better predictability would reduce unnecessary maintenance and prevent unexpected asset failures. One interviewee commented:

> *"The most important area of development in asset management is predictability of failures. There is still a lot of work to do in order to predict failures early enough."*

Another common problem in asset management is that there is only a limited amount of money given for maintenance. Thus, asset management is continuous prioritizing what assets should be repaired next. Because of that criticality classification of assets and maintenance tasks plays major role in prioritizing. One of the interviewees said:

"We have done different types of criticality assessments for about 30 000 assets. ... You never get enough money to repair all assets. We prioritize maintenance tasks all the time."

5 Information Needs in Maintenance and Investment Planning

5.1 Maintenance Specific Challenges Related to Data Utilization

There are generic prioritized objectives for data utilization in asset management, such as prioritization and scheduling of the maintenance work, however, the challenges are often case specific and depend on the specific production environment and maintenance schemes. It seems that there is a great interest in developing better capabilities for maintenance planning with respect to lifecycle costs and profits, instead of pure cost optimization. This emphasizes the role of data and knowledge in the context of maintenance programme optimization. Criticality analyses are in common use as a part of this, however, there is a need for procedures that help to incorporate the asset criticality to the daily routines.

EAM and CMM-systems are still largely focused on supporting the daily maintenance management activities. If the lifecycle data is then managed in a number of systems, there is a risk of data being fragmented, poorly managed and also poorly exploited. Holistic management of the asset lifecycle data is thus a significant challenge. Refinement of data into actionable insight is challenging particularly at production line level while collection of relevant insight is better managed at asset level. Increased amount of data has not so far resulted in a step change of the efficiency but instead optimization and fine tuning of processes here and there.

According to the interviews, many of the current challenges are related to the quality of the event data caused by incomplete reporting of the failure, process disturbance, debottlenecking experiences, mistakes made, learnings and observations. Furthermore, all the interviewees referred to the fact that data and information are widespread in different information systems and many of them stated that all the relevant information should be available in one individual system. We conclude that instead of merely integrating information systems based on case-specific needs, the asset management processes described in terms of how data is utilized in different use cases, will be used as a guideline for integration. Maintenance event data is still inadequately utilized in the development of the maintenance activities. However, it can be concluded that utilization of the event data is more dependent of the proper processes and models being implemented rather than due to inadequate analytics methodologies. The processes of collecting data should be improved in order to more harmonized data in terms of maintenance terminology and format.

5.2 Data Utilization in Maintenance Prioritization

Prioritization of the maintenance activities relies too often and too much on the decisions made at operational level, with insufficient evidence collected and based on single events or biased opinions. Thus, no sufficient tools or models exist for prioritization and therefore experience-based and often biased selections are made. Prioritization is an example of an activity requiring cross-disciplinary collaboration and information sharing.

The interviews revealed that since no adequate evidence exists that could support strong prioritization, maintenance resources are not optimally allocated. The interviews thus support the conclusion that there is a need for an approach for maintenance and investment planning that is data-driven and utilizes the expert knowledge effectively.

6 Concept for Hierarchical Replacement and Maintenance Investment Elaboration

We could argue that there is no "world-class maintenance". Instead, best practices depend on the equipment to be maintained and the production environment. We consider asset management as an integrated approach, where capacity, investments, operation and upkeep of assets are optimized in the specific environment.

There is a need to thoroughly understand the failure behavior of the assets and to analyze the impacts of different maintenance strategies in order to select the suitable and most cost-effective items for the maintenance programme. It is concluded that there is a need for a process that better uses lifecycle data as an integral part of lifecycle decision-making. Here we focus on the aspects of strategic and tactical asset management where development targets are identified for maintenance actions and investments.

A hierarchical approach is proposed in Fig. 1 where the following key process elements address the effective combination of expert knowledge and maintenance and process event data. Criticality assessment and dependability analysis are proposed as a dynamic part of the process where expert knowledge and event data is continuously used for updates. Timing of the updates of the maintenance programme or updates of the maintenance resource plan as well as the definition of the investment plan for the future are defined by the organization and they should be supported by proper tools.

Here the first two phases are used as input information and a particular analysis for maintenance resource allocation and investment planning are proposed.

Criticality Assessment: Criticality assessments are widely implemented in industry, however, their utilization as an integral part of asset lifecycle decision-making is still limited. In this process, criticality assessment can be done hierarchically so that the results can be used as a guideline for allocating resources for more detailed assessments. The results of the criticality assessment can, in many cases, be used directly but the results can be complemented with the help of other analysis methods. Data-based approaches are preferred, however, the availability of relevant data typically guides the balance between use of asset data and expert judgment in making the assessments.

Hierarchical Data-Based Dependability Analysis: The hierarchical approach continues in the availability performance analyses where the failure behaviour of the systems is more thoroughly analysed than in the criticality assessment phase. The failure behaviour of production systems is affected by various factors, such as the lifecycle phase of the production equipment, maintenance and investment history, stress and process conditions (for instance, temperature and humidity) and overloads. This information makes it

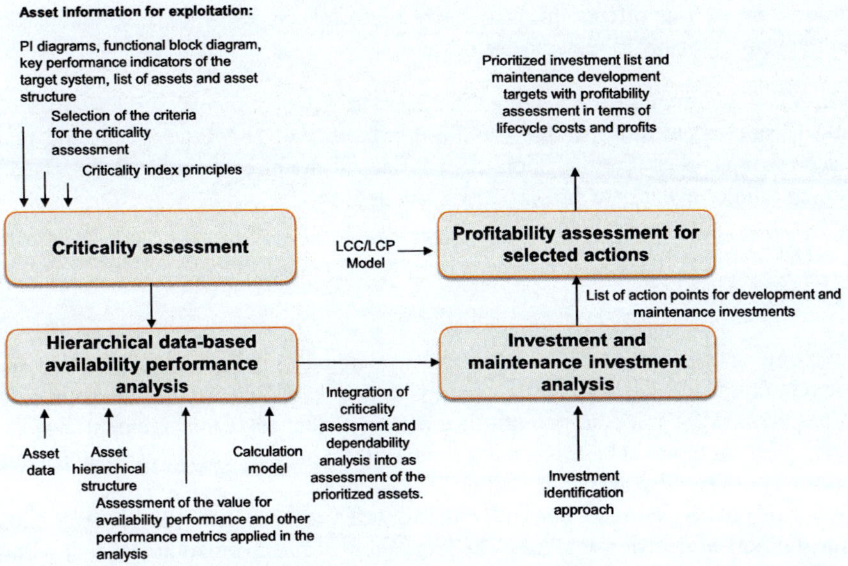

Fig. 1. A process for hierarchical asset assessment for investment and maintenance development target planning.

possible to plan a balanced maintenance programme. It is known that failure analyses can be very resource intensive, however, the hierarchical approach aims to invest most resources to the most significant system parts so that the results from upper level analyses are used to make focusing decisions on the lower levels.

Investment and Maintenance Investment Analysis: RCM decision logic can be used for reviewing the asset specific task planning and this phase results in the most profitable selections for the on-going maintenance programme of the system. However, the needs for larger investments are not considered in this analysis. Therefore, an investment analysis is carried out where identification and reasoning of the investments are made. This phase should integrate cross-disciplinary information gathered from operations and maintenance to assess, for instance, lifecycle phase and remaining lifetime of the assets, effects of the investments on the failure behaviour and reduced risks. In practice, we propose that the hierarchical data-based dependability analysis and a synthesis of the most significant items in criticality assessment is used together with a list of investment proposals from operations and maintenance functions as a basis for the identification and assessment phases. This ensures a more thorough analysis of the failure-related risk or expected bottlenecks. The investments should then be analysed with respect to their most profitable scope. Thus, the related functions and system parts are also analysed with respect to their criticality, expected failure behaviour, capacity performance and lifecycle phase. A coarse and semi-quantitative cost-based assessment can be used at this stage, where the costs and impacts are analysed. In addition to the optimized scope, the analysis should include estimates on the investment costs, effects of investment shut-down as well as effects on availability performance, capacity, maintenance and other relevant case-specific operations costs.

Profitability Assessment for Selected Actions: A lifecycle cost and profit model should be applied in order to analyse the profitability of each investment and maintenance resource allocation decision. Assessment of the effects can be done based on the integration of event data and expert knowledge.

7 Conclusions

Technology development and companies' investments in digital technologies have resulted in rapid increase in the amount and variety of available data in process industry. This has created a significant potential for more efficient asset management, however, the stepwise development based on the technological advancements have created a range of systems criticized by users due to their inefficiency.

A lot of challenges have been identified related to how data can be exploited as an integral part of decision-making. The large number of information systems without effective integration and fragmented and unmanageable data have been identified as the primary generic challenges.

Current paper proposes a process that could tie together expert knowledge and the different pieces of information produced from data and use that according to the asset hierarchy. In practice, paper presents the steps needed to compose an understanding of the criticality and failure behavior of the system and finally to make a coherent plan of development actions regarding maintenance and maintenance investments.

References

1. Crespo Márquez, A., Macchi, M., Parlikad, A.K. (eds.): Value Based and Intelligent Asset Management. Springer, Cham (2020). https://doi.org/10.1007/978-3-030-20704-5
2. Ouertani, M., Parlikad, A., McFarlane, D.: Towards an approach to select an asset information management strategy. Int. J. Comput. Sci. Appl. **5**(3), 25–44 (2008)
3. Betz, M.: The secret life of machines – boundary objects in maintenance, repair and overhaul. In: Floréen, P., Krüger, A., Spasojevic, M. (eds.) Pervasive 2010. LNCS, vol. 6030, pp. 174–191. Springer, Heidelberg (2010). https://doi.org/10.1007/978-3-642-12654-3_11
4. Mahlamäki, K., Nieminen, M.: Analysis of manual data collection in maintenance context. J. Qual. Maint. Eng. **26**(1), 104–119 (2019). https://doi.org/10.1108/JQME-12-2017-0091
5. Kortelainen, H., Komonen, K., Laitinen, J., Valkokari, P., Hanski, J.: Knowledge-based life cycle management. Kunnossapitoyhdistys Promaint ry. 241 p. (2021). https://cris.vtt.fi/en/publications/tiet%C3%A4mysperusteinen-elinjakson-hallinta (in Finnish)
6. Kortelainen, H., Kunttu, S., Valkokari, P., Ahonen, T.: Asset management decisions—based on system thinking and data analysis. In: Tse, P.W., Mathew, J., Wong, K., Lam, R., Ko, C.N. (eds.) Engineering Asset Management - Systems, Professional Practices and Certification. LNME, pp. 1083–1093. Springer, Cham (2015). https://doi.org/10.1007/978-3-319-09507-3_92
7. McAfee, A., Brynjolfsson, E.: Big data: the management revolution. Spotlight on big data. Harvard Bus. Rev. **90**, 60–68 (2012)

8. Barton, D., Court, D.: Making advanced analytics work for you – a practical guide to capitalizing on big data. Spotlight on big data. Harvard Bus. Rev. **90**, 78–83 (2012)
9. Nonaka, I., Takeuchi, H.: The Knowledge-Creating Company. Oxford University Press, Oxford (1995)
10. Redding, L.E., Hockley, C.J., Roy, R., Menhen, J.: The role of maintenance, repair, and overhaul (MRO) knowledge in facilitating service led design: a nozzle guide vane case study. In: Amadi-Echendu, J., Hoohlo, C., Mathew, J. (eds.) 9th WCEAM Research Papers. LNME, pp. 379–395. Springer, Cham (2015). https://doi.org/10.1007/978-3-319-15536-4_31
11. Amadi-Echendu, J., de Smidt, M.: Integrating tacit knowledge for condition assessment of continuous mining machines. In: Amadi-Echendu, J., Hoohlo, C., Mathew, J. (eds.) 9th WCEAM Research Papers. LNME, pp. 119–130. Springer, Cham (2015). https://doi.org/10.1007/978-3-319-15536-4_10
12. Camarinha-Matos, L.M., Afsarmanesh, H., Ortiz, A. (eds.): PRO-VE 2020. IAICT, vol. 598. Springer, Cham (2020). https://doi.org/10.1007/978-3-030-62412-5
13. Hastings, N.A.J.: Physical Asset Management With an Introduction to ISO55000. Springer, Cham (2015). https://doi.org/10.1007/978-3-319-14777-2
14. McCall, J.J.: Operating characteristics of opportunistic replacement and inspection policies. Manage. Sci. **10**(1), 85–97 (1963)
15. Nakajima, S.: Introduction to Total Productive Maintenance (TPM). Productivity Press, Portland (1988)
16. ISO 55000–2. Asset Management - Overview, principles and terminology', BSI. https://doi.org/10.1007/978-3-7908-2720-0_12 (2014)
17. Rosqvist, T. Laakso, K., Reunanen, M.: Value-driven maintenance planning for a production plant. Reliab. Eng. Syst. Saf. **94**(1), 97–110 (2009)
18. Kortelainen, H., Salmikuukka, J., Pursio, S.: Reliability simulation model for systems with multiple intermediate storages. In: Annual Reliability and Maintainability Symposium. 2000 Proceedings. International Symposium on Product Quality and Integrity, 8th edn. IEEE Institute of Electrical and Electronic Engineers. Proceedings of the Annual Reliability and Maintainability Symposium, vol. 2, pp. 65–70 (2000). https://doi.org/10.1109/RAMS.2000.816285
19. Gopalakrishnan, M., Skoogh, A., Salonen, A., Asp, M.: Machine criticality assessment for productivity improvement. Int. J. Product. Perform. Manag. **68**(5), 858–878 (2019)
20. Lopes, I.S., Figueiredo, M.C., Sá, V.: Criticality evaluation to support maintenance management of manufacturing systems. Int. J. Ind. Eng. Manage. **11**(1), 3–18 (2020)
21. Crespo, A. Moreu de Leon, P. Gomez Fernandez, J.F. Parra Marquez, C., Lopez Campos, M.: The maintenance management framework: a practical view to maintenance management. J. Qual. Maintenance Eng. **15**(2), 167–178 (2009)
22. Kunttu, S., Ahonen, T., Kortelainen, H., Reunanen, M.: An approach to improve the cost-effectiveness of a maintenance programme. In: Proceedings of the 2nd World Congress on Engineering Asset Management (EAM), pp. 1109–1116 (2007)
23. Sharma, A., Yadava, S.G., Deshmukh, S.G.: A literature review and future perspectives on maintenance optimization. J. Qual. Maint. Eng. **7**(1), 5–25 (2011)
24. Räikkönen, M., et al.: Towards risk-conscious investment decision-making and value creation. VTT Technical Research Centre of Finland. VTT Technology No. 282 (2017). http://www.vtt.fi/inf/pdf/technology/2017/T282.pdf
25. Patton, M.Q.: How to Use Qualitative Methods in Evaluation. Sage, Newbury Park (1987). https://doi.org/10.1017/CBO9781107415324.004

Analysis of the Impact of Changeover Time and Priority Rules on the Timely Execution of Customer Orders

Paulina Rewers[1(✉)], Anna Karwasz[1], and Marta Czaja[2]

[1] Poznan University of Technology, M. Sklodowskiej-Curie 5 sq., 60-965 Poznan, Poland
paulina.rewers@put.poznan.pl

[2] Intitek Polska Sp. z o. o., Baraniaka 88C/Building D Street, 61-131 Poznan, Poland

Abstract. The article is the preliminary research results determining the influence of selected criteria on the timeliness of customer orders, the workstation load and the required size of the finished products warehouse. The criteria included in this paper are the changeover time and the priority rule. The literature analysis covers issues related to the priority rules and machine changeover. The following part of the work presents the adopted research methodology, production process model and input data. Production plans were prepared successively and 9 model variants were designated, with different changeover times and priority rules applied. The work also simulated the production process in the FlexSim program, using previously prepared production plans. For the evaluation of the models, a summary of the obtained results was taken into account, showing the timely execution of customer orders, the workstation load and the required size of the finished products warehouse. The article ended with conclusions, in which a variant was indicated, in which a combination of factors allows to achieve the highest punctuality of order fulfillment, the smallest size of the warehouse and the optimal load of workstations (the shortest inactivity, the longest implementation of the production process).

Keywords: Priority rules · Production scheduling · Computer simulation

1 Introduction

In the era of constantly growing customer requirements and increasing competition, production companies must constantly improve the quality of manufactured products, reduce production costs and increase the timeliness of orders. Unstable demand causes frequent changes in the production schedule, which may, on the one hand, cause a sharp increase in inventories of finished products, generating additional costs for the company (e.g. storage costs) or, on the other hand, difficulties in timely delivery of products to the customer [1]. In order to solve the above problem, enterprises often decide to implement various methods and production management tools. The analysis of the literature on the subject indicates that solutions supporting decision-making processes [2–5], quality management [6–9] and lean manufacturing [10, 11] are most often implemented.

J. Machado et al. (Eds.): icieng 2021, LNME, pp. 12–24, 2022.
https://doi.org/10.1007/978-3-030-78170-5_2

The concept of Lean Manufacturing in its assumptions focuses on the elimination or significant reduction of waste, ie activities that do not increase the value of the product in the eyes of the customer, and therefore do not add value to them [12]. One of the activities that brings no value and at the same time reduces the available production time is the changeover of the machine or device. Long changeover times make it necessary to produce in large series, which significantly reduces production flexibility (understood as the ability to quickly respond to customer needs) and increases the stock of finished products [13, 14]. The use of the SMED (Single Minute Exchange of Dies) method, which is part of the Lean Manufacturing concept, allows for a significant reduction in changeover times for machines and devices. As indicated in the literature on the subject, the use of the SMED method allows in the first stage of implementation to reduce the changeover time by about 30–50%, and in the next stages by another 20–30% [15]. Moreover, the application of the SMED method may increase production flexibility [16], reduce production costs and storage costs [17].

In addition to the improvement of production processes, an important aspect that determines the timely completion of customer orders or the proper load on workstations is production planning. In the planning process, key decisions are made regarding the efficient implementation of production, solving existing problems or determining the necessary measures to achieve the set goals. An important element is to provide complete and reliable input data that enable the creation of a production plan that is realistic [18, 19]. The literature on the subject indicates many methods and algorithms for planning and scheduling production. Most often, however, enterprises decide to use heuristic methods that do not guarantee an optimal solution, but generate an acceptable solution within an acceptable time. These methods are based on priority rules [20, 21].

Improving processes and proper production planning can also contribute to reducing the number of rejects, better use of production resources and increasing production efficiency, while reducing the consumption of raw materials, i.e. material, energy, etc. Thanks to this, the company is able to reduce the amount of pollution and waste generated, which is in line with the trend of sustainable development.

The aim of the article is to develop and conduct simulation tests in the FlexSim program, which will allow to determine the impact of selected factors: changeover time and the priority rule, on the timely execution of customer orders, workstation load and the required size of the finished products warehouse. To do this, it is necessary to check for which combination of selected factors, the timeliness of customer orders fulfillment and machine load will reach the highest level, and the required warehouse size will be the smallest.

2 Selection of Priority Rules

Priority rules are decision rules used to determine the order in which tasks are performed on specific workstations in a production unit [22]. There are many rules of priority in the literature. In practice, the most frequently used are [23–27]: shortest task time, longest task time, longest next task time, shortest processing time, longest processing time, shortest preparation and finishing time, nearest task due date, longest preparation and finishing time, first in first out.

The article presents the research results for three priority rules:

- shortest processing time (SPT), determines the sequence of operations according to their labor intensity, in this case the task with the shortest processing time is performed first [28],
- longest processing time (LPT), determines the sequence of operations according to their labor consumption, but in this case the task with the longest processing time is performed first [22],
- first in first out (FIFO), determines the order of tasks depending on the order of entry of orders to production. This means that the job received earliest is performed first. [22].

The entire planning and scheduling process depends on the proper selection of the priority rules, which allows to meet the given criteria. The above priority rules are one of the most frequently used in manufacturing companies due to the minimization of the production cycle [29, 30]. One of the key factors of competitiveness and meeting customer needs is the ability to quickly respond to customer needs.

3 Research Methodology

The simulation study is designed to determine the impact of changeover times and priority rules on the timeliness of customer orders fulfillment, workstation load and the required size of the finished products warehouse. All input data used to perform the test are simulation data generated for the test.

The research methodology presented in Fig. 1 was adopted to conduct the study.

The following sections present a detailed description and implementation of the stages of the research methodology.

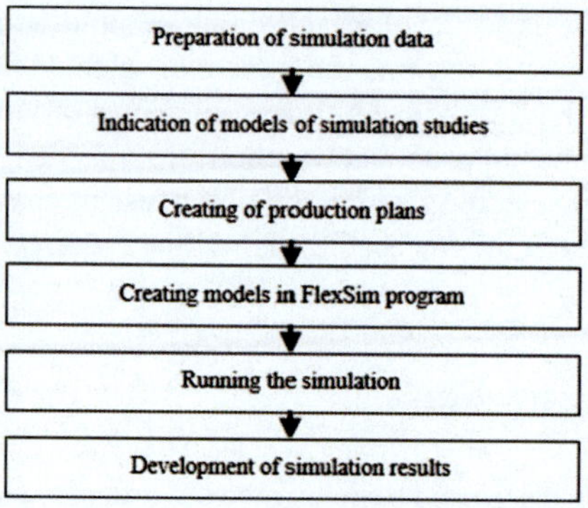

Fig. 1. Research methodology.

3.1 Preparation of Simulation Data

Two product families were accepted for the study, each family contains 8 products. The products W1-W8 belong to the first family of products (F1), while the products W9-W16 belong to the second family (F2). Historical data of customer orders was generated for each of the products. They provided information on the number of orders in a period of one year, the minimum and maximum number of orders for a specific product during one day, the minimum number of items in an order, the maximum number of items in an order, and the average number of items in an order. The data made it possible to calculate the arithmetic mean and standard deviation of the number of pieces in the order, thanks to which the range of order sizes was obtained, i.e. the minimum and maximum number of orders for a given product by the customer during the simulation testBy analyzing how many days an order for a given product was recorded, the arithmetic mean and standard deviation were also calculated, which determined the minimum and maximum number of days for the order in the simulation study, i.e. the frequency of customer orders.

The next step was to develop a manufacturing process for both product families. It was assumed that each family will be produced on one production line. In the case of F1, 7 technological operations and quality control should be performed, and in the case of F2, 3 technological operations and quality control should be performed. It was assumed that the quality control for both production lines is 100%. The product quality control station W1-W8 has a defect rate of 5%, and the product quality control station W9-W16 has a defect rate of 4%.

In the next stage, production operation times and basic changeover times for all manufactured products were generated in the program prepared by the authors. On the basis of the above data, calculations were made regarding: the product production program (taking into account the deficit coefficient described above), the pace and cycle of production and the required number of machines for each production operation. For the calculations, it was assumed that the production takes place in a two-shift system, where the number of hours per shift is 8 h, and the time is 30 min for each shift.

Figure 2 shows a diagram of the manufacturing process for the F1 product family, and Fig. 3 for the F2 product family.

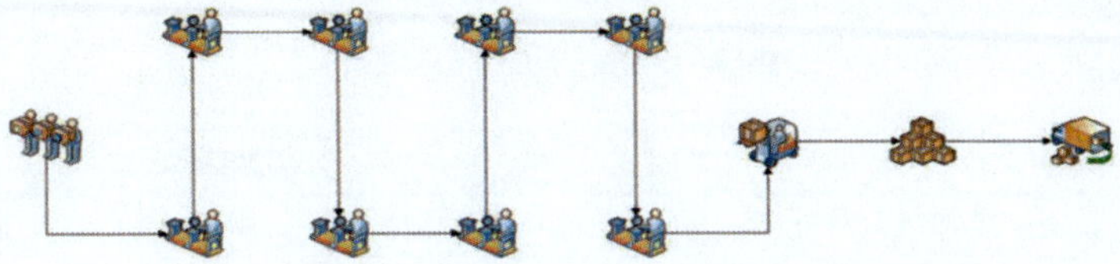

Fig. 2. Diagram of the F1 family production process.

The last element of data preparation was generating random customer orders for individual products, in accordance with the previously calculated values of the arithmetic mean and standard deviation (frequency of orders, range of order sizes). The fragment with the generated orders is shown in Fig. 4.

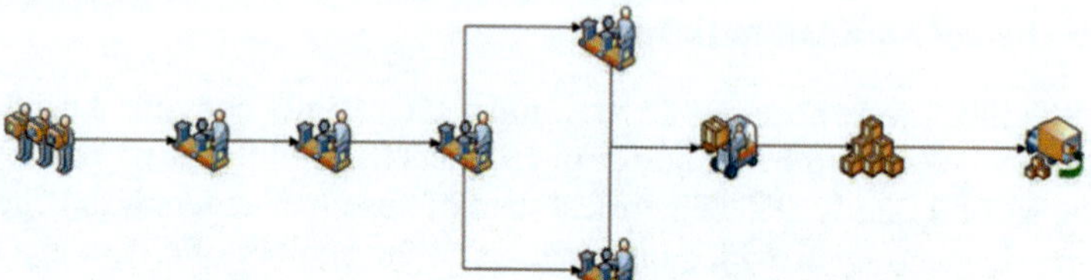

Fig. 3. Diagram of the F2 family production process.

Day	W1		W2		W3	
	Order?	Pieces	Order?	Pieces	Order?	Pieces
1	yes	129			yes	99
2	yes	79	yes	55		
3	yes	194			yes	100
4	yes	83				
5	yes	130	yes	131	yes	118
Minimum		78		40		69
Maximum		194		206		165
Average		135		149		113

Fig. 4. Fragment of the orders table.

3.2 Indication of Models of Simulation Studies

Nine variants of the combination of changeover times and priority rules were adopted for the simulation tests. As previously indicated, the 3 most common priority rules were adopted: STP, LTP and FIFO. There were also 3 variants of changeover times T1, T2, T3. The first variant of T1 covered the base values of the changeover times, which are the longest. In the second variant, the values of changeover times (T2) were reduced in relation to the first variant by about 1/3, and the third variant determines the changeover times (TPZ3), reduced by about 2/3, compared to the first variant. Table 1 shows the factor combinations and the corresponding simulation model numbers.

Table 1. Simulation models.

	FIFO	LTP	STP
T1	Model T1-FIFO (M1)	Model T1-LTP (M4)	Model T1-STP (M7)
T2	Model T2-FIFO (M2)	Model T2-LTP (M5)	Model T2-STP (M8)
T3	Model T3-FIFO (M3)	Model T3-LTP (M6)	Model T3-STP (M9)

In the case of T2 and T3 times, in accordance with the SMED rules, the preparatory activities were divided into external and internal. In the first time variant (T1), it was assumed that the worker has to leave the machine, e.g. to pick up the appropriate tools. For the second option (T2), the times are reduced compared to T1 as it is as-sumed that the worker is not leaving the workplace to pick up the tools. The worker's task is only to assemble the required tool for further work and install in the machine tool. In the third time variant (T3), it was assumed that the worker has prepared as-sembled tools, which allows for a significant reduction of the changeover time - his task is only to install the finished tool in the machine tool .

The presented variants will be assessed in terms of: timeliness of customer orders, machine load and the required size of the warehouse, the most important criterion being the timeliness criterion and the least important requirement for the size of the warehouse.

3.3 Creation of Production Plans

The production plans were created according to the previously adopted priority rules and performed calculations, in particular regarding the production program. The table below (Table 2) presents the order of manufacture of products for the M4-M6 models.

Table 2. The order to produce products according to the LTP rule priorities.

F1 family production line		F2 family production line	
Sequence	Product	Sequence	Product
1	W6	1	W15
2	W7	2	W14
3	W3	3	W16
4	W5	4	W11
5	W2	5	W9
6	W4	6	W13
7	W8	7	W10
8	W1	8	W12

3.4 Creating Models in FlexSim Program

The next stage of the research was the preparation of 9 simulation models in the FlexSim program and the implementation of previously prepared production data concerning e.g. operation times, changeover times, production plan or customer orders. The models differ in production data, while the layout of machines and stations remains unchanged. As a result, the applied simulation is deterministic and discrete.

The simulation model for the line producing the F1 family and the line producing the F2 family consists of (Fig. 5):

- sources of raw materials (ENTRANCE),
- three buffers (BUFFER 1, 2, 3),
- ten machines (MACHINE -10) and three quality control stations (K1-K3),

- finished products warehouse (WAREHOUSE),
- an element enabling leaving products from the warehouse (WAREHOUSE_EXIT),
- customer order acceptor (ORDER),
- element issuing production orders (ORDER _EXIT).

The production plan, with operation times, changeovers times, the order in which the products are made, and the number of items to be manufactured, is implemented in the "Entrance" element. This is a source of raw materials that allows you to determine the specific number of products produced, and also to determine the time when the raw materials needed for production are to appear on the buffer. "Buffer" is an interstate magazine. When the machine finishes pro-ducing one piece of product, it takes another from the buffer.

- The element defined as "Warehouse_exit" allows you to release products from the supermarket and control of the number of pieces of all products that were collected to carry out the customer's order. The "Orders" element specifies the size of the order, as well as the time after which they are carried out (orders appear once a day, at the end of the business day). The last element "Orders_exit" allows you to complete orders and control the number of completed orders.

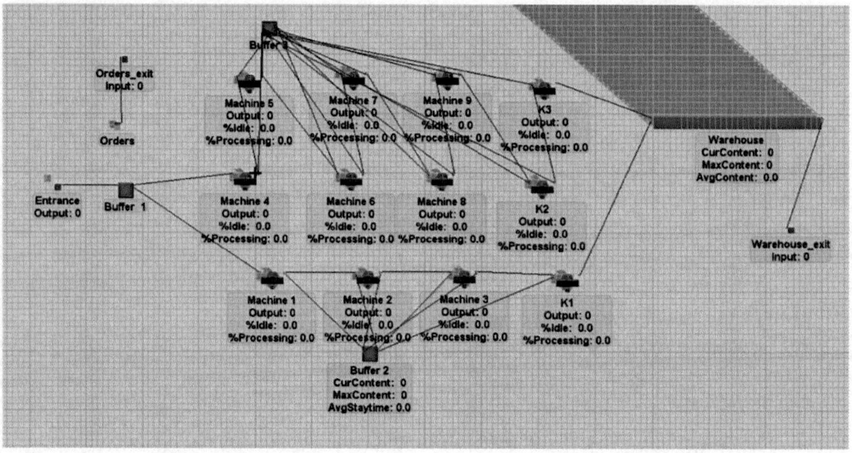

Fig. 5. Model of the manufacturing process in FlexSim program.

3.5 Running the Simulation

The simulation was carried out for 9,720,000 s, or 180 working days with 2 shifts each day. The break time was 30 min for each work shift, therefore the given simulation time is the time in which the sum of break times was omitted (if the program had included the break time, the simulation would have lasted 10,368,000 s). The program set the end of the simulation time to 9,720,000.01 s, as it allowed for the correct completion of the simulation.

3.6 Development of the Results

Introduction

For the purposes of the study, it was assumed that the timely execution of customer orders means sending a ready (complete) order within 5 working days of placing the order by the customer. It should be emphasized that the customer receives his order only when it is complete. This means that if there are not enough items in the warehouse to fulfill a given order, the customer receives them only when there are enough items in the warehouse to fulfill the order.

The customer may only receive a complete order, if there are not enough items to complete the order, it is defined as: in progress. An explanation of the terms used for contract fulfillment is provided below:

- completed, i.e. orders that have been completed, regardless of whether the customer received them on time or late,
- in progress, i.e. orders that by the end of the simulation were not fulfilled, but were accepted for execution and would have been fulfilled if the simulation had not ended.

The completed orders were divided into orders timely and orders untimely, defined in the study as:

- completed on time, i.e. according to the principle that the customer orders, the customer gets (immediately after ordering or up to 5 working days),
- completed after the deadline, i.e. the customer received a complete order, but with a delay, the customer waited for the order for more than 5 working days.

Timely Execution of Customer Orders

Table 3 shows the simulation results in relation to the timely execution of customer orders. The largest number of orders in total was made in the models that used the FIFO method. However, considering in which case more orders were processed on time, more orders were processed on time for models that implement the process according to the STP rule. The results also show that most orders were realized when the changeover times were the shortest (T3), and the fewest orders were realized when the changeover times were the longest.

Workplace Load

Table 4 shows the results for each model, showing the average station load for the F1 family production line, and Table 5 shows the results for the F2 family production line. The data shows that in the case of the F1 line, the highest percentage of the total production time in each model is for the process. The best (in both production lines) are models with the FIFO rule applied. Thus, in models using the FIFO rule, the average idle time is the shortest. On the other hand, taking into account the changeover time, the most favorable (due to the shortest time) are models in which the LPT rule was applied. Due to the fact that the failure time in each simulation study was the same, also its share in the total simulation time does not change for individual models.

Table 3. Summary of the results of customer order fulfillment [pieces].

Model	Completed	Completed on time	Completed after the deadline	In progress
T1-FIFO	2068	661	1407	385
T2-FIFO	2182	889	1293	271
T3-FIFO	2213	1006	1207	240
SUM	6463	2556	3907	896
T1-LTP	2014	613	1401	439
T2-LTP	2167	848	1319	286
T3-LTP	2199	987	1212	254
SUM	6380	2448	3932	979
T1-STP	2068	662	1406	385
T2-STP	2181	891	1290	272
T3-STP	2212	1091	1193	241
SUM	6461	2572	2889	898

Table 4. Summary of the results of the average machine load for the F1 family production line [%].

Model	Work	Changeover	Idleness	Accident
T1-FIFO	76.51	21.19	4.73	0.16
T2-FIFO	76.90	14.63	8.31	0.16
T3-FIFO	77.49	7.93	14.44	0.16
AVERAGE	76.97	14.58	9.16	0.16
T1-LTP	71.09	45.51	5.3	0.16
T2-LTP	76.01	16.78	7.03	0.16
T3-LTP	77.36	9.25	13.21	0.16
AVERAGE	74.82	23.85	8.51	0.16
T1-STP	73.94	21.15	4.78	0.16
T2-STP	76.90	14.61	8.31	0.16
T3-STP	77.49	7.93	14.45	0.16
AVERAGE	76.11	14.56	9.18	0.16

The Size of the Finished Products Warehouse

Table 6 presents data on the average and maximum number of finished products items in the finished goods warehouse. The table above shows that the greatest demand for storage space was in the case of the T3-STP model, and the lowest for the T1-LTP model.

Table 5. Summary of the results of the average machine load for the F1 family production line [%].

Model	Work	Changeover	Idleness	Accident
T1-FIFO	78.88	17.76	3.28	0.08
T2-FIFO	83.24	12.56	4.06	0.08
T3-FIFO	85.64	7.02	7.26	0.08
AVERAGE	82.59	12.45	4.87	0.08
T1-LTP	78.82	17.76	3.34	0.08
T2-LTP	83.24	12.56	4.06	0.08
T3-LTP	85.64	7.02	7.26	0.08
AVERAGE	82.57	12.45	4.89	0.08
T1-STP	78.82	17.76	3.34	0.08
T2-STP	83.24	12.56	4.06	0.08
T3-STP	85.64	7.02	7.26	0.08
AVERAGE	82.57	12.45	4.89	0.08

Table 6. Summary results of the number of items in stock [pieces].

Model	Average value	Maximum value
T1-FIFO	1200	2700
T2-FIFO	3500	6000
T3-FIFO	6500	7000
T1-LTP	2500	2750
T2-LTP	3550	4500
T3-LTP	4500	13000
T1-STP	1020	2200
T2-STP	3500	4000
T3-STP	8500	13500

4 Conclusions

This paper presents the results of preliminary simulation studies concerning the indication of the impact of changeover time and the priority rule on the timely execution of customer orders, workstation load and the required size of the finished products warehouse. Computer simulation was carried out in FlexSim for 9 simulation models: T1-FIFO, T2-FIFO, T3-FIFO, T1-LTP, T2-LTP, T3-LTP, T1-STP, T2-STP and T3-STP. The models were created by combining the various variants of the examined factors: three different variants of changeover times and three different variants of selected priority rules. The results were compiled taking into account three criteria:

- the highest timely execution of customer orders,
- the highest percentage of machines and workstations,
- the smallest size of the finished products warehouse required.

Taking into account the timely execution of orders, i.e. orders that were completed on time, the best results were obtained in the T3-STP model. In the same model, the smallest number of orders were obtained late, i.e. within 5 days from the date of placing the order.

In the case of workstation load, both the F1 and F2 family production line load should be taken into account. The best result, i.e. the highest percentage of production lead time, was obtained for the T3-FIFO model. However, the model also has the highest percentage of idle time, which may be due to the shortest changeover time among all models. Thus, it can be concluded that the time saved on changeover has become a time of machine idle time.

When analyzing the results for the demand for the size of the finished goods warehouse, it can be seen that the smallest results, i.e. the least warehouse space, are needed in the T1-SPT model. The longest changeover times in this case resulted in less frequent delivery of products to the warehouse, thus causing an increased number of orders not completed on time, which can be seen in Table 3.

In summary, the models were identified that best meet established criteria and they are:

- T3-STP – timely execution of customer orders,
- T3-FIFO – load of machines and workstations,
- T1-STP – the required size of the products goods warehouse.

However, taking into account the fact that in order to increase competitiveness and meet customer requirements, the company must absolutely meet the deadlines for order fulfillment, and considering this criterion as the most important, it was assumed that the T3-STP model is the best among all the tested models.

Acknowledgements. The studies were realized with a support from statutory activity financed by Polish Ministry of Science and Higher Education (0613/SBAD/4677).

References

1. Varela, M.L.R., Trojanowska, J., Carmo-Silva, S., Costa, N.M.L., Machado, J.: Comparative simulation study of production scheduling in the hybrid and the parallel flow. Manage. Prod. Eng. Rev. **1**(2), 10–20 (2010)
2. Kłos, S., Patalas-Maliszewska, J.: Throughput analysis of automatic production lines based on simulation methods. In: Jackowski, K., Burduk, R., Walkowiak, K., Woźniak, M., Yin, H. (eds.) IDEAL 2015. LNCS, vol. 9375, pp. 181–190. Springer, Cham (2015). https://doi.org/10.1007/978-3-319-24834-9_22

3. Kunz, G., Machado, J., Perondi, E.: Using timed automata for modeling, simulating and verifying networked systems controller's specifications. Neural Comput. Appl. **28**(5), 1031–1041 (2015). https://doi.org/10.1007/s00521-015-2115-5
4. Sika, R., Hajkowski, J.: Synergy of modeling processes in the area of soft and hard modeling. In: 8th International Conference on Manufacturing Science and Education (MSE), Trends in New Industrial Revolution. EDP Sciences (2017)
5. Sika, R., Rogalewicz, M.: Demerit Control Chart as a decision support tool in quality control of ductile cast-iron casting process. In: 8th International Conference on Manufacturing Science and Education (MSE), Trends in New Industrial Revolution. EDP Sciences (2017)
6. Hamrol. A.: Strategies and practices of efficient operation. LEAN SIX SIGMA and other. PWN, Warsaw (2015). (in Polish)
7. Diering, M., Dyczkowski, K.: Assessing the raters agreement in the diagnostic catheter tube connector production process using novel fuzzy similarity coefficient. In: IEEE International Conference on Industrial Engineering and Engineering Management, pp. 228–232 (2016)
8. Jasarevic, S., Diering, M., Brdarevic, S.: Opinions of the consultants and certification houses regarding the quality factors and achieved effects of the introduced quality system. Tehnicki Vjesnik-Technical Gazette **19**(2), 211–220 (2012)
9. Borkowski, S., Knop, K.: Challenges faced in modern quality inspection. Manage. Prod. Eng. Rev. **7**(3), 11–22 (2016)
10. Trojanowska, J., Kolinski, A., Galusik, D., Varela, M.L.R., Machado, J.: A methodology of improvement of manufacturing productivity through increasing operational efficiency of the production process. In: Hamrol, A., Ciszak, O., Legutko, S., Jurczyk, M. (eds.) Advances in Manufacturing. LNME, pp. 23–32. Springer, Cham (2018). https://doi.org/10.1007/978-3-319-68619-6_3
11. Liker, J.K.: The Toyota Way, 14 Management Principles from the Worlds Greatest Manufacturer. 1st edn. McGraw-Hill, London (2004)
12. Concalves, T.I.G., Sousa, P.S.A., Moreira, M.R.A.: Does lean practices implementation impact on company performance? A meta-analytical research. Manage. Prod. Eng. Rev. **10**(4), 11–24 (2019)
13. Korytkowski, P., Grimaud, F., Dolgiu, A.: Exponential smoothing for multi-product lot-sizing with heijunka and varying demand. Manage. Prod. Eng. Rev. **5**(2), 20–26 (2014)
14. Dennis, P.: Lean Production Simplified, 2nd edn. Productivity Press, New York (2007)
15. Shingo, S.: A Revolution in Manufacturing: The SMED System. Productivity Press, Stanford (1985)
16. Ulutas, B.: An application of SMED methodology. Int. J. Innov. Sci. Res. 1194–1197 (2011)
17. Sarkar, B., Sarkar, M.: Application of Optimalization in Production, Logistics, Inventory, Supply Chain Management and Block Chain, MDPI, Printed Edition of the Special Issue Published in Mathematics (2020)
18. Graves, S.C.: A review of production scheduling. Oper. Res. **29**(4), 646–675 (1981)
19. Fuchigami, H.Y., Rangel, S.: A survey of case studies in production scheduling: analysis and perspectives. J. Comput. Sci. **25**, 425–436 (2018)
20. Vanhoucke, M.: Optimizing regular scheduling objcctives: schedule generation schemes. PM Knowledge Center (2012). http://www.pmknowledgecenter.com/dynamic_scheduling/baseline/optimizing-regular-scheduling-objectives-schedule-generation-schemes. Accessed 15 Dec 2019
21. Chakrabortty, R.K., Rahman, H.F., Ryan,M.J.: Efficient priority rules for project scheduling under dynamic environments: a heuristic approach. Comput. Ind. Eng. **140**, 106287 (2020)
22. Wróblewski, K.J., Krawczyński, R., Kosieradzka, A., Kasprzyk, S.: Priority Rules in Manufacturing Flow Control. Scientific and Technical Publishing House, Warsaw (1984). (in Polish)

23. Binchao, C., Matis, T.I.: A flexible dispatching rule for minimizing tardiness in job shop scheduling. Int. J. Prod. Econ. **141**(1), 360–365 (2013)
24. Błażewicz, J., Ecker, K.H., Pesch, E., Schmidt, G., Węglarz, J.: Scheduling in Computer and Manufacturing Processes. Springer, Heidelberg (1996). https://doi.org/10.1007/978-3-662-03217-6
25. Jardzioch, A., Bulwan, K.: The prioritisation of production orders under the bee colony algorithm. Adv. Manuf. Sci. Technol. **37**(4), 49–59 (2013)
26. Krenczyk, D., Skołud, B.: Transient states of cyclic production planning and control. Appl. Mech. Mater. **657**, 961–965 (2014)
27. Trojanowska, J., Varela, M.L.R., Machado, J.: The tool supporting decision making process in area of job-shop scheduling. In: Rocha, Á., Correia, A.M., Adeli, H., Reis, L.P., Costanzo, S. (eds.) WorldCIST 2017. AISC, vol. 571, pp. 490–498. Springer, Cham (2017). https://doi.org/10.1007/978-3-319-56541-5_50
28. Baker, K.R., Dan, T.D.: Principles of Sequencing and Scheduling. Wiley, New Jersey (2009)
29. Vieira, G.G., Varela, M.L.R., Putnik, G.D., Machado, J.M., Trojanowska, J.: Integrated platform for real-time control and production and productivity monitoring and analysis, Romanian review precision mechanics. Opt. Mechatron. (Mecahitech'16) **50**, 119–127 (2016)
30. Rewers, P., Trojanowska, J., Diakun, J., Rocha, A., Reis, L.P.: A study of priority rules for a levelled production plan. In: Hamrol, A., Ciszak, O., Legutko, S., Jurczyk, M. (eds.) Advances in Manufacturing. LNME, pp. 111–120. Springer, Cham (2018). https://doi.org/10.1007/978-3-319-68619-6_11

Study of the Portuguese Challenges in the Context of European Union to Identify Adaptation Strategies for the Industry 4.0

Cristiano de Jesus(✉) and Rui M. Lima

Department of Production and Systems, ALGORITMI Center, School of Engineering, University of Minho, 4800-058 Guimarães, Portugal
b12121@algoritmi.uminho.pt

Abstract. The adaptation of business models to the Industry 4.0 is an urgent demand and a clear concern that emerges from economic relations. However, the concepts of the Fourth Industrial Revolution and the ways of implementation are not yet assimilated at all by the agents of the productive sector and the scientific researches carried out, in many cases, present solutions that are too generic, so that the companies, in turn, find it difficult to recognize them as applicable in their particular reality. The aim of this article is to identify and describe, through research in literature and analysis of economic reports, the specificities of Portugal in the context of the European Union regarding the barriers and the conditions of adhesion to the Industry 4.0. The perspective is that the challenges presented points to priority aspects and serves as a parameter for future research work aimed at the design and development of implementation strategies that have greater adherence to the challenges faced by Portuguese industry and, therefore, have greater potential for success. The topic of Industry 4.0 is quite broad and the article may help researchers and companies to locate themselves in the universe of discussions and proposals of solutions, as well as to be oriented in the decision making about, above all, the adoption or construction of tools for the evaluation of results and performance indicators of projects, in order to increase the chances of success in the context of Portugal in the European Union environment.

Keywords: Industry 4.0 · Industrial Engineering and Management · Portugal · European Union

1 Introduction

The implementation models of the Industry 4.0 proposed so far are generic and differ between them, lack a unicity in terms of direction, and this makes the application difficult.

In this study, it is intended to address this gap through a systematic document analysis and investigate factors that can guide the design of a management model for projects implementing the Industry 4.0, which may be suitable to the Portuguese challenges in the context of the European Union (EU).

J. Machado et al. (Eds.): icieng 2021, LNME, pp. 25–35, 2022.
https://doi.org/10.1007/978-3-030-78170-5_3

The study is conducted with the objective of answering the following questions:

1. What is the economic challenge in Portugal in the context of the European Union as regards the conditions for adapting to technological changes?
2. What strategic factors related to Industry 4.0 present shortfalls or potential for development in Portugal in the context of the European Union?
3. What are the opportunities for scientific research and how can they be conducted so that they better meet the most pressing needs in the context of Portugal?

2 Theoretical Background

The factors for the adoption of management models are very diverse and respond differently depending on the context of the environment and the size of the company, however, this cannot be an impediment to achieving a satisfactory approach in terms of best practices and implementation guidelines.

Stefan et al. (2018) [11] affirm that small and medium enterprises are not always able to perceive any relationship between their activities and the conventional definition of Industry 4.0, so they consider that guides and models are important instruments of guidance for implementation. Despite this need, as argued by Stefan et al. (2018) [11] consider that the existing models have in common the fact that they are not geared to specific situations, so that they can only be applied by hiring an external consultancy.

Another aspect is that many of these solutions have greater emphasis on technology and organizational structure. To obey this general perspective, Schuh et al. (2017) [9], from Acatech, state that the objective of a project to implement the Industry 4.0 should make a company agile in a changing environment and a maturity model serves in this context, according to the authors, to guide companies to perform this transformation.

In the same way, Tonelli et al. (2016) [12] consider processes to be important factors impacting the competitiveness of an organization and therefore suggest attention to Business Process Improvement (BPI). The authors mention BPM – Business Process Modelling as an important tool for process integration and consider the ANSI/ISA-95 Enterprise/Control standard as support for this purpose.

For the monitoring of process performance, they present as a solution the maturity indexes. They affirm that there are several frameworks that organize these instruments with a view to a purpose, as already mentioned, and were developed by professionals, academics and international consortia, however, most of them are based on the CMM – Capability Maturity Model, or CMMI – Capability Maturity Model Integration.

3 An Analysis of the Portuguese Challenges in the Context of European Union for Implementation of Industry 4.0

This section is dedicated to presenting a set of conditions, barriers and drivers, for the implementation of Industry 4.0 (I4.0) in Portugal, considering the EU context.

3.1 Digital Maturity to the I4.0

Following the steps taken to prepare the productive sector for the technological transformations in Industry 4.0, Portugal has a program called "National Strategy Portugal 2030 – National Strategy for Portugal Post 2020" [3].

The program provides a self-diagnosis of the digital maturity of the Portuguese economy in order to better support the definition of roadmaps for technological transformation, and to ensure that the knowledge generated from the experiments and implementation experiences is widely disseminated and shared.

The diagnosis of digital maturity is carried out through consultation of companies and institutions. In relation to companies, it can be seen that the number of companies that systematically develop and put into practice disruptive ideas is below average in the European Union, ideas such as joint development between companies and entities, application of digital skills, reconfiguration of the operational and business model, adoption of Industry 4.0 technologies, adherence to technological and innovation ecosystems.

In the events called "open days", working groups, interventions and debates organized by COTEC Portugal – Business Association for Innovation, it was identified that the context of industry 4.0 in Portugal is marked by the existence of distinct groups of companies regarding the level of assimilation of the concepts related to Industry 4.0, including autonomous and self-sufficient companies in skills and resources in implementation, companies with some level of sensitivity to the theme and that conduct experiments in a way that obtains some results and, finally, companies that are not sensitized and do not have internally resources and skills to perform the necessary changes by themselves.

A qualitative representation of this reality is also reflected in quantitative surveys, as can be seen in Fig. 1.

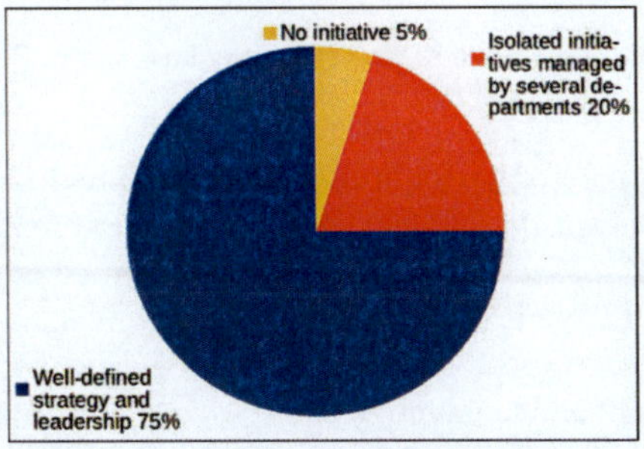

Fig. 1. Digital transformation in large Portuguese enterprises [1]

Companies recognize Industry 4.0 has the potential to leverage competitiveness, but their concerns are the inadequacy of current management mechanisms and business models, requiring interoperability, the integration of the value chain and digital skills in human resources. There are also difficulties in accessing funding programs to support investment in digitization projects [3].

According to the 2019 Edition of the Annual Study of the Economy and the Digital Society of Portugal, the main barriers are represented in Fig. 2.

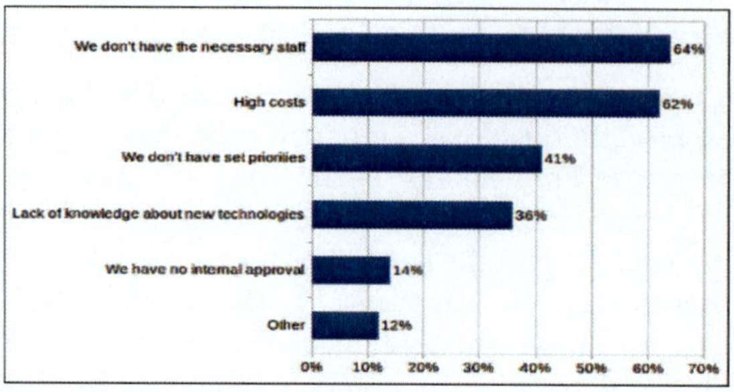

Fig. 2. Main barriers to the implementation of digital transformation initiatives in large Portuguese enterprises [1]

More data can be complemented to the information showed in Fig. 2. The main problems raised by the World Economic Forum to do business in Portugal are plotted in the chart illustrated in the Fig. 3.

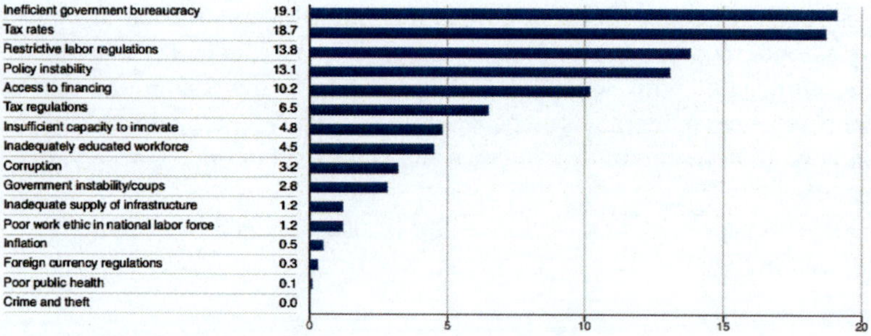

Fig. 3. Main problematic factors for doing business in Portugal [15]

The WEF data give more details about the costs pointed in the previous graph as a significant problem. As for the qualification of the workforce, the two graphs show different aspects. It does not seem that the supply of qualified labor is the only problem and neither the biggest of them in this category. It may be that the greatest challenge lies in the training of the current workforce and that they have been trained according to old paradigms of management and technology.

According to data from the World Economic Forum, value-added manufacturing companies grow by 0.2% in Portugal and of these, 25.7% adopt medium or high technology [13]. According the analysis presented by Silva (2018) [10], Portugal's level of competitiveness has remained stable over the last ten years in part because it also does not improve the rates of innovation and sophistication factors in business.

In his study on the competitiveness of Portuguese companies in the European Union, Silva (2018) [10] notes that in fact the level of adoption of technology is not a problem, nor is the quality of education, which makes Portugal a country that has a reasonable level of technological readiness for Industry 4.0. However, high competitiveness is not only defined by these factors, it is necessary to improve the development of the financial market, and the efficiency and productivity rates.

3.2 Current Business Conditions that May Influence the Transformation for the I4.0?

It is important to consider that the cost of a 1 GB of Internet in Portugal is one of the largest in the EU. While in Portugal 1 GB of Internet coast $13.98, in Finland, for example, coast $1.16 [2]. Other indices presented by the World Economic Forum on the relative condition of Portugal in the EU, concerned to dimensions as technology, innovation, labor force, product market and business dynamism, suggest relevant topics for understanding the maturity of the economy and Portuguese firms for Industry 4.0.

According to the reports, Portugal has a medium to good performance in most indicators. The country stands out in the capacity of the companies to absorb technology and the impact of Information and Communication Technology (ICT) on new services and products is meaningful in comparison with their peers in the EU. In number of subscriptions of Internet, Fixed-broadband and Fiber, Portugal is ahead of more than 120 countries out of a total of 141, but is an intermediate position in mobile-broadband subscriptions. However, the perspective of grow in many aspects is positive if the analysis is focused in indicators as venture capital deal volume per size of economy – position 17 out of a total of 100 countries, and the quality of research institutions – 27 out of a total of 141 [13–16].

About labor force, in percentage of working population, Portugal ranks ahead the most countries in manufacturing employment (15/100), availability of scientists and engineers (27/100) and skillset of graduates (27/141). In terms of Education and measures to the future labor force, stands out indicators concerned to quality of math and science education and ease of hiring foreign labor. In the other hand, Portugal needs to improve the capacity to attract and retain talent, mean years of schooling years, the practices of hiring and firing and, mainly, the internal labor mobility, an indicator which Portugal occupies the 116 position of a total of 140 [13–16].

Labor mobility refers to the capacity of personnel able to move between positions, organizations and regions, within an economy and between different economies. This indicator is important because it affects growth and production [5].

According Pordata [6], Portugal has one of the lowest rates in labor productivity, a special factor that contribute to the economic growth and social welfare. Portugal is behind 21 countries out of 27 in this indicator. Portugal's level of competitiveness measured by the World Economic Forum compared to that of 141 other countries is in the number 34 position [15].

Regarding product marketing and business dynamism, Portugal is strong in competition in services and has efficiency of the clearance process, has good potential in growth of innovative companies and the amount of companies which embraces disruptive ideas, but needs to improve in attitudes toward entrepreneurial risk [13–16].

Pourabdollahian et al. (2016) [8] present detailed data on the technology adoption of Portuguese companies in relation to their peers in the European Union. High performance manufacturing corresponds to the adoption of high-performance technology that enables resources such as systems combining flexibility, precision and zero-defect mechanisms (e.g. high precision machine tools, advanced sensors, 3D printers). Figure 4 show the level of adoption of high-performance manufacturing technologies in Portugal compared to some other countries in European Union.

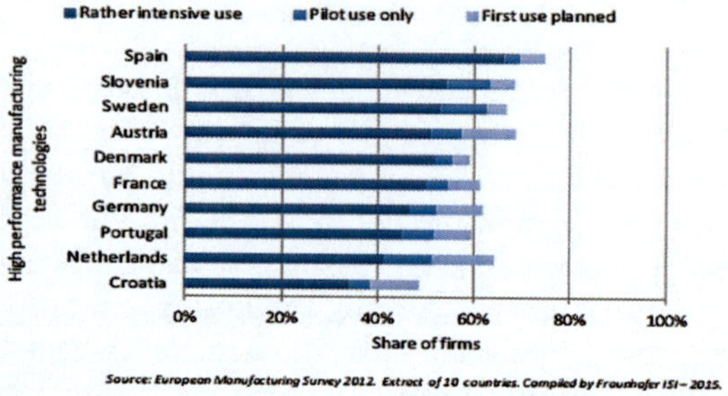

Fig. 4. Companies using high-performance manufacturing technologies, by country [8]

ICT-enabled smart manufacturing are Information and Communication Technology that enables digital integration technologies implemented in productive systems. As can be seen in Fig. 5 Portugal stands out in access and employment of ICT, but not necessarily, as can be seen in Fig. 4, to enable the smart manufacturing.

It is estimated that 16.9% of Portuguese workers work in manufacturing companies [13]. However, according to data from the Francisco Manuel dos Santos Foundation, the largest employers in Portugal are Small and Medium-sized Enterprises (SME), 78% of workers are in the SME [7]. Pourabdollahian et al. (2016) [8] present numbers extracted from a survey with European companies from 10 countries which identify that in European Union, technologies are adopted in the most time by large enterprises in the various levels of intensive use. It is reasonable to suppose that Portugal follow this same trend. The scenario designed by the graphs shows that there is a bigger potential of development in SME with more impact in the expansion of the social benefices.

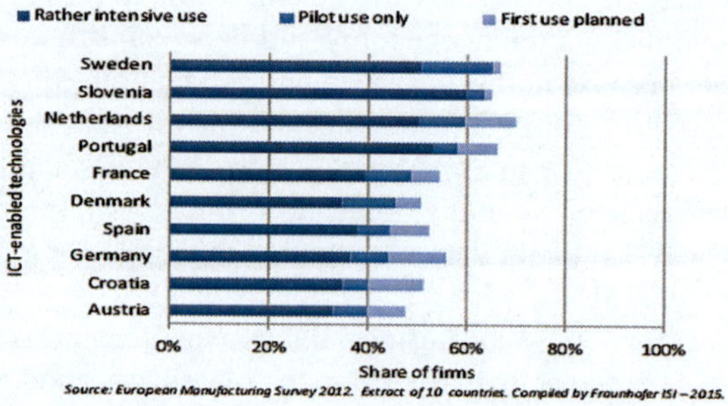

Fig. 5. Companies using ICT-enabled technologies, by country [8]

Pourabdollahian et al. (2016) [8] present an analysis of the drivers, barriers and factors to prepare EU companies for the adoption of the advanced manufacturing model and consider that countries need to develop policies to strengthen small and medium-sized enterprises, to promote the adoption of new technologies and to support companies in reconfiguring their business models. The study deals with ICT-enabled intelligent manufacturing that corresponds to the integration of technologies and processes to enable the concept of intelligent manufacturing, advanced technologies and sustainable technologies. The latter will not be commented on in this article because it is not the focus of the objectives. According to the authors, the main barriers for the adoption of ICT-enabled intelligent manufacturing and the advanced manufacturing model are as follows: changes in the rules and forces that define the value chain, difficulties in the management of the political environment, the regulatory, legal and tax environment of work, in the definition of requirements, in the evaluation of the cost-benefit of investments, in the conciliation of motivations and purposes between partner companies and in the alignment of processes, resistance to changes, lack of experts, lack of qualified labor, lack of education and training programs aimed at advanced management practices, lack of innovation in the training process, difficulties in developing solutions on an industrial scale, in accessing and replicating data among partners, and lack of shared tools in a very dynamic context of change.

3.3 Future Perspective and Opportunities on Industry 4.0

An overall view of European competitiveness in the context of Industry 4.0 is illustrated in the Fig. 6. Scoreboard is a tool designed to identify national and European competitiveness in the context of Industry 4.0. The countries analyzed are grouped into three groups: “lagging”, “mid-tier” and “leading”. Portugal occupies the last position of the mid-tier group, as can be seen in Fig. 6 and the 12th in the global set. It is projected that an increase of 1 point in this i4.0 index may result in incremental GDP growth of between 2.1% and 2.4% per year over 5 years [3].

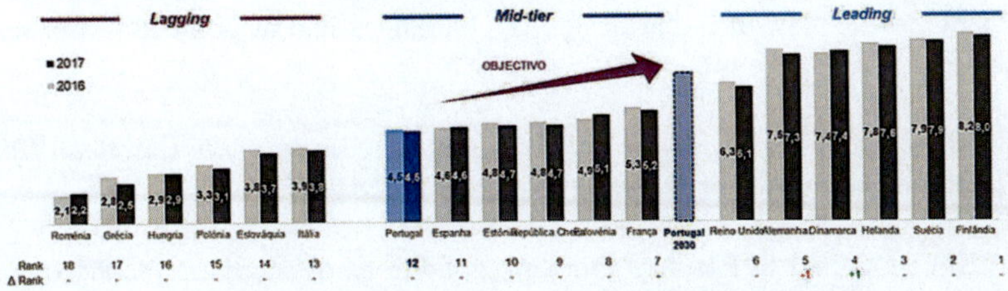

Fig. 6. i4.0 index scores in 2016 and 2017 [3]

Although Portugal is well positioned in several aspects, it presents challenging indices on sensitive factors for the digital market such as delay in the application of connectivity technology, demands in the field of training of labor and the adoption of good management practices. This assessment also appears in other economic reports as can be seen in the Fig. 7.

On the other hand, existing programs and actions already carried out place Portugal in a condition of potential of progress and development, as can be seen in the Fig. 8.

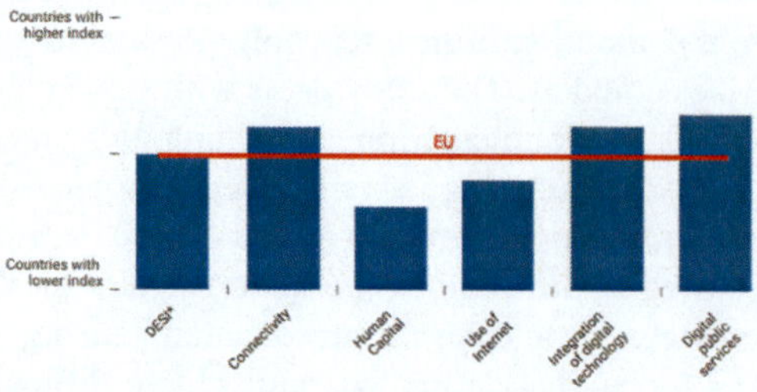

Fig. 7. Portugal's relative position in terms of digital competences in the DESI 2017 Index of the European Commission. Source DESI: Digital Economy & Society Index [4]

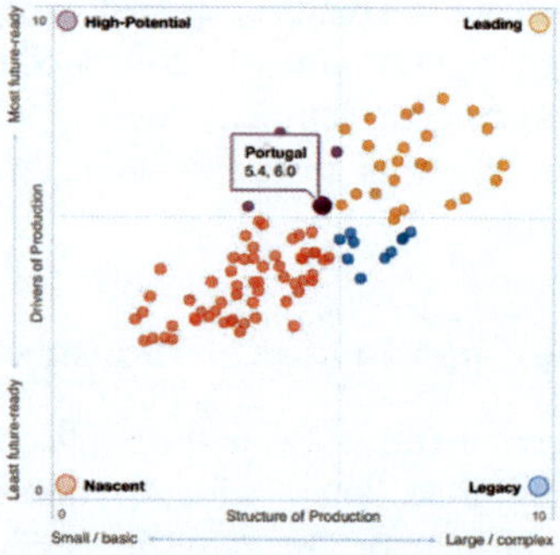

Fig. 8. Readiness for the Future of Production Assessment [15]

4 Discussion

Regarding the issues proposed for investigation in this study, with the data raised so far it is possible to identify an initial design of challenges and to point to future research projects.

1. *What is the economic challenge in Portugal in the context of the European Union as regards the conditions for adapting to technological changes?*

The data show that in Portugal there are conditions of access to technology, it has development and modernization programs and, although there are barriers that prevent more accelerated progress, the country has representations of all sectors that conduct initiatives to minimize the factors that limit the competitiveness and productivity of companies. As could be seen in Fig. 1, only 5% of Large Portuguese Enterprises do not dedicate actions for Industry 4.0. Although technological modernization is a concern of Portuguese companies, the reports indicate a significant discrepancy between the use of technology in the productive system and actions to develop the concept of smart manufacturing. However, the scenario designed by the study shows that Portugal has good conditions to be inserted in the industry development agenda for the 4.0 Industry model. In addition to a technological infrastructure in full development, an increasing opening of financing lines and incentive programs, the country also stands out in the

EU in the quality of education in math and sciences, and also in scientific research and innovation.

2. *What strategic factors related to Industry 4.0 present shortfalls or potential for development in Portugal in the context of the European Union?*

As seen in Fig. 2, the companies have difficulties with the qualification of workforce in spite of good positions of Portugal in quality of math and science Education and research institutions. Data presented by the World Economic Forum show that two factors can be responsible for this which need to be investigate with more details: (1) universities has the mission to prepare by themselves by updating your curricula and education methodologies in specific areas as Engineering and Business Administration for responding the need to develop new skills; (2) the companies need to be willing for professionalizing not only your operations but also your strategies and concepts of management, employing qualified professionals and developing a new organizational culture by adopting policies of permanent develop of new competences dedicated to current staff.

The surveyed data also shows that small and medium enterprises have a very significant social and economic importance because they employ most of the workforce, currently around 78% according to statistics presented by the Francisco Manuel dos Santos Foundation. However, it is worth noting that most companies in this sector of the economy have difficulties in visualizing the adherence of their business environment to the Industry 4.0 and are unable to apply the available implementation frameworks because they are too generic. Thus, developing Industry 4.0 frameworks that can be easily adapted for the specificities of SMEs may contribute to overcome this obstacle. Additionally, such framework could act as model for the implementation of management best practices and in this way, improve the productivity performance. It should be noted that such a framework must include culture adaptation issues so that they become more open to changes in their business model and management practices.

Therefore, in a first analysis, the hypothesis is raised that the cause of the low productivity of the workforce in Portugal in comparison with other European Union countries is the distance between the management and Industrial Engineering models developed and perfected in the academic and scientific environment of those employed by the companies, especially by small and medium enterprises, either because the supply of skilled labor suffers the challenge of its expansion, or also because companies need to modernize not only in the technological aspect, but also in the organizational culture so that they become more open to changes in their business model and management practices.

3. *What are the opportunities for scientific research and how can they be conducted so that they better meet the most pressing needs in the context of Portugal?*

This study shows that the development of a methodology to implement the production standard called Industry 4.0 is an opportunity of research. Moreover, this implementation methodology should be suitable to the challenges of Portugal in the context of the European Union, with special attention to the needs of SME-Small and Medium-Sized Enterprises. The basis for the construction of this model can be the existing generic models for organizational reconfiguration projects. This could be adapted through methodologies for construction of specific models, which draws up on surveys, interviews and case studies. This model can also be built by rereading the traditional Industrial Engineering concepts more sensitive to the environment of Industry 4.0 in order to draw attention to the need for action that favors systemic thinking, Lean philosophy, organizational culture, higher level skills, innovation and sustainability, and others, to achieve a research proposal which explore concepts like strategies frameworks, maturity models, roadmap implementation and continuous improvement enabled by assessment readiness, always directed at Portuguese Small and Medium-Sized Enterprises.

5 Conclusions

The Industry 4.0 depends on a chain of cooperation because it is a technological revolution. Thus, the lack of a national policies enabling the insertion of this concepts in the productive system can be a problem. Despite that Portugal has such policies, namely an Industry 4.0 development agenda, there are several structural challenges that can be seen by companies as obstacles to individual actions. Exploratory research to characterize the context of Portugal in its specific conditions and in relation to the European Union, identifies that a strategy of an implementation project of Industry 4.0 can achieve better results if it focuses on small and medium enterprises, due to the social and economic importance of SMEs in the country and because they face the greatest difficulties to obtain guidance in structuring competitive strategies, business models and organizational architectures. Another approach that should be considered to support this transformation, is the management of competences as a way to continuously adapt and improve the availability of a qualified workforce. Such approaches would be potentiated by a strong cooperation between universities and companies. These activities are not enough in long-term if they are not aligned with a strong proposal for developing a new organizational culture focused in the valuation of professionalization of the operations and management approach. Is expected that this article can make a contribution to guide research-action intentions with the proposal to develop complete frameworks composed by tools as assessment readiness techniques, maturity model, roadmap for combining management by processes and information systems, for designing a project for implementation of Industry 4.0 in Portuguese SMEs.

Acknowledgments. This work has been supported by FCT – Fundação para a Ciência e Tecnologia within the R&D Units Project Scope: UIDB/00319/2020.

References

1. ACEPI, & IDC. Estudo Anual da Economia e da Sociedade Digital—Edição 2019. ACEPI - Digital Economy Portugal; IDC - Analyse de Future. http://m.acepi.pt/download.php?f=ACEPI%20-%20Estudo%20da%20Economia%20Digital%20em%20Portugal,%20edi%E7%E3o%202019.pdf. Accessed 10 Dec 2019

2. Cable.co.uk. Worldwide Mobile Data Pricing League | Cost of 1GB in 230 countries—Cable.co.uk. Cable. https://www.cable.co.uk/mobiles/worldwide-data-pricing/. Accessed 11 Dec 2019
3. COTEC. Programa Portugal Indústria 4.0—Fase II. https://drive.google.com/file/d/1-guCKhDJXCwkcdAVSc-zfW6sBDbPuog2/view?usp=sharing. Accessed 10 Dec 2019
4. FCT. Portugal INCoDe 2030—Um Programa Integrado de Competências Digitais para Portugal, 2017–2030. FCT – Fundação para a Ciência e a Tecnologia, Portugal. http://www.incode2030.gov.pt/sites/default/files/incode2030_pt_0.pdf. Accessed 10 Dec 2019
5. Fellini, I., Ferro, A., Fullin, G.: Recruitment processes and labor mobility: the construction industry in Europe. Work Employ Soc. **21**(2), 277–298 (2007). https://doi.org/10.1177/0950017007076635
6. Pordata. Retrato de Portugal na Europa—Edição 2019. Fundação Francisco Manuel dos Santos. https://www.pordata.pt/Retratos/2019/Retrato+de+Portugal+na+Europa-84. Accessed 10 Dec 2019
7. Pordata. Retrato de Portugal—Edição 2019. Fundação Francisco Manuel dos Santos. https://www.pordata.pt/Retratos/2019/Retrato+de+Portugal-83. Accessed 12 Dec 2019
8. Pourabdollahian, G., et al.: IDEA Consult, Irish Translators' and Interpreters' Association, VTT, Fraunhofer, European Commission, & Directorate-General for Internal Market, I., Entrepreneurship and SMEs. An analysis of drivers, barriers and readiness factors of EU companies for adopting advanced manufacturing products and technologies. Publications Office (2016)
9. Schuh, G., Anderl, R., Gausemeier, J., Wahlster, W. Industrie 4.0 Maturity Index - Managing the Digital Transformation of Companies (acatech STUDY). Herbert Utz Verlag (2017)
10. da Silva, R.: Por que não cresce a economia portuguesa? Uma viagem à competitividade das boas empresas de Portugal. Caleidoscópio (2018)
11. Stefan, L., Thom, W., Dominik, L., Dieter, K., Bernd, K.: Concept for an evolutionary maturity based Industrie 4.0 migration model. Procedia CIRP **72**, 404–409 (2018). https://doi.org/10.1016/j.procir.2018.03.155
12. Tonelli, F., Demartini, M., Loleo, A., Testa, C.: A novel methodology for manufacturing firms value modeling and mapping to improve operational performance in the Industry 4.0 Era. Procedia CIRP **57**, 122–127 (2016). https://doi.org/10.1016/j.procir.2016.11.022
13. WEF. The Inclusive Growth and Development Report. World Economic Forum. https://www.weforum.org/reports. Accessed 10 Dec 2019
14. WEF. Readiness for the Future of Production Assessment 2018 edition. World Economic Forum. https://www.weforum.org/reports. Accessed 10 Dec 2019
15. WEF. The Global Competitiveness Index 2017–2018 edition. World Economic Forum. https://www.weforum.org/reports. Accessed 10 Dec 2019
16. WEF. Global Competitiveness Index 4.0 2019 edition. World Economic Forum. https://www.weforum.org/reports. Accessed 10 Dec 2019

Application of Linear Regression for Evaluation of Production Processes Effectiveness

Małgorzata Grzelak, Anna Borucka(✉), and Patrycja Guzanek

Military University of Technology, 00406 Warsaw, Poland
{malgorzata.grzelak,anna.borucka}@wat.edu.pl,
patrycja.guzanek@student.wat.edu.pl

Abstract. The effectiveness of production processes is a measure of the proper control of an enterprise, its development and effective meeting the market's needs. Production capacity planning and process scheduling is undoubtedly the key to success. However, in many cases, it is first of all necessary to analyse the current situation, eliminate waste and look for areas to improve business operations. Such actions are supported by ongoing analyses of the tasks being performed and assessment of the accuracy of decisions. In enterprises without modern IT systems, enabling permanent analysis and assessment of production processes, mathematical tools and methods can be helpful. One of them, the linear regression, is presented in this article, highlighting the factors which, among the subjects studied, significantly affect the modelled phenomenon, and the degree of such impact. The analysis was carried out on the example of a manufacturing company from medical industry.

Keywords: Multifactorial regression · Production processes effectiveness · Operations management

1 Introduction

The activities of manufacturing companies are profit-driven through the effective implementation of their objectives. They can be classified in two main dimensions, the economic and material ones. The first is to ensure good financial condition of the company by improving the efficiency of its operations. In the second one, the activities are aimed at increasing market share and meeting its needs by improving productivity. The guarantee of effective accomplishment is the implementation of an appropriate method of production management, enabling the reconciliation of two orientations: the market one focused on the customer, and the efficiency orientation, serving to improve the effectiveness of operations. In the era of the industrial revolution 4.0, the wide use of technical equipment controllers, IT software (e.g. MES or ERP) and industrial robots allows for ongoing diagnostics and control of the production processes being realised, leading to an increase of effectiveness. Moreover, it allows for their full autonomy, both in terms of organisation and the ability to quickly respond to changes in market demand.

The reasons behind the fourth generation of the industrial revolution are the availability of large information resources and the ability to archive, process and make right

J. Machado et al. (Eds.): icieng 2021, LNME, pp. 36–47, 2022.
https://doi.org/10.1007/978-3-030-78170-5_4

decisions based on them. However, analyses of the Polish producers' market indicate that only 15% of Polish factories are automated, and the main methods of data collection are still dominated by the manual creation of electronic databases (59% of entities surveyed) and only paper documentation (16% of companies) [1]. Furthermore, the research conducted reveals that only 14% of the respondents have specialised analytical teams in their structures, while 53% of companies do not have significant skills in this area, using basic skills of individual employees on an ad hoc basis [2]. Therefore, from a practical point of view, it is still important to develop simple mathematical tools to support decision making in the production planning process, allowing the quantitative and qualitative assessments of the implemented activities to be made based on the available information resources. Advanced mathematical analysis tools are not useful in such cases, hence there is a need for the application of uncomplicated, utilitarian methods. One of them is described in this article. The multifactorial regression model was chosen. The study was based on data of a manufacturing process of laboratory tips in the medical industry. The purpose of the publication is to present the possibility of using the regression model to assess factors affecting the effectiveness of the production process, measured in this case by the performance of technical equipment (machinery), as well as the diagnostics of the analysed company in this respect.

2 The State of the Problem - Literature Analysis

The development of production systems and increased competition on the market requires entrepreneurs to permanently analyse their production processes to identify downtime and eliminate waste. Effective assessment is possible only through the use of appropriate methods of supervising and controlling production processes, including the operation of machinery and technical equipment. It should be based on reliable information on the condition of the entire system. It is also important to choose the right measures to assess the effectiveness of the company equipment, adapted to the data that can be obtained, reflecting the mode of operation of individual machines, as well as being in line with the objectives set by the organisation [3]. The analysis of the literature showed many possibilities of assessing the effectiveness of technical equipment. Economic (e.g. profitability index, fixed and variable costs of machine maintenance), technical and operational (e.g. machinery reliability index), information and economic (e.g. repair service intensity indicator) or safety indicators (number of accidents at service and using machines) can be distinguished [4].

The above indicators can be used at various levels of the organisation to assess the effectiveness of both, the entire production process and the selected production line or a single machine. The method of selection for analysing the production system performance is defined in the PN-EN:15347:2019-12 standard. It provides a list of key performance indicators (KPIs) which enable comprehensive assessment and improvement of the efficiency and durability of technical equipment maintenance, based on both, external (e.g. location of the entity, national labour policy, market and economy or environmental conditions), and internal factors (e.g. size of the organisation, its structure and objectives, complexity of implemented processes or age of technical equipment). This document, as a method of streamlining production processes, recommends using the Deming cycle, which includes four basic steps: defining the organisation objectives and identification of KPIs, measuring the current KPI values using qualitative and quantitative assessment methods, comparing current KPIs with their nominal values, and analysing differences, as well as the implementation of activities to improve production processes [5].

In the era of industrial revolution 4.0, numerous companies have modern machine parks in which, by integrating digitally controlled machines with information technologies, it is possible to read indicators supporting the supervision of production processes in real time [6, 7]. However, particularly in smaller enterprises (also in the one being the subject of the study), the records are based on hand-drawn data sheets in Excel. This hinders the ongoing monitoring, but does not prevent an ad hoc analysis and performance evaluation [4]. However, it requires the use of appropriate mathematical methods.

High production effectiveness is primarily determined by the reliability of the machines and their ease of maintenance. Literature analysis proved that the above reliability models and operational efficiency models were most commonly used for the above purpose. Reliability models allow a temporary assessment of the functioning of technical facilities. The MTBF (Mean Time Between Failure) and MTTR (Mean Time To Repair) values are most commonly calculated [8, 9]. The former is used to measure reliability based on the average time between failures and their frequency [10], while the latter is based on the measurement of the average real time of repair from the moment of reporting to removing the fault [11].

The operational effectiveness model for measuring the performance of a production system uses primarily the OEE indicator (Overall Equipment Effectiveness) [12, 13]. It is the product of three elements related to the availability, effectiveness of the machine park and the quality of manufactured products [14, 15].The above model enables the assessment of the entire production process, taking into account the time of machine availability, their performance and the quality index of products manufactured. The essence of OEE is based on a comparison of the actual use of the machine with its performance indicators in ideal conditions, i.e. when production and its preparation is carried out according to schedule. Following the literature, it should be assumed that the desirable reference point is the level of global production proposed by Seichi Nakajime, for which the OEE is 85.41% [4].

For the examined enterprise, one of the parameters of the OEE indicator, i.e. the performance index, was indicated as the most desirable in the assessment of the effectiveness of implemented activities. It is measured as the quotient of the actual to optimal production (350 pieces/day in the examined enterprise) as per the relationship (1):

$$W_Q = \frac{Q_r}{Q_O} \tag{1}$$

where:
W_Q – machine operational effectiveness index;
Q_R – actual production volume [pcs./24 h];
Q_O – required production volume [pcs./24 h].

Calculated that way, it illustrates the degree of effective use of technical equipment, enabling the identification of areas which need improvement.

In the literature analysed, to assess the production process effectiveness, the most frequently used are the measures regarding the degree of effective use of operations and workplace [16], as well as those related to unproductive time [17]. However, multiple regression models are not widely used. Among the items available, Turóczy et al. point to the possibility of its application in the chemical industry to increase competitiveness, flexibility and adaptability in terms of the cost of operations [18]. However, Anayeche et. al. developed a comparison of the effectiveness of the assessment of the performance of production processes using multiple regression and artificial neural networks. They pointed out that the latter tool may provide better results, but it is more complicated in use and interpretation, and requires more complex input data [19].

The article consists of three main parts. The first briefly reviews the literature in terms of the available indicators for assessing the effectiveness of machines and technical equipment in production processes. Then, the subject of the study is described and the multifactorial regression model is discussed. The third part presents the results obtained, indicating the factors which significantly affect the volume of production. The entire elaboration ends with conclusions and recommends the directions for further research.

3 Methodology and Subject of the Study

The subject of the study is a medical industry company that specialises in the production of laboratory, sterile and non-sterile tips, with a capacity from 0.5 to 1,200 μl. The production is based on injection machines which work in conjunction with an automatic packaging system. Stainless steel moulds are used in the manufacturing process. The manufacturing process consists of three phases. Plasticised plastic is injected into a mould cavity closed in the machine, forming a moulding which is then pressed in the mould and cooled. After the pressing phase, the plastic is drawn from a hopper and plasticised, and the unit is moved away from the mould. In the last step, the moulding is removed from the mould cavity and automatically packaged on trays.

The study covered the effectiveness of the production process of disposable medical pipette tips manufactured by a selected machine. The data collected dated back to 2018 (from January to December) and concerned the records of each of two (12 h) shifts completed in the continuous production cycle. That way, a sheet containing 504 observations was obtained, including variables such as shift number, tip type, number of man-hours, tray type, employee number, and number of trays.

Linear multifactorial regression was applied for the study, which allows a qualitative approach to the relationship between many independent variables and a dependent variable. The general form of the regression model is represented by the Eq. (2):

$$Y = F\left(X_1, X_{2,\ldots,} X_3,\ \varepsilon_t\right) \tag{2}$$

- where:
- Y – dependent variable;
- $X_1, X_{2,\ldots,} X_3$ – independent variables;
- ε_t – internal disturbances.

An estimation of the regression function Y in relation to X in general population is the regression function y in relation to x in a random sample, called an approximate with the form:

$$\hat{y}_i = a_1 x + a_0 + u_i \quad (3)$$

where:

$$\mathrm{u_i} = \mathrm{y_i} - \hat{\mathrm{y}}_\mathrm{i} \quad (4)$$

Model parameters are estimated using the method of least squares (LSM). Diagnosis of the regression model includes statistical verification of the model, checking its significance and significance of individual parameters, as well as the residual analysis. The following assumptions are made for the linear regression models:

- the variance of the random component is equal for all observations: $D^2(\varepsilon_i) = \sigma^2$ for $i = (1,\ 2, \ldots,\ n)$;
- there is no significant impact of factors not included in the model on the average value of the dependent variable: $E(\varepsilon_i) = 0$ for $i = (1,\ 2, \ldots,\ n)$;
- there is no autocorrelation of random components;
- random component ε_i has a distribution $N(0, \sigma)$ for $i = (1,\ 2, \ldots,\ n)$ [20].

At the first stage of the study, the collected data was evaluated. In the examined enterprise, 80% was considered a satisfactory level of performance, and its variability is shown in Fig. 1. An analysis of the box-plot and basic descriptive statistics indicates that most observations are below the accepted level of satisfaction. Therefore, it is important to identify factors which significantly affect the production processes and to point out the areas which can be improved.

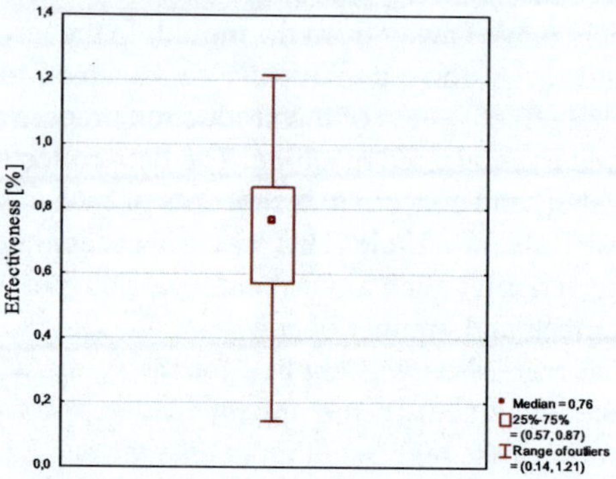

Fig. 1. Production performance box-plot

Based on the selection of information contained in the worksheets, explanatory variables were specified and their impact on the effectiveness of the production process was assessed. A quantitative variable was selected: *the number of man-hours* and qualitative ones: *shift (first, second), calendar month, employee, type of tip* (10 μl, 1000 μl) and *type of tray (FR flat red, DR deep red, FW flat white, DB deep blue).*

The Anova variance analysis makes it possible to assess the significance of the impact of qualitative factors on the dependent variable. However, it requires that the dependent variable in individual groups have a normal distribution. Therefore, in the first step, the compliance of the variable *performance* distribution with the normal distribution was checked in groups related to the *month*, the *employee operating the machine*, the *type of tray*. To this end, the Shapiro-Wilk test was applied in which the null hypothesis assumes that the distribution of the examined feature is consistent with the normal distribution, while according to the alternative hypothesis, the distribution of the examined feature is different from the normal distribution. The results obtained demonstrate that the distributions examined do not comply with the normal distribution, therefore there are no grounds to apply the variance analysis. In such a case, it is possible to use a non-parametric equivalent, i.e. the Kruskal-Wallis test, the results of which in individual groups are presented in Table 1.

Table 1. Kruskal-Wallis test results.

Variable	K-W test statistic value	*p*-value
Month	145.88	0.00
Employee	20.89	0.00
Tray type	114.89	0.00
Shift	4.88	0.02
Tip type	160.58	0.00

The results of the Kruskal-Wallis test indicate a significant impact of the variables studied on the modelled phenomenon. The value p in each case it is lower than the adopted level of significance $\alpha = 0.05$.

For the qualitative variable *number of man-hours*, the correlation coefficient was examined with the variable *performance*. It was obtained $r = 0{,}21$, the value of t-student's test statistics was 4.47, and the p-value 0.00 which means a weak but statistically significant relationship between variables.

Later, detailed analyses of individual groups were carried out to minimise the number of categories. Box-plots were applied (examples for the month and type of tray are shown in Fig. 2 and 3) and multiple comparisons of average ranks for all samples.

The analysis made it possible to combine variables which do not differ significantly from each other into groups. For the months, 4 groups were proposed including the registered observations: in October (1st Group), in March, May and November (2nd Group), in February, April and August (3rd Group), in January, June, July, September and December (4th Group). Employees were divided according to the performance achieved. Only one of them clearly stood out from the others (employee 6), therefore two subgroups were created: the first one including the employee no. 6 and the second including the remaining staff. For the type of trays, the analysis showed no significant differences

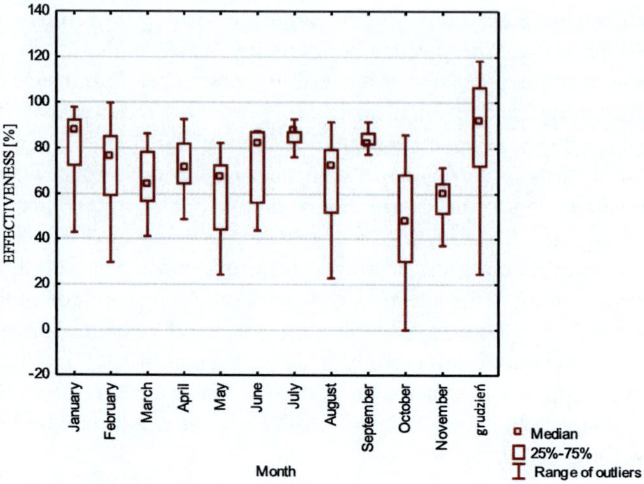

Fig. 2. Performance box-plot for individual months

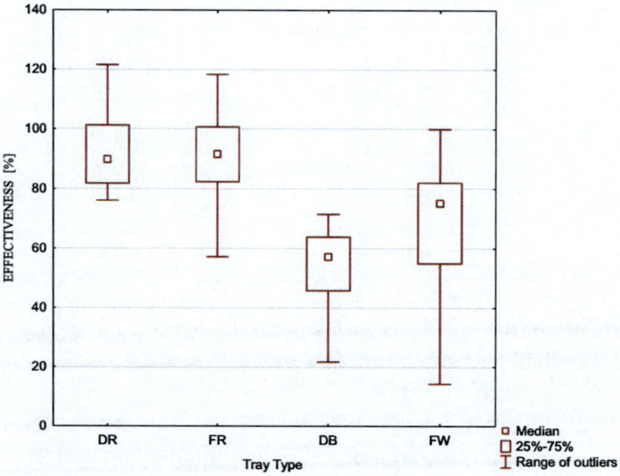

Fig. 3. Performance box-plots depending on the tray type

in the effectiveness when using deep red (DR) and flat red (FR) trays, therefore it was decided to combine them, while the others, i.e. the flat white (FW) and deep blue (DB) trays were left unchanged.

4 Multiple Regression Model

Distinguished regressors were used to build a multifactorial regression model. Table 2 presents the estimated parameters of functions.

Table 2. Estimated parameters of the regression model.

Corrected $R^2 = 0.33$ $F(9.49) = 28.99$, $p < 0.00$, Std. estimation error = 0.21 N = 504	b	Std. error	t(494)	p
Man-hours (mh)	0.01	0.00	4.56	0.00
DR/FR tray type (DR/FR)	0.13	0.04	3.61	0.00
FW tray type (FW)	−0.05	0.02	−2.53	0.01
Tip 10 (t.10)	0.11	0.04	2.81	0.01
First shift (fs)	0.15	0.04	3.58	0.00
Remaining staff (rs)	0.23	0.04	5.43	0.00
Months: 2nd group (m2)	0.08	0.03	2.30	0.02
Months: 3rd group (m3)	0.19	0.03	6.80	0.00
Months: 4th group (m4)	0.13	0.04	2.94	0.00

All estimated parameters are statistically significant, which is proven by the calculated *p* value of the test statistics which, in each case, is below the assumed significance level $\alpha = 0.05$.

They are interpreted as follows. Individual increase in the number of man-hours results in a 1% increase in the effectiveness. Production on the first shift leads to effectiveness increase by 15%. The use of deep and flat red trays in the production process increases the effectiveness by 13%, with a 5% decrease when using flat white trays. Effectiveness in the production of tips with a capacity of 10 μl increases the productivity by 11% compared to the 1,000 μl tips. In a group of employees, the reference point was the sixth employee with the lowest performance. When the machine is operated by other employees, the process effectiveness increases by 23%. In the group of months, the reference point assumed was October, with the lowest average effectiveness value. Production in months of the group 2 (March, May, November) increases the effectiveness by 8%, in the group 3 months (February, April, August) the above indicator increases by 19% and by 13% during work in months of the group 4 (January, June, July, September, December).

The regression equation takes the following form (3):

$$y = 0.01 \cdot x_{mh} + 0.13 \cdot x_{DF/FR} - 0.05 \cdot x_{FW} + 0.11 \cdot x_{t.10} + 0.15 \cdot x_{fs} + 0.23 \cdot x_{rs.} + 0.08 \cdot x_{m2} + 0.19 \cdot x_{m3} + 0.13 \cdot x_{m4} \quad (5)$$

where:
mh – man-hours, DF/FR – DR/FR tray type, FW – FW tray type, t.10 – tip 10, fs – first shift, rs – remaining Staff, m2 – months: 2nd group, m3 – months: 3rd group, m4 – months: 4th group.

Model verification was based on residual analysis. The compliance of their distribution with the normal distribution was verified. The Shapiro-Wilk test statistics was

$S - W = 0, 825$ and the value $p = 0.00$, and it is below the adopted level of significance $\alpha = 0.05$, indicating that there were no grounds for adopting the null hypothesis of compliance with the normal distribution.

The next step was to examine the the independence of observation errors. The Durin-Watson test statistics was calculated which was DW = 1.53 and indicated the existence of a positively correlated random component. This is confirmed by the autocorrelation plot shown in Fig. 4.

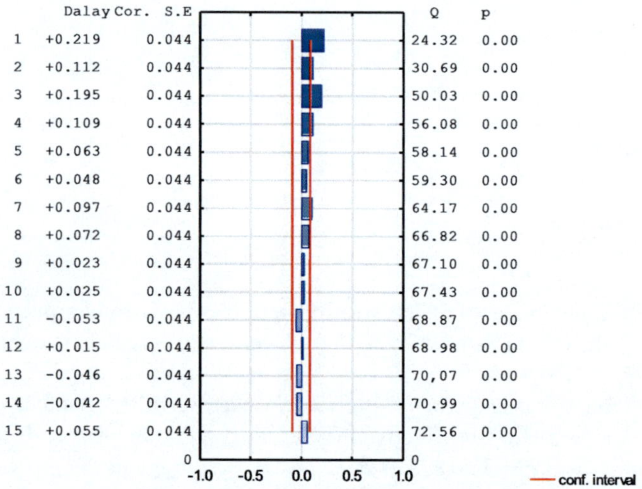

Fig. 4. Model residual autocorrelation function plot

The above results indicate that the model did not fully reflect all relationships present in the process. The determination coefficient obtained is not high $R^2 = 33\%$ and the residuals do not meet the required assumptions. Nevertheless, the proposed model meets its function in qualitative analyses, indicating areas which require modification. It is primarily inefficient staff management (the efficiency of employees of the second shift is significantly lower than that of those working during the day) and problems in the packaging process (placing laboratory tips in any tray other than the red one) that reduces the effectiveness of the process.

The reason for only satisfactory matching of the model is the machine downtime. In Fig. 5 showing the plot of observed and predicted values, it is clearly seen that the regression function does not reflect non-production situations which affects the value of the model residuals, but in the others it well reflects the variability of the series. The calculated values of ex post errors presented in Table 3 are also satisfactory.

The mean squares residual (MSE) is close to zero, while the mean absolute error (MAE) indicates that the actual implementations of the performance variable will deviate by approximately 14% from the predicted values. This allows for the model to be considered satisfactory. Furthermore, it indicates that the developed tool can effectively support decision making in the production management process.

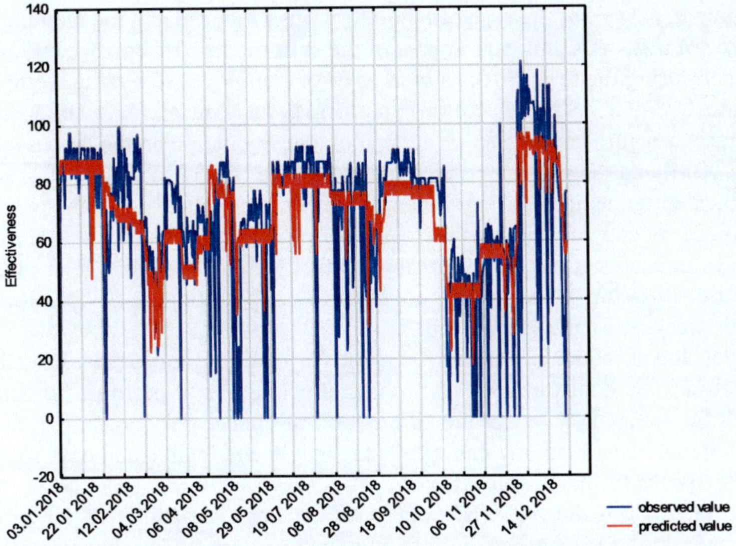

Fig. 5. . A plot of values observed and predicted by the model

Table 3. Quality of predictions of the developed model

Mean squares residuals (MSE)	0.04
Mean absolute error	0.14
Relative mean squared error	0.12
Relative average deviation (MRAE)	0.23

5 Conclusions

The study demonstrated that inefficient staff management significantly reduces performance. The productivity of employees on the second shift is significantly lower than that of those working during the day, and the use of machinery by operator no. 6 results in a decrease in the volume of actual production. Moreover, the laboratory tip packaging system should be subjected to detailed analysis. It is necessary to implement improvement measures aimed at identifying the reasons for the decrease in effectiveness when packaging products in trays other than red.

The nature of the business, the method of data collection and the lack of a specialist, analytical team prompted the search for simple mathematical tools which could support decision making processes in the company. A multifactorial regression model was proposed which is a simple and accessible tool enabling the assessment of implemented processes by non-specialist personnel in advanced data analysis. The dependent variable was the effectiveness, and the variables explaining the number of man-hours assigned to product making, type of shift, month of production, number of the employee operating the equipment, as well as the type of laboratory tips produced and the method of packaging (type of tray). A significant impact of the shift on the performance indicator (it is 15% higher during the first shift), as well as of the employee (operating equipment by operator no. 6 results in a 23% decrease in performance) was identified. The analysis also demonstrated a significant impact of the manufacturing month on the process effect (in February, April and August it increased by 19%). The type of laboratory tips produced and the packaging method also significantly affect the effectiveness of the process. The production of smaller capacity tips increases the productivity by 11%, while packaging in red trays increases it by 13%. The above results indicate a clear need to take corrective actions leading, among others to improve the performance of operator no. 6 or employees working on the second shift.

The developed model is primarily suitable for qualitative analyses. It well reflects the impact of individual variables on process effectiveness. It performs its function in the search for areas allowing for the elimination of unnecessary downtime. Moreover, it is versatile and usable in any enterprise, the results are legible and easy to interpret, which also speaks for its usefulness.

However, since it did not fully cope with the large variability of the process, in particular when the performance drops to zero, in the future studies other models will be considered that may be applicable to this type of empirical data, for example the logistic regression model.

The objective of the article was to present the possibility of using the multifactorial regression model to assess the effectiveness of the laboratory tip production process in terms of factors affecting it. Preliminary statistical analysis demonstrated that in the enterprise analysed, most observations related to situations in which the effectiveness is below the level of 80% assumed by the company. Therefore, the main factors affecting the implemented processes were identified and the degree of their impact was determined, ultimately aiming to identify areas and operations which need improvement.

References

1. Raport "Drogi do przemysłu 4.0. Robotyzacja na świecie i lekcje dla Polski", Polski Instytut Ekonomiczny, Warszawa (2019)
2. Raport "Smart industry Polska 2019 – inżynierowie w dobie czwartej rewolucji przemysłowej", Ministerstwo Przedsiębiorczości i Technologii/Simens, Warszawa (2019)
3. Świderski, A., Borucka, A., Grzelak, M., Gil, L.: Evaluation of the machinery readiness using semi-Markov processes. Appl. Sci. **10**(4), 1541 (2020)
4. Antosz, K.: Metodyka modelowania, oceny i doskonalenia koncepcji Lean Maintenance. Oficyna Wydawnicza Politechniki Rzeszowskiej, Rzeszów (2019)
5. Szwedzka, K., Jasiulewicz-Kaczmarek, M.: Determining maintenance services using production performance indicators. Res. Logist. Prod. **6**, 361–374 (2016)
6. Timofiejczuk, A., Brodny, J., Loska, A.: Exploitation policy in the aspect of Industry 4.0 concept - overview of selected research. Multi. Aspects Prod. Eng. **1**(1), 353–359 (2018)
7. Jasiulewicz-Kaczmarek, M., Gola, A.: Maintenance 4.0 technologies for sustainable manufacturing – an overview. IFAC PapersOnLine **52–10**, 91–96 (2019)
8. Jasiulewicz-Kaczmarek, M.: Identification of maintenance factors influencing the development of sustainable production processes–a pilot study. IOP Conf. Ser. Mater. Sci. Eng. **400**, 062014 (2018)

9. Antosz, K.: Maintenance – identification and analysis of the competency gap. Eksploatacja i Niezawodnosc – Maintenance Reliab. **20**(3), 484–494 (2018)
10. Ratnayake, R.M.C., Antosz, K.: Risk-based maintenance assessment in the manufacturing industry: minimisation of suboptimal prioritization. Manage. Prod. Eng. Rev. **8**(1), 38–45 (2017)
11. Sobaszek, Ł, Gola, A.: Survival analysis method as a tool for predicting failures. Actual Probl. Econ. **177**(3), 421–428 (2016)
12. Maohammadi, M., Rai, P., Gupta, S.: Performance evaluation of bucket based excavating, loading and transport (BELT) equipment–an OEE approach. Arch. Min. Sci. **62**(1), 105–120 (2017)
13. Antosz, K., Pasko, L., Gola, A.: The use of intelligent systems to support the decision-making process in lean maintenance management. IFAC PapersOnLine **52–10**, 148–153 (2019)
14. Gola, A., Pastuszak, Z., Relich, M., Sobaszek, Ł., Szwarc, E.: Scalability analysis of selected structures of a reconfigurable manufacturing system taking into account a reduction in machine tools reliability. Eksploatacja i Niezawodnosc – Maintenance Reliab. **23**(2), 242–252 (2021)
15. Kosicka, E., Gola, A., Pawlak, J.: Application-based support of machine maintenance. IFAC-PapersOnLine **52–10**, 131–135 (2019)
16. Miao, R., Khanna, M.: Effectiveness of the biomass crop assistance program: roles of behavioral factors, credit constraint, and program design. Appl. Econ. Perspect. Policy **39**(4), 584–608 (2017)
17. Jasiulewicz–Kaczmarek, M., Antosz, K., Żywica, P., Mazurkiewicz, D., Sun, B., Ren, Y.: Framework of machine criticality assessment with criteria interactions. Eksploatacja i Niezawodnosc – Maintenance Reliab. **23**(2), 207–220 (2021)
18. Turóczy, Z., Liviu, M.: Multiple regression analysis of performance indicators in the ceramic industry. Procedia Econ. Financ. **3**, 509–514 (2012)
19. Anayaeche, C., Ighravewe, D.E.: Predicting performance measures using linear regression and neural network: a comparison. Afr. J. Eng. Res. **1**(3), 84–89 (2013)
20. Dette, H., Munk, A.: Validation of linear regression models. Ann. Stat. 778–800 (1998)

A Novel Discrete Particle Swarm Optimization Algorithm for the Travelling Salesman Problems

José A. Sequeiros[1], Rui Silva[1], André S. Santos[1], J. Bastos[1], M. L. R. Varela[3,4](✉), and A. M. Madureira[1,2]

[1] Institute of Engineering, Polytechnic Institute of Porto - ISEP/IPP, Porto, Portugal
{1180162,1180146,abg,jab,amd}@isep.ipp.pt
[2] Interdisciplinary Studies Research Center (ISRC), Porto, Portugal
[3] University of Minho, Braga, Portugal
leonilde@dps.uminho.pt
[4] Algoritmi Research Centre, Guimarães, Portugal

Abstract. There are Optimization Problems that are too complex to be solved efficiently by deterministic methods. For these problems, where deterministic methods have proven to be inefficient, if not completely unusable, it is common to use approximate methods, that is, optimization methods that solve the problems quickly, regardless of their size or complexity, even if they do not guarantee optimal solutions. In other words, methods that find "acceptable" solutions, efficiently. One particular type of approximate method, which is particularly effective in complex problems, are metaheuristics. Particle Swarm Optimization is a population-based metaheuristic, which has been particularly successful. In order to broaden the application and overcome the limitation of Particle Swarm Optimization, a discrete version of the metaheuristics is proposed. The Discrete Particle Swarm Optimization, DPSO, will change the PSO algorithm so it can be applied to discrete optimization problems. This alteration will focus on the velocity update equation. The DPSO was tested in an instance of the Traveling Salesman Problem, att48, 48 points problems proposed by Padberg and Rinaldi, which showed some promising results.

Keywords: Metaheuristics · Particle Swarm Optimization · Discrete optimization

1 Introduction

Optimization problems are transversal to many knowledge areas, as science, engineering, management or business [1–11]. There are some that are defined as hard optimization problems, which cannot be solved optimally by a deterministic method, within a practical time frame [12]. There are several categories for these problems, as depending on if they are continuous or discrete, constrained or unconstrained, mono or multi-objective, static or dynamic [13]. In this paper, a discrete version of a well-known metaheuristic is proposed, particularly, a discrete version of the Particle Swarm Optimization. The procedure of this discrete version of Particle Swarm Optimization, DPSO, will be

J. Machado et al. (Eds.): icieng 2021, LNME, pp. 48–55, 2022.
https://doi.org/10.1007/978-3-030-78170-5_5

demonstrated with a Route Planning Problem, specifically Travelling Salesman Problem (TSP). Route planning is a type of problem which that aims to determine the shortest, most cost-effective, available route from point (x) to point (y) on a map, stopping at each given point (city). A TSP is a route planning problem in which a tour visits each city exactly once [14]. It is a problem which is easy to define but, as the number of points increases, the difficulty increases exponentially [15]. This paper will be divided in five parts. In Sect. 2, a Literature Review which will focus on metaheuristics and Particle Swarm Optimization, in Sect. 3, Discrete PSO will be explored, Sect. 4 will focus on the Proposal, Sect. 7, Preliminary results and Sect. 8, Conclusions.

2 Literature Review

2.1 Metaheuristics

In the last three decades, a great interest and attention have been payed to metaheuristics [16]. There are Optimization Problems that are too complex to be solved efficiently by deterministic methods. For these problems, where deterministic methods have proven to be inefficient, if not completely unusable, it is common to use approximate methods, that is, optimization methods that solve the problems quickly, regardless of their size or complexity, even if they do not guarantee optimal solutions. In other words, methods that find "acceptable" solutions, efficiently. One particular type of approximate method, which is particularly effective in complex problems, are metaheuristics. A metaheuristic is an algorithm designed to solve a variety of hard optimization problems without the need of great adjustments on each problem [12, 14]. As such, there is a wide range of different metaheuristics which will vary according to the problem. There can be single-solution based metaheuristics such as local search [17], simulated annealing [17, 18] or tabu search [19], among others, or Population-Based Metaheuristics as Evolutionary Algorithms [20] and Swarm Intelligence [21], in which PSO and DPSO is included [14]. According to Boussaïd, metaheuristics share characteristics as their inspiration by some nature principles. These principles are borrowed from physics, biology or ethology and they have several parameters that need to be fitted to the problem [16].

2.2 Particle Swarm Optimization

The Particle Swarm Optimization (PSO) is a population-based optimization method first proposed by Kennedy and Eberhart [22]. A swarm, in this context, is a population of several solutions of a given problem, and each solution is seen as an organism in a social environment, also called particle. This method attempts to mimic the behaviour of a swarm of individuals moving on the solution space. Its main inspiration is the behaviour of a school of fish or a flock of birds. It can be used to solve a wide variety of problems, including neural network training and function minimization [23]. The most important features of the PSO are the simplicity of implementation and the absence of information requirements for the gradient [24]. PSO algorithm is an extremely effective optimization technique to solve multimodal continuous optimization problems [21]. As said before, the implementation is very simple as it generates a random initial population of particles, in which every individual particle keeps memory of three parameters: position, velocity and fitness, which provides a solution of the problem. At every iteration, each individual particle updates its position based on the velocity calculated considering two factors; its previous best position (the best solution obtained by that specific particle) and the best one among all the particles in the population (the best solution obtained by the swarm). The algorithm will iterate continuously until the stopping criteria is reached [22, 25].

```
Initialize a population of particles with random positions and velocities
  Repeat
        Evaluate f (x_i) ;
        For all particles i
           Adapt velocities using equation (1):
                  (1)   v_i(t) = v_i(t − 1) + ρ1 × (p_i − x_i(t − 1)) + ρ2 × (p_g − x_i(t − 1));
           Update position using equation (2):
                  (2)   x_i(t) = x_i(t − 1) + v_i(t) ;
           If f (x_i) < f (p_besti) Then p_besti = x_i ;
           If f (x_i) < f (g_best) Then g_best = x_i ;
           Update(x_i, v_i) ;
        EndFor
Until Stopping criteria
```

Fig. 1. Pseudocode of the particle swarm optimization algorithm, adapted from [14, 16]

With this method, PSO has been attained great results and applied on numerous continuous and combinatorial optimization problems [25]. Even though, it has been reported that this algorithm as some drawbacks as premature convergence or getting trapped in a local optima [16, 26]. There is considerable research work conducted into the original formulation of PSO, in both continuous and discrete problems, in order to reduce these restraints.

3 Discrete PSO

In order to be applied to discrete optimization problems, the PSO algorithm needs some reformulation. In their work of 1997, Kennedy & Eberhart proposed a new interpretation of PSO to solve these discrete problems. Both particles position and velocities are defined by changing probabilities that a bit is set to 0 or 1 [27]. This probability is set by a function of two factors, individual and social memory, which will allow the particles to move in a state space restricted to 0 and 1 [21]. Lately, several papers had described some changes to the initial PSO algorithm in order to solve TSP. For instance, Zhong et al., introduced a new parameter, called mutation factor, and it helped to keep balance between exploitation and exploration [28]. Additionally, different pairwise exchanging operations might be used, 2-swap or 2-exchange, for instance [29]. In this paper, it is proposed a new approach to DPSO, altering the velocity update equation.

4 Proposal

As said before, a swarm is a population of individuals seen as an organism is a social environment. As such, for each iteration, a neighbourhood must be defined for each individual. There are many possibilities to define such a neighbourhood. Conventionally, two methods are used, g_{best} or l_{best} [14]. In this paper, it has been chosen the g_{best} neighbourhood, in which every particle is allured to the best solution found by any other member of the swarm [30]. As said by Kennedy, this type of neighbourhood is equivalent to a fully connected social network where every particle is capable of compare performances with the other members of population. The best solution found by any member of the population represents the leader of the swarm, the one used to guide the group toward better solutions.

Each particle results in a combination of three vectors:

- The x-vector which keeps record of the current position (location),
- The p-vector which keeps record of the location of the best solution found by any other particle in the swarm,
- The v-vector which contains information for the direction, gradient, in which particle will move along.

In order to adapt an algorithm used to continuous problems, some adaptations to the algorithm in Fig. 1 must be made for discrete optimization problems. There are two main differences, which are the mapping between particle positions and velocity models [14]. There were made some changes from the model proposed by Talbi in his book [14], represented by Eq. 1 and 2, Velocity and particle update position, respectively.

$$v[n]_i(t+1) = Int\left[c_1 \times v[n]_i(t) \times Rnd() + c_2 \times \left(p[n]_i^{best} - p[n]_i(t)\right) \times Rnd()\right] \quad (1)$$

$$p[n]_i(t+1) = p[n]_i(t) + v[n]_i(t+1) \quad (2)$$

The main difference is:

- The random factor was substituted by a 10% probability of contribution of each social component for the velocity update, Eq. 3.

$$\begin{aligned} v[n]_i(t+1) = {} & Int(0{,}1 \times c_1 + Rnd()) \times v[n]_i(t) + Int(0{,}1 \times c_2 + Rnd()) \\ & \times \left(p[n]_i^{best} - p[n]_i(t)\right) + Int(0{,}1 \times c_3 + Rnd()) \times \left(p[n]_{gbest} - p[n]_i(t)\right) \end{aligned} \quad (3)$$

In which:

- c_1 is the Inertia factor, making the particle move with a position and velocity influenced by the previous state.
- c_2 is the influence of the historic best, which makes the algorithm more conservative.
- c_3 is the influence of the g_{best}.

The neighbourhood is generated by all pairwise interchange method, and the position update equation is the same as proposed by Talbi [14].

5 Preliminary Results

A preliminary study was carried out using att48, 48 points problems proposed by Padberg and Rinaldi, and the instances are available at [31].

For the experiments of DPSO, there were tested five parameters at three different levels. The initial values are based on the work of Odili, Peker and Van Den Bergh [15, 23, 32], and adapted for this paper accordingly. There have been performed three Taguchi experimental designs in order to identify the best parameters to solve the problem in study. For the design of experiments of DPSO, it was used a Taguchi L18 orthogonal design, since there were tested five parameters at three different levels. The initial values are described on Table 1, based on the work of Odili, Peker and Van Den Bergh [15, 23, 32], and adapted for this paper accordingly.

One important part of the study is the parameter optimization systemization which might be used widely to test a greater number of experiments in less time. The best parameters for DPSO were obtained 18 trials instead of 35. On Table 1 are the parameters which had the best results solving the att48 problem.

Table 1. Parameters Swarm size, Inertia weight, Historic best weight, G_{best} weight and Iterations without improvement with the best results solving att48.

Swarm size	Inertia weight	Historic best weight	G_{best} weight	Iterations Without improvement
162	1,5	2	0,25	1875

The best solutions were obtained using the parameters above, in which the sequence for the problem in study is:

1 8 9 38 31 44 18 7 28 6 37 19 27 17 43 30 36 46 33 20 47 21 39 48 5. 29 2 42 26 4 35 45 10 24 32 13 25 14 23 11 12 15 40 3 34 41 16 22.

This solution has a cost of 34385, 3% higher than the optimal 33522. The full results for the run are on the Table 2.

Table 2. Results for the att48 problem using DPSO with the optimized parameters.

Run	N° of iterations to best solution	Best solution	Deviation to optimal
1	1068228	36915	10%
2	1053648	37568	12%
3	1643814	42249	26%
4	1440342	34385	3%
5	1179198	36054	8%

Even though the results are few, the first results are promising.

6 Conclusions

Particle swarm optimization is simple algorithm that can solve effectively a wide range of problems. Although, it needs an adaptation in order to be applied to discrete optimization problems. With this paper, the initial algorithm was adapted to solve discrete problems and to minimize the probability of premature convergence and trapping in local optima.

Some changes are proposed, namely, the inclusion of a Inertia factor, influenced by the previous state and by a 10% probability of contribution of each social component for the velocity update, instead of the previous random factor. In addition, the neighbourhood is generated by all pairwise interchange method, with no changes in the position update equation proposed by Talbi.

The initial results are promising even though there is some space to improve. The algorithm takes a lot of time to run due to the fact that it had been developed in a VBA environment. The algorithm improved with the tuning parameter, even though a few more run should be needed.

As future work, it is proposed the optimization of coding in order to improve the efficiency and the time needed of each run and the use of more sophisticated operations on the neighbourhood search. With this, a new design of experiments should be run to tune up parameters.

Acknowledgement. This work has been supported by national funds through FCT – Fundação para a Ciência e Tecnologia within the project ref. UID/CEC/00319/2019.

References

1. Varela, M.L.R., et al.: Collaborative paradigm for single-machine scheduling under just-in-time principles: total holding-tardiness cost problem. Manage. Prod. Eng. Rev. **9**(1), 90–103 (2018)
2. Gangala, C., Modi, M., Manupati, V.K., Varela, M.L.R., Machado, J., Trojanowska, J.: Cycle time reduction in deck roller assembly production unit with value stream mapping analysis. In: Rocha, Á., Correia, A.M., Adeli, H., Reis, L.P., Costanzo, S. (eds.) WorldCIST 2017. AISC, vol. 571, pp. 509–518. Springer, Cham (2017). https://doi.org/10.1007/978-3-319-56541-5_52
3. Pavlenko, I., Trojanowska, J., Gusak, O., Ivanov, V., Pitel, J., Pavlenko, V.: Estimation of the reliability of automatic axial-balancing devices for multistage centrifugal pumps. Periodica Polytechnica Mechanical Engineering **63**(1), 52–56 (2019). https://doi.org/10.3311/PPme.12801
4. Varela, M.L.R., Putnik, G.D., Manupati, V.K., Rajyalakshmi, G., Trojanowska, J., Machado J.: Collaborative manufacturing based on cloud, and on other I4.0 oriented principles and technologies: a systematic literature review and reflections. Manage. Prod. Eng. Rev. **9**(3), 90–99 (2018)
5. Araújo, A.F., Varela, M.L.R., Gomes, M.S., Barreto, R.C.C., Trojanowska, J.: Development of an intelligent and automated system for lean industrial production, adding maximum productivity and efficiency in the production process. In: Hamrol, A., Ciszak, O., Legutko, S., Jurczyk, M. (eds.) Advances in Manufacturing. LNME, pp. 131–140. Springer, Cham (2018). https://doi.org/10.1007/978-3-319-68619-6_13

6. Starzyńska, B., Szajkowska, K., Diering, M.: A study of raters' agreement in quality inspection with the participation of hearing disabled employees – continuation. In: Diering, M., Wieczorowski, M., Brown, C.A. (eds.) MANUFACTURING 2019. LNME, pp. 212–220. Springer, Cham (2019). https://doi.org/10.1007/978-3-030-18682-1_17
7. Kujawinska, A., Vogt, K., Diering, M., Rogalewicz, M., Waigaonkar, S.D.: Organization of visual inspection and its impact on the effectiveness of inspection. In: Hamrol, A., Ciszak, O., Legutko, S., Jurczyk, M. (eds.) Advances in Manufacturing. LNME, pp. 899–909. Springer, Cham (2018). https://doi.org/10.1007/978-3-319-68619-6_87
8. Antosz, K. Maintenance – identification and analysis of the competency gap, Eksploatacja i Niezawodnosc – Maintenance Reliab. **20**(3), 484–494 (2018). http://dx.doi.org/10.17531/ein.2018.3.19
9. Ribeiro, R.A. Varela, L.R.: Fuzzy optimization using simulated annealing: an example set. In: Verdegay, J.L. (eds.) Fuzzy Sets Based Heuristics for Optimization, pp. 159–180. Springer, Heidelberg (2003). https://doi.org/10.1007/978-3-540-36461-0_11
10. Kunz, G., Machado, J., Perondi, E., Vyatkin, V.: A formal methodology for accomplishing IEC 61850 real-time communication requirements. IEEE Trans. Ind. Electron. **64**(8), 6582–6590 (2017). art. no. 7878522. https://doi.org/10.1109/TIE.2017.2682042
11. Leão, C.P., et al.: Web-assisted laboratory for control education: remote and virtual environments. In: Uckelmann, D., Scholz-Reiter, B., Rügge, I., Hong, B., Rizzi, A. (eds.) ImViReLL 2012. CCIS, vol. 282, pp. 62–72. Springer, Heidelberg (2012). https://doi.org/10.1007/978-3-642-28816-6_7
12. Hansen, P., Mladenovi, N., Brimberg, J.: Variable neighborhood search. In: Handbook of Metaheuristics, vol 57, pp. 61–86 (2003)
13. Ghoreishi, S.N. Clausen, A., Joergensen, B.N.: Termination criteria in evolutionary algorithms: a survey. In: IJCCI, pp. 373–384 (2017)
14. Talbi, E.-G.: Metaheuristics - From Design to Implementation (2009)
15. Peker, M., Şen, B., Kumru, P.Y.: An efficient solving of the traveling salesman problem: the ant colony system having parameters optimized by the Taguchi method. Turkish J. Electr. Eng. Comput. Sci. **21**(SUPPL. 1), 2015–2036 (2013)
16. Boussaïd, I., Lepagnot, J., Siarry, P.: A survey on optimization metaheuristics. Inf. Sci. (Ny) **237**, 82–117 (2013)
17. Henderson, D., Jacobson, S.H., Johnson, A.W.: The theory and practice of simulated annealing. In: Glover, F., Kochenberger, G.A. (eds.) Handbook of Metaheuristics, pp. 287–319. Springer, Boston (2003). https://doi.org/10.1007/0-306-48056-5_10
18. Nourani, Y., Andresen, B.: A comparison of simulated annealing cooling strategies. J. Phys. A. Math. Gen. **1**(41), 8373–8385 (1998)
19. Zhou, A.H., et al.: Traveling-salesman-problem algorithm based on simulated annealing and gene-expression programming. Information **10**(1), 7 (2018)
20. Eiben, A.E., Smit, S.K.: Evolutionary algorithm parameters and methods to tune them. In: Hamadi, Y., Monfroy, E., Saubion, F. (eds.) Autonomous search, pp. 15–36. Springer, Heidelberg (2011). https://doi.org/10.1007/978-3-642-21434-9_2
21. Shelokar, P.S., Siarry, P., Jayaraman, V.K., Kulkarni, B.D.: Particle swarm and ant colony algorithms hybridized for improved continuous optimization. Appl. Math. Comput. **188**(1), 129–142 (2007)
22. Kennedy, J., Eberhart, R.: Particle Swarm Optimization, p. 7, IEEE (1995)
23. Van Den Bergh, F.: An analysis of particle swarm optimizers, University of Pretoria (2001)
24. Pedersen, M. Good parameters for particle swarm optimization. Technical report HL1001, Hvass Lab., vol. HL1001, pp. 1–12 (2010)
25. Akhand, M., Hossain, S., Akter, S.: A comparative study of prominent particle swarm optimization based methods to solve traveling salesman problem. Int. J. Swarm Intell. Evol. Comput. **5**(3), 11 (2016)

26. Tewolde, G.S. Hanna, D.M., Haskell, R.E.: Enhancing performance of PSO with automatic parameter tuning technique. In: 2009 IEEE Swarm Intelligence Symposium SIS 2009 - Proceedings, no. 1, pp. 67–73 (2009)
27. Fan, H.: Discrete particle swarm optimization for TSP based on neighborhood. J. Comput. Inf. Syst. **6**(10), 3407–3414 (2010)
28. Zhong, W.L., Zhang, J., Chen, W.N.: A novel discrete particle swarm optimization to solve traveling salesman problem. In: 2007 IEEE Congress on Evolutionary Computation CEC 2007, pp. 3283–3287 (2007)
29. Tasgetiren, M.F., Kizilay, D., Pan, Q.K., Suganthan, P.N.: Iterated greedy algorithms for the blocking flowshop scheduling problem with makespan criterion. Comput. Oper. Res. **77**, 111–126 (2017)
30. Kennedy, J.: Small worlds and mega-minds: effects of neighborhood topology. In: Congress on Evolutionary Computation, pp. 1931–1938 (1999)
31. Discrete and Combinatorial Optimization, Universität Heidelberg. http://comopt.ifi.uni-heidelberg.de/software/TSPLIB95/
32. Odili, J.B.. Mohmad Kahar, M.N.: Solving the traveling salesman's problem using the African buffalo optimization. Comput. Intell. Neurosci. **2016**, 1–12 (2016)

The Algorithms for Robust Scheduling of Production Jobs Under Machine Failure and Variable Technological Operation Times

Łukasz Sobaszek(✉), Arkadiusz Gola, and Antoni Świć

Faculty of Mechanical Engineering, Department of Production Computerisation and Robotisation, Lublin University of Technology, Lublin, Poland
{l.sobaszek,a.gola,a.swic}@pollub.pl

Abstract. Robust scheduling of production jobs has been extensively studied. New methods are being sought to support effective production scheduling in the presence of disruptions caused by uncertain events. The range, extent and variety of uncertainty factors cannot be fully anticipated, which is why the solutions developed should necessarily allow for handling several disruptions simultaneously. Therefore, this study proposes a robust scheduling method that considers two factors contributing to the uncertainty of the schedule – machine failure and variable technological operation times. In particular the algorithms for robust scheduling using prediction tools: for prediction of machine failure, estimation and implementation of redundant service time buffers and for the prediction of variable technological operation times are presented.

Keywords: Robust scheduling · Uncertainty factors · Machine failures · Real technological operation times

1 Introduction

Minimising makespan is the major objective function of modern manufacturing systems [1]. To maintain high credibility with customers, manufacturers must execute production processes efficiently, but most essentially, they must maintain the timeliness of order completion [2, 3]. During production, numerous disruptions can occur that negatively impact on the process itself and on meeting contractual deadlines [4]. Therefore, effective planning and control of production flow, whose key element is scheduling, becomes crucial [5, 6]. Scheduling provides a breadth of information about the current and future state of executed processes [7–9].

1.1 Production Scheduling Under Uncertainty

Multiple disturbances negatively impact on the nominal operating conditions. Therefore, a well-developed robust schedule will take into account not only production processes but also other uncertainty factors, among which the most prominent are [10–14]:

J. Machado et al. (Eds.): icieng 2021, LNME, pp. 56–67, 2022.
https://doi.org/10.1007/978-3-030-78170-5_6

- resource uncertainty (machine/robot failure),
- order uncertainty (new orders or due dates)
- quality and supply uncertainty (material shortage, poor product quality),
- operation yield uncertainty (incorrect estimation of operation times),
- labour uncertainty (absence or malaise of employees, reduced or extended operation times).

A process subject to constant changes becomes disorganised. Hence, the current and growing trend for robust scheduling studies in production management studies [15, 16]. This approach aims to produce a schedule that will be resistant to potential disturbances, *i.e.* the one that accounts for the variability of production system parameters. A robust scheduling process consists of the following levels [2, 17]:

1. **Predictive scheduling** (*off-line*) of:

- nominal schedule – based on nominal system parameters,
- robust schedule – considering uncertainty and variability as part of an executed process.

2. **Reactive scheduling** in which the schedule is subject to modification after the occurrence of uncertainty, *i.e.* in production (*on-line*) – implemented process changes create an alternative version of the schedule.

The most commonly employed robust scheduling techniques include [15, 16]: redundancy-based techniques, contingent scheduling, partial reactive approaches and sensitivity analysis.

1.2 Practical Applicability of Robust Scheduling

Robust schedules in production environments have been intensively studied in recent years. This approach to the construction of schedules is strongly associated with complex problems – typically in flow-shop and job-shop production systems, which are widely used forms of production organisation in real manufacturing systems [5].

The literature on robust scheduling in the context of flow-shop environments is rather sparse. The authors tend to combine various techniques to solve flow-shop problems with a robust approach. The overwhelming majority of publications investigate predictive-reactive scheduling problems, mainly focusing on the reactive level. Several algorithms are typically selected for constructing schedules for flow-shop problems: genetic [18, 19], hybrid genetic [20, 21], simulated annealing [22] or local search algorithms [23]. Integer programming models have also been reported to handle job-shop problems [24]. Major challenges of scheduling are in terms of uncertainty of processing times and modification of orders, whereas the availability of equipment is considered as a secondary issue in research.

In comparison, a greater deal of previous research into robust scheduling has focused on job-shop systems. This model provides a better reflection of typical production environments, where the order of jobs is imposed by technological documentation. Researchers have been emphasising the need for devising a different approach than the theoretical one, thus voicing the need for a new trend to emerge in job-shop problem research, including a predictive and reactive approach. Predictive scheduling tends to remain somewhat in the background, whereas it is the reactive approach that is focusing the attention of researchers. In job-shop conditions, job scheduling is predominantly handled by: genetic algorithms and their hybrids [25–27], artificial immune systems [28], stochastic programming [29], knowledge-based expert systems [30] or simple technical solutions (*e.g.* shifting operations) [31].

The main uncertainty factor recognised in job-shop is the uncertainty of resources – machine failure, tool damage, employee turnover *etc.* Uncertainty of operation times is sporadically considered in job-shop scheduling investigations, whereas uncertainty associated with orders is given negligible attention.

Existing research on robust scheduling appears to omit the predictive scheduling level, as the attention of researchers is focused on investigating effective methods of reactive scheduling. In addition, even if solutions for the latter are presented, they usually address a single uncertainty factor.

Therefore, this work sets out to present an original method for designing a robust production schedule that accounts for two-factor uncertainty: (1) technological machine failure, (2) variable technological operation times.

A further benefit of the proposed solution is that it allows for the use of historical data, which are subsequently implemented in the predictive production scheduling model. In addition, historical data enable the analysis of relevant uncertainty factors.

2 Mathematical Model for Robust Job-Shop Scheduling Under Two-Factor Uncertainty

The job-shop scheduling problem consists in assigning jobs from the set of jobs $J = \{J_1, J_2, \ldots, J_n\}$ to the set of available machines $M = \{M_1, M_2, \ldots, M_m\}$ so that the schedule is optimised in terms of the objective function. Processing job J_i on machine M_j is described as operation. Simultaneously, we must consider the technology of processes, which is described by the matrix of machine orders $MO = [o_{ij}]$. Times of particular operations are described by the matrix of processing times $PT = [pt_{ij}]$. The size of matrices MO and PT is $m \times n$.

Machine failure uncertainty is determined from the following sets:

- $FT_{Mj} = \{ft_{Mj1}, ft_{Mj2}, \ldots, ft_{Mjn}\}$ describing potential failure times of machines (expressed in hours),
- $P_{Mj} = \{p_{Mj1}, p_{Mj2}, \ldots, p_{Mjn}\}$ describing the probability of machine failure,
- $TB_{Mj} = \{tb_{Mj1}, tb_{Mj2}, \ldots, tb_{Mjn}\}$ defining time buffers (length of potential service times in minutes).

Uncertainty of processing times is given by:

- matrix $PPT = [pt_{ij} + \Delta pt_{ij}]$ of pre-emptive processing times (the size of matrices PPT is $m \times n$).

The elements of sets FT_{Mj},P_{Mj}, TB_{Mj} and matrix *PPT* are computed by means of the robust scheduling method with prediction tools.

3 Algorithms for Robust Scheduling with Prediction Tools

Figure 1 presents the flowchart of the method, derived from the computational model. The two main components of the process are job scheduling processes and the prediction algorithms.

The scheduling was performed on a standard set of typical production data – the number of technological machines, the number of jobs and the machining technology. The data provided the input for developing a nominal production schedule, which was subsequently subjected to modifications using the prediction algorithms:

1. Modification I – machine failure constraint – time buffers introduced in case of potential failure of technological machines.
2. Modification II – variable technological operation time constraint – nominal operation times are updated according to the values obtained from prediction.

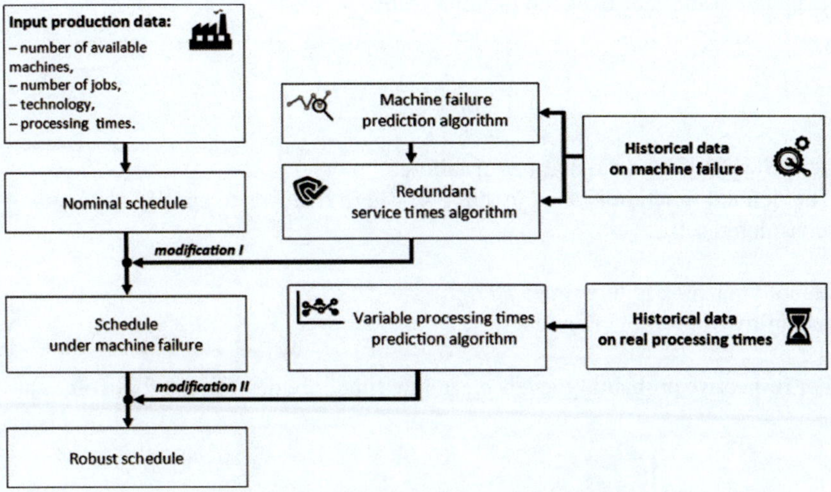

Fig. 1. The flowchart of the proposed method

Machine failure prediction involved the use of Survival Analysis, typically employed when a duration of a particular object or phenomenon is sought. On the other hand, it is used because it provides an important insight into examined processes, which, also in this case, becomes the basis for predicting the process behaviour [28]. The approach followed in the prediction of variable technological operation times assumed their times as variables. As a result, the probability calculus could be used, which allowed us to fit the empirical data to theoretical random variable distributions, thus providing the basis for the prediction of variable job processing times.

3.1 Algorithm for Predicting Machine Failure

The algorithm predicting machine failure determines potential times of failure ft_{Mji} and the probability of occurrence p_{Mji}. In order to estimate these parameters, it is necessary to implement the subsequent stages of the proposed algorithm.

In **stage 1** of the algorithm, the machine for which the prediction process will be carried out is designated, and transfer the historical data from the set T_{Mj}:

$$T_{Mj} = \{t_1, t_2, \ldots, t_n\} \text{ [hours]}, \tag{1}$$

where: t_i – i-th time of failure.

In **stage 2**, the failure data for machine M_j are sorted in the ascending order and saved by means of the sequence:

$$\{(t_i, d_i)\}_{1 \le k \le n},\ t_i \in T_{Mj}, \tag{2}$$

where: t_i – time between failures, d_i – number of failures.

The data are subsequently filtered so as to remove outliers (atypical values).

Stage 3 of the algorithm is crucial for further inference because it is at this stage that the Survival Function for an analysed machine is determined. This is performed by the Kaplan-Meier estimator from the formula below:

$$\hat{S}(t) = \begin{cases} 1, & for\ \ t < t_1 \\ \prod_{t_i \le t} \frac{r_i - d_i}{r_i}, & for\ t_1 < t \end{cases}, \tag{3}$$

where: $r_i = \sum_{j=i}^{k} d_j$ – total number of failures.

The defined function is used in **stage 4**, where it serves to specify elements of the two considered sets:

- potential machine failure times FT_{Mj},
- probability of machine failure P_{Mj}.

For respective probability levels p_i, failure times are determined ft_{Mji} (Fig. 2).

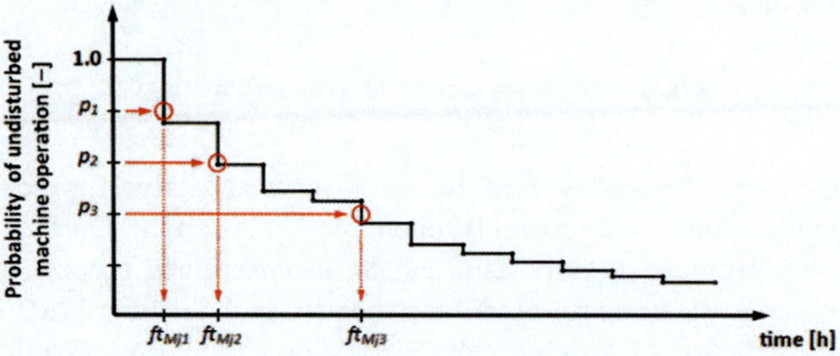

Fig. 2. Machine failure times derived from Survival Function

From the Survival Function given by the Kaplan-Meier estimator, the probability p_i of undisturbed machine processing time ft_{Mji} is determined from the specific historical data. Machine failure probability is obtained from:

$$p_{Mji} = 1 - p_i, \tag{4}$$

where: p_{Mji} – probability of failure, p_i – probability of undisturbed processing.

As a result, the algorithm establishes the failure times and failure probability expressed by (p_{Mji}, ft_{Mji}) and stored in the respective sets P_{Mj} and FT_{Mj}.

3.2 Algorithm for Estimating and Implementing Redundant Time Buffers

To produce a robust schedule, it is first necessary to estimate service times using historical data from set RT_{Mj}:

$$RT_{Mj} = \{rt_1, rt_2, \ldots, rt_n\}\,[\text{min}], \tag{5}$$

where: $rt_i \in \langle 0;\ 480\rangle$ – i-th service time (duration).

Stage 1 commences with loading, filtering and ordering historical data in ascending order. Thus obtained, the data are divided into appropriate subsets:

$$RT_{Mj} = \left\{RT_{Mj1}, RT_{Mj2}, \ldots, RT_{Mj8}\right\}, \tag{6}$$

where: RT_{Mji} – subset of service times.

The subsets obtained in stage 1 are used in **stage 2** of the algorithm. In this stage, typical times of machine failure are defined – the cardinality of each subset is determined, which in turn enables the visualisation of results and simplifies their analysis (Fig. 3). In stage 2, auxiliary sets are determined; to establish service times, the subset of highest cardinality from set RT_{Mji} is found. To this end, two auxiliary subsets are determined RT'_{Mj} and TB'_{Mj}:

$$RT'_{Mj} = \left\{\overline{\overline{RT}}_{Mj1}, \overline{\overline{RT}}_{Mj2}, \ldots, \overline{\overline{RT}}_{Mj8}\right\}. \tag{7}$$

then:

$$\bigvee_{\max\left(RT'_{Mj}\right)} TB'_{Mj} = RT_{Mji};\ TB'_{Mj} = \left\{rt_i \in \left\langle \min\left(RT_{Mji}\right); \max\left(RT_{Mji}\right)\right\rangle\right\}, \tag{8}$$

where: RT_{Mji} – subset of highest cardinality.

Stage 3 of the algorithm divides the estimated time buffers so as to estimate their size prior to implementation. Set TB_{Mj} is derived from:

$$TB_{Mj} = \left\{tb_{Mj1}, tb_{Mj2}, \ldots, tb_{Mjn}\right\}, \tag{9}$$

and particular elements of sets are determined from the relationship:

$$tb_{Mji} = \frac{\max\left(TB'_{Mj}\right)}{n_p} \cdot i, \tag{10}$$

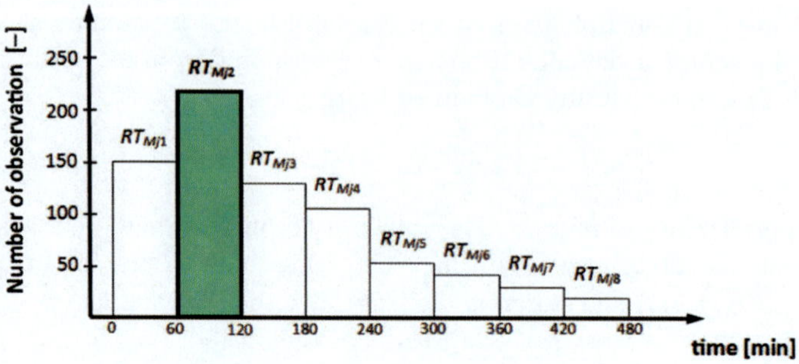

Fig. 3. Historiogram of subsets

where: i – the number of an element from set TB_{Mj},
n_p – the number of probability levels (elements of set P_{Mj}).

Depending on the number of probability levels considered by the user (elements of set P_{Mj}), stage 3 is iteratively repeated until satisfactory results are produced – set TB_{Mj}. The algorithm is executed for each machine included in the schedule that operates under failure constraint.

In **stage 4,** service time buffers are chosen using the algorithm below:

```
IF p_Mji ≤ p_kMj1
    choose time buffer tb_Mj1 (minimal)
WHEREAS IF p_Mji > p_kM1 and p_Mji ≤ p_kMj2
    choose time buffer tb_Mj2 (where tb_Mj2 > tb_Mj1)
    ...
WHEREAS IF p_Mji > p_kMj(n-1) and p_Mji ≤ p_kMjn
  choose time buffer tb_Mjn (where tb_Mjn > tb_Mj(n-1))
END OF CONDITION
```

In **stage 5** of the algorithm, service time buffers are implemented into the production schedule as indicated by the elements of set FT_{Mj} of machine. Any interfering operation is shifted right in the machine order of jobs (Fig. 4).

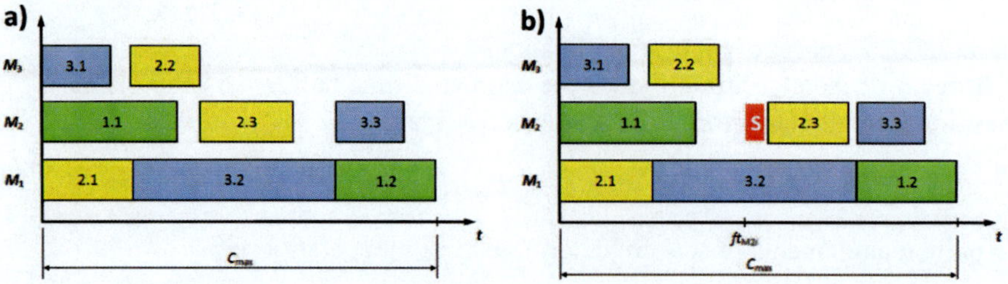

Fig. 4. Job schedule before (a) and after (b) the implementation of the service time buffer

3.3 Algorithm for Predicting Variable Processing Times of Job

By predicting variable technological operation times, pre-emptive processing times of operations, ppt_{ij} of particular jobs are established, and as a result, changes in the operation processing times Δpt_{ij} (elements of the matrix PPT).

In **stage 1** of the algorithm, the number of operations of a specific job is defined and historical data contained in set RPT_{oij} are loaded:

$$RPT_{oij} = \left[rpt_1, rpt_2, \ldots, rpt_n\right], \tag{11}$$

where: RPT_{oij} – real set of processing times of operation j of job i,
rpt_n – observed real processing time of operation j of job i.

In **stage 2**, the data is prepared for further analysis. They are sorted in the ascending order, and subjected to initial statistical analysis, *i.e.* filtering to exclude outliers, and subsequently to determine the minimum, maximum, average, range and quartile values.

Stage 3 of the algorithm is the initial identification of the hypothetical probability distribution of times of operations. For this purpose, the empirical data are used to determine asymmetry coefficient – skewness γ_1 and concentration coefficient – kurtosis γ_2, determined from the k-th moment of a continuous random variable RPT_{oij}:

$$\mu_k = \frac{1}{N}\sum\nolimits_{i=1}^{N} (rpt_i - \mu)^k, \quad \mu = \frac{1}{N}\sum\nolimits_{i=1}^{N} rpt_i\,, \tag{12}$$

$$\gamma_1 = \frac{\mu_3}{(\mu_2)^{\frac{3}{2}}}, \quad \gamma_2 = \frac{\mu_4}{(\mu_2)^2} - 3. \tag{13}$$

where: μ_k – central moment of k–th order, N – number of observations,
rpt_i – real processing time, μ – average value.

Determining $(\gamma_1)^2$ and γ_2 of the empirical probability distribution enables preliminary identification of the shape of hypothetical probability distribution using the Cullen-Frey graph.

Stage 4 consists in approximating the parameters of the initially selected distribution to determine whether the theoretical distribution fits to describe the occurrence of variable processing times of a given technological operation (whether it can be employed for prediction). In the next step, the non-parametric Kolmogorov-Smirnov test is employed to verify distributional adequacy of results using the empirical distribution function. The procedure of the Kolmogorov-Smirnov test includes the verification of the hypothesis concerning the compatibility of the empirical distribution $F_n(t)$ from the n-element sample with the cumulative distribution $F(t)$ for the selected theoretical distribution. At the level of significance $\alpha \in (0, 1)$, we set a null hypothesis:

$$H_0 : F_n(rpt_i) = F_0(rpt_i) \tag{14}$$

against the alternative hypothesis:

$$H_1 : F_n(rpt_i) \neq F_0(rpt_i) \tag{15}$$

In order to verify the working hypothesis, we determine the biggest difference between the cumulative distributors, as in Fig. 5:

$$D_n = \sup_{rpt} |F_n(rpt) - F(rpt)| \tag{16}$$

Subsequently, based on the tabulated data for n observations and the assumed level of significance α, we determine the critical value $d_n(1-\alpha)$ for statistics D_n. Given that $D_n < d_n(1-\alpha)$, then at the level of significance α, there are no grounds to reject the hypothesis H_0, and therefore the distribution of operation times is assumed to be equal to the theoretical distribution $F(rpt)$, which is then implemented in the prediction of future times of given technological operations.

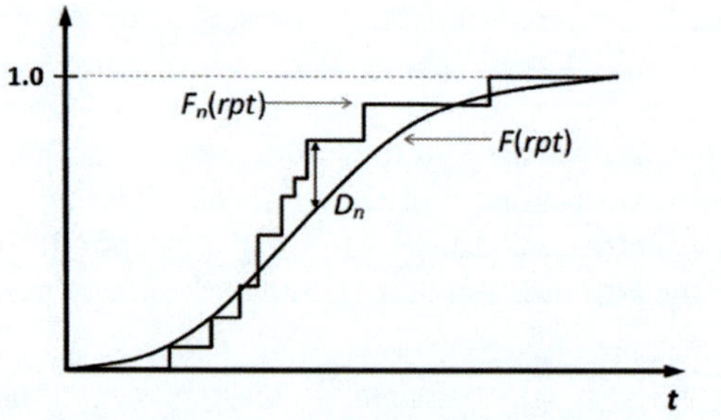

Fig. 5. Difference between empirical and hypothetical cumulative distribution function

Additionally, for a given probability, the range of execution values of operation j for job i is determined (at the significance level α, the estimated value $\mathcal{E}RPT_{oij}$replaces pre-emptive processing times ppt_{ij}).

Stage 5 of the algorithm is executed depending on the results of the statistical inference. If the data exhibits an atypical distribution then the interval-linear nonparametric estimation of the cumulative distribution function is performed. The results from the estimation are then employed to predict the time variability. The interval-linear distribution function $\widetilde{F}_n(rpt)$ is determined from:

$$\widetilde{F}_n(rpt_i) = \begin{cases} 0, \ rpt < 0 \\ \frac{F_n(rpt_1)}{2} \cdot \frac{rpt}{F_n(rpt_1)}, \ 0 < rpt < rpt_1 \\ g(rpt), \ rpt_i < rpt < rpt_{i+1}, \ 1 < i \le k-1 \\ F_n(rpt_{k-1}) + \frac{1-F_n(rpt_{k-1})}{2} \cdot \left(1 + \frac{rpt - rpt_k}{rpt_{k+1} - rpt_k}\right), \ rpt_k < rpt < 2 \cdot rpt_k - rpt_{k-1}, \\ 1, \ rpt < 2 \cdot rpt_k - rpt_{k-1}, \end{cases} \tag{17}$$

$$g(rpt) = F_n(rpt_{i-1}) + \frac{F_n(rpt_i) - F_n(rpt_{i-1})}{2} + \frac{F_n(rpt_{i+1}) - F_n(rpt_{i-1})}{2} \cdot \frac{rpt - rpt_i}{rpt_{i+1} - rpt_i}.$$

Based on the designated distribution function, the values of future operation times can be predicted (Fig. 6).

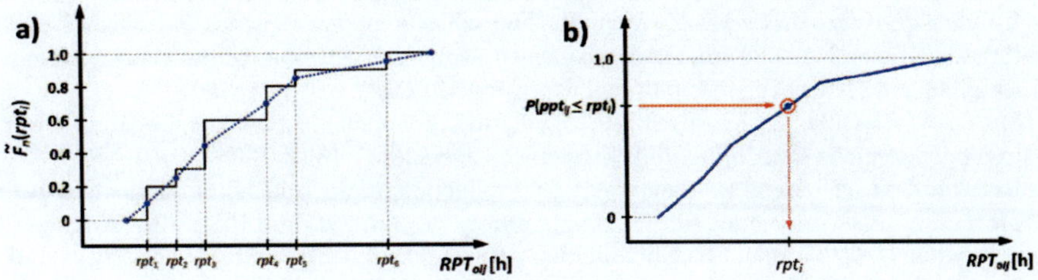

Fig. 6. Estimated cumulative distribution function (a) prediction (b)

The execution of the presented algorithm provides information about the predicted processing times of an investigated operation ppt_{ij}, and the change in technological operation times Δpt_{ij} – that must be added in order to obtain a robust schedule.

4 Summary and Conclusions

The robust scheduling method proposed in this paper is designated for application in schedules created under two-factor uncertainty constraint: machine failure and variable processing times. The motivation behind the conducted works is two-fold. On the one hand, the algorithmic method was developed in response to the paucity of similar solutions that has been reported widely in the literature, on the other hand, it stems from the experience gained in cooperation with industry. The primary aim of the method is to produce a predictive production schedule that will compensate for the effects of the considered uncertainty factors. The executed prediction algorithm computes a robust production schedule, whose makespan shows little discrepancy from the real project duration. Each of the proposed algorithms has been individually verified using a set of historical production data. In further studies, the prediction algorithm will be tested against a complex production environment composed of multiple machines performing multiple processes.

References

1. Trojanowska, J., Varela, M.L.R., Machado, J.: The tool supporting decision making process in area of job-shop scheduling. In: Rocha, Á., Correia, A.M., Adeli, H., Reis, L.P., Costanzo, S. (eds.) WorldCIST 2017. AISC, vol. 571, pp. 490–498. Springer, Cham (2017). https://doi.org/10.1007/978-3-319-56541-5_50
2. Hong, G.: Bulding Robust Schedules using Temporal Protection – An Empirical Study of Constraint Based Scheduling Under Machine Failure Uncertainty. Toronto, Ontario (1996)
3. Gola, A., Kłosowski, G.: Development of computer-controlled material handling model by means of fuzzy logic and genetic algorithms. Neurocomputing **338**, 381–392 (2019)
4. Sobaszek., Ł, Gola, A., Świć, A.: Time-based machine failure prediction in multi-machine manufacturing systems. Eksploatacja i Niezawodnosc – Maintenance Reliab. **22**(1), 52–62 (2020)
5. Jian, X., Li-ning, X., Ying-wu, C.: Robust scheduling for multi-objective flexible job-shop problems with random machine break-downs. Int. J. Prod. Econ. **141**(1), 112–126 (2013)

6. Jasiulewicz-Kaczmarek, M., Żywica, P.: The concept of maintenance sustainability performance assessment by integrating balanced scorecard with non-additive fuzzy integral. Eksploatacja I Niezawodnosc – Maintenance Reliab. **20**(4), 650–661 (2018)
7. Sobaszek, Ł., Gola, A., Kozłowski, E.: Prediction of variable technological operation times in production jobs scheduling. IFAC-PapersOnLine **52**(13), 1301–1306 (2019)
8. Bocewicz, G., Nielsen, I., Banaszak, Z.: Production flows scheduling subject to fuzzy processing time constraints. Int. J. Comput. Integr. Manuf. **29**(10), 1105–1127 (2016)
9. Bocewicz, G.K., Nielsen, I.E., Smutnicki, C., Banaszak, Z.A.: Towards the leveling of multi-product batch production flows. A multimodal networks perspective. IFAC-PapersOnLine **51**(11), 1434–1441 (2018)
10. Seybold, L., Witczak, M., Majdzik, P., Stetter, R.: Towards robust predictive fault-tolerant control for a battery assembly system. Int. J. Appl. Math. Comput. Sci. **25**(4), 849–862 (2015)
11. Jensen, M.T.: Generating robust and flexible job shop schedules using genetic algorithms. IEEE Trans. Evol. Comput. **7**(3), 275 (2003)
12. Kujawinska, A., Vogt, K., Diering, M., Rogalewicz, M., Waigaonkar, S.D.: Organization of visual inspection and its impact on the effectiveness of inspection. In: Hamrol, A., Ciszak, O., Legutko, S., Jurczyk, M. (eds.) Advances in Manufacturing. LNME, pp. 899–909. Springer, Cham (2018). https://doi.org/10.1007/978-3-319-68619-6_87
13. Kujawińska, A., Diering, M., Rogalewicz, M., Żywicki, K., Hetman, Ł.: Soft modelling-based methodology of raw material waste estimation. In: Intelligent Systems in Production Engineering and Maintenance – ISPEM 2017, Advances in Intelligent Systems and Computing, vol. 637, pp. 407–417 (2017)
14. Antosz, K., Ratnayake, R.M.C.: Spare parts' criticality assessment and prioritization for enhancing manufacturing systems' availability and reliability. J. Manuf. Syst. **50**(2), 212–225 (2019)
15. Deepu, P.: Robust Schedules and Disruption Management for Job Shops, Bozeman, Montana (2008)
16. Klimek, M.: Techniques of generating schedules for the problem of financial optimization of multi-stage project. Appl. Comput. Sci. **15**(1), 18–35 (2019)
17. Sobaszek, Ł, Gola, A., Kozłowski, E.: Job-shop scheduling with machine breakdown prediction under completion time constraint. Ann. Comput. Sci. Inf. Syst. **15**, 437–440 (2018)
18. Fuad, R.H., Sarker, R., Essam, D.: A real-time order acceptance and scheduling approach for permutation flow shop problems. Eur. J. Oper. Res. **247**(2), 488–503 (2015)
19. Choi, S.H., Wang, K.: Flexible flow shop scheduling with stochastic processing times: a decomposition-based approach. Comput. Ind. Eng. **63**(2), 362–373 (2012)
20. Sungbum, J., Jinwoo, P.: A hybrid genetic algorithm for the hybrid flow shop scheduling problem with nighttime work and simultaneous work constraints: a case study from the transformer industry. Expert Syst. Appl. **42**(15–16), 6196–6204 (2015)
21. Kianfar, K., Fatemi Ghomi, S.M.T., Oroojlooy, J.A.: Study of stochastic sequence-dependent flexible flow shop via developing a dispatching rule and a hybrid GA. Eng. Appl. Artif. Intell. **25**(3), 494–506 (2012)
22. Chung-Cheng, L., Shih-Wei, L., Kuo-Ching, Y.: Robust scheduling on a single machine to minimize total flow time. Comput. Oper. Res. **39**(7), 1682–1691 (2012)
23. Almeder, C., Hartl, R.F.: A metaheuristic optimization approach for a real-world stochastic flexible flow shop problem with limited buffer. Int. J. Prod. Econ. **145**(1), 88–95 (2013)
24. Chung-Cheng, L., Kuo-Ching, Y., Shih-Wei, L.: Robust single machine scheduling for minimizing total flow time in the presence of uncertain processing times. Comput. Ind. Eng. **74**, 102–110 (2014)
25. Varela, M.L.R., et al.: Collaborative paradigm for single-machine scheduling under just-in-time principles: total holding-tardiness cost problem. Manage. Prod. Eng. Rev. **9**(1), 90–103 (2018)

26. Jian, X., Li-Ning, X., Ying-Wu, C.: Robust scheduling for multi-objective flexible job-shop problems with random machine breakdowns. Int. J. Prod. Econ. **141**(1), 112–126 (2013)
27. Bierwirth, C., Mattfeld, D.C.: Production scheduling and rescheduling with genetic algorithms. Evol. Comput. **7**(1), 1–17 (1999)
28. Xingquan, Z., Hongwei, M., Jianping, W.: A robust scheduling method based on a multi-objective immune algorithm. Inf. Sci. **179**, 3359–3369 (2009)
29. Janak, S.L., Lin, X., Floudas, C.A.: A new robust optimization approach for scheduling under uncertainty – II uncertainty with known probability distribution. Comput. Chem. Eng. **31**, 171–195 (2007)
30. Henning, G.P., Cerda, J.: Knowledge-based predictive and reactive scheduling in industrial environments. Comput. Chem. Eng. **24**, 2315–2338 (2000)
31. Mehta, S.V., Uzsoy, R.M.: Predictable scheduling of a job shop subject to breakdowns. IEEE Trans. Robot. Autom. **14**(3), 365–378 (1998)
32. Lawless, J.F.: Statistical Models and Methods for Lifetime Data. Wiley, Hoboken (2003)

Prioritizing Internal Production on MRI Waiting List Management: An Optimization Model

Leandro Barretiri[1(✉)], Bruno S. Gonçalves[1,2], Rui M. Lima[1], and José Dinis-Carvalho[1]

[1] Department of Production and Systems, ALGORITMI Center, School of Engineering, University of Minho, 4800-058 Guimarães, Portugal
a89843@alunos.uminho.pt
[2] School of Management and Technology of the Polytechnic of Porto, Porto, Portugal

Abstract. This research intends to estimate the optimal monthly Magnetic Resonance Imaging (MRI) production in a Portuguese public hospital. A time horizon of three years, containing detailed information about the ideal amount of daily production, is included in the evaluation. The objective is to minimize yearly MRI exams cost and maximize the internal production to gradually reduce the response time, which will consequently decrease the waiting list for MRI exams. The current waiting list in the hospital under-study is around 11,4 months, a long response time compared to the 90 days maximum waiting time pre-established by the Portuguese National Health Service (NHS). The results obtained through the developed linear programing model revealed that the under-study hospital must keep the internal and external production on its maximum level for 17 months to reach the expected waiting list length of 90 days. Given that the demand is higher than production capacity, the model exposed that external production will not be totally extinct for a period longer than three months during the analyzed period.

Keywords: Operations management · Waiting list management · Capacity planning · Optimization model

1 Introduction

There is an evident global increase in demand for medical imaging techniques for the diagnosis of medical pathologies. Magnetic Resonance Imaging (MRI) is currently the best medical technique for body scanning with high image quality results. This aspect represents a higher cost compared to other medical imaging techniques. Another complication caused by the growing demand for MRI scans is that it overwhelms hospitals' imaging sectors, which are unable to meet the demand due to lack of equipment, lack of qualified personnel, and due to the fact that the response time is much higher than the stipulated by the NHS, resulting in patient dissatisfaction.

Operations management techniques may be used as part of a hospital's success factors [1]. As an example, a properly capacity management is part of a lean service model pathway for waste reduction, value-added increasing, patient flow [2], and patient satisfaction.

J. Machado et al. (Eds.): icieng 2021, LNME, pp. 68–78, 2022.
https://doi.org/10.1007/978-3-030-78170-5_7

The imaging sector of the hospital where the project was developed need to manage the capacity of internal and external production to reduce the waiting list. Thus, this work aims implementing an optimization model capable of maximizing the internal production by minimizing monthly cost in order to reduce the waiting list.

The present paper is divided in five sections. The first one is the project introduction. The literature review on the use of mathematical models for capacity planning is presented in the second one. Then, the problem description is made, followed by the model formulation. The results obtained after the implementation are shown in the fourth chapter. In the last section, the work conclusions are exposed.

2 Literature Review

Vieira et al. (2020) stress that, besides the poor-quality health service experienced by patients, health organizations may suffer penalties for not complying with the response deadlines stipulated by the Central Administration of the Portuguese NHS. They also concluded, through a case study performed in a public hospital, that personal engagement is the most important input for continuous problem-solving methodologies such as mathematical model implementation.

According to Conforti et al. (2008), the push to curb spending experienced by health organization results in a management mindset with greater concern in maximizing the use of available resources, thus being able to provide treatment in a timely manner and consequently guarantee the flow of patients. The authors emphasize that the hospital patient scheduling system plays an important role in the healthcare organizations overall performance and that it is decisive for an efficient patient flow. They argue that mathematical optimization models can result in significant contributions to the hospital operations management progress.

Healthcare waiting list management and outpatient scheduling has been object of academic research for over 50 years. Priyan & Uthayakumar (2016) advocate that practicing organizations management scientific methods is an outstanding approach for operations improvements in healthcare environment. For the authors, healthcare operations planning has to be taken as priority for nations where the importance of providing a proper and satisfactory service is primordial, given that operations research can support hospitals for problem solving through modeling approaches.

Addressing the same health operational management field, Farughi et al. (2020) exposed the relevance of complementing the optimization models performance with the use of independent algorithms. They declare that mathematical models in parallel with agent-based and quantitative procedures has greater capacity to be converted into benefits when compared to heuristic and empirical solutions.

Focused on capacity planning and demand forecasting, Sharifi (2014) declare that these are complex tasks for various health services. The main aspects that translate this particularity are the permanent uncertainty, the high users' volume, and the complexity of associated interconnection.

Given that capacity planning is associated with a high financial cost of specific resources and that it can also interfere positively or negatively in clinical cases with risk of death, capacity planning administration is essential in the health care industry [7].

Improved operations management through relevant approaches, such as "lean philosophy" adapted to the health industry, can assist in the implementation of a culture of continuous improvement for an improved management focused on patients flow [8].

3 Problem Statement

This paper is the result of a one-year case study (Feb 2020–Feb 2021) developed in a large Portuguese public hospital in which the imaging sector works every business day from 8:00 to 20:00. The average number of monthly requests for MRI exams is, during the study period, equal to 1550, against an internal capacity, considering days worked as weekdays, of 840 MRI exams per month.

The imaging department of the under-study hospital works with the possibility of running overtime during the weekdays and at the weekends, making it possible for the internal production to reach a maximum average production of 1480 MRIs per month, which would bring the average monthly internal production closer to the demand. By internal production, this article refers to all MRI scans performed on weekdays and overtime, including weekends.

Considering that the current waiting list for MRI exams is 10070, the average response time for MRI scans is 6.8 months. If only the weekdays are considered, with an average production of 840 MRs per month, the response time goes up to 11.4 months, which means that a patient in need of medical diagnosis through MRI exam, with low priority level, may have to wait for almost a year for this exam to be performed.

The priority levels in the imaging sector of the concerned hospital can differ from high, medium, and low priority. The Portuguese government stipulated that 90 days is the time limit for patients to wait, in case of no urgency, for MRI exams to be performed, taking into account the date of doctor requisition [9].

If the hospital is unable to meet this deadline, the patient has to be routed to other national health service entities or private units with agreements or conventions. However, the possibility of referring patients to another public entity is unlikely since they are also unable to respond to internal demand and the outsourcing to private units represents a high cost to public coffers.

The fact that the capacity restrictions, at the under-study hospital, are currently calculated using annual averages, implies that results are presented in a non-absolute form, which can be interpreted in an ambiguous way. Another particularity is the fact that the number of holidays and working days varies monthly and from year to year, implying variations that significantly alter the results obtained. Also, the fact that the personnel vacation period (July 15th to September 15th) indicates, due to the lack of professionals, the impossibility for overtime MRI performing during this period, is a restriction that, if not considered, results in misinterpretation of results.

This ambiguous interpretation can lead to wrong decision from the hospital management, considering that it is based only on average data, which may represent an unreal scenario. Early investment in equipment or mistaken planning of future budgets, wrong personnel scheduling, and unnecessary outsourcing of services, are the standing out among erroneous decisions that can be taken by the hospital administrative sector due to lack of precision in the information provided with respect to the response time for waiting list reduction.

The use of a mathematical model capable of calculating the ideal monthly internal and external production for cost reduction aims to mitigate this problem.

Due to the high cost associated with performing MRI, these errors can lead to high financial losses for the health organizations. In the hospital where this work was carried out, the possibility of reducing the long waiting list for MRIs is in the out contracting of the service for one external supplier selected through public tender. Because of the budget constraints, this externalization has a maximum monthly limit (500 MRIs) that makes it impossible to carry out a high number of MRIs by the external supplier.

Currently, in the under-study hospital, the monitoring of the waiting list management for MRI exams is done in an Excel spreadsheet with limited functionalities, where the average monthly production quantities (internal and external) need to be manually inserted. This is a time-consuming activity and does not guarantee a reliable result, given that this procedure does not consider the working days variation in each month of the year neither the monthly holidays nor the personal vacation.

The research team's role is to implement improvements in the imaging sector of the hospital in question to be converted into benefits for patients. After analyzing the current waiting list management, we developed a linear optimization model to minimize the cost associated with MRI exams performed internally and externally with the aim of maximizing internal production, considering that its costs are much lower.

Linear programing is applicable when the objective and constraints of the problem can be translated into linear functions, which is the case of the problem identified in this research work.

4 Optimization Model

To a better understanding how the developed linear optimization model works, this section describes the problem modeling. The objective is to reach and maintain the waiting list close to the pre-stipulated 90 days by guaranteeing the minimum MRI external production, thus ensuring that internal production is maximized, with priority for the weekdays and, if possible, without practicing overtime.

Among all the problem constraints, the monthly capacity constraint is the one that varies every month, simply because the number of working days on each month of the year is not constant. Also, during the holiday period, there are restrictions that influence the final result of the monthly production capacity.

Bearing in mind that the hospital has the objective of reducing the waiting list as quickly as possible, the optimization is done globally (three years period), but the initial months are prioritized through algorithms by considering all internal and external capacity constraints.

In short, the system restrictions are, monthly capacity constraints, with variations in production quantities from weekdays to weekends. There are also restrictions on specific holidays when the imaging sector does not work (Good Friday, Easter, Christmas, and New Year's Eve).

The model aims to assist the hospital in reducing the waiting list by maximizing internal production before contracting external service in case of a waiting list higher than 90 days. In this sense, the model was developed to work dynamically in the data relationship between different months through algorithms inclusion.

Another aspect is the constraints imposed by the annual vacation period, which, in July and September, reduces the overtime weekdays and weekends production capacity to 50% and, in August, reduces the overtime production in 100%. These restrictions are automatically imposed through an algorithm, developed by the research team, that identifies and restricts to zero the amount of production in extraordinary hours in the period from 15 July to 15 August.

A relevant feature of this tool is the fact that it dynamically generates the monthly tables (three years period from the starting month), so that the optimization model can distribute, based on the financial cost, the optimal quantity to be internally produced on each day of the month and also the optimal quantity to be externally produced if necessary. Thus, the algorithms' function is to automatically generate the existing constraints, according to each month conditions, conditions that can vary according to the month length, the existence of holidays in each month, and the possibility of reduction in production due to staff vacations.

Another restriction that is automatically generated by another algorithm developed by the research team, is the minimum quantity to be produced in each month. This algorithm calculates, based on demand, in the waiting list, and in the maximum waiting time pre-stipulated objective, the minimum production quantity for each month. The same algorithm also considers, through the dynamic tables, the number of working days of each month, taking into account the holidays that the imaging sector does not work.

Thus, the proposed mathematical model, combined with all developed algorithms, can minimize the monthly production cost, since the unit cost per MRI internally performed is considerably lower than the external cost production.

With the maximization of the internal MRI production, the hospital can gradually reduce the waiting list until reaching an average response time of three months. After reaching this goal, the model must, based on the monthly demand, establish monthly production capable of sustaining the waiting list near to the limit stipulated by the Portuguese NHS.

As it is an optimization problem within the operational research area, the mathematical notation of the parameters, the decision variables, the objective function, and the problem constraints must be defined. Thus, the mathematical notation is characterized next:

Parameters

The parameters are of 3 types:

1 - To indicate the cost of each MRI produced:

$\mathbf{WV}_j$ (weekday value): during the business days in month j.

OV_j (overtime value): in overtime in month j.
SV_j (Saturday value): on Saturdays in month j.
CV_j (Sunday value): on Sundays in month j.
EV_j (external value): externally produced on a specific month j.

2 - To indicate the quantity of MRIs that can be produced:
WC_j (weekday capacity): during the business days in month j.
OC_j (overtime capacity): in overtime in month j.
SC_j (Saturday capacity): on Saturdays in month j.
CC_j (Sunday capacity): on Sundays in month j.
EC_j (external capacity): externally produced in month j.

3 - To indicate the necessary month production:
MD_j (month demand): in a specific month j.
The **MD_j** parameter is calculated by the algorithm that determines the minimum quantity to be produced in each month based on the beginning of the month Waiting List (WL), in the average monthly Demand (D), which is calculated based on the production of the previous year, in the Monthly Production (MP) capacity, that is automatically calculated independently for each month, and in the desired Response Time (RT).

Indexes

$i = \text{weekdays}\{1, \textit{number of business days in month } j\}$
$j = \text{months}\{1, 36\}$
$k = \text{Saturdays}\{1, \textit{number of Saturdays in month } j\}$
$l = \text{Sundays}\{1, \textit{number of Sundays in month } j\}.$

Decision Variables

$$x_{ij} = \text{number of exams performed on weekday i in month j.} \tag{1}$$

$$y_{ij} = \text{number of exams overtime performed on weekday i in month j.} \tag{2}$$

$$s_{kj} = \text{number of exams performed on Saturday k in month j.} \tag{3}$$

$$c_{lj} = \text{number of exams performed on Sunday l in month j.} \tag{4}$$

$$ep_j = \text{number of exams outsourced performed in month j.} \tag{5}$$

$$x_{ij},\ y_{ij},\ s_{kj},\ c_{lj},\ ep_j \geq 0$$

Objective Function
The objective function is to minimize the cost of analyzed period production, as expressed by the Eq. (6):

$$Z = Min\sum_{j=1}^{36}\left[\sum_{i=1}^{m}(x_{ij} \times WV_j + y_{ij} \times OV_j) + \sum_{k=1}^{n}(s_{kj} \times SV_j) + \sum_{l=1}^{t}(c_{lj} \times CV_j) + ep_j \times EV_j\right] \quad (6)$$

m = number of business days in month j.
n = number of Saturdays in month j.
t = number of Sundays in month j.
$m \in \{1, \textit{number of business days}\}$
$n \in \{1, \textit{number of saturdays}\}$
$t \in \{1, \textit{number of sundays}\}$.

Constraints

$$\sum_{i=1}^{m} x_{ij} \le WC_j \quad (7)$$

$$\sum_{i=1}^{n} y_{ij} \le OC_j \quad (8)$$

$$\sum_{i=1}^{k} s_{kj} \le SC_j \quad (9)$$

$$\sum_{i=1}^{l} c_{ij} \le CC_j \quad (10)$$

$$ep_j \le EC_j \quad (11)$$

$$\sum_{i=1}^{m} x_{ij} + \sum_{i=1}^{n} y_{ij} + \sum_{i=1}^{k} s_{kj} + \sum_{i=1}^{l} c_{ij} + ep_j \ge MD_j \quad (12)$$

$$WC_j,\ OC_j,\ SC_j,\ CC_j,\ EC_j \ge 0\,\forall\, j\ \in \{1,\ 36\}$$

The model is constrained by the weekday production capacity (7), by the overtime capacity (8), the Saturday capacity (9), the Sunday capacity (10), by the external capacity (11), and by the monthly demand (12). The case study results are presented and discussed next.

5 Results and Discussion

By using a Microsoft Excel spreadsheet, all information necessary to parameterize and solve the problem was organized. The OpenSolver toll with the Simplex method was used as optimization engine.

The model parameters data are presented in Table 1:

Table 1. Problem parameters

Model parameters					
$\mathbf{WV}_j$	Weekday Value	1 × V (EUR)	$\mathbf{WC}_j$	Weekday Capacity	40 (MRIs)
$\mathbf{OV}_j$	Overtime Value	4 × V (EUR)	$\mathbf{OC}_j$	Overtime Capacity	15 (MRIs)
$\mathbf{SV}_j$	Saturday Value	4 × V (EUR)	$\mathbf{SC}_j$	Saturday Capacity	45 (MRIs)
$\mathbf{CV}_j$	Sunday Value	4 × V (EUR)	$\mathbf{CC}_j$	Sunday Capacity	45 (MRIs)
$\mathbf{EV}_j$	External Value	6 × V (EUR)	$\mathbf{EC}_j$	External Capacity	500 (MRIs)
$\mathbf{MD}_j$	Month Demand	Algorithmically calculated (MRIs)			
Month demand algorithm parameters					
WL	Waiting List	10070 (MRIs)			
D	Demand	1550 (MRIs)			
RT	Response Time	3 (months)			
MP	Month Production	Varies monthly (MRIs)			

The actual costs of performing MRIs were not reported to the research team. The only information provided about the costs is related to the occasion when the exam is performed. Thus, we know that performing MRI internally on business days represents the lowest cost. As for the same exam, the cost is four times higher if executed in overtime and on weekends, and six times higher if externally performed. With this information, we use fictitious values to find the optimal solution, however, a cost analysis will not be displayed in this case study.

Posterior to data collection and to the analysis of the current scenario at the under-study hospital imaging sector, the model exposed that even using the maximum external production that is limited to 500 MRIs per month, it is necessary 17 months of maximum internal and external production for the waiting list to reach the response time of 90 days stipulated by the Portuguese health ministry.

The model additionally exposes that will be possible to reduce external production in 62,1% in the remaining 18 months of the analysis and still maintaining the queue close to the expected limit. The model also shows that the weekend production can be also reduced in some months that external production is not needed. Figure 1 exposes the necessary MRIs production during the three years analyzed period.

To better understand how it will impact the response time, Fig. 2 displays the waiting list behavior in the analyzed period. The orange line in the graph represents the hospital's target (in months) for the waiting list.

It is also possible to use de model to analyze what the impact that hiring more technicians or the acquisition of a new MRI machine can generate in the internal production and consequently in the reduction of the waiting list. Other scenarios can also be analyzed by changing the decision variables, for instance, increasing or decreasing the number of overtime hours or by changing the monthly demand.

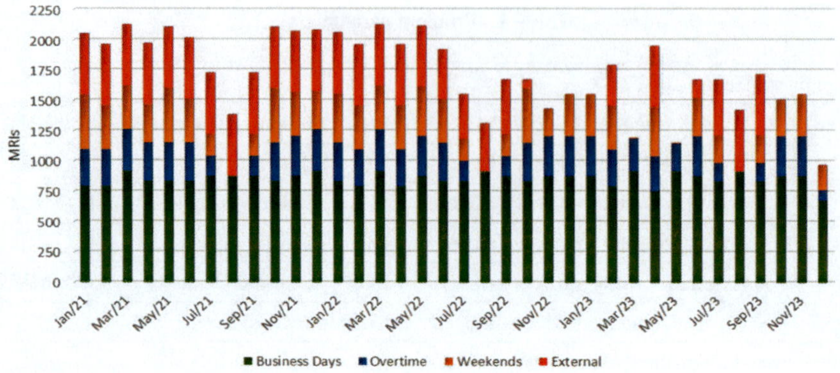

Fig. 1. Required MRIs production

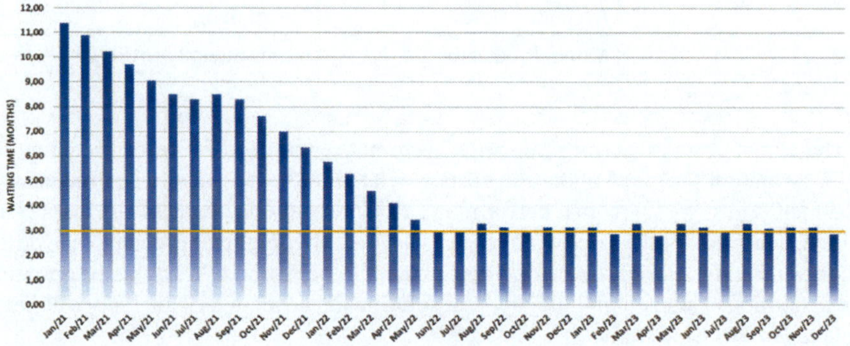

Fig. 2. Waiting list behavior

The capacity planning problem presented, reinforced by the lack of tools such as a model with a greater level of detail and that performs dynamically the waiting list control, brings to light the challenges faced by hospital managers, planners, and physicians.

One of the possible reasons for this condition is related to the fact that the capital available to health service providers is more controlled than before. This factor, associated with a greater concern on the part of citizens regarding health, causes overcrowding in health units that do not have sufficient resources.

The developed model makes the waiting list control and the capacity planning process simpler for the imaging sector management. The dynamism of this tool makes it unlimited in a matter of future dates to be analyzed with a view that automatically presents three years ahead based on the month chosen as initial.

Thus, with the fulfilment of the decision variables, the algorithms present in the Excel tables calculate the minimum quantities to be monthly produced so that the optimization model can distributes the production, with a lower cost priority and respecting the constraints, among business days, overtime, weekends, and external production to achieve the goal of reducing the waiting list efficiently.

6 Conclusions

According to the feedback from the hospital management, the model will be of great use as it addresses capacity planning in a useful and dynamic way, expressing exactly what the monthly quantity of internal and external production is, an approach that was not possible before the model implementation. This toll can positively impact the hospital's performance in agility for data analysis and to assist decision making based on the expected response time.

The outcomes obtained from the proposed model revealed, by meeting all problem constraints, the period of time needed to reach the 90 days waiting list target within the current scenario. Seventeen months is the necessary time for the goal to be achieved, which may be considered a high period regarding patients' notion of a valuable service. Anyway, this is the reality of many hospitals, considering the capacity restrictions, amid intense pressure to reduce monetary expenses.

Besides of exposing a long response time by the imaging sector of the under-study hospital, an urgent need for increased production capacity was identified, since the demand for MRI scans tends to continue rising. A quick solution to increase the production would be to invest in other MRI equipment, or alternatively to find a higher level of external service.

Based on this reality, a methodology capable of reducing waste and increase value-added in the service performed by utilizing the existing resources is a credible action for internal improvements that can culminate in faster response time and greater patient satisfaction.

The developed tool represents an initial evolution in terms of improvements, but as the need for a new project is clear, the implementation of the Overall Equipment Effectiveness (OEE) [10] is a possibility for future work. That would be a way for uncover wastes in the MRI exams performing process, which could then be addressed and reduced to increase the overall effectiveness of MRI equipment.

Acknowledgments. This work has been supported by projects UIDCEC003192019 and POCI-01-0145-FEDER-030299, from Fundação para a Ciência e Tecnologia (FCT), Portugal.

References

1. Lima, R.M., Dinis-Carvalho, J., Souza, T.A., Vieira, E., Gonçalves, B.: Implementation of lean in health care environments: an update of systematic reviews. Int. J. Lean Six Sigma **12** (2020)
2. Sharifi, S., Saberi, K.: Capacity planning in hospital management: an overview. Indian J. Fund. Appl. Life Sci. **4**(2), 515–521 (2014)
3. Vieira, E., Gonçalves, B.S., Lima, R.M., Dinis-Carvalho, J.: Optimization model for waiting list management and service continuous improvement. In: International Conference on Quality Engineering and Management, 2020-September, pp. 807–819 (2020)
4. Conforti, D., Guerriero, F., Guido, R.: Optimization models for radiotherapy patient scheduling. 4OR **6**, 263–278 (2008)
5. Priyan, S., Uthayakumar, R.: Operations research in healthcare supply chain management under fuzzy-stochastic environment. Supply Chain Log. Manage., 1833–1875 (2016)

6. Farughi, H., Tavana, M., Mostafayi, S., Santos Arteaga, F.J.: A novel optimization model for designing compact, balanced, and contiguous healthcare districts. J. Oper. Res. Soc. **71**(11), 1740–1759 (2020)
7. Hall, R.: Handbook of Healthcare System Scheduling (International Series in Operations Research & Management Science 168). Springer, New York (2011). https://doi.org/10.1007/978-1-4614-1734-7
8. Rechel, B., Wright, S., Barlow, J., McKee, M.: Hospital capacity planning: from measuring stocks to modelling flows. Bull. World Health Organ. **88**(8), 632–636 (2010)
9. Ministério da Saúde: Portaria n.º 153/2017 de 4 de maio. Diário da República 1ª série, Portugal (2017)
10. Souza, T.A., Roehe Vaccaro, G.L., Lima, R.M.: Operating room effectiveness: a lean healthcare performance indicator

Conceptual Model of Production Engineers' Actions to Monitor Workers' Exposure to Occupational Risks

Arminda Pata[1], José Carlos Sá[2,3](✉), Francisco José Gomes da Silva[2], Olga Silva[3], Luís Barreto[3], and Luís Pinto Ferreira[2]

[1] D. Dinis Higher Institute (ISDOM), Marinha Grande, Portugal
arminda.pata@isdom.pt
[2] Polytechnic of Porto, School of Engineering (ISEP), Porto, Portugal
cvs@isep.ipp.pt
[3] Instituto Politécnico de Viana do Castelo, Rua Escola Industrial e Comercial Nun Alvares, 4900-347 Viana do Castelo, Portugal

Abstract. The purpose of the present study consists in the development of a conceptual model to optimize assignments between staff's and workstations with manual assembly processes. This model is planned to satisfy customers' needs without compromising workers' health. The development of occupational risk measurement tools is urgent and essential to prevent health problems in work teams. Staff managers should establish appropriate assignments to the characteristics of available and existing resources to complete the tasks without compromising productivity, safety, and health. A life free of work-related musculoskeletal disorders can extend career length and the number of active workers. Therefore, industrial managers need to integrate knowledge from different scientific fields into their decision-making process. They need to identify, specify and focus their efforts on the changes that characterize their productive systems (e.g., layout, workload, cycle time, occupational diseases), and follow the evolution of demographic changes, in order to adjust, with greater celerity, their actions to different dynamic problems. Hence, the implementation of an instrument of measure can contribute to increasing the wellbeing of people at industrial workplaces. Consequently, workers' health is less compromised over the span of an entire professional career.

Keywords: Scheduling optimization · Occupational health safety · Work-related musculoskeletal disorders

1 Introduction

When analyzing Occupational Health and Safety (OHS) within a globalized, diversified and competitive market, it is very important to ensure future healthy generations. The increasing use of ever-tightening rigorous standards (e.g., demand, deadlines, assembly quality, differentiation models) regarding costumers' needs can change very quickly.

J. Machado et al. (Eds.): icieng 2021, LNME, pp. 79–87, 2022.
https://doi.org/10.1007/978-3-030-78170-5_8

Thus, it is very important to prepare flexible manual assembly processes to cope with the demanding variety of market requirements, but without compromising employees' health. In warehouse management systems, where a large quantity of manual handling is done, batch-sizing decisions can have an important influence on workload and human performance. It can also have a strong impact on ergonomic parameters and therefore on workers' welfare [1]. A lot of scheduling problems (e.g., bicycle assembling, fish handling) can be reduced by using appropriate decision support systems (DSS) on the assembly processes [2–10]. DSS enable more efficient solutions compared to the use of traditional methods (e.g., experience of managers, experienced workers). The results from the scheduling of working hours can minimize hypothetical disorders if the exposure is balanced regarding all the causes of injuries [11]. The optimum schedule of production systems with manual tasks is generated by defining the most suitable workload for the work team. The individual work schedule is constructed with the lowest number of resources, throuhgout the imposed takt-time [11, 12]. The exactness and consistency of the results generated by DSS can encourage managers to question steps, routines, and explore different scenarios and work models. These efficient ways of predicting and solving problems improve the efficiency of the decisions. These systems can allow the decision-maker to explore a greater variety of admissible solutions and, consequently, to choose the best one. Therefore, in the work planning and programming phases, the market's active changes can force organizations to implement a vast variety of adjustments in decision-making. A balanced workload in accordance with assumptions, constraints, variables, and constants can lead the managers to respond more quickly and more efficiently to the high uncertainty of supply and demand. The computational results can demonstrate that a polynomial algorithm is highly efficient for large-scale problems, which offers a competent solution for practical applications [13]. In addition to time and resources, the mathematical programming models are subject to precedence, non-concurrency, lead and lag times, and zonal constraints, being capable of capturing multi-resource and multi-mode activities [12]. Some linear programming models are extended for hybrid approaches [14] to explore different variants that include a lot of characteristics that the industrial environment contains.

2 Literature Review

The planning and programming of workers' activities and the management of machines and equipment at production systems are iterative and dynamic processes [11]. Using a problem approach that incorporates diverse types of information about a production system can result in a combinatorial optimization of problems that may have different applications in real work context (e.g., scheduling [12], sequencing, human resources management, task assignment [15]), and add value to support daily decisions. Some authors made mathematical models, such as the mathematical model for resource allocation in emergency situations with the co-operation of non-governmental organizations (NGOs) under uncertainty [16], mathematical models for location of temporary relief centers and dynamic routing of aerial rescue vehicles [17–19], mathematical models for flight-to-gate reassignment with passenger flows [20]. Thus, there is a growing body of literature that recognizes the importance of combinatorial optimization for the resolution of programming problems [6, 11, 21]. However, despite the measures implemented to manage the workload in industries [22], labour conditions to perform a task is mostly not taken into consideration. Such criteria (e.g., age, gender, health condition, exposure time, capacities of work resolution, the physical effort at the workstation) have a great impact on workers' health and safety and are still problems under-explored problems and less referred at the current literature, in addition to the certification of management systems [23]. Many organizations integrate their Management Systems [24–28], namely the Quality Management System [29–35], the Environmental Management System [36–40] and the Hygiene, Safety and Health at Work Management System [41–45], among others, seeking to save human, economic and financial resources [46–53].

This case study creates a dynamic evaluation to solve the daily problems of OHS in industrial environments with manual assembly processes. It is estimated to strengthen scientific knowledge and to improve the industrial practices with a methodology of planning to support decision-making in those work contexts. It is also intended to contribute to the improvement of a DSS able to plan the workload for repetitive manual tasks, thus helping to prevent workers' injuries and potential stops or slowdowns in production.

3 Conceptual Model

To respond to the literature gap regarding the absence of limit values, it has been created a conceptual model as a pathway to monitor workers' real exposure at any industrial environment with manual assembling processes.

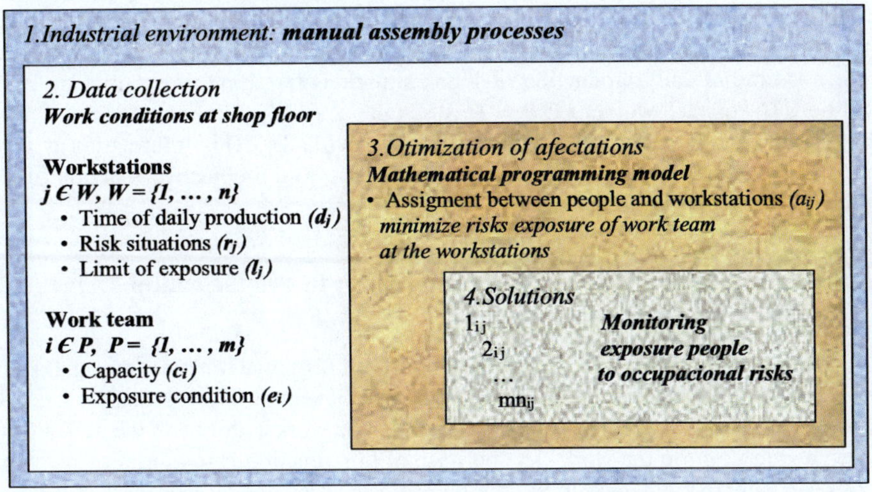

Fig. 1. Pathways to monitor the workers' real exposure at any industrial environment

The conceptual model (Fig. 1) to evaluate criteria that minimize the work exposure risks of a whole work team has four pathways. The model was tested and validated by CPLEX® optimization software and OPL language, using test instances to generate the solution to the problem presented in a time format (i.e. scheduling each worker's map), based on a mathematical model to monitor workers' exposure to occupational risks in manual assembly processes. This conceptual model is estimated to contribute to scientific advancement within the scope of Occupational Health and Safety (OHS). The approach presented encourages the formulation of new mathematical models designed to monitor workers' exposure to work conditions.

3.1 Industrial Environment – Manual Assembling Processes: (Pathway 1)

Workers recurrently stay at the same workstation, exposed to the same working conditions for long periods of time, even though they are able to perform work at other workstations. Sometimes they change to another workstation but exposure is not monitored. When workers undergo cumulative exposure to occupational risks that affect their health, the production systems are susceptibles to failures and stops. People have limits. There are no standard values to ensure that workers are exposed to safe work conditions, similar to what happens regarding vibrations and noise. The industrial environment can compromise workers' OHS. Thus, it is crucial to generate engineers' actions to monitor workers' exposure to occupational risks, mainly in manual assembling processes.

3.2 Data Collection – Workstations and Work Team: (Pathway 2)

It is impossible to control what it is not known. Thus, to monitor workers' exposure to the conditions of performing manual work, it is essential to collect some data. These data can consist of existing information (e.g., a database, historical data, predetermined standard times) or data that has to be investigated using methodologies (e.g., time study, work sampling). The purpose of this data is to characterize each workstation $\boldsymbol{j} \in \boldsymbol{W} = \boldsymbol{1} \ldots \boldsymbol{n}$: time of daily production $(\boldsymbol{d_j})$, risk situations $(\boldsymbol{r_j})$, limit of exposure to work conditions $(\boldsymbol{l_j})$ of each worker $\boldsymbol{i} \in \boldsymbol{P} = \boldsymbol{1} \ldots \boldsymbol{m}$: (Capacity $(\boldsymbol{c_i})$, Exposure condition $(\boldsymbol{e_i})$), which defines and characterizes each operator individually. This information will be stored in a database that will later be used according to the objective function of the problem to be analyzed.

3.3 Optimization of Affections – Development of the Mathematical Model: (Pathway 3)

First, the formulation of the mathematical model (i.e., objective function) that characterizes the problem of monitoring the exposure of a work team is developed. This function allows to optimize the effects established between the workstations and the person who is assigned to perform the work. The purpose of this function is to minimize the work team's exposure risks, by monitoring the exposure accumulated by each worker. A list of input data (Table 1) is considered to formulate a mathematical model. The mathematical model uses limits established by the decision-maker (e.g., industrial manager).

Table 1. List of input data of mathematical programming model.

Description		
Number of work posts	Units	Workstations
Daily production time	Minutes	
Risk of effort occurrence	OCRA scale	
Number of workers = Dimension of work team	Units	People
Capacity to perform work	Minutes	
Statement of exposure to work conditions	Scale of Katz	
Time frame of planning and programming activities	Minutes	Scheduling

3.4 Solutions – Actions to Monitor Workers' Exposure to Occupational Risks: (Pathway 4)

Workers are assigned to workstations without surpassing previously established limits, thus enabling scenarios to be compared. This allows minimizing longer exposure to work conditions.

This mindset allows recognizing the real risk management, profitability, and resource availability (i.e., human and technical). All principles must be in harmony with customer satisfaction without harming workers' health and safety where safety can be integrated with quality and environment to improve workers' quality of living. The conceptual model (Fig. 1) helps production engineers to develop a tool for a wide range of scientific and industrial processes, where a strong education and new ideas are always necessary towards sustainable development.

Thus, the result generated by the objective function guides the decision-maker in making decisions. Since the result exposes the optimization of the effects established between workstations and workers, this allows monitoring individual exposure of the work team and, consequently, contributing to the extension of professional careers and healthy future generations.

4 Conclusion

The most important asset of the conceptual model is the mathematical formulation, allowing to: extract the information about real work conditions, optimize assignments and also generate admissible solutions in a short period of time. All solutions present a balanced exposure regarding to the work team. This allows minimizing the penalization associated with the frequent use of the body at work posts. Lastly, workers' exposure is balanced to the industrial environment under analysis.

At the real context of the shop floor, it is important to know the processing system and also to have instruments that help to quickly make decisions. It is also essential to previously prepare the manual assembling processes. At the start of the line, there will exist a certain downtime at each workstation, while a produced unit does not arrive at the end of the line. When the first workstations do not guarantee the work of the following workstations, they are inactive. Then, the desired production planning cannot be satisfied within the appointed time. Normally, this depends on the demand for the products or the absence of one or more workers, requiring to adjust the workload in a balanced way. Consequently, workers' health is affected to ensure the output because they stay all the time at the same workstation.

The main limitations found in the development of this study result from gaps found in the literature review, such as the absence of the occupational risks list and their exposure limits. As there is no standard list or scientific documents that can serve as a reference to allow properly filling in the matrices, the decision maker must create the list of individual characteristics to be considered in the definition and characterization of people and workstations. It is based on the information contained in the matrices that the mathematical model establishes the assignments, minimizing people's exposure to occupational risk. As a result of the gaps found in the literature, the criteria used to define the input data can be considered a limitation. The formulation of the mathematical model allows the reading of any information represented in the matrices, so it would be important to create lists of occupational risks and their exposure limits, covering new occupational diseases that result from the emergence of industry 4.0. The last limitation concerns the culture itself rooted in organizations in relation to industry 4.0, due to the challenges and opportunities that this may allow in terms of people management.

Corporate culture must be aligned with the 4th Industrial Revolution in order to adhere to new working methods that will keep or make companies competitive. Naturally, without harming people's quality of living and reconciling work with life.

The contribution of this work may stimulate future developments in extracting zettabyte value from stored information. This information extracted from the databases may be focused on keeping people active, healthy, safe and with quality of living, and future work should analyze information to make decisions based on historical data: aligned with the elimination of occupational accidents, adjusted to the training [54] with the real needs of the work organization and focused on eliminating unwanted work-related musculoskeletal disorders.

References

1. Battini, D., Glock, C.H., Grosse, E.H., Persona, A., Sgarbossa, F.: Ergo-lot-sizing: an approach to integrate ergonomic and economic objectives in manual materials handling. Int. J. Prod. Econ. **185**, 230–239 (2017)
2. Kucukkoc, I., Zhang, D.: Balancing of parallel U-shaped assembly lines. Comput. Oper. Res. **64**, 233–244 (2015)
3. Ogan, D., Azizoglu, M.: A branch and bound method for the line balancing problem in U-shaped assembly lines with equipment requirements. J. Manuf. Syst. **36**, 46–54 (2015)
4. Chen, Y.-Y.: A hybrid algorithm for allocating tasks, operators, and workstations in multi-manned assembly lines. J. Manuf. Syst. **16**, 196–209 (2017)
5. Sener, A., Atabak, E., Tolga, B.: Combinatorial Benders cuts for assembly line balancing problems with setups. Eur. J. Oper. Res. **259**, 527–537 (2017)
6. Blum, C., Puchinger, J., Raidl, G., Roli, A.: Hybrid metaheuristics in combinatorial optimization: a survey. Appl. Soft Comput. **11**, 4135–4151 (2011)

7. Qiuhua, T., Zixiang, L., LiPing, Z., Chaoyong, Z.: Balancing stochastic two-sided assembly line with multiple constraints using hybrid teaching-learning-based optimization algorithm. Comput. Oper. Res. **82**, 102–113 (2017)
8. Savino, M.M., Brun, A., Mazza, A.: Dynamic workforce allocation in a constrained flow shop with multi-agent system. Comput. Ind. **65**, 967–975 (2014)
9. Gonca, T., Seyda, T.: Assembly line balancing with positional constraints, task assignment restrictions and station paralleling: a case in an electronics company. Comput. Ind. Eng. **64**, 602–609 (2013)
10. Otto, A., Scholl, A.: Incorporating ergonomic risks into assembly line balancing. Eur. J. Oper. Res. **212**, 277–286 (2011)
11. Pata, A., Moura, A.: Applying metaheuristics to minimize work-related musculoskeletal disorders. Int. J. Technol. Hum. Interact. **14**, 17–34 (2018)
12. Russell, A., Taghipour, S.: Multi-objective optimization of complex scheduling problems in low-volume low-variety production systems. Int. J. Prod. Econ. **208**, 1–16 (2019)
13. Zhang, R.-Q., Yi, M., Wang, Q.-Q., Xiang, C.: Polynomial algorithm of inventory model with complete backordering and correlated demand caused by cross-selling. Int. J. Prod. Econ. **199**, 193–198 (2018)
14. Byrne, M.D., Hossain, M.M.: Production planning: an improved hybrid approach. Int. J. Prod. Econ. **93–94**, 225–229 (2005)
15. Shigeno, M., Saruwatarib, M.: An algorithm for linear fractional assigment problems. Discret. Appl. Math. **56**, 333–343 (1995)
16. Sarma, D., Bera, U.K., Das, A.: A mathematical model for resource allocation in emergency situations with the co-operation of NGOs under uncertainty. Comput. Ind. Eng. **137**, 106000 (2019)
17. Alinaghian, M., Aghaie, M., Sabbagh, M.S.: A mathematical model for location of temporary relief centers and dynamic routing of aerial rescue vehicles. Comput. Ind. Eng. **131**, 227–241 (2019)
18. Aryanezhad, M., Deljoo, V., Mirzapour Al-e-hashem, S.: Dynamic cell formation and the worker assignment problem: a new model. Int. J. Adv. Manuf. Technol. **41**, 329–342 (2009)
19. Ramos, T., Gomes, M., Barbosa-Póvoa, A.: Solving a multi-product, multi-depot vehicle routing problem by a hybrid method. In: IO2011 (2011)
20. Pternea, M., Haghani, A.: Mathematical models for flight-to-gate reassignment with passenger flows: state-of-the-art comparative analysis, formulation improvement, and a new multidimensional assignment model. Comput. Ind. Eng. **123**, 103–118 (2018)
21. Saraç, T., Sipahioglu, A.: Generalized quadratic multiple knapsack problem and two solution approaches. Comput. Oper. Res. **43**, 78–89 (2014)
22. Othman, M., Bhuiyan, N., Gouw, G.: Integrating workers' differences into workforce planning. Comput. Ind. Eng. **63**, 1096–1106 (2012)
23. Morgado, L., Silva, F.J.G., Fonseca, L.M.: Mapping occupational health and safety management systems in Portugal: outlook for ISO 45001:2018 adoption. Proc. Manuf. (2020, in Press)
24. Talapatra, S., Santos, G., Uddin, K., Carvalho, F.: Main benefits of Integrated Management Systems through literature review. Int. J. Qual. Res. **13**(4), 1037–1054 (2019)
25. Santos, G., Mendes, F., Barbosa, J.: Certification and integration of management systems: the experience of Portuguese small and medium enterprises. J. Clean. Prod. **19**(17–18), 1965–1974 (2011)
26. Barbosa, L.C.F., Oliveira, O.J., Santos, G.: Proposition for the alignment of the integrated management system (quality, environmental and safety) with the business strategy. Int. J. Qual. Res. **12**(4), 925–940 (2018)

27. Carvalho, F., Santos, G., Gonçalves, J.: Critical analysis of information about integrated management systems and environmental policy on the Portuguese firms' website, towards sustainable development. Corp. Soc. Respons. Environ. Manage., 1–20 (in Press)
28. Carvalho, F., Santos, G., Gonçalves, J.: The disclosure of information on Sustainable Development on the corporate website of the certified Portuguese organizations. Int. J. Qual. Res. **12**(1), 253–276 (2018)
29. Santos, G., Murmura, F., Bravi, L.: Developing a model of vendor rating to manage quality in the supply chain. Int. J. Qual. Serv. Sci. **11**(1), 34–52 (2019)
30. Sá, J.C., Amaral., A., Barreto, L., Carvalho, F., Santos, G.: Perception of the importance to implement ISSO 9001 in organizations related to people linked to quality - an empirical study. Int. J. Qual. Res. **13**(4) 1055–1070 (2019)
31. Rodrigues, J., Sá, J.C., Ferreira, L.P., Silva, F.J.G., Santos, G.: Lean management "quick-wins": results of implementation. a case study. Qual. Innov. Prosp. J. **23**(3), 3–21 (2019)
32. Santos, G., Barbosa, J.: QUALIFOUND - a modular tool developed for quality improvement in foundries. J. Manuf. Technol. Manag. **17**(3), 351–362 (2006)
33. Bravi, L., Murmura, F., Santos, G.: The ISO 9001:2015 quality management system standard: companies' drivers, benefits and barriers to its implementation. Qual. Innov. Prosp. J. **23**(2), 64–82 (2019)
34. Araújo, R., Santos, G., Costa, J., Sá, J.C.: The quality management system as a driver of organizational culture: an empirical study in the Portuguese textile industry. Qual. Innov. Prosp. J. **23**(1), 1–24 (2019)
35. Costa, A.R., Barbosa, C., Santos, G., Alves, M.R.: Six sigma: main metrics and R based software for training purposes and practical industrial quality control. Qual. Innov. Prosp. J. **23**(2), 83–99 (2019)
36. Félix, M.J., Santos, G., Barroso, A., Silva, P.: The transformation of wasted space in urban vertical gardens with the contribution of design to improving the quality of life. Int. J. Qual. Res. **12**(4), 803–822 (2018)
37. Santos, G., Rebelo, M., Lopes, N., Alves, M.R., Silva, R.: Implementing and certifying ISO 14001 in Portugal: motives, difficulties and benefits after ISO 9001 certification. Total Qual. Manag. Bus. Excell. **27**(11–12), 1211–1223 (2015)
38. Santos, D., Rebelo, M., Santos, G.: The integration of certified management systems. Case study – organizations located at the district of Braga, Portugal. Procedia Manuf. **13**, 964–971 (2017)
39. Santos, G., Sá, J.C., Oliveira, J., Ramos, D., Ferreira, C.: Quality and Safety Continuous Improvement Through Lean Tools. Chapter 8 of the Book "Lean Manufacturing – Implementation, Opportunities and Challanges", coordenado por Francisco Silva & Luis Ferreira (Inst.Sup.Eng.Porto). Nova Science, New York (2019)
40. Silva, F.J.G., Gouveia, R.M.: Cleaner Production - Toward a Better Future. Springer, Cham (2020). https://doi.org/10.1007/978-3-030-23165-1
41. Ferreira, N., Santos, G., Silva, R.: Risk level reduction in construction sites: towards a computer aided methodology - a case study. Appl. Comput. Inform. **15**, 136–143 (2019)
42. Gonçalves, I., Sá, J.C., Santos, G., Gonçalves, M.: Safety stream mapping—a new tool applied to the textile company as a case study. Stud. Syst. Decis. Control **202**, 71–79 (2019)
43. Santos, G., Barros, S., Mendes, F., Lopes, N.: The main benefits associated with health and safety management systems certification in Portuguese small and medium enterprises post quality management system certification. Saf. Sci. **51**(1), 29–36 (2013)
44. Santos, G., Rebelo, M., Ramos, S., Silva, R., Pereira M., Ramos, G.N.L.: Developments regarding the integration of the occupational safety and health with quality and environment management systems. In: Kavouras, I.G., Chalbot, M.-C.G. (eds.) Developments Regarding the Integration of the Occupational Safety and Health with Quality and Environment Management Systems, pp. 113–146. Nova Science, EUA, New York (2014)

45. Ferreira, N., Santos, G.: Simultaneity of risk in the civil site construction: an analysis of accumulated risk. In: Occupational Safety and Hygiene II - Selected Extended and Revised Contributions from the International Symposium Occupational Safety and Hygiene (SHO 2014), pp. 259–264 (2014)
46. Bravi, L., Murmura, F., Santos, G.: Manufacturing Labs: where digital technologies help improve life quality. Int. J. Qual. Res. **12**(4), 957–974 (2018)
47. Santos, G., et al.: Value creation through quality and innovation – a case study on Portugal. TQM J. **31**(6), 928–947 (2019)
48. Marques, C., Lopes, N., Santos, G., Delgado, I., Delgado, P.: Improving operator evaluation skills for defect classification using training strategy supported by attribute agreement analysis. Measurement **119**, 129–141 (2018)
49. Santos, G., Afonseca, J., Murmura, F., Félix, M.J., Lopes, N.: Critical success factors in the management of ideas as an essential component of innovation and business excellence. Int. J. Qual. Serv. Sci. **3**(3), 214–232 (2018)
50. Santos, G., Doiro, M., Mandado, E., Silva, R.: Engineering learning objectives and computer assisted tools. Eur. J. Eng. Educ. **44**(4), 616–628 (2019)
51. Santos, G., Bravi, L., Murmura, F.: SA 8000 as a tool for a sustainable development strategy. Corp. Soc. Responsib. Environ. Manag. **25**, 95–105 (2018)
52. Bravi, L., Murmura, F., Santos, G.: Attitudes and behaviours of Italian 3D prosumer in the era of additive manufacturing. Proc. Manuf. **13**, 980–986 (2017)
53. Pinto, B., Silva, F.J.G., Costa, T., Campilho, R.D.S.G., Pereira, M.T.: A strategic model to take the first step towards Industry 4.0 in SMEs. Proc. Manuf. **38**, 637–645 (2019)
54. Bastos, A., Sá, J.C., Silva, O., Fernandes, M.C.: A study on the reality of Portuguese companies about work health and safety. Occup. Safety Hygiene II, 687–691 (2014)

Economic Analysis of a Modular Storage System Designed for Railway Industry Use

Marilena Stoica(✉) and Lucian Cucu

University Politehnica of Bucharest, 060042 Bucharest, Romania
marilena.stoica@upb.ro

Abstract. The present study proposes the product design of an innovative external luggage storage system, analyzed from economical point of view which can be digitalized by developing the attached module in order to organize the distribution of the passengers and luggage in the same time. It has been developed a case study of an industrial product of which assembly uses standardized structural materials among side modular subassemblies used in railway industry. The predicted cost effectiveness of a product is an essential part of the design process that will influence the entire workflow of the product design cycle. It consists in detailed structure analysis of each part, manufacturing process and assembly methods from a cost point of view. The product studied is a modular structure that can provide secure user-friendly storage space in public spaces. The cost analysis will provide a full insight in the feasibility of the product from the customer point of view. The product contains multiple modules in order to respect the ergonomic requirements and to increase the comfort of the passengers. The main advantage of this innovative baggage system is that it was designed to be fixed on the external structure of the wagon. Thus, the passenger concerns about the safety of the luggage ends at the beginning of the travel trip; the insufficient storage space designed in interior of the wagon is assured outside, it also solves the problem of difficult access for the elderly and kids accessing the storage area inside the wagon, especially when the train is moving and decreases the time required for embarking and disembarking for passenger with very big luggage. Thus, such a product would fluidize the travel time flow of the train.

Keywords: Storage system · Railway · Cost · Structural materials

1 Introduction

Automatization represents an advancement in technology with the purpose to deliver a more easy and relaxed way of living on a daily base life. The process of automatization varies across the engineering fields and it consists in combining successfully different technologies in order to make it work. Depending of the designation the automatization process it can involve mechanical and software engineering. The correct correlation of the two is a must in order to obtain a solid functional automatization process independent of the field its designed for, industry automatization or consumer goods automatization. Public transportation is a necessity for most of the developed countries and could represent a viable ecological transportation alternative for overcrowded cities [1].

J. Machado et al. (Eds.): icieng 2021, LNME, pp. 88–96, 2022.
https://doi.org/10.1007/978-3-030-78170-5_9

One of the most important aspect of public transportation, beside safety and a good tight schedule is comfort [2]. The comfort level provided by the transportation method of the passengers is a decisive factor in the transport choice of the user for a daily commute or leisure travels. After an in-depth study of the rail passenger transportation situation in Europe, most of the users where not pleased with the comfort of the train carriage, especially with the storage capability of the train carriage [3]. Research shows that 15 to 20% of travelers travel with large luggage and they don't have enough storage space inside the train carriage. In order to overcome this negative situation, they deposit their luggage on the corridor, thus obstructing the people who move inside the train [4]. The solution described in this paper involves a new innovative concept located at the ends of the train carriage on the exterior – External storage System – ESS. The ESS concept consists in a carousel type storage device that allows those 15–20% of travelers to store their large luggage in a safe manner without obstructing passageways and also to improve the train time table, optimizing passenger flow at the train station. The ESS is automated, using the train ticket's bar code in order to allocate storage space for the users that really needs it, thus additional personnel will not be necessary for the operation. Being automated the storage system will provide a safer way of travel, registering the users automatically, thus it will monitor the user and luggage data, in order to deny transportation of dangerous items [5].

The cost of such product- ESS- will have a decisive factor in the implementation decision, thus is imperative to determine the approximate fabrication and assembly costs of the concept. This study will offer an analysis over the production costs of a subassembly of the concept and will offer an insight over the production costs of the final product. The subassembly studied is one of the storage units for the luggage and its components through which is attached to the carrying system. The storage unit is designed to accommodate the largest hand carried luggage on the market, and to offer good accessibility from both sides depending on its position in the carousel. The outer shell of the storage unit is designed to be user friendly, made from hard plastic in order not to deteriorate the user's luggage. The plastic shell of the unit is supported by a metal frame that is designed to take the loads and it is attached to the carrying structure system of the carousel. The entire storage unit assembly can move from side to side inside the carousel in order to allow access from either side of the train but also from inside of the carousel as well, through an access corridor. Subsequent paragraphs, however, are indented.

2 Design and Optimization

2.1 Concept

Due to the limited internal train carriage space and European passenger train requirements the solution to offer more efficient storage space was to move it to the exterior of the train. This concept required detailed analysis due to the strict train dynamics regulations, thus the optimal solution was to attach it at the ends of the train carriage [6]. An important aspect of the passenger railway system is the timetable the train must respect. The current solution for large storage in the railway system is a separate storage train carriage. The benefit of the separate train carriage is that it will offer large amounts of storage, but this advantage fails in comparison to the time needed for the passengers to access the storage train carriage at embarkation and disembarkation, thus negatively impacting the train timetable [7].

Our solution divides the storage space from one big unit into separate units for each train carriage, providing better accessibility for passengers, thus shortening the timetable, thus offering more comfort and a more cost-effective train journey. This solution unlike the separate storage train carriage offers the possibility for the passengers to access at any time their luggage [8]. The layout for the new external storage system provides access and coupling for the train carriages (see Fig. 1).

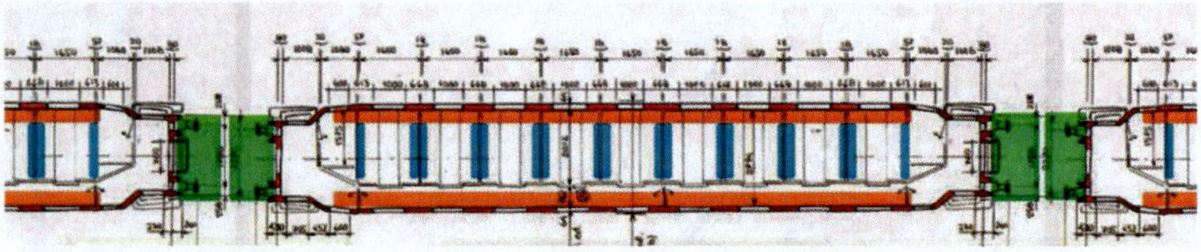

Fig. 1. New External Storage System – ESS (marked green)

This aspect meant to couple the train wagons through the new storage system concept, so a new coupling system was developed. Another criterion was that the new storage system needs a passthrough corridor to enable people to pass through safely form one train carriage to another.

2.2 Modular Structure ESS

This new modular structure ESS was described in [9]. It consists of a metal structure protected by a composite outer shell, attached through a metal exoskeleton and the coupling system to the train carriage (see Fig. 2). The whole assembly is modular and can be attached at either end of the train carriage.

The ESS is designed to allow the passengers access to the storage capsule on both sides, regardless of the train station placement. It is also designed to optimize the passengers flow, at embarkation and disembarkation, through the ESS ticket scan access system that allows only one person at a time to remove their luggage, thus avoiding crowding at the ESS access doors. The coupling system is similar to the standard one, using a slightly modified coupling arm assuring the same functionality and safety. The ESS module can be further developed to be implemented into other public transportation methods or as a stationary storage solution and it would offer the possibility to monitor luggage storage in public spaces, thus avoiding multiple dangerous scenarios through luggage scanning and a user registration platform.

The metal structure has been analyzed in [8] on flexion studies and has been optimized in order to achieve better stiffness. The choice for material is, Stainless steel E295 with a safety factor of 3.76 (see Fig. 3). The safety factor value is critical in an economic analysis, influencing the choice of material for manufacturing, thus the price of materials used, so an optimal safety factor was studied and achieved for this type of application considering multiple material variants.

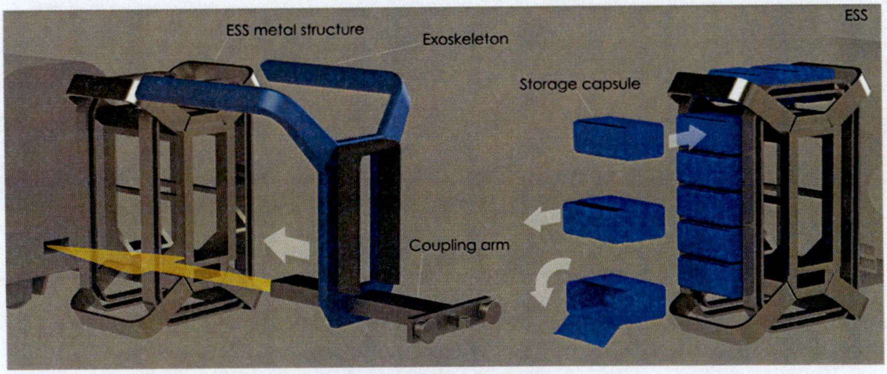

Fig. 2. The ESS assembly

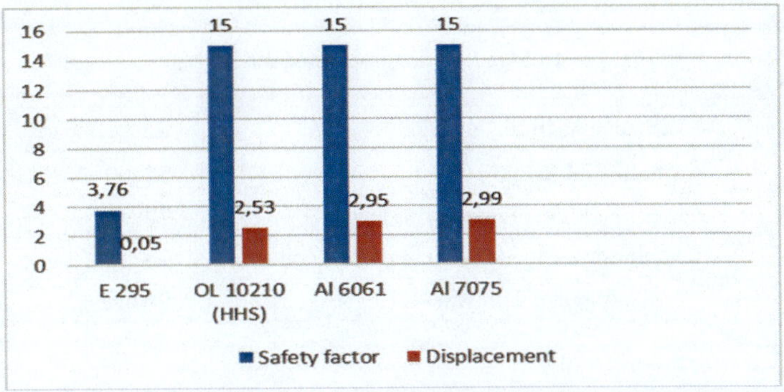

Fig. 3. Safety factor comparison for the selected materials

The storage capsule dimensions (see Fig. 4) can allow storage for the largest hand carried luggage on the market. Based on mechanical properties [10], the chosen material for the ESS storage capsule could be either High-Density Polyethylene (HDPE) or Polycarbonate (PC300). Most manufacturers of plastic containers use high density polyethylene (HDPE or PEHD) as manufacturing material, being a thermoplastic made from petroleum. These are sometimes called "alkathene" or "Polyethylene" when used for pipes. With a high strength-density ratio, HDPE is used in the production of plastic bottles, corrosion-resistant pipes, geo-membranes and plastic timber. HDPE density can range from 0.93 g/cm^3 to 0.97 g/cm^3. Although HDPE density is higher than that of PC 300, HDPE has a lower branching, giving it strong intermolecular forces and a tensile strength higher than PC 300. The difference in resistance is greater than the difference in density, giving HDPE a higher specific resistance.

Starting with the standard dimensions of the luggage, presented in Table 1, the ESS was designed in order to include 14 storage capsules inside the module. The capsule transportation inside the module is assured by a chain transmission, C2/62 H, according to standard [11]. The cinematics study of the structure revealed stability issues with the capsules. The solution was to design guiding elements (see Fig. 5).

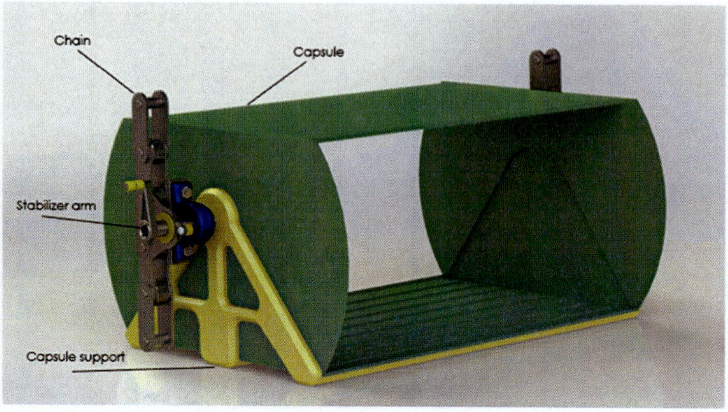

Fig. 4. ESS capsule assembly

Table 1. Standard dimensions of the luggage

Days of travel	Dimensions of the luggage	Recommended height
1–4	Small hand luggage	50–60 cm
4–7	Medium hand luggage with wheels	70 cm
7–14	Medium to large hand luggage with wheels	76 cm
14–21	Large luggage with wheels (biggest size)	90 cm

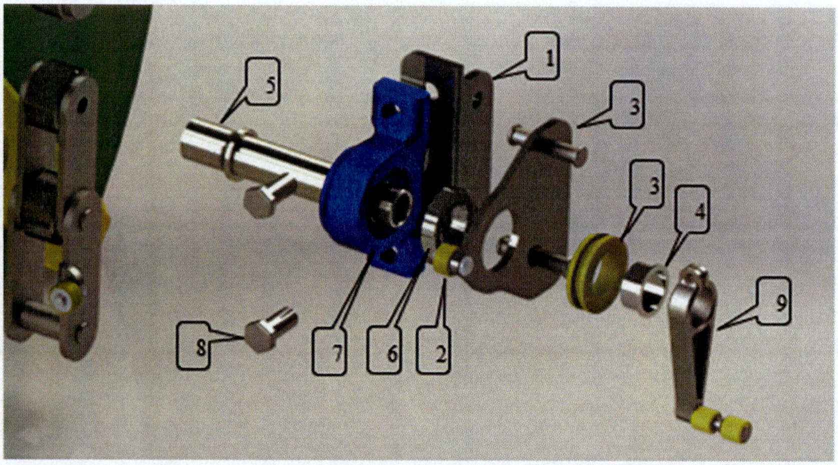

Fig. 5. Capsule transportation and stabilization assembly

The carrier and stabilizer assembly are formed by a connecting flange (1), one guide roller (2), a support flange (3), rubber and metal bushings (3, 4), a metal shaft (5), one ball bearing (6), a bearing housing (7), two hex bolts with nuts (8) and a stabilizer arm (9).

3 Economic Analysis

Cost calculation wise, the study considers the recommendations from [12] which consists in a product development process, respectively four stages namely, analyzing, concepting, designing and finalizing. All phases have many steps, and the core activity is designing and concepting. Hence, the analyze is very important in the design process.

According to calculus of the ESS, the study was divided in two section. In the first section, the storage capsule, that is custom designed (see Table 1) will be analyzed, together with the carrier and stabilization assembly and the second is the main structure containing these capsules.

In order to provide a comprehensive cost estimation for ESS, the following relation was used with different types of material and fabrication methods. The relation was based on a well known study in systematic development of Solution [12]. The material data library is based on the mechanical properties, dimensions, shapes and relative costs.

$$M = V_b \cdot k_v^* \cdot k_{v0} \cdot \left(1 + g_{w/z}\right) \tag{1}$$

Where: V_b - gross materials volume, k_v^* - relative costs, k_{v0} - relative material costs based on volume, $\left(1 + g_{w/z}\right)$ – overall maintenance/ subcontracting costs - $w \cong z$.

The gross material volume was provided by the CAD files of each component.

For the first study, the storage capsule costs were determined for two different materials using injection moulding. We can observe that the biggest percentage of the total material costs is attributed to the injection moulding process (see Fig. 6, 7).

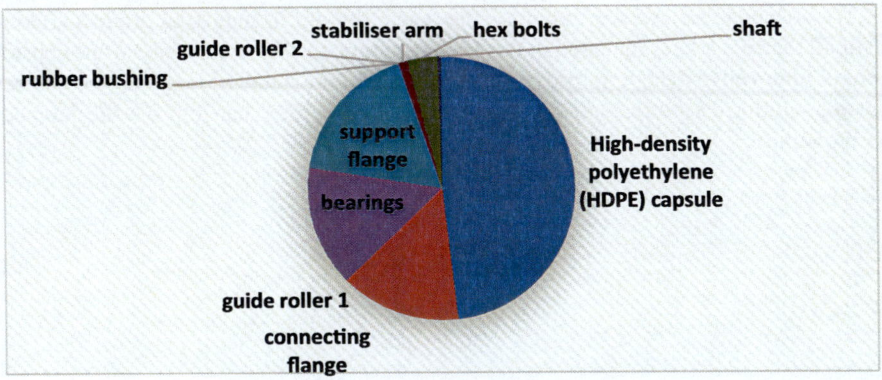

Fig. 6. ESS capsules materials cost [euro]

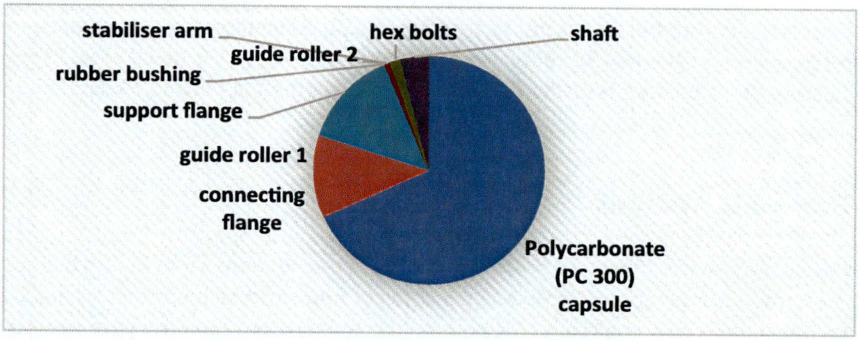

Fig. 7. ESS capsules materials cost [euro]

The study shows that using HDPE, as a base material for the storage capsules it reduces the final costs with over 20% from the total material costs.

The ESS structure design uses standard structural materials. The study shows that using this material offers better materials/manufacturing costs.

The gross material volume of the structure is similar with the gross volume of the 14 storage capsules. Thus, the optimal choice for ESS storage capsule manufacturing is HDPE, because the material costs of the capsules and structure are proportional, from economical point of view.

Additional to material costs, the analysis contains also the manufacturing costs. This considers the losses, storage and manipulation costs. The relation between the two types of costs is in Eq. 2

$$H = \frac{M}{M'} \tag{2}$$

Where: H - total manufacturing costs and M' – material costs share from the product manufacturing costs (Fig. 8).

The cost analysis gives an insight in the feasibility of this industrial product. Taking act of all the advantages included in this ESS structure and the modularity of the capsule storage (innovative design, improved cost, safety factor, standard luggage) we can say that this solution can be adapted in the railway industry because can improve the demands in ergonomics.

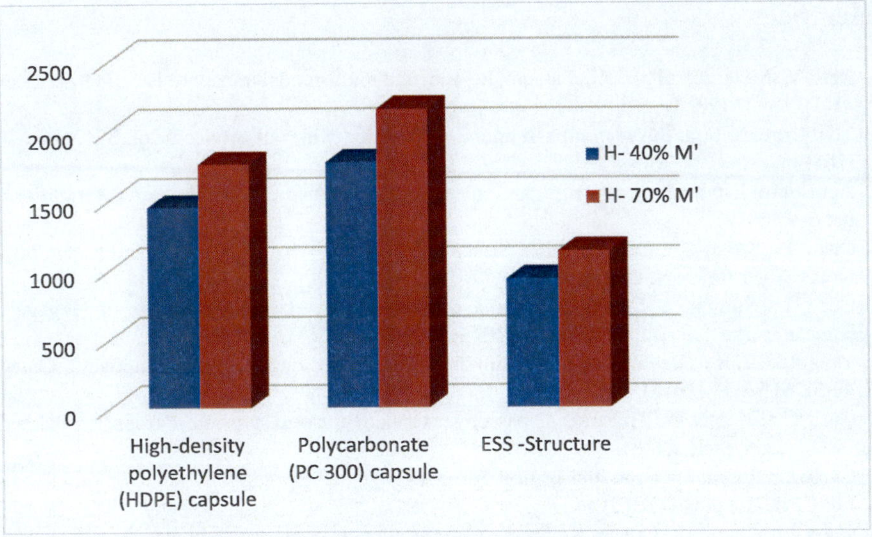

Fig. 8. The capsule/structure manufacturing costs

4 Conclusions

The study shows the importance of an economic analysis in the product development cycle and how the design process can be optimized based on simple but efficient calculations regarding the manufacturing possibilities. The economic analysis shows that regardless of manufacturing coefficient cost, the total ESS module cost is mostly influenced by the storage capsule material choice, especially the material density. This is shown in the higher manufacturing cost for the capsule with PC 300. Although the higher costs for PC 300, the material is not representative for a higher product quality and functionality. After the analysis we can conclude that the material type influences the total cost of ESS with a percentage of 15%.

The ESS assembly can be obtained at the lowest cost, considering a manufacturing ratio between 40%–70%, with 2400–3300 Euro.

This study allows the designer to optimize the material choice and manufacturing process for this type of application reducing the manufacturing costs and complexity of the final product and can be used for an estimation of costs in a feasibility study in order to develop the ESS production business plan.

Acknowledgement. This work has been funded by the European Social Fund from the Sectoral Operational Programme Human Capital 2014–2020, through the Financial Agreement with the title "Scholarships for entrepreneurial education among doctoral students and postdoctoral researchers (Be Antreprenor!)", Contract no. 51680/09.07.2019 - SMIS code: 124539.

References

1. Reis, V., Meier, J.F., Pace, G., Palacin, R.: Rail and multi-modal transport. Res. Transp. Econ. **41**(1), 18–22 (2013)
2. T.G. Organization: Survey on passengers' satisfaction with rail services, vol. 326, pp. 1–13 (2011)
3. Accenture: Establishing new rules and roles for public transportation. Beyond the traditional, pp. 6–9 (2012)
4. Cucu, L., Simion, I., Cazorla, M.P., Stoica G.F., Cananau, S.: Ergonomics and accessibility in passenger trains. S. Bull. Ed. **78**(D) (2016)
5. Cucu, L., Simion, I., Stoica G.F., Cananau, S.: Modular design for storage of luggage in passenger trains. S. Bull. Ed. **77**(D) (2015)
6. Yang, L., Li, K., Gao, Z., Li, X.: Optimizing trains movement on a railway network. Omega **40**(5), 619–633 (2012)
7. Nash, C.: European rail reform and passenger services, the next steps. Res. Trans. Econ. Leeds Inst. Transp. Stud. **22**, 61–70 (2010)
8. Cucu, L.: Research on modularization of the luggage transportation by rail. Doctoral Thesis, Ed. UPB, Bucuresti (2019)
9. Cucu, L., Stoica, M., Crisan, N., Stoica, G.F.: Design of an innovative luggage storage system for passenger trains. Int. Sci. J. Sci. Tech. Union Mech. Eng. "Industry-4.0" **7**(4), 21–24, (2019)
10. http://www.mtu.edu/materials/k12/experiments/tensile. Accessed 1 Nov 2018
11. Standard ISO 1977:2006, ISO 286-2.
12. VDI Guideline 2222: Systematic Development of Solution Principles. Engineering Design Methodology, Berlin, Germany (1997)

Robust Supply Chain Network Design Under Facility Disruption by Consideration of Risk Propagation

V. K. Manupati[1,2], Shukalya Akash[3], K. Illaiah[1,2], E. Suresh Babu[1,2], and M. L. R. Varela[4(✉)]

[1] Mechanical Engineering Department, NIT Warangal, Warangal, Telangana, India
[2] Computers Science Engineering Department, NIT Warangal, Warangal, Telangana, India
[3] Blooms System Pvt Ltd, Mumbai, Maharashtra, India
[4] University of Minho, Guimarães, Portugal
leonilde@dps.uminho.pt

Abstract. A supply chain consists of facilities that are subjected to various types of disruption. When one or more facility in the supply chain is disrupted, its impact propagates and reduce the performance of the supply chain. This research paper takes into consideration the probability of a disruption scenario to design a robust network for a supply chain. A stochastic mathematical formulation has been presented as a mixed-integer linear programming model for maximizing the concerned profit. For measuring the supply chain's robustness, the final quantity delivered after risk propagation in the disruption scenario as a robustness index (RI) is utilized. A numerical simulation is conducted to calculate the robustness index at different linking intensity and node threshold. The robustness index is compared under different disruption scenarios to obtain the optimal combination concerning linking intensity and node threshold.

Keywords: Risk propagation · Multi-echelon disruption · Supply chain robustness · Supply chain design

1 Introduction

The supply chain faces the risk of getting disrupted by natural and human-made disasters, strikes, legal contract disputes, terrorism, etc. Regardless of type, disruption in most circumstances results in unpleasant effects such as halting various functionalities within the network, forcing realignment with respect to logistics strategy among many others. This turns out to be costlier in the long run. No facility is entirely immune from predicted and unpredicted disruption. Yet, most of the research considers only a few disruption scenarios or worst-case scenarios. In general, the primary aim of designing the supply chain network is either minimizing the total cost or maximization of profit. If all disruption scenario is not considered, then sometimes loss or opportunity loss in case of disruption is significant.

J. Machado et al. (Eds.): icieng 2021, LNME, pp. 97–107, 2022.
https://doi.org/10.1007/978-3-030-78170-5_10

In March of 2000, a semiconductor plant operating in United States under Philips ceased to be in operation for around 3 out of the 4 quarters in a year due to an eight-minute fire event leading to sale loss estimated at $40 million directly and a $2.34 billion loss to Ericsson's mobile phone unit, as referred by Sheffi et al., in 2005 [1]. The terrorist attack dated 11th September 2001 led to all US borders being cordoned off and cancellation of various flights. This circumstance coerced majority proportion of assembly lines of Ford to lie idle due to the dearth of equipment which in turn were provided from abroad [2]. In 2009, strike at two General Motor part manufacturers prompted the shutdowns of 26 assembly manufacturer plant, which brought about the production loss of 5,00,000 vehicles and a loss of 809 million dollars to General Motor [2]. Earthquake in Gujarat, India in January of 2001, Japan Tsunami dated March 2011, Typhoon in the Philippines dated November 2013 are some of the instances where natural disasters have led to disruption of various global supply chains.

Simultaneously predicting the disruption along with understanding its behaviour is becoming possible due to the recent advancement of technology. In this research paper, a robust supply chain network is designed by considering all disruption scenarios, including multi-echelon disruption. This model focuses on obtaining a risk-neutral supply chain network with the prominent feature of being robust. Further, this robustness parameter is calculated through robustness index (RI) and can be defined as the ability of the supply chain for performing well during uncertainty.

Recent research done on supply chain design and robustness are analysed in Sect. 2. The modelling approach for calculating probability, supply chain design, and robustness index calculation is explained in detail in Sect. 3. This methodology is used in Sect. 4 to optimize the network of supply chain and simulation of RI at different linking intensity and node threshold. The probability of disruption scenario, supply chain network design, Robustness Index is explained briefly in Sect. 3.1. Section 5 draws this paper to a conclusion and deliberates some avenues for future research.

2 Literature Review

Ahmadi-Javid and Seddighi in 2013, presented the location-routing problem with a network of producer–distributors that create and circulate a single item to specific clients in a supply-chain [3]. Li et al. in 2013 review two models are for facility location design i.e., reliable P-median problem (RPMP) and reliable incapacitated fixed charge location (RUFL) problem [2]. Azad et al. in 2014, did provide a model for capacitated supply chain network design (SCND) to demonstrate under occasional interruptions in facility and logistics [4]. Klibi and Martel in 2013 proposed a methodology for the supply chain network (SCN) under unreliability [5]. Mari et al. in 2014, developed a model to optimize network design for a manageable and reliable inventory using resilience and sustainability [6]. Jeong et al. in 2014, did present an integrated system to plan crisis co-ordinations systems emergency logistics networks (ELNs) given robustness risk and efficiency measurements [7]. Marufuzzaman et al. in 2014, suggested a scientific model that designs a dependable multi-modular transportation network, where multi-purpose hubs are liable to site-dependent probabilistic disturbances [8]. Hatefi and Jolai in 2014, proposed a robust and reliable model for an incorporated forward-reverse logistics network [9].

Hatefi et al. in 2015(a), considered programming model with credibility constraint of an integrated forward-reverse logistics network with uncertain disruptions in the facility [10]. Hatefi et al. 2015(b), suggested a fuzzy possibilistic programming approach to plan a dependable forward-reverse transport network equipped with hybrid facilities taking into consideration uncertain and random aberration causing disruption in the facility [11]. Sadghiani et al. in 2015, developed a multiple deterministic sets covering model is proposed and stretched out to a possibilistic situation based robust display by disruption profiling and scenario generation [12]. Tang et al. in 2016, conducted a numerical simulation to measure robustness in different disruption scenarios [13]. Hasani and Khosrojerdi in 2016, presented a non-linear mixed-integer model for robust supply chain network design under uncertainty, pointing out six strategies to alleviate the danger of associated disruptions [14]. Table 1 give the abbreviations for uncertain parameters and other parameters involved in objective function for designing the supply chain network.

Table 1. Abbreviation for uncertain parameters in supply chain network design.

Abbreviation	Parameter with uncertainty
D	Demand
C	Activity Cost (e.g., transportation, production)
S	Supply quantity of various facilities within network
AF	Availability of network facilities
AT	Availability of transportation links/modes between network's entities
CA	The capacity of network facilities/transportation links
DC	Disrupted products/supply/commodities in SC facilities
TT	Transportation time through entities of Supply chain network
A	Annual no. of the vehicle visit

3 Theoretical Risk Model

For delineating the risk propagation, a four-tier supply chain considered comprising of the supplier, production center, distribution center, and customer. If the node of the supply chain is disrupted or failed, the risk will propagate through connectivity links.

In Fig. 1, there is a simultaneous disruption at supplier1, production center2, and distribution center1. In the case of supplier1 disruption, 'first propagation' is from supplier1 to production center1; 'second propagation' is from production center1 to distribution

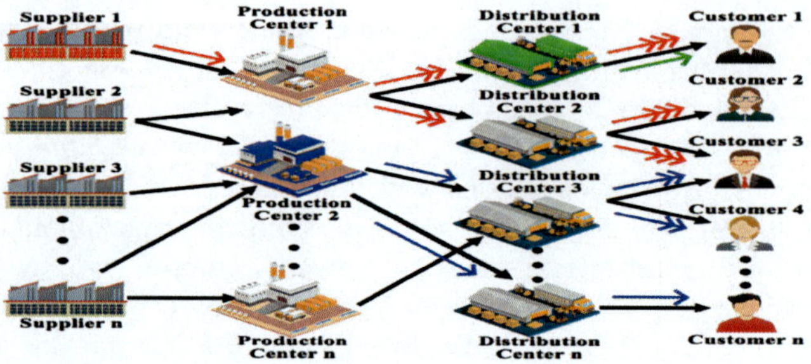

Fig. 1. Facility disruption and propagation of risk between different levels.

center1 and distribution center2 and 'third propagation' is from distribution center1 to customer1 and distribution center2 to customer2 and customer3. There is some hypothesis followed while designing the supply chain network. The supplier can only connect to the manufacturer, and the manufacturer can only connect to the retailer. Each node is subjected to random disruption. Disruption can also be regional, affecting a set of nodes simultaneously.

Each supplier has maximum capacity CAP_f, beyond which it cannot supply. Similarly, each manufacturer and retailer have a maximum capacity of receiving MA_m and MA_r, respectively. The retailer k is subjected to demand, which is less than MA_m. Retailer k sell the product at selling price SP_r. When the quantity delivered QD_r to retailer r is more than the demand D_r, the excess is sold at a discount price with the loss of DC_r. If the quantity deliver QD_r to retailer r is less than the demand D_r, the unmet demand is lost, which incorporates shortage cost SC_r. Mixed-integer linear programming (MILP) has been implemented to design a supply chain optimized network.

Parameters

P_{sl}: Probability of disruption scenario l.
D_{rk}: Demand of retailer k.
D_{sfli}: Binary setting: 1 if supplier i in scenario l is in disruption; 0 otherwise.
D_{smlj}: Binary setting: 1 if manufacturer j in scenario l is in disruption; 0 otherwise.
D_{srlk}: Binary setting: 1 if retailer k in scenario l is in disruption; 0 otherwise.
Y_{fi}: 1 if supplier i is selected; 0 otherwise.
Y_{mj}: 1 if manufacturer j is selected; 0 otherwise.
Y_{ri}: 1 if retailer i is selected; 0 otherwise.
Q_{mrjk}: Unit of product ordered from manufacturer j to retailer k.
Q_{fmij}: Unit of product ordered from supplier i to manufacturer j.
LST_{srlk}: Level of storage in scenario l from retailer k.
LSH_{srlk}: Level of shortage for retailer k in scenario l.
QD_{srlk}: Quantity delivered to retailer k in scenario l.

The mathematical formulation of our problem is presented below:

$$Max\ Z = \sum_{k=1}^{r}\sum_{l=1}^{s} SP_k * QD_{kl} * P_{sl}(1 - D_{kl})$$
$$- \left\{ \left[\sum_{i=1}^{f} FC_i * Y_i + \sum_{j=1}^{m} FC_j * Y_j + \sum_{k=1}^{r} FC_k * Y_k \right] \right.$$
$$+ \left[\sum_{j=1}^{m}\sum_{k=1}^{r}\sum_{l=1}^{s} (1 - D_{jl}) * Q_{jk} * VC_{jk} * P_{sl} + \sum_{i=1}^{f}\sum_{j=1}^{m}\sum_{l=1}^{s} (1 - D_{il}) * Q_{ij} * VC_{ij} * P_{sl} \right]$$
$$\left. + \left[\sum_{k=1}^{r}\sum_{l=1}^{s} LST_{kl} * DC_k * P_{sl} \right] + \left[\sum_{k=1}^{r}\sum_{l=1}^{s} LSH_{kl} * SC_k * P_{sl} \right] \right\} \quad (1)$$

Subjected to

$$QD_{kl} = \sum_{j=1}^{m} (1 - D_{jl}) * Q_{jk}\ \forall k \quad (2)$$

$$QD_{kl} = DR_k + LST_{kl} - LSH_{kl}\ \forall k \quad (3)$$

$$QD_{kl} \le MA_k\ \forall k \quad (4)$$

$$\sum_{i=1}^{f} Q_{ij} - \sum_{k=1}^{r} Q_{jk} \ge 0\ \forall j \quad (5)$$

$$\sum_{i=1}^{f} Q_{ij} \le MA_j * Y_j\ \forall j \quad (6)$$

$$\sum_{j=1}^{m} Q_{ij} \le CAP_i * Y_i\ \forall i \quad (7)$$

$$QD_{kl}, Q_{ij}, Q_{jk}, LST_{kl}, LSH_{kl} > 0\ \forall i, j, k \quad (8)$$

The objective of the problem shown in Eq. (1), maximizes the expected profit, which is obtained by subtracting the total cost from the revenue. Here, the total cost is calculated as the sum fixed of each and respective, supplier manufacturer and retailer; the variable cost of delivering product from supplier to manufacturer and manufacturer to retailer; shortage cost, and discount cost. Equation (2) shows that the quantity delivered to retailer k in scenario l is the sum of quantity deliver from all the manufacturer to retailer k. Equation (3) shows that the quantity delivered to each retailer is the sum of demand and storage minus the shortage, where demand, storage, and shortage are non-negative integers. Equation (4) and (5) shows that the quantity delivered to each retailer and manufacturer is less than their maximum acceptable delivery. Equation (6) shows that the total quantity delivered to manufacturer j is more than quantity delivered from manufacturer j. In Eq. (7), the quantity delivered from each supplier should not exceed its maximum capacity. Equation (8) assures that the quantity supplied is not harmful.

3.1 Model Construction

The supply chain network optimized in this research paper consists of six suppliers, six manufacturers, and six retailers, as shown in Sect. 3.3. *1S*, *2S*, *3S*, *1M*, *2M*, *3M*, *1R*, *2R*, and *3R* are internal nodes. *4S*, *5S*, *6S*, *4M*, *5M*, *6M*, *4R*, *5R* and *6R* are international nodes. The MILP model is solved using CPLEX. Table 2, Table 3, Table 4, Table 5 are the input parameter of the mathematical model illustrated in Sect. 3.2 and Sect. 3.3. Table 2 gives the simultaneous disruption probability of domestic and international suppliers, manufacturers, and retailers. Table 3 presents the retailer side parameter: Maximum acceptable delivery, demand, selling price before discount, shortage cost, discount-cost, fixed cost, and the local probability of disruption. Table 4 presents manufacturer side parameter: Variable cost per product from manufacturer *j* to retailer *k* of each product, fixed cost, maximum acceptable delivery, and the local probability of disruption. Table 5 gives supplier side parameter: Variable cost per product from supplier *i* to manufacturer *j* of each product, fixed cost, maximum capacity, and the local probability of disruption.

Table 2 demonstrates the simultaneous domestic, and international disruption probability of retailers is 0.00003 and 0.0001, respectively. Table 3 shows that the maximum acceptable delivery of retailer *1R* is 500, and the demand is 350. Selling price before discount, shortage cost, discount-cost per product is 43, 8 and 25 respectively, and fixed cost of opening retailer *1R* is 700. The local disruption probability of retailer 1R is 0.04. Table 4 presents that fixed cost, maximum acceptable deliver, and local disruption probability of manufacturer *1M* is 1400, 1000, and 0.02, respectively. The variable cost of delivering per product from manufacturer *1M* to retailer *1R* is 16. Similarly, in Table 5, fixed cost, maximum capacity, and local disruption probability of supplier *1F* is 1400, 1500, and 0.01, respectively. The variable cost of delivering per product from supplier *1F* to manufacturer *1M* is 14.

Table 2. Simultaneous disruption probability.

α^*_{1f}	α^*_{2f}	α^*_{1m}	α^*_{2m}	α^*_{1r}	α^*_{2r}
0.00003	0.0001	0.00005	0.0004	0.00008	0.0006

Table 3. Retailer side parameter.

Retailer	*1R*	*2R*	*3R*	*4R*	*5R*	*6R*
Max. delivery	500	700	600	550	800	650
Demand	350	500	500	400	600	400
Selling price	43	49	40	52	46	54
Shortage cost	8	9	7	9	8	10
Discount cost	25	27	24	26	24	29
Fixed cost	700	900	800	1200	1000	1100
Local disr. prob.	0.04	0.03	0.02	0.07	0.12	0.11

Table 4. Manufacturer side parameter

j	*k*								
	1R	*2R*	*3R*	*4R*	*5R*	*6R*	*Fixed cost*	*Max. delivery*	*Local disr. prob.*
1M	16	11	10	14	11	12	1400	1000	0.02
2M	5	18	14	6	12	17	1200	1200	0.04
3M	6	6	12	18	16	13	1500	800	0.03
4M	15	13	17	16	16	13	1000	1100	0.07
5M	8	13	15	11	16	15	800	900	0.09
6M	5	18	8	15	5	16	1600	1300	0.06

Table 5. Supplier side parameter

i	j								
	1M	*2M*	*3M*	*4M*	*5M*	*6M*	*Fixed cost*	*Max. capacity*	*Local disr. prob.*
1F	14	15	13	5	7	15	1400	1500	0.01
2F	16	11	15	19	5	19	1200	1600	0.05
3F	6	18	11	11	6	17	2300	2000	0.04
4F	17	9	7	17	19	16	1300	1700	0.11
5F	11	7	18	17	19	8	1500	1300	0.06
6F	8	19	12	11	9	14	1800	1800	0.08

4 Computational Results

Initially, we have considered 18 nodes, but the final result shows that the optimal supply chain consists of 3 suppliers, 4 manufacturers, and 6 retailers. Table 6 shows the quantity supplied from supplier *i* to manufacturer *j*, and Table 7 presents the quantity supplied from manufacturer *j* to retailer *k*.

Table 6 shows that supplier *1F*, *2F*, and *6F* are not selected, and the quantity deliver from supplier *3F* to manufacturer *1M* is 1000. Similarly, Table 7 shows that manufacturers *4M* and *5M* are not selected, and the quantity delivered from manufacturer *1M* to retailer *3R* is 350. With the given input data, the expected profit considering all the disruption scenario is equal to 67133.75. When there is no disruption, the profit will increase to 81050.

A numerical simulation is conducted to calculate RI at different linking intensity and node threshold, and the results obtained are shown in Figs. 2, 3, 4, 5 and 6. According to these results presented, both linking intensity increases, and the node threshold increases with the step size of 0.1 from 0.3 to 1 and from 0 to 1, respectively. Linking intensity of all connectivity links and thresholds of all nodes is assumed to be equal while calculating RI. Two cases are considered; in the first case, exactly one node is disrupted, and in the second case, exactly two nodes are disrupted. The total number of disruption scenarios is the first and second cases are ${}^{13}C_1$ and ${}^{13}C_2$, respectively.

To find RI at random node removal, we have conducted a simulation on a different number of node removal. The RI value is based on the average of 10 simulations.

Table 6. Quantity supplied from supplier to manufacturer.

	1M	*2M*	*3M*	*4M*	*5M*	*6M*
1F	0	0	0	0	0	0
2F	0	0	0	0	0	0
3F	1000	0	0	0	0	0
4F	0	600	800	0	0	0
5F	0	350	0	0	0	950
6F	0	0	0	0	0	0

Table 7. Quantity supplied from manufacturer to retailer.

	1R	*2R*	*3R*	*4R*	*5R*	*6R*
1M	0	0	350	0	0	650
2M	400	0	0	550	0	0
3M	100	700	0	0	0	0
4M	0	0	0	0	0	0
5M	0	0	0	0	0	0
6M	0	0	150	0	800	0

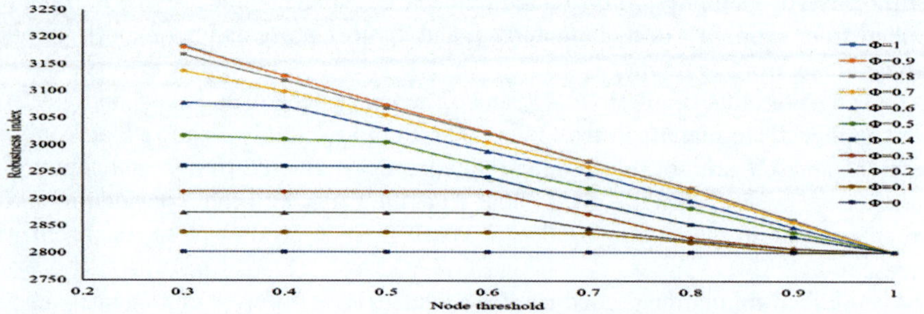

Fig. 2. RI at the different linking intensity.

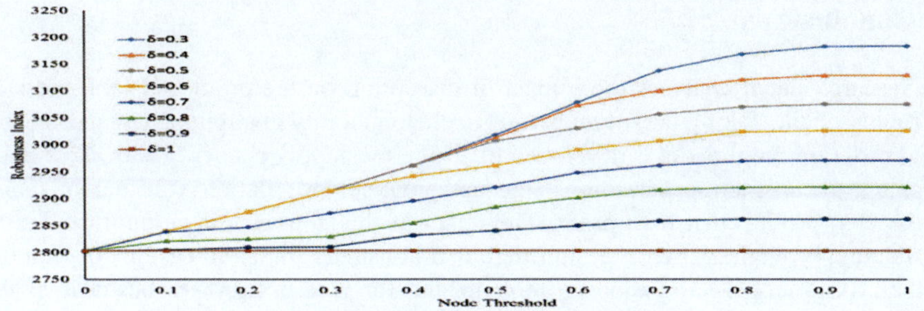

Fig. 3. RI at different node threshold.

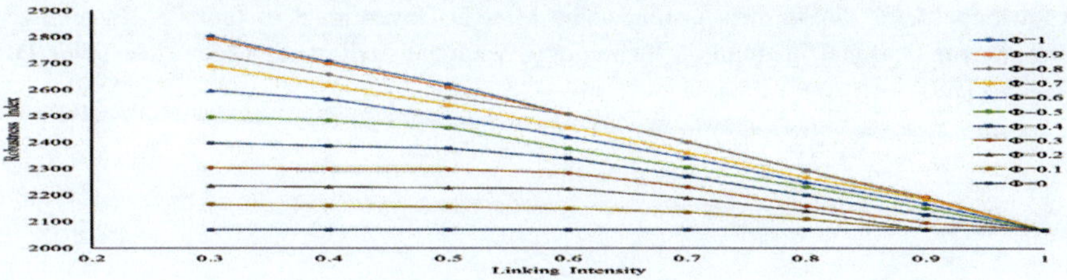

Fig. 4. RI at the different linking intensity.

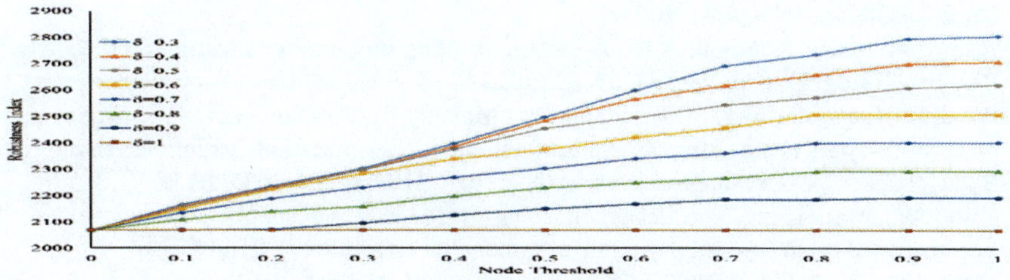

Fig. 5. RI at different node threshold.

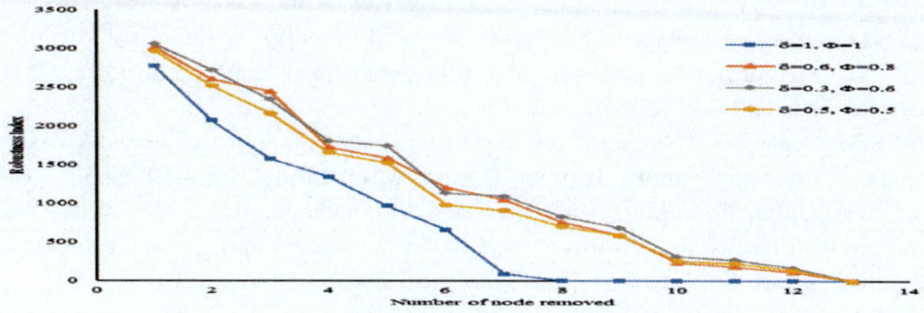

Fig. 6. RI at different number of node removal.

5 Conclusions

The research paper analyses the impact of disruption on the profit and robustness of the supply chain. Taking into account multi-echelon facility disruption, a mixed-integer linear programming model is developed to design the supply chain network. This main advantage of considering disruption in the mathematical formulation is the maximization of expected profit rather than profit. The result of this approach is an optimized risk-neutral supply chain network design that also considers losses in case of disruption. Further, a numerical simulation is implemented for measuring the robustness of the supply chain in different disruption scenarios. The robustness index is expressed in terms of the final quantity deliver by all the retailers.

Acknowledgement. The project is funded by MHRD Government of India as research seed money scheme National Institute of Technology Warangal, India with Order No NITW/ DIR/ 2018/478/1138.

References

1. Sheffi, Y., Rice Jr., J.: A supply chain view of the resilient enterprise. MIT Sloan Manag. Rev. **47**, 41 (2005)
2. Li, Q., Zeng, B., Savachkin, A.: Reliable facility location design under disruptions. Comput. Oper. Res. **40**(4), 901–909 (2013)
3. AhmadiJavid, A., Seddighi, A.H.: A location-routing problem with disruption risk. Transp. Res. Part E Log. Transp. Rev. **53**, 63–82 (2013)
4. Azad, N., Saharidis, G.K., Davoudpour, H., Malekly, H., Yektamaram, S.A.: Strategies for protecting supply chain networks against facility and transportation disruptions: an improved Benders decomposition approach. Ann. Oper. Res. **210**(1), 125–163 (2013)
5. Klibi, W., Martel, A.: The design of robust value-creating supply chain networks. OR Spectrum **35**(4), 867–903 (2013). https://doi.org/10.1007/s00291-013-0327-6
6. Mari, S.I., Lee, Y.H., Memon, M.S.: Sustainable and resilient supply chain network design under disruption risks. Sustainability **6**(10), 6666–6686 (2014)
7. Jeong, K.Y., Hong, J.D., Xie, Y.: Design of emergency logistics networks, taking efficiency, risk and robustness into consideration. Int. J. Log. Res. Appl. **17**(1), 1–22 (2014)
8. Marufuzzaman, M., Eksioglu, S.D., Li, X., Wang, J.: Analyzing the impact of intermodal-related risk to the design and management of biofuel supply chain. Transp. Res. Part E Log. Transp. Rev. **69**, 122–145 (2014)
9. Hatefi, S.M., Jolai, F.: Robust and reliable forward–reverse logistics network design under demand uncertainty and facility disruptions. Appl. Math. Model. **38**(9–10), 2630–2647 (2014)
10. Hatefi, S.M., Jolai, F., Torabi, S.A., Tavakkoli-Moghaddam, R.: A credibility-constrained programming for reliable forward–reverse logistics network design under uncertainty and facility disruptions. Int. J. Comput. Integr. Manuf. **28**(6), 664–678 (2015)
11. Hatefi, S.M., Jolai, F., Torabi, S.A., Tavakkoli-Moghaddam, R.: Reliable design of an integrated forward-revere logistics network under uncertainty and facility disruptions: a fuzzy possibilistic programing model. KSCE J. Civ. Eng. **19**(4), 1117–1128 (2015). https://doi.org/10.1007/s12205-013-0340-y
12. Sadghiani, N.S., Torabi, S.A., Sahebjamnia, N.: Retail supply chain network design under operational and disruption risks. Transp. Res. Part E Log. Transp. Rev. **75**, 95–114 (2015)

13. Tang, L., Jing, K., He, J., Stanley, H.E.: Robustness of assembly supply chain networks by considering risk propagation and cascading failure. Phys. A **459**, 129–139 (2016)
14. Hasani, A., Khosrojerdi, A.: Robust global supply chain network design under disruption and uncertainty considering resilience strategies: a parallel memetic algorithm for a real-life case study. Transp. Res. Part E Log. Transp. Rev. **87**, 20–52 (2016)

Impact of Governmental Support for the Implementation of Industry 4.0 in Portugal

Bruno Silva Faria[1], Ana Correia Simões[2(✉)], and José Coelho Rodrigues[1,2]

[1] Faculty of Economy, University of Porto, Porto, Portugal
up201701318@fep.up.pt
[2] INESC TEC, Rua Dr. Roberto Frias, Porto, Portugal
ana.c.simoes@inesctec.pt

Abstract. Technological breakthroughs, such as the Internet of Things, Big Data repositories, artificial intelligence or additive manufacturing, are triggering a Fourth Industrial Revolution. This new revolution, also known as Industry 4.0, is characterized by the combination of physical and digital worlds in digital ecosystems that connect the different members in the value chain from clients to suppliers and distributors. Companies are redefining their strategies based on this new paradigm to obtain a competitive advantage. They aim to achieve more efficient and flexible productive processes that can produce high-quality products at low costs, investing on mass customization to satisfy their clients. Accordingly, governments are implementing support programs that create a suitable environment for the adoption of technological innovation strategies by the companies. Although some programs may diverge in some objectives, they all aim to promote workers' skills adaptation, technological supply development, and business modernization. The Portuguese Government also released its program for Industry 4.0 support, known as Portugal i4.0, which is intended to stimulate Portuguese economy digitalization. Furthermore, in latest years, it has been supporting projects through European funds mobilizations from Portugal 2020 program. The present study analyses whether companies that received financial support from Portuguese government to implement innovative projects, within the Industry 4.0 paradigm, were able to improve economic and financial performance and competitivity gains. For such purpose, it was applied an inference statistical method to analyse the differences verified in economics and financial indicators between the periods before and after projects implementation in a selected group of companies.

Keywords: Industry 4.0 · Digitalization · Technology innovation · Competitiveness · Economic performance · Financial performance

1 Introduction

The 4th Industrial Revolution paradigm is a current priority on the executive committees' agenda seeking to identify new opportunities to diversify markets and increase revenues

J. Machado et al. (Eds.): icieng 2021, LNME, pp. 108–120, 2022.
https://doi.org/10.1007/978-3-030-78170-5_11

[1]. By adopting new technology tools, companies seek to transform operations throughout the value chain, increasing efficiency, flexibility and reliability to make them more agile [1]. The digitalization of business and production processes is critical as it enables better communication and cooperation throughout the value chain. The constant information sharing with customers enables companies to offer customized products, which can be obtained at more competitive prices due to the advances in manufacturing processes [2]. Similarly, the direct link with different elements of the supply chain makes its management much more efficient.

The new industrial revolution is a result of the technological advances of recent years. The next-generation information networks emergence, the Internet of Things (IoT) application to the industrial sector (Industrial IoT - IIoT), the radio frequency identification (RFID) use, and the Big Data analysis and Cloud Computing use have allowed data collection and analysis make the processes more productive flexible, faster, and efficient, capable of manufacturing products with higher quality at reduced costs [3].

Therefore, it is crucial that companies assess their current digital maturity and define an appropriate strategy in order to meet the challenges of this new technological context. According to a PWC study, published in 2016, based on responses from more than 2000 companies from 26 countries, at that time only 33% of the respondents considered their company digitally advanced (i.e., had a high degree of their products, processes and services digitalized) [6].

In this context, it is crucial that governments implement measures that promote the creation of favorable conditions for the industrial sector digital evolution. These conditions encompass numerous factors, such as the creation of financing channels, tax incentives, technological infrastructures, access to global markets, proximity to teaching and research centers and the training of human resources [5]. Therefore, the aim of this study is to analyse whether the government financial support that is being invested in Portuguese companies in the industrial sector, to implement projects within the Industry 4.0 paradigm, are allowing them to improve economic and financial performance and competitivity gains.

2 Industry 4.0 Impacts and Challenges

After the evolution enabled by the steam engine, electricity, mass production, electronics and automated production, the next phase of manufacturing evolution is now being discussed towards a complete production system digitalization, interconnecting machines, people and processes [2]. Industry 4.0 aims to take advantage of emerging technological advances to improve the industrial sector performance, enabling organizations to remain competitive. The purpose of this new paradigm is to combine the physical and digital worlds, namely through the creation of Cyber-Physical Systems (CPS), which allow full digitalization of the physical assets and their integration into digital ecosystems [1]. This principle is fundamental to the development of intelligent production processes composed by sensorized machines and devices, with the ability to communicate in real-time with each other and with other elements of the value chain in order to adapt to different scenarios [6].

The previous industrial revolutions had a strong impact, especially on the factory floor and production processes. However, the 4th Industrial Revolution is expected to have a broader impact, changing organizations' versatility. In addition to anticipating an improvement in the production processes with the introduction of new technologies, this new paradigm has implications throughout the value chain, changing the way that organizations relate to their suppliers and customers [2]. For example, according to the study of Simões et al. (2019) [7], one of the main expected impacts of collaborative robot's adoption on assembly production lines is to obtain operational efficiency gains. Participants in this study noted that the introduction of this technology improves product quality and productivity, increases flexibility, and optimizes workspace utilization.

On the other hand, with IoT use in equipment monitoring it becomes possible to detect and predict anomalies, through the processing of the collected information [8]. Another advantage of the use of IoT is energy efficiency improvement. Using production cycle analysis, it is possible to make decisions that improve production lines' planning towards an energy consumption minimization [8]. Through the adoption of disruptive technologies and new business models developed towards the Industry 4.0 paradigm, approximately 57% of Portuguese industrial companies expect an average revenue increase of up to 10% [4]. While around 55% of them expect cost reductions above 10% and roughly 70% expect efficiency gains above 10% [4].

With the sophistication of the productive and organizational processes and their interconnectedness, there are several changes in the products' life cycle, working environment and labor market, leading to the emergence of new challenges [1]. Thus, in order to achieve better competitive levels, it is crucial that companies are prepared for digital transformation by adopting the resources, strategies and frameworks best suitable to face those challenges [9].

Along with capital investment, investment in human resources will be a critical factor in this transformation process. According to Eurostat data, about 40% of European workers have insufficient digital skills [10]. Jobs in this new context will require a combination of multidisciplinary skills, which the current education and training system cannot answer. According to Maresova et al., this is a field that should receive greatest attention from companies and governments, these entities should combine efforts to train workers in order to fit their skills with new technological trends and opportunities [11].

As the number of connected equipment and products increases, issues related to reliability, system availability, and operational security become major challenges in organizations' management. Therefore, investments in cybersecurity also become essential to prevent data manipulation situations, unauthorized remote access, loss of intellectual property, or in extreme cases "sabotage" of the facilities [12].

Table 1 lists the major impacts and challenges of implementing a pro-Industry 4.0 strategy that emerge from the previous literature review.

Table 1. Key impacts and challenges of Industry 4.0 [1–4, 7, 9, 11]

Impacts	Challenges
Operational efficiency	Alignment with strategy
Increased productivity	Resistance to change
Cost reduction	Infrastructure investment
Bulk customization	Human resources training
Supply chain optimization	Standards creation
Labor market change	Cybersecurity

3 Industry 4.0 Support Programs

Governmental institutions in various countries have, over the last few years, developed strategic programs aimed at supporting the transformation of their industrial sectors. Conscious of the potential impacts that the adoption of new digital technologies can have on the efficiency and competitiveness of their economies, governments have been implementing policies that seek to capitalize on the strengths of their industries but also to mitigate their gaps.

The strategic programs of the different countries seek to adapt their context to the 4th Industrial Revolution and undertake different approaches depending on their political traditions, degree of centralization, institutional infrastructures, and type of technologies [13]. However, there are significant commonalities between the different strategic programs, particularly in what concerns workers' skills adaptation, technological supply development and companies' modernization [14]. Furthermore, there is an effort of international cooperation towards a joint priority regarding the standardization of digital technologies and the transference of scientific and technological knowledge [2]. Some programs that are being implemented are Platform Industry 4.0 from Germany, Alliance pour l'Industrie du Futur from France, Advanced Manufacturing Partnership from the USA, Made in China 2025 from China, Make in India from India, and Industry Innovation Movement 3.0 from South Korea.

The 4th Industrial Revolution is a notable opportunity for Portugal to overcome some of its competitive barriers, such as lack of internal market scale and peripheral location. According to the 2016 Digital Economy & Society Index of the European Commission, Portugal was in 15th position, in terms of the level of digital competitiveness, above the European Union (EU) average. In another UBS study, which analysed 45 countries, Portugal was the 23rd most well-prepared economy to adopt Industry 4.0, with its infrastructures, innovation capacity and general competences as its main added value. As a conclusion of these studies, it can be inferred that although the country is reasonably well prepared for this transformation, that preparation is not yet reflected in its current level of competitiveness [15].

With the aim of generating the conditions for the development of industry and national services in the digital age and promoting the adoption of new technologies, the Ministry of Economy launched in 2017 the first phase of the program "Portugal i4.0".

For the elaboration of this program, interviews, workshops, and auditions were carried out with the main stakeholders from different sectors of the Portuguese economy, with the participation of more than 200 entities and companies, including multinational companies such as Bosch, Volkswagen, Altice-PT, Siemens, Microsoft, Google, Huawei or Deloitte [16]. The strategic plan consisted of 60 public and private measures grouped into six major areas of priority action: Human Resources Training; Cooperation Ecosystem; StartUp i4; Financing and investment support; Internationalization; Legal and regulatory adaptation. These areas aimed to achieve three main objectives: accelerate the adoption of Industry 4.0 technologies and concepts in the Portuguese companies; promote internationally the Portuguese technology companies; and, make Portugal an attractive hub for investment in the Industry 4.0 context [17].

In order to accelerate the investments and adhesion by the national companies, the Government has forecast the injection of up to 4.500 million euros in the Portuguese economy in the following years, with about half of that amount (up to 2.26 billion) being mobilized from the European Structural and Investment Funds through Portugal 2020. After two years of implementation of the first phase of the Portugal i4.0 program, which was mainly mobilizing and demonstrating, 95% of the measures were implemented, with over 24,000 companies and 550,000 people covered by the different initiatives [18].

In April 2019, the second phase of the program was launched, featuring a more transformative logic, designed to expand the number of companies that truly capitalize on the benefits of Industry 4.0 [18]. Analyzing the results of the first phase, the companies that benefited most from the initiatives implemented were those with a greater awareness of the themes and benefits related to Industry 4.0 and with greater availability of resources and skills [18]. Therefore, for Phase 2 to truly have the intended transformative impact, it would be necessary to mobilize companies with a lower level of maturity in terms of digitalization. To this end, it was necessary to implement measures aiming to mitigate the difficulties of companies in the implementation of this type of project, usually associated with a smaller financial and human resources capacity, and it was also fundamental to promote the sharing of resources and knowledge among SMEs [19].

In order to bridge the identified gaps and leverage a generalized transition to Industry 4.0, the 2nd phase of Portugal i4.0 considers it necessary to act on three strategic lines: Generalize i4.0, Enable i4.0, Assimilate i4 .0 [18]. In this phase it is estimated a mobilization of public and private investments in the amount of 600 million euros over the following two years. The goal is to involve more than 20,000 companies in various initiatives, train more than 200,000 workers, and finance more than 350 manufacturing projects [20].

4 Method

Companies and Governments have over the past few years joined forces to improve the digital capabilities of the Portuguese industry. The definition of appropriate policies and strategies are crucial to ensure a successful transition to the digital age. Taking this into consideration, this study aims at understanding whether the government financial support provided for the Portuguese companies in the industrial sector to implement projects within the Industry 4.0 paradigm in recent years are allowing them to improve economic and financial performance and competitivity gains.

4.1 Sample Selection

To conduct this study the companies in the Portuguese industrial sector that had implemented projects within the Industry 4.0 paradigm were considered. The sample of companies in this study consists of those that received financial support to conduct the projects, under the Portugal 2020 program. The database of Portugal 2020 program (available on the official website) was used to identify the companies with approved projects between January 2014 and September 2018. Since the aim of this study is to analyse the impacts resulting from the implementation of the Industry 4.0 paradigm, a minimum of 2 years in the time gap was considered for each company. Therefore, in this study, only companies with projects approved before 2017 were considered.

The projects in the Portugal 2020 official database were not typified within the Industry 4.0 paradigm. To overcome this difficulty, it was decided to select projects that promote the digitalization, the use of information and communication technologies, and the development of intelligent and interconnected systems in the productive and business context. The projects were selected searching for keywords related to Industry 4.0 paradigm in the descriptive texts, in the titles, and the synthesis of the project proposals, similar to the technique applied in the study of the Agência Nacional de Inovação, in its study about R&D projects related to Industry 4.0, co-financed in QREN and PT2020. The keywords used were: Smart Factory; Business intelligence; Digitalization; Robotics; Automation; Internet of Things; Cloud computing; Artificial intelligence; Big data; Machine learning; Data mining; Additive Manufacturing; Advanced Materials; Industry 4.0; i4.0.

Based on the previous criteria, a first sample of 89 projects was obtained. However, this first selection of projects, included projects implemented by companies, as well as Universities, Research Institutions, Business Associations or Municipalities. Since the scope of this study is focused on projects implemented in companies, a new filter was applied and a final sample of 54 projects were selected (those conducted in companies). As referred before, the database is organized by projects, and a company can have more than one project. The 54 projects selected were implemented in 41 companies. A detailed analysis on the project description of these 54 projects allowed to exclude 12 projects that did not fit the study objective. After this detailed analysis, the final sample resulted in 42 projects, corresponding to 31 companies (Fig. 1).

4.2 Variables Selection

A literature review was performed to identify the variables to use that best represent the measures aimed for this study: economic and financial performance, and competitivity gains. However, there is no consensus regarding the variables that should be used [22]. Therefore, the variables to conduct the analysis were defined based on their capacity to measure the economic and financial performance within the context of this study. In the following paragraphs the selected variables are presented.

Turnover (NR), which allows companies to measure corporate growth, while their competitive position in the global market may be evaluated through the Exports variable (EXP). One of the most anticipated effects of the implementation of Industry 4.0 is the impact on business productivity [2]. Thus, by assessing the evolution of Gross Value

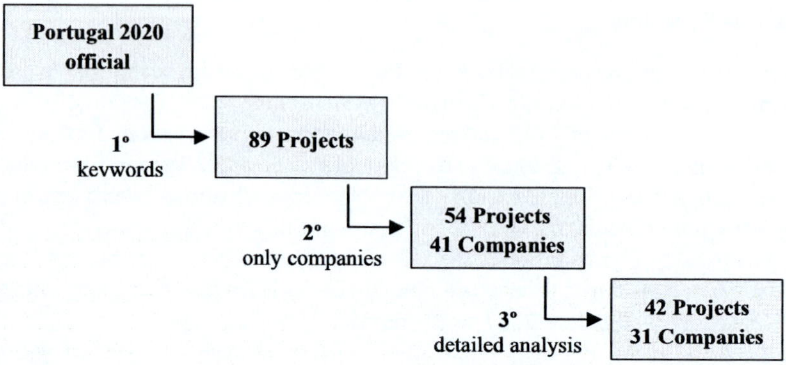

Fig. 1. Selection of the study sample

Added (GVA), one can see if there was an exponential impact on resource use optimization [23]. Similarly, by relating the GVA and the average number of workers with the Apparent Labor Productivity (ALP) variable, one can evaluate the efficiency of the use of human resources in productivity.

The evolution of Industry 4.0 paradigm requires companies to make high investments. Through the Return on Assets (ROA) analysis, it is possible to measure the return on capital of the company, since this variable allows us to evaluate the capacity of companies to generate profit by making investments [24]. However, it is important that these investments do not compromise the company's ability to pay off its debts in the short term, so this study will also consider developments in the liquidity ratio (LQD), namely the overall liquidity ratio [23]. Similarly, the financial health of companies can also be assessed through Financial Autonomy (FIN AUT) [25]. Regarding the long term, it is advisable to resort to the degree of solvency (SOLV), since it allows to measure, for long periods of time, the firm's solidity against the foreign capital invested in the company [26].

Having defined the companies' sample and the variables to use, the data (values for the variables) were collected using the Sabi database [https://sabi.bvdinfo.com/]. This is a private database that contains historical detailed financial information for the Portuguese companies. Based on the financial data collected in Sabi, on the two time periods of the analysis, the values of the economic and financial variables were calculated.

4.3 Data Analysis

This study aims to analyse whether companies that received financial support from Portuguese government to implement innovative projects, within the Industry 4.0 paradigm, were able to improve economic and financial performance and competitivity gains. In this way, it is intended to assess the differences in economic and financial performance between the ex-ante and ex-post periods. To conduct the data analysis a paired samples method was applied, since it is intended to analyse the behavior of quantitative variables of the same group of companies at two different instants in time, between which occurred events that may have influenced the evolution of variables [21]. The analysis of variables differences in these two instants in time, aims to assess the existence of statistically significant differences between the preceding period and the period after the financial incentive application.

In this type of studies, it is usual to use the parametric test, t-test, as it is a test that allows verifying if there are statistically significant differences between the variables average, considering the two time periods, through a test of hypothesis (t-test) (Eq. 1) [21]. However, given the small size of the sample (31 companies), it is recommended, before applying the t-test, to test the variables for the normal distribution. If the test rejects a normal distribution of the variables, a nonparametric test, such as the Wilcoxon test, should be used [21].

$$\begin{cases} H_0 : \mu_{t-1} = \mu_{t+1} \\ H_1 : \mu_{t-1} < \mu_{t+1} \end{cases} \tag{1}$$

5 Results Presentation and Discussion

The sample consists of 7 large, 10 medium, and 14 small companies. In terms of location (NUTS II), the 31 companies are mainly located in the North region, in a total of 17 companies, followed by the Center region with 10 companies. Overall, the 31 companies had a total eligible expenditure allocated to the operation of € 82 330 950, resulting in a total approved funding of € 36 728 950. Regarding the funding, the projects were financed by European Structural and Investment Funds, more specifically, 35 funded by the European Regional Development Fund (ERDF), and 7 by the European Social Fund (ESF).

5.1 Hypothesis Test Results

For the data analysis, the advanced statistical analysis software, SPSS (Statistical Package for the Social Sciences), version 25, was used. Since, the sample contains only 31 values for each variable, the Shapiro-Wilk test was used to assess the normal distribution of the variables.

After this test, it was possible to conclude that only the "financial autonomy" variable follow a normal distribution (for both periods, before and after the project implementation) with a significance value higher than 0.05. The t-test was only used for this variable. For the remaining variables, the Wilcoxon test was applied.

These two tests were applied to the variables according to their distribution, in order to assess if there were statistically significant differences between the values in the two periods, ex-ante and ex-post the allocation of the structural fund. Table 2 presents, for each variable, the type of test that was applied, the Hypothesis test results and the significance value. In order to understand the evolution direction of the variables, it was decided to calculate the growth rates for each of the variables.

Table 2. Hypothesis test results

Variables	Test	Result	Significance	Growth rate
NR	Wilcoxon	Rejected	0,000	28,17
EXP	Wilcoxon	Rejected	0,004	25,93
GVA	Wilcoxon	Rejected	0,006	38,89
ALP	Wilcoxon	Accepted	0,739	0,96
ROA	Wilcoxon	Accepted	0,327	−15,87
LQD	Wilcoxon	Accepted	0,845	−3,70
FIN AUT	Teste t	Accepted	0,417	5,80
SOLV	Wilcoxon	Accepted	0,170	23,93

5.2 Hypothesis Test Results

The hypothesis test results allows to conclude that only three variables, turnover (NR), gross value added (GVA) and exports (EXP), presented statistically significant differences between the two periods. As such, only for these three variables is possible to conclude that the technology/process implementation have statistically significant differences regarding the Community funds application.

By analyzing the growth rate of these three variables, it can be seen that they evolved favorably, which corroborates the expectation on improving operational efficiency and competitiveness gains associated with the implementation of projects within the Industry 4.0 paradigm. The results obtained are significant, especially when compared with the aggregate results of the manufacturing sector for the period 2014–2017. For example, in the case of turnover, between this period, it appears that the sector grew 12,07% [27], significantly lower than the 28,17% growth rate for the companies in the sample. Similarly, in the case of the value-added and exports variables, there are significant differences. In the case of gross value added, the sector grew by 25,35% [28], also below the 38,89% of the companies in the sample. In the case of the value of exports, it was estimated that the sector grew by 8% [29], considerably below to the 25,93% for the companies in the sample.

On the other hand, it is also important to understand why the remaining five variables (apparent labor productivity, return on assets, liquidity, financial autonomy, solvency) did not yield statistically significant results. The first explanation could be the presence of "other non-considered factors", that go beyond the implementation of co-financed projects, which could influence the evolution of variables. For example, for the apparent labor productivity one possible explanation could be the lack of adequate skills of human resources to take full advantage of the technology, or the lack of investment in training the human resources to optimize the use of new technologies. In the case of the return on assets variable, the results may have been influenced by the short period considered in the study. Therefore, it may suggest considering a longer post-implementation period for analysis. Although this suggestion may also be considered to justify the negative results on the financial variables (ROA and LQD), the results may also be justified by the financial situation of each company prior to project implementation.

5.3 Discussion

During the research, it was not possible to find quantitative studies that assessed the impact of Industry 4.0 strategy on the companies' performance variables. However, there are several studies that, although not focusing directly on Industry 4.0 technologies implementation impact, analyse the performance of companies that have applied innovation strategies. The results of those studies can be useful to compare with the results of this study, since the last ones also involve innovation.

For example, the study "Destination: Growth and Innovation" conducted by COTEC Portugal (2017) [30] concludes that the most innovative SMEs tend to outperform other SMEs. This study indicates that when comparing these two groups of companies, the most innovative SMEs have a net result 7.8 times higher and a turnover 3.7 times higher. Our results are in line with the results obtained in this study, since the companies analysed had a higher turnover than their sector. Similarly, the results of Correia & Costa (2016) [26], shows that for the total sample of companies that implemented co-financed projects, aiming at innovation, technological development, qualification, and internationalization, all the variables, except apparent productivity of the work, presented similar results.

On the other hand, the results of our study can also be compared with the perspectives that entities and entrepreneurs have on the implementation of Industry 4.0 related projects. For example, according to a study conducted by PWC [4], 43% of Portuguese industrial companies are expected to achieve revenue increases of over 10% by implementing Industry 4.0 projects, in addition to an improvement in resource efficiency and cost reduction. This study also estimates that production times can be accelerated by about 120% and the time to put the products on the market can be reduced by about 70% [4]. The results obtained for the evolution of gross added value, turnover and exports allow us to conclude that expectations about efficiency gains, cost reduction and increased revenues can be achieved through a strategy of technological updating.

Another conclusion from this PWC study is that more than half (60%) of the national companies surveyed expect that investments in Industry 4.0 can be paid back within 2 years or less through an investment of approximately 5% of annual revenues. A similar value (56%) was presented in this PWC study for the overall companies on the other countries surveyed [4]. However, in our study, the hypothesis test results for the asset return variable did not allow to verify the existence of statistically significant differences, so, as already mentioned, it will be expected that a period of more than two years will be necessary in order to truly return on the investments made in this area. Therefore, the optimistic expectations of most respondents in the PWC study may not be true.

6 Conclusions

The implementation of the Industry 4.0 paradigm in the industrial sectors is expected to improve companies' performance and competitivity gains. The financial support provided by governments is crucial for the ability of making such investments, and, consequently, improve local economies. This study aimed to analyse whether government financial support invested in Portuguese companies of the industrial sector to implement projects within the Industry 4.0 paradigm are allowing those companies to improve economic and financial performance. The results show that turnover, gross value added, and exports of companies supported by the government in projects within the Industry 4.0 paradigm present statistically significant differences for the results before and after the implementation of the projects. Given the high growth rates for turnover and exports, it can be argued that the expectations of increased revenue and competitiveness associated with the implementation of Industry 4.0 were met. Similarly, the increase in gross value-added shows that the adoption of new technologies and new organizational models makes possible to optimize the use of resources, and, therefore, increase operational efficiency and cost savings.

Taking into account the results of this study, to achieve the full potential of Industry 4.0, it will be necessary that Governments promote measures that allow the industrial sector to digital update itself. The current industrial revolution requires that the various elements of the value chain (suppliers, producers, distributors, retailers) be at similar digital maturity levels, in order to guarantee a fluid and efficient exchange of information, capable of promoting the optimization of processes and cost reduction.

The short period of analysis may have been one of the major study's limitations. This study only considered projects targeted at Industry 4.0 that were implemented with support from the Portugal 2020 program. In order to have data to compare the performance and competitiveness of companies before and after the project, it was decided to limit the analysis to companies that received funding between 2014 and 2016. Since the nature of this type of projects, in general, implies a significant transformation of production processes and management practices, which are assumed to take a long time to operationalize and realize the expected performance gains, the time window analysed may be too short to evaluate the impact in some performance and competitive dimensions, such as apparent labor productivity, return on assets, liquidity, financial autonomy, and solvency, which did not yield statistically significant results in this study. Another limitation of the present study is the difficulty in isolating the origin of the impacts analysed. In other words, it is assumed that the financial support and the nature of the implemented projects were responsible for improving the competitiveness indicators that proved to be statistically significant (turnover, gross value added and exports). However, one cannot completely exclude the possibility that other factors related to the companies' context could also have impact on these results.

This study could be complemented, in the future, with interviews with companies' managers in order to understand and contextualize the quantitative results. These interviews can be used to understand how the projects were implemented, what synergies were created within the company and with the other elements of the value chain, what major changes occurred at the level of structures and organization, what were the major obstacles and challenges they faced during the implementation.

Acknowledgments. This work is financed by the ERDF – European Regional Development Fund through the Operational Programme for Competitiveness and Internationalisation - COMPETE 2020 Programme and by National Funds through the Portuguese funding agency, FCT - Fundação para a Ciência e a Tecnologia within project POCI-01-0145-FEDER- 016418 and this project was also financed by the Portuguese funding agency, FCT - Fundação para a Ciência e a Tecnologia, through national funds, and co-funded by the FEDER, where applicable.

References

1. Kagermann, H., Wahlster, W., Helbig, J.: Recommendations for implementing the strategic initiative INDUSTRIE 4.0. In: Final report of the Industry 4.0. Working Group. Acatech, Frankfurt am Main, Germany (2013)
2. Pereira, A.C., Romero, F.: A review of the meanings and the implications of the Industry 4.0 concept. Procedia Manuf. **13**, 1206–1214 (2017)
3. Rüßmann, M., et al.: Industry 4.0: the future of productivity and growth in manufacturing industries. Boston Consult. Group **9**(1), 54–89 (2015)
4. Correia, A.B., Deus, P., Baptista, J.R.: Indústria 4.0: Construir a empresa digital. PricewaterhouseCoopers Portugal, vol. 1, pp. 1–30 (2016)
5. Balasingham, K.: Industry 4.0: securing the future for German manufacturing companies (Master's thesis, University of Twente) (2016)
6. Chen, Y.: Integrated and intelligent manufacturing: perspectives and enablers. Engineering **3**(5), 588–595 (2017)
7. Simões, A.C., Lucas Soares, A., Barros, A.C.: Drivers impacting cobots adoption in manufacturing context: a qualitative study. In: Trojanowska, J., Ciszak, O., Machado, J.M., Pavlenko, I. (eds.) MANUFACTURING 2019. LNME, pp. 203–212. Springer, Cham (2019). https://doi.org/10.1007/978-3-030-18715-6_17
8. Nagy, J., Oláh, J., Erdei, E., Máté, D., Popp, J.: The role and impact of Industry 4.0 and the internet of things on the business strategy of the value chain the case of Hungary. Sustainability. **10**(10), 3491 (2018)
9. Erol, S., Jäger, A., Hold, P., Ott, K., Sihn, W.: Tangible Industry 4.0: a scenario-based approach to learning for the future of production. Procedia CiRp **54**, 13–18 (2016)
10. Comissão Europeia.: Digitising European Industry - Reaping the full benefits of a Digital Single Market (2016). https://ec.europa.eu/digital-single-market/en/policies/digitising-european-industry
11. Maresova, P., et al.: Consequences of Industry 4.0 in business and economics. Economies. **6**(3), 46 (2018)
12. Thames, L., Schaefer, D. (eds.): Cybersecurity for Industry 4.0. SSAM, Springer, Cham (2017). https://doi.org/10.1007/978-3-319-50660-9
13. IEDI.: Estratégias nacionais para a Indústria 4.0: Brasil. São Paulo, IEDI (2018)
14. Bidet-Mayer, T., Ciet, N.: L'industrie du futur: une compétition mondiale. La Fabrique de l'Industrie (2016)
15. COTEC.: Programa Portugal i4.0 (2018). https://www.industria4-0.cotec.pt/programa/medidas/
16. República Portuguesa.: Portugal i4.0 (2018). http://www.dgae.gov.pt/comunicar-as-empresas/financiamento/industria-40.aspx
17. IAPMEI.: Indústria 4.0 (2018). https://www.iapmei.pt/Paginas/Industria-4-0.aspx
18. KPMG Portugal.: Portugal i4.0 (Indústria 4.0) – Fase II (2019)
19. IAPMEI.: Indústria 4.0 - Governo lança Fase II do programa (2019). https://www.iapmei.pt/NOTICIAS/Industria-4-0-Governo-lanca-2-fase-do-programa.aspx
20. República Portuguesa.: Apresentação na nova fase do i4.0 (2019). https://www.portugal.gov.pt/pt/gc21/comunicacao/comunicado?i=primeiro-ministro-e-ministro-adjunto-e-da-economia-na-apresentacao-na-nova-fase-do-i40

21. Hall, A., Neves, C., Pereira, A.: Grande Maratona de Estatística no SPSS: Escolar Editora (2011)
22. Neves, J. C. D.: Avaliação e gestão da performance estratégica da empresa. Texto Editora, Lisboa (2011)
23. Martins, A.: Introdução à Análise Financeira de Empresas. 2ªEdição. Porto: Grupo Editorial Vida Económica (2004)
24. Marco, R.: Gender and economic performance: evidence from the Spanish hotel industry. Int. J. Hosp. Manag. **31**(3), 981–989 (2012)
25. Moreira, J.: Análise Financeira de Empresas - da Teoria à Prática. 4ªEdição. Porto: Associação da Bolsa de Derivados do Porto (2001)
26. Correia, C., Costa, V.: Avaliação do Impacto dos Fundos do QREN no Desempenho Económico e Financeiro das PME: O Caso da Indústria Transformadora do Concelho de Barcelos. Portuguese J. Financ. Manag. Accounting **2**(4) (2016)
27. PORDATA.: Volume de negócios das empresas: total e por sector de actividade económica (2019a). https://www.pordata.pt/Portugal/Volume+de+negócios+das+empresas+total+e+por+sector+de+actividade+económica-2913
28. PORDATA: Valor acrescentado bruto das empresas: total e por sector de actividade económica (2019b). https://www.pordata.pt/Portugal/Valor+acrescentado+bruto+das+empresas+total+e+por+sector+de+actividade+econ%C3%B3mica-2915
29. INE.: Estatísticas da Produção Industrial 2017, Instituto Nacional de Estatística (2018)
30. COTEC: Destino: Crescimento e inovação, O impacto da inovação na performance económico-financeira das PME e no seu crescimento (2017). http://www.cotecportugal.pt/imagem/Relatorios/20170523_Relato%CC%81rio%20agregadoEstudoDestino.pdf

Exponential Inequalities over the Parameters of a Strongly Regular Graph

Luís Almeida Vieira[1,2(✉)] and Vasco Moço Mano[3]

[1] Faculty of Engineering of Porto, Section of Mathematics, Porto, Portugal
[2] Center of Mathematics of University of Porto, University of Porto, Porto, Portugal
lvieira@fe.up.pt
[3] University of Porto, Porto, Portugal

Abstract. In this paper we work, in the context of Euclidean Jordan algebras, on the problem of finding admissibility inequalities for primitive strongly regular graphs. We connect each strongly regular graph to a particular Euclidean Jordan algebra $\mathcal{A}$ spanned by the natural powers of its adjacency matrix and by the identity matrix. By creating an exponential series with the elements of a Jordan frame of $\mathcal{A}$, we are able to deduce some necessary conditions for the existence of a strongly regular graph.

Keywords: Euclidean Jordan algebras · Graph theory · Strongly regular graphs

1 Introduction

In this paper we establish a couple of necessary conditions for the existence of strongly regular graphs. To study the existence of these graphs constitutes a difficult and challenging problem in the field of Combinatorics.

In the background of our research, we consider a particular Euclidean Jordan algebra related to a strongly regular graph. We are then able to establish some asymptotic inequalities involving the elements of the spectrum and the parameters of the strongly regular graph. Failing the compliance of these inequalities renders the existence of such a graph impossible.

The presentation of our results is divided in the following manner. In Sect. 2 we present the most important theoretical concepts and results for the development of our work. Strongly regular graphs are addressed in Subsect. 2.1 and Euclidean Jordan algebras are surveyed in Subsect. 2.2. In Sect. 3, we associate a an Euclidean Jordan algebra $\mathcal{A}$ to a strongly regular graph via its adjacency matrix and we are able to deduce some exponential inequalities involving the elements of the spectrum and the parameter set of a strongly regular graph.

2 Theoretical Background

Along this section we will describe the basic concepts about strongly regular graphs and Euclidean Jordan algebras needed for our paper. Our aim is to

J. Machado et al. (Eds.): icieng 2021, LNME, pp. 121–133, 2022.
https://doi.org/10.1007/978-3-030-78170-5_12

establish admissibility conditions over the spectrum and over the parameters of strongly regular graphs.

2.1 Strongly Regular Graphs

Herein, we assume that the reader is familiarized with the basics of Graph Theory. See [1] for a detailed survey.

A graph is called *simple* if it doesn't contain *multiple edges*, that is, different edges connecting the same pair of vertices, or *loops*, which are edges that connect a vertex to itself. A graph that permits multiple edges is called a *multigraph*. More precisely, a simple graph X is a pair $(V(X), E(X))$ where $V(X)$ is a non empty set of elements called *vertices* and $E(X)$ is a set formed by ordered pairs of distinct elements of $V(X)$ called *edges*.

A graph X such that every pair of distinct vertices is adjacent is called a *complete graph*. The *degree* of a vertice x of a graph X is the number of edges of $E(X)$ incident to x. A graph X is called $p-$*regular* if all the vertices of X have the same degree p.

Given two graphs X_1 and X_2 such that $V(X_1) \cap V(X_2) = \emptyset$, the *union* $X_1 \cup X_2$ is a graph such that the $V(X_1 \cup X_2) = V(X_1) \cup V(X_2)$ and such that $E(X_1 \cup X_2) = E(X_1) \cup E(X_2)$. A graph that can't be expressed as the union of two graphs is called a *connected* graph, otherwise, it is called a *disconnected* graph.

A *strongly regular graph* (srg), X, is a simple regular graph such that the number of common neighbors to any pair of vertices in X is fixed and depends only on the nature of adjacency of that pair of vertices. This special class of regular graphs was introduced for the first time introduced by R. C. Bose in the paper [2].

Moreover, a $(n; p, a, c)$-srg is a p-regular graph with n vertices in which any pair of adjacent vertices has a common neighbors and any pair of non-adjacent vertices has c common neighbors. The set $(n; p, a, c)$ is normally referred to as the *parameter set* of G.

The adjacency matrix of a $(n; p, a, c)$-srg G is a 01-matrix of order n, where each entry is 1 if the corresponding vertices are neighbors and 0 otherwise. It is known (see [1]) that this adjacency matrix A satisfies the equation

$$A^2 = pI_n + aA + c(J_n - A - I_n),$$

where I_n denotes the identity matrix and J_n is the all ones matrix, both of order n. Also, the eigenvalues of A, frequently referred to as the eigenvalues of G, are p, r and s, where r and s are such that:

$$r = \frac{a - c + \sqrt{(a-c)^2 + 4(p-c)}}{2},$$

$$s = \frac{a - c - \sqrt{(a-c)^2 + 4(p-c)}}{2}.$$

It is known as well that, if X is a $(n; p, a, c)$-srg, then any eigenvalue of X is in modulus less or equal than p.

We finish this very short survey on strongly regular graphs with the interesting relationship of these graphs with their complement graphs. Indeed, if X is a $(n; p, a, c)$-srg, then its complement, $\overline{G}$, is a $(\overline{n}; \overline{p}, \overline{a}, \overline{c})$-srg where $\overline{n} = n, \overline{p} = n - p - 1, \overline{a} = n - 2 - 2p + c$ and $\overline{c} = n - 2p + a$.

A srg X is *primitive* if and only if X and $\overline{X}$ are connected. A $(n; p, a, c)$-srg is not primitive or *imprimitive* if and only if $c = p$ or $c = 0$. To exclude trivial cases, we consider X to be a primitive srg and also not empty and not complete, which implies that its parameter set must satisfy the condition $0 < c < p < n - 1$.

Given a parameter set, it is not trivial to decide if there is a corresponding srg. In fact, there are many parameter sets called unfeasible because there isn't a corresponding srg. To decide if a given parameter set is feasible or *unfeasible* is a challenging problem in the field of Combinatorics. Among the relevant admissibility conditions on these graphs parameter sets we can refer the Krein inequalities, presented in [3] by Scott.

2.2 Euclidean Jordan Algebras

To deduce our feasibility conditions over the spectrum and over the parameters of a srg G we will make use of a particular Euclidean Jordan algebra associated to G.

A *Jordan algebra*, $\mathcal{A}$, is an algebra over a field $\mathbb{K}$ with finite dimension endowed with a product of vectors $\lozenge : (a, b) \mapsto a\lozenge b$ that is bilinear mapping, from $\mathcal{A} \times \mathcal{A}$ to $\mathcal{A}$, such that we have $a\lozenge b = b\lozenge a$ and $a\lozenge(a^{2\lozenge}\lozenge b) = a^{2\lozenge}\lozenge(a\lozenge b)$, for any vectors a and b of $\mathcal{A}$ where $a^{2\lozenge} = a\lozenge a$.

A non zero vector of $\mathcal{A}$, $\mathbf{e}$, with the property $\mathbf{e}\lozenge a = a\lozenge \mathbf{e} = a$ for all a in $\mathcal{A}$, is the *unit* vector of the algebra $\mathcal{A}$.

From now on, we will suppose that the field $\mathbb{K}$ is the field of the real numbers as we will only consider real Jordan algebras. Furthermore, in the remaining of this text, we will only consider finite dimensional Jordan algebras containing a unit element $\mathbf{e}$.

Example 1. Let $\mathcal{A} = \mathrm{Sym}(n, \mathbb{R})$ be the algebra over the field $\mathbb{R}$ of real symmetric matrices of order n, with the operation $a\lozenge b = \frac{ab+ba}{2}$, where a and b are vectors of $\mathcal{A}$. We will show that $\mathcal{A}$ is a Jordan algebra.

Let a be a vector of $\mathcal{A}$. We firstly, define the powers of a as: $a^{0\lozenge} = \mathbf{e}, a^{1\lozenge} = a$, $a^{2\lozenge} = a\lozenge a$ and $a^{m\lozenge} = a\lozenge a^{(m-1)\lozenge}$, for m in $\mathbb{N}$. But since $a^{2\lozenge} = a\lozenge a = \frac{aa+aa}{2} = \frac{a^2+a^2}{2} = a^2$ then, using induction over m, we obtain $a^{m\lozenge} = a^m$ for any m in $\mathbb{N}$.

Now, we will prove that $\mathcal{A}$ is a Jordan algebra. Indeed, we have, for any a and b in $\mathcal{A}$, $a\lozenge b = \frac{ab+ba}{2} = \frac{ba+ab}{2} = b\lozenge a$ and

$$
\begin{aligned}
a\Diamond(a^{2\Diamond}\Diamond b) &= a\Diamond(a^2\Diamond v)\\
&= a\Diamond\left(\frac{a^2b+ba^2}{2}\right)\\
&= \frac{a\left(\frac{a^2b+ba^2}{2}\right)+\left(\frac{a^2b+ba^2}{2}\right)a}{2}\\
&= \frac{\frac{a^3b+aba^2+a^2ba+ba^3}{2}}{2}\\
&= \frac{\frac{a^3b+a^2ba+ba^3+aba^2}{2}}{2}\\
&= \frac{\frac{a^2(ab+ba)+(ba+ab)a^2}{2}}{2}\\
&= \frac{\frac{a^2(ab+ba)+(ab+ba)a^2}{2}}{2}\\
&= \frac{a^2(a\Diamond b)+(a\Diamond b)a^2}{2}\\
&= \frac{a^{2\Diamond}(a\Diamond b)+(a\Diamond b)a^{2\Diamond}}{2}\\
&= a^{2\Diamond}\Diamond(a\Diamond b).
\end{aligned}
$$

Example 2. Let $\mathcal{A}=\mathbb{R}^m$. Then $\mathcal{A}$ equipped with the componentwise multiplication of elements of $\mathcal{A}$ defined by

$$c\Diamond d=(c_1,c_2,\cdots,c_m)\Diamond(d_1,d_2,\cdots,d_m)=(c_1d_1,c_2d_2,\cdots,c_md_m),$$

where $c=(c_1,c_2,\cdots,c_m)$ and $d=(d_1,d_2,\cdots,d_m)$ are vectors of $\mathcal{A}$, is a Jordan algebra.

Indeed, for $c=(c_1,c_2,\cdots,c_m)$ and for $d=(d_1,d_2,\cdots,d_m)$ in $\mathcal{A}$ we have

$$
\begin{aligned}
c\Diamond d &= (c_1,c_2,\cdots,c_m)\Diamond(d_1,d_2,\cdots,d_m)\\
&= (c_1d_1,c_2d_2,\cdots,c_md_m)\\
&= (d_1c_1,d_2c_2,\cdots,d_mc_m)\\
&= (d_1,d_2,\cdots,d_m)\Diamond(c_1,c_2,\cdots,c_m)\\
&= d\Diamond c
\end{aligned}
$$

and, also,

$$
\begin{aligned}
c\Diamond(c^{2\Diamond}\Diamond d) &= (c_1,c_2,\cdots,c_m)\Diamond((c_1^2,c_2^2,\cdots,c_m^2)\Diamond(d_1,d_2,\cdots,d_m))\\
&= (c_1,c_2,\cdots,c_m)\Diamond(c_1^2d_1,c_2^2d_2,\cdots,c_m^2d_m)\\
&= (c_1^3d_1,c_2^3d_2,\cdots,c_m^3d_m)\\
&= (c_1^2c_1d_1,c_2^2c_2d_2,\cdots,c_m^2c_md_m)\\
&= (c_1^2,c_2^2,\cdots,c_m^2)\Diamond(c_1d_1,c_2d_2,\cdots,c_md_m)\\
&= c^{2\Diamond}\Diamond(c\Diamond d).
\end{aligned}
$$

Note that the unit vector of the Jordan algebra $\mathcal{A}$ is the vector $\mathbf{e} = (1, 1, \cdots, 1)$.

Example 3. In this example we consider the subspace $\mathcal{B}$ of the Jordan algebra $\mathcal{A} = \mathrm{Sym}(n, \mathbb{R})$ of Example 2 generated by $B_0, B_1, \cdots, B_{l-1}$ and B_l of $\mathcal{A}$ whose entries are zero or one such that:

i) $B_0 + B_1 + B_2 + \cdots + B_l = J_n$ where J_n is the all ones matrix of $\mathcal{A}$;
ii) $B_0 = I_n$;
iii) $B_i B_j = \sum_{k=0}^{l} \alpha_{ij}^k B_k$.

Here we must observe that, for c and d in $\mathcal{B}$, we have that $cd = dc$ where cd and dc are the usual multiplication of the matrix c by d and the multiplication of the d by c respectively. We, firstly must note that

$$\begin{aligned} B_j B_i &= (B_i B_j)^T \\ &= \left(\sum_{k=0}^{l} \alpha_{ij}^k B_k \right)^T \\ &= \sum_{k=0}^{l} \alpha_{ij}^k B_k^T \\ &= \sum_{k=0}^{l} \alpha_{ij}^k B_k \\ &= B_i B_j \end{aligned}$$

and, secondly, that, for c and d two elements of $\mathcal{B}$, we have $c = \sum_{i=0}^{l} \beta_i B_i$ and $d = \sum_{j=0}^{l} \gamma_j B_j$. Therefore, we have:

$$\begin{aligned} cd &= \left(\sum_{i=0}^{l} \beta_i B_i \right) \left(\sum_{j=0}^{l} \gamma_j B_j \right) \\ &= \sum_{i=0}^{l} \sum_{j=0}^{l} \beta_i \gamma_j B_i B_j \\ &= \sum_{i=0}^{l} \sum_{j=0}^{l} \beta_i \gamma_j B_j B_i \\ &= \sum_{i=0}^{l} \sum_{j=0}^{l} \gamma_j \beta_i B_j B_i \\ &= \sum_{j=0}^{l} \sum_{i=0}^{l} \gamma_j \beta_i B_j B_i \\ &= dc. \end{aligned}$$

Therefore, $c\Diamond d = cd$. Note that

$$\begin{aligned} c\Diamond d &= \frac{cd+dc}{2} \\ &= \frac{cd+cd}{2} \\ &= \frac{2cd}{2} \\ &= cd. \end{aligned}$$

Now, we have that $\mathcal{B}$ is closed under the operation $\Diamond$ since is closed for the operation cd. Indeed, we have

$$\begin{aligned} cd &= \left(\sum_{i=0}^{l} \beta_i B_i\right)\left(\sum_{j=0}^{l} \gamma_j B_j\right) \\ &= \sum_{i=0}^{l}\sum_{j=0}^{l} \beta_i\gamma_j B_i B_j \\ &= \sum_{i=0}^{l}\sum_{j=0}^{l} \beta_i\gamma_j \sum_{m=0}^{l} \alpha_{ij}^m B_m \\ &= \sum_{m=0}^{l}\left(\sum_{i=0}^{l}\sum_{j=0}^{l} \beta_i\gamma_j\alpha_{ij}^m\right) B_m. \end{aligned}$$

Therefore, since cd is in $\mathcal{B}$, then $c\Diamond d$ is in $\mathcal{B}$. So, we conclude that $\mathcal{B}$ is a subalgebra of $\mathcal{A}$ and, therefore, it is a Jordan subalgebra of $\mathcal{A}$. And the unit of $\mathcal{B}$ is the identity I_n.

Example 4. Let m be a natural number and consider the real vector space $\mathcal{A}_{m+1} = \mathbb{R}^{m+1}$ equipped with the product $\Diamond$ defined, for a and $b \in \mathcal{A}_{m+1}$, by the equality

$$a\Diamond b = \begin{bmatrix} a|b \\ a_1\overline{b} + b_1\overline{a} \end{bmatrix},$$

where $a = \begin{bmatrix} a_1 \\ a_2 \\ \vdots \\ a_m \\ a_{m+1} \end{bmatrix}$, $b = \begin{bmatrix} b_1 \\ b_2 \\ \vdots \\ b_m \\ b_{m+1} \end{bmatrix}$, $\overline{a} = \begin{bmatrix} a_2 \\ a_3 \\ \vdots \\ a_m \\ a_{m+1} \end{bmatrix}$ and $\overline{b} = \begin{bmatrix} b_2 \\ b_3 \\ \vdots \\ b_m \\ b_{m+1} \end{bmatrix}$.

Now, we will prove that $a\Diamond b = b\Diamond a$ and that $a\Diamond(a^{2\Diamond}\Diamond b) = a^{2\Diamond}\Diamond(a\Diamond b)$. Indeed, we have the following calculations:

$$a\Diamond b = \begin{bmatrix} a|b \\ a_1\overline{b} + b_1\overline{a} \end{bmatrix} = \begin{bmatrix} b|a \\ b_1\overline{a} + a_1\overline{b} \end{bmatrix} = b\Diamond a,$$

$$\begin{aligned}
a\Diamond(a^{2\Diamond}\Diamond b) &= a\Diamond((a\Diamond a)\Diamond b) \\
&= a\Diamond\left(\begin{bmatrix} ||a||^2 \\ 2a_1\overline{a} \end{bmatrix} \Diamond \begin{bmatrix} b_1 \\ \overline{b} \end{bmatrix}\right) \\
&= \begin{bmatrix} a_1 \\ \overline{a} \end{bmatrix} \Diamond \begin{bmatrix} ||a||^2 b_1 + 2a_1\overline{a}|\overline{b} \\ ||a||^2\overline{b} + 2b_1 a_1\overline{a} \end{bmatrix} \\
&= \begin{bmatrix} a_1 b_1(||a||^2 + 2||\overline{a}||^2) + (||a||^2 + 2a_1^2)\overline{a}|\overline{b} \\ a_1||a||^2\overline{b} + (2a_1^2 b_1 + ||a||^2 b_1 + 2a_1\overline{a}|\overline{b})\overline{a} \end{bmatrix}
\end{aligned}$$

and

$$\begin{aligned}
a^{2\Diamond}\Diamond(a\Diamond b) &= \left(\begin{bmatrix} a_1 \\ \overline{a} \end{bmatrix} \Diamond \begin{bmatrix} a_1 \\ \overline{a} \end{bmatrix}\right) \Diamond \left(\begin{bmatrix} a_1 \\ \overline{a} \end{bmatrix} \Diamond \begin{bmatrix} b_1 \\ \overline{b} \end{bmatrix}\right) \\
&= \begin{bmatrix} ||a||^2 \\ 2a_1\overline{a} \end{bmatrix} \Diamond \begin{bmatrix} a_1 b_1 + \overline{a}|\overline{b} \\ a_1\overline{b} + b_1\overline{a} \end{bmatrix} \\
&= \begin{bmatrix} a_1 b_1(||a||^2 + 2||\overline{a}||^2) + (||a||^2 + 2a_1^2)\overline{a}|\overline{b} \\ a_1||a||^2\overline{b} + (2a_1^2 b_1 + ||a||^2 b_1 + 2a_1\overline{a}|\overline{b})\overline{a} \end{bmatrix}.
\end{aligned}$$

Hence, we deduced that $a\Diamond(a^{2\Diamond}\Diamond b) = a^{2\Diamond}\Diamond(a\Diamond b)$. We have shown that $\mathcal{A}$ is a Jordan algebra. The unit element of $\mathcal{A}$ is the element $\mathbf{e} = \begin{bmatrix} 1 \\ \overline{0} \end{bmatrix}$. Indeed, we have

$$\begin{aligned}
\mathbf{e}\Diamond a &= \begin{bmatrix} 1 \\ \overline{0} \end{bmatrix} \Diamond \begin{bmatrix} a_1 \\ \overline{a} \end{bmatrix} \\
&= \begin{bmatrix} a_1 \\ 1\overline{a} + a_1\overline{0} \end{bmatrix} \\
&= \begin{bmatrix} a_1 \\ \overline{a} \end{bmatrix} \\
&= a.
\end{aligned}$$

Since $\mathcal{A}_{m+1}$ is a commutative algebra then the unit of $\mathcal{A}_{m+1}$ is the element $\mathbf{e} = \begin{bmatrix} 1 \\ \overline{0} \end{bmatrix}$.

An *Euclidean Jordan algebra* (EJA), $\mathcal{A}$, is a Jordan algebra endowed with an inner product $\cdot|\cdot$ with the property $(a\Diamond b)|c = b|(a\Diamond c)$, for all a, b and c in $\mathcal{A}$.

Example 5. The Jordan algebra $\mathcal{A} = \mathrm{Sym}(n, \mathbb{R})$ endowed with the inner product $\cdot|\cdot$ defined by $a|b = \mathrm{trace}(a\Diamond b)$, for all a and b in $\mathcal{A}$, is an EJA (see [4]).

An EJA is called *simple* if and only if it has only trivial ideals and any non trivial EJA is a direct sum of simple Euclidean Jordan algebras.

Example 6. The algebra $\mathcal{A} = \mathbb{R}^m$ endowed with the component wise multiplication of vectors and with the usual inner product is an EJA. Indeed, for $e = (e_1, e_2, \cdots, e_m)$, $f = (f_1, f_2, \cdots, f_m)$ and $g = (g_1, g_2, \cdots, g_m)$ in $\mathcal{A}$ we have that:

$$
\begin{aligned}
(e\diamondsuit f)|g &= ((e_1, e_2, \cdots, e_m)\diamondsuit(f_1, f_2, \cdots, f_m))\,|(g_1, g_2, \cdots, g_m) \\
&= (e_1 f_1, e_2 f_2, \cdots, e_m f_m)|(g_1, g_2, \cdots, g_m) \\
&= \sum_{i=1}^{m} e_i f_i g_i
\end{aligned}
$$

and that

$$
\begin{aligned}
f|(e\diamondsuit g) &= (f_1, f_2, \cdots, f_m)|\,((e_1, e_2, \cdots, e_m)\diamondsuit(g_1, g_2, \cdots, g_m)) \\
&= (f_1, f_2, \cdots, f_m)|e_1 g_1, e_2 g_2, \cdots, e_m g_m) \\
&= \sum_{i=1}^{m} f_i e_i g_i \\
&= \sum_{i=1}^{m} e_i f_i g_i.
\end{aligned}
$$

Therefore, since we have $(e\diamondsuit f)|g = f|(e\diamondsuit g)$ for any e, f and g in $\mathcal{A}$, then the Jordan algebra $\mathcal{A}$ is an EJA.

Euclidean Jordan algebras have many applications on several fields of Mathematics, namely on establishing the formalism of quantum mechanics [5], on combinatorics [6,7], on constructing the theoretical foundations for the interior-point methods [8–10], as well as in applications to statistics, [11,12], and to robotics, [13]. We would also like to mention the generalization of the Cauchy interlacing theorem over hermitian matrices to simple Euclidean Jordan algebras and the generalization of the Fischer, Hadamard, Bergstrom and Oppenheim inequalities and others over determinants in matrix theory to Euclidean Jordan algebras, see [14,15]. Regarding the theory of interior point methods, we would also like to refer the interesting and clear works of Farid Alizadeh, presented in the chapter *An introduction to formally real Jordan algebras and their applications in optimization* of the book [16].

For thorough surveys on Euclidean Jordan algebras, the reading of the monographs [1,17–19] is suggested.

The vector a in $\mathcal{A}$ is an *idempotent* of $\mathcal{A}$ if $a^{2\diamondsuit} = a$. Two vectors a and b of $\mathcal{A}$ are called *orthogonal* if $a\diamondsuit b = 0$. A set $\{e_1, e_2, \ldots, e_k\}$ is called a *complete system of non-zero orthogonal idempotents* (CSOI) if each element of the set is an idempotent, each two different elements of the set are orthogonal and $e_1 + e_2 + \cdots + e_k = \mathbf{e}$.

Example 7. Consider the EJA $\mathcal{A} = \mathrm{Sym}(n, \mathbb{R})$ and let E_{ii}, for $i = 1, \cdots, n$, be the matrix such that $(E_{ii})_{kj} = 0$ if $k \neq i$ or $j \neq i$ and $(E_{ii})_{ii} = 1$. Let l be in $\mathbb{N}$ such that $1 < l \leq n - 1$. The set

$$
\left\{ \sum_{i=1}^{l} E_{ii}, \sum_{i=l+1}^{n} E_{ii} \right\}
$$

is a $CSOI$ of $\mathcal{A}$.

Example 8. Consider the EJA $\mathcal{A}_{m+1}$. Then, the set

$$\left\{ \begin{bmatrix} \frac{1}{2} \\ -\frac{\overline{a}}{2\|\overline{a}\|} \end{bmatrix}, \begin{bmatrix} \frac{1}{2} \\ +\frac{\overline{a}}{2\|\overline{a}\|} \end{bmatrix} \right\}$$

is *CSOI* of $\mathcal{A}_{m+1}$, where $\overline{a}$ is a non zero vector of $\mathbb{R}^m$.

Example 9. Let m and l be in $\mathbb{N}$ such that $1 < l \leq m-1$, and let's consider the canonical basis

$$<e_1, e_2, \cdots, e_m> = <(1, 0, \cdots, 0), (0, 1, \cdots, 0), \cdots, (0, 0, \cdots, 1)>$$

of $\mathbb{R}^m$. Then

$$\left\{ \sum_{i=1}^{l} e_i, \sum_{i=l+1}^{m} e_i \right\}$$

is a *CSOI* of the EJA $\mathcal{A} = \mathbb{R}^m$ equipped with the componentwise operation of two vectors $a = (a_1, a_2, \cdots, a_m)$ and $(b = (b_1, b_2, \cdots, b_m)$ of $\mathbb{R}^m$ defined by $a \Diamond b = (a_1 b_1, a_2 b_2, \cdots, a_m b_m)$.

The *rank* of an element v in $\mathcal{A}$ is the least natural number k, such that the set $\{e, , v^{1\Diamond}, v^{2\Diamond}, \ldots, v^{k\Diamond}\}$ is linearly dependent and we denote it by $\text{rank}(v) = k$. The rank of the algebra $\mathcal{A}$ is denoted by $\text{rank}(\mathcal{A})$ and is defined through the equality $\text{rank}(\mathcal{A}) = \max\{\text{rank}(v) : v \in \mathcal{A}\}$. The vectors of $\mathcal{A}$ whose rank is equal to the rank of $\mathcal{A}$ are called *regular*. Note that the set of the regular vectors of $\mathcal{A}$ is an open and dense set in $\mathcal{A}$.

Let's consider a regular vector v of $\mathcal{A}$ and let's consider the natural number $r = \text{rank}(v)$. Then the set $\{\mathbf{e}, v^{1\Diamond}, v^{2\Diamond}, \cdots, v^{r\Diamond}\}$ is a linearly dependent set of $\mathcal{A}$ and the set $\{\mathbf{e}, v^{1\Diamond}, v^{2\Diamond}, \ldots, v^{(r-1)\Diamond}\}$ is linearly independent set of $\mathcal{A}$. Hence, there are reals $a_1(v),_2(v), \ldots, a_{r-1}(v)$ and $a_r(v)$ such that $v^{r\Diamond} - a_1(v)v^{r-1\Diamond} + \cdots + (-1)^r a_r(v)\mathbf{e} = 0$, where 0 is the zero vector of $\mathcal{A}$. The polynomial in λ

$$p_v(\lambda) = \lambda^r - a_1(v)\lambda^{r-1} + \cdots + (-1)^r a_r(v)$$

can be, therefore, deduced which is designated as the *characteristic polynomial* of v. Observe that each a_i is a homogeneous polynomial of degree i in the coordinates of the vector v in a given basis of $\mathcal{A}$. Though p_v is deduced for a regular element of $\mathcal{A}$, the definition can be extended to any element of $\mathcal{A}$, considering that each a_i is homogeneous and the already referred property of the set of regular elements of $\mathcal{A}$ which is dense in $\mathcal{A}$. The *eigenvalues* of v are defined as the roots of the p_v, $\lambda_1, \lambda_2, \ldots, \lambda_{r-1}$ and λ_r. Finally, the numbers $a_1(v)$ and $a_r(v)$ of p_v are called the *trace* and the *determinant* of v, respectively.

For each element v in an EJA $\mathcal{A}$, there are unique real numbers $\alpha_1(v), \alpha_2(v)$, $\ldots, \alpha_{k-1}(v)$ and $\alpha_k(v)$, all distinct, and a unique CSOI $\{v_1, v_2, \ldots, v_k\}$, such that

$$v = \alpha_1(v)v_1 + \alpha_2(v)v_2 + \cdots + \alpha_k(v)v_k. \quad (1)$$

The decomposition of v presented in (1) is called the *first spectral decomposition* of v and the numbers $\alpha_j(v)$'s appearing in (1) correspond to the eigenvalues of v.

An idempotent u is primitive if u is not the zero vector of $\mathcal{A}$ and is an idempotent of $\mathcal{A}$ and if the writing of it as the sum of two non-zero idempotents is impossible. A CSOI such that any element of it is primitive is called a *Jordan frame* (JF).

Example 10. The set

$$\left\{\begin{bmatrix} \frac{1}{2} \\ -\frac{\overline{a}}{2\|\overline{a}\|} \end{bmatrix}, \begin{bmatrix} \frac{1}{2} \\ +\frac{\overline{a}}{2\|\overline{a}\|} \end{bmatrix}\right\}$$

is a JF of the EJA $\mathcal{A}_{m+1}$, where $\overline{a}$ is a non zero vector of $\mathbb{R}^n$.

Example 11. The set

$$\{e_1, e_2, \cdots, e_m\},$$

where $<e_1, e_2, \cdots, e_m>$ is the canonical basis of $\mathbb{R}^m$, is a JF of $\mathcal{A} = \mathbb{R}^m$ endowed with component wise multiplication of vectors and the usual inner product of $\mathbb{R}^m$

For each v in an EJA $\mathcal{A}$, there is a JF $\{v_1, v_2, \cdots, v_r\}$ and real numbers $\beta_1, \beta_2, \cdots, \beta_{r-1}$ and β_r such that

$$v = \beta_1(v)v_1 + \beta_2(v)e_2 + \cdots + \beta_r(v)v_r. \tag{2}$$

The decomposition (2) is called the second spectral decomposition of v.

We finish this subsection on Euclidean Jordan algebras with a pertinent observation. Let n be a natural number, $\mathcal{A}$ be a real EJA with dimension and rank equal to n and v be a regular element of $\mathcal{A}$. Since the set of regular vectors of $\mathcal{A}$ is dense in $\mathcal{A}$, then there is a unique CSOI, $\mathcal{S} = \{e_1, e_2, \cdots, e_n\}$, such that $v = \sum_{i=1}^n \lambda_i e_i$. But this means that $\mathcal{S}$ is a linearly independent set of $\mathcal{A}$ and, therefore, S is a basis of $\mathcal{A}$.

3 Exponential Inequalities over the Spectrum of a Strongly Regular Graph

From this point onwards, we are considering G to be a $(n; p, a, c)$-srg, such that $n - 1 > p > c > 0$, with adjacency matrix A with the three distinct eigenvalues, p, r and s. Observe that in our notation p and r are the non-negative eigenvalues of G and s is the negative eigenvalue of G.

In order to deduce our admissibility inequalities for G we will start by connecting a real EJA with dimension and rank equal to 3 to A.

We consider the EJA of the symmetric matrices of order n, denoted by $\text{Sym}(n, \mathbb{R})$, equipped with the map

$$a \Diamond b = \frac{ab + ba}{2}$$

endowed with the inner product $a|b = \text{trace}(a \Diamond b)$, representing ab and ba the usual products of matrices the matrices a by b and of b by a. The algebra

$\mathrm{Sym}(n, \mathbb{R})$ has a subalgebra, $\mathcal{A}$, that is generated by the identity matrix, I_n, and the natural powers of A. The subalgebra $\mathcal{A}$ is an EJA with dimension and rank equal to 3 due to the fact that A has three distinct eigenvalues. We denote by $\mathcal{B} = \{E_1, E_2, E_3\}$ the unique JS of $\mathcal{A}$ associated to A, whose elements are given by

$$E_1 = \frac{1}{n}I_n + \frac{1}{n}A + \frac{1}{n}(J_n - A - I_n),$$
$$E_2 = \frac{|s|n + s - p}{n(r-s)}I_n + \frac{n+s-p}{n(r-s)}A + \frac{s-p}{n(r-s)}(J_n - A - I_n),$$
$$E_3 = \frac{rn + p - r}{n(r-s)}I_n + \frac{-n+p-r}{n(r-s)}A + \frac{p-r}{n(r-s)}(J_n - A - I_n).$$

For the following part of our work some notation needs to be introduced. The set of real matrices of order n will be denoted by $M_n(\mathbb{R})$. For any matrices $G = [g_{ij}]$, $H = [g_{ij}]$ in $M_n(\mathbb{R})$, we denote by $G \circ H = [g_{ij}h_{ij}]$ the *entrywise product* of the matrices G and H and by $G \otimes H = [g_{ij}H]$ the *Kronecker product* of G and H. Additionally, for any non-negative integer number m and for any $D \in M_n(\mathbb{R})$ we define the powers $D^{m\circ}$ with the equalities: $D^{0\circ} = J_n$, $D^{1\circ} = D$ and for $m \geq 2$ we define $D^{m\circ} = D^{(m-1)\circ} \circ D$ (see [20]).

We are now in conditions to construct the following series of entrywise powers of the elements of $\mathcal{B}$. For $\epsilon > 0$, consider the exponential series, S_ϵ, defined as

$$S_\epsilon = \sum_{k=0}^{+\infty} \frac{(\epsilon E_3^{2\circ})^{k\circ}}{k!}.$$

After some algebraic manipulation we can rewrite S_ϵ as

$$S_\epsilon = \exp\left(\epsilon\left(\frac{(rn+p-r)^2}{(n(r-s))^2}\right)\right) I_n + \exp\left(\epsilon\left(\frac{(-n+p-r)^2}{(n(r-s))^2}\right)\right) A$$
$$+ \exp\left(\epsilon\left(\frac{(p-r)^2}{(n(r-s))^2}\right)\right)(J_n - A - I_n). \tag{3}$$

At this point, we consider the element $S_{3\epsilon} = E_3 \circ S_\epsilon$. We, then, have that

$$S_{3\epsilon} = \sum_{i=1}^{3} q_{3\epsilon}^i E_i$$

is the spectral decomposition of $S_{3\epsilon}$ with respect to the JF $\mathcal{B}$ of $\mathcal{A}$. Note that $q_{3\epsilon}^i \geq 0$ for $i \in \{1, 2, 3\}$. We establish the inequality (5) of Theorem 1 by an algebraic analysis of the eigenvalues $q_{3\epsilon}^1$, making use of the equality (4)

$$\frac{rn+p-r}{n(r-s)} + \frac{-n+p-r}{n(r-s)}p + \frac{p-r}{n(r-s)}(n-p-1) = 0, \tag{4}$$

noting that for $p < \frac{n}{3}$ we have that

$$\frac{n-p+r}{rn+p-r} > \frac{2}{3r+1},$$

and supposing that $q_{3\epsilon}^1 \geq 0$.

Theorem 1. *Let a, c, p and n be natural numbers such that $0 < c < p < n-1$ and let G be a $(n; p, a, c)$-srg and with the eigenvalues p, r and s. If $p < \frac{n}{3}$ then we have that*

$$(3r+1)(3r+2)r > 2p. \tag{5}$$

Now, making an analogous study of the element

$$S'_\epsilon = \sum_{k=0}^{+\infty} \frac{(\epsilon E_2^{2\circ})^{k\circ}}{k!}$$

as we did before and by the spectral analysis of the eigenvalue $q^1_{2\epsilon}$ of spectral decomposition

$$E_2 \circ S'_\epsilon = \sum_{i=1}^{3} q^i_{2\epsilon} E_i,$$

we deduce the next inequality.

Theorem 2. *Let a, c, p and n be natural numbers such that $0 < c < p < n-1$ and G be a $(n; p, a, c)$-srg with the eigenvalues p, r and s. If $p > \frac{2n}{3}$ then we have that*

$$|s|(3|s|-1)(|s|-1) > \frac{2}{3}(n-p-1). \tag{6}$$

The results obtained in Theorem 1 and Theorem 2 provide two admissibility conditions for the existence of an imprimitive srg. Using these conditions we are testing parameter sets for their existence but we can already conclude, from inequalities (5) and (6), that the positive eigenvalue r can't be too small compared with the regularity p of a srg of order n when $p < \frac{n}{3}$ and the negative eigenvalue s can't be too small compared to the regularity $n-p-1$ of the complement graph when $p > \frac{2n}{3}$.

Acknowledgments. Luís Vieira in this paper was partially financed by the Center of Research of Mathematics of University of Porto (UID/MAT/00144/2013), which is supported by the Foundation for Science and Technology of Portugal, and with the Ministry for Science, Technology and Higher Education and the European structural funds through the programs FEDER, under the partnership agreement PT2020.

References

1. Mano, V.M., Vieira, L.A.: Strongly regular graphs and Euclidean Jordan algebras. Revelations within an unusual relashionship. Lap Lambert Academic Publishing, Berlin (2016)
2. Bose, R.C.: Strongly regular graphs, partial geometries and partially balanced designs. Pacific J. Math. **13**, 389–419 (1963)
3. Scott, L.L.: A condition on Higman parameters. Notices. Amer. Math. Soc. **20**, A-97 (1973)

4. Vieira, L.: Euclidean Jordan algebras and some conditions over the spectra of a strongly regular graph. 4open **2**, 1–19 (2019)
5. Jordan, P., Neumann, J.V., Wigner, E.: On an algebraic generalization of the quantum mechanical formalism. Ann. Math. **35**, 29–64 (1934)
6. Mano, V.M., Martins, E.A., Vieira, L.A.: On generalized binomial series and strongly regular graphs. Proyecciones J. Math. **4**, 393–408 (2013)
7. Vieira, L.A.: Euclidean Jordan algebras, strongly regular graphs and Cauchy Schwarz inequalities. Appl. Math. Inf. Sci. **13**(2), 437–44 (2019)
8. Cardoso, D.M., Vieira, L.A.: On the optimal parameter of a self-concordant barrier over a symmetric cone. Eur. J. Oper. Res. **169**, 1148–1157 (2006)
9. Faybusovich, L.: Several Jordan-algebraic aspects of optimization. Optimization **57**(3), 379–393 (2008)
10. Faybusovich, L.: Infinite-dimensional quadratic optimization: interior-point methods and control applications. Appl. Math. Optim. **36**(1), 43–66 (1997)
11. Massam, H., Neher, E.: Estimation and testing for lattice condicional independence models on Euclidean Jordan algebras. Ann. Stat. **26**, 1051–1082 (1998)
12. Malley, J.D.: Statistical Applications of Jordan Algebras. Springer Verlag, New York (1994)
13. Helmke, U., Ricardo, S., Yoshizawa, S.: Newton's algorithm in Euclidean Jordan algebras with applications to robotics **2**(3), 283–298 (2002)
14. Gowda, M.S., Tao, J.: The Cauchy interlacing theorem in simple Euclidean Jordan algebras. Linear Multiliear Algebra **59**(1), 65–86 (2011)
15. Gowda, M.S., Tao, J.: Some inequalities involving determinants, eigenvalues and Schur complements in Euclidean Jordan algebras. Positivity **15**, 381–399 (2011)
16. Farid, A.: An introduction to formally real Euclidean Jordan algebras and their applications in optimization. In: Anjos, M.F., Lasserre, J.B. (eds.) The Handbook on Semidefinite, Conic and Polynomial Optimization,pp. 229–337. Springer, Boston (2012)
17. Faraut, J., Korányi, A.: Analysis on Symmetric Cones. Oxford Mathematical Monographs. Clarendon Press, Oxford (1994)
18. Jacobson, N.: Structure and Representation of Jordan Algebras, volume XXXIX of Colloquium Publications American Mathematical Society, Providence, Rhode Island (1968)
19. Baes, M.: Spectral functions and smoothing techniques on Jordan algebras. Ph.D. Thesis, University Catholique of Louvain (2006)
20. Horn, R., Johnson, A.: Matrix Analysis. Cambridge University Press, Cambridge, C.R (2013)

Stakeholder Management in University-Industry Collaboration Programs: A Case Study

Gabriela Fernandes[1](✉), Marlene Capitão[2], Anabela Tereso[2], José Oliveira[3], and Eduardo B. Pinto[4]

[1] Univ Coimbra, CEMMPRE, Department of Mechanical Engineering, Rua Luís Reis Santos, Coimbra, Portugal
gabriela.fernandes@dem.uc.pt
[2] ALGORITMI, Engineering School, University of Minho, Campus de Azurém, Guimarães, Portugal
[3] BOSCH Car Multimedia, S.A, Rua Max Grunding 35, Braga, Portugal
[4] CCG/ZGDV Institute, Campus de Azurém, Guimarães, Portugal

Abstract. Poor stakeholder management repeatedly comes up as one of the main reasons for unsuccessful projects. This issue takes a higher dimension when we look at the inter-organizational context, where traditional stakeholder management practices may be insufficient. This paper investigates stakeholder management applied to the context of collaborative research and development (R&D) programs between university and industry, a particular form of inter-organizational relationships. A large case study involving an R&D collaboration between a university and an industry in Portugal, collaborating with each other for more than seven years, with 170 million euros of investment, is analyzed based on participant observation and document analysis. A methodology for stakeholder management in collaborative programs between university and industry is proposed, which includes several systematic practices. For example, in the process of identify stakeholders, it is suggested to identify the expectations, required contributions and previous experience of stakeholders, and in the process of monitor stakeholder engagement comes, to define KPIs for measuring their engagement evolution. This is a methodology that considers the subjective aspects of human action and, therefore, applies tools and techniques that fully involve the stakeholder, such as surveys to monitor their self-engagement. The methodology can be used as a starting point for the systematic development of stakeholder management processes in any large collaborative project and program.

Keywords: Stakeholder management · University-industry collaboration · Program and Project stakeholders · Inter-organizational relationships

1 Introduction

In spite of the perceived advantages of university-industry collaborations (UICs), there are also inherent challenges, namely different cultures, different short, medium, and long-term objectives, and initiatives complexity [1]. Barnes, Pashby, and Gibbons [2]

J. Machado et al. (Eds.): icieng 2021, LNME, pp. 134–147, 2022.
https://doi.org/10.1007/978-3-030-78170-5_13

suggest that good project management minimizes most of the problems associated with the culture gap in UIC. Fernandes et al. [3] demonstrated the value of project management for different UIC stakeholders. Nevertheless, projects continue to fail as results do not meet stakeholder expectations [4]. Looking at projects, they all have stakeholders who are impacted or can impact the project positively or negatively, and some stakeholders may have a limited ability to influence the work or outcomes of the project, while others may have a significant influence on the project and its expected outcomes [5]. The unique nature of collaborative projects makes many traditional project management practices insufficient [6]. The boundaries of a collaborative project become even wider and the knowledge more diversified [7], which means that coordinating stakeholders in these conditions represents a critical issue that needs to be addressed to increase project success [8].

This paper focuses on stakeholder management in inter-organizational programs, with a view to the success of collaborative projects. More specifically this paper seeks to answer the following question: *How can organizations effectively manage stakeholders of large UIC programs*? To address this question, the authors analyzed the existing literature on stakeholder management practices and conducted a longitudinal case study to develop a stakeholder management methodology for this particular context of inter-organizational collaborations. The case study analyzes a collaboration between a university and a major industry in Portugal, the Bosch-UMinho collaboration. The Bosch-UMinho collaboration has lasted for more than seven years and represents an investment of more than 170 million euros. The proposed stakeholder management methodology is intended to give an important contribution to inter-organizational collaborative large programs and projects, by the systematization of stakeholder management processes, and to benefit future research in this field.

This paper begins with a review of the research literature on collaborative programs between university and industry and stakeholder management. Next, the case study is presented, starting by presenting the Bosch-UMinho collaboration and the stakeholder management practices under the collaboration. Then, the proposed methodology for stakeholder management in UIC programs is presented, and the paper closes with the research conclusions, where the limitations and next steps are also identified.

2 Background

This section focuses on a literature review of UIC programs and stakeholder management. In the context of UIC programs, the benefits, conflicts of interest, and possible solutions are analyzed. On stakeholder management, four processes are analyzed: identify stakeholders, plan stakeholder engagement, manage stakeholder engagement, and monitor stakeholder engagement. Next, the main differences in the management of stakeholders in traditional and collaborative R&D projects and the main practices applied to this context are discussed.

2.1 University-Industry Collaboration R&D Programs

University-industry knowledge transfer is a broad concept that identifies a wide set of interactions between companies and universities that aim at the exchange of knowledge related to research, science, and technology [9]. From a general perspective, the exchange of knowledge and skills between university and industry contributes positively to reduce market failures and realizing the full social benefits of R&D investments [10]. Fernandes and O'Sullivan [11] emphasize several benefits of R&D collaborations for companies, universities, and society. According to the authors, through these collaborations, the companies can acquire knowledge capable of generating additional profit, as well as improving the knowledge and skills of their scientific work. As for the benefits for universities, the additional income resulting from the collaboration can be used to improve their research conditions (e.g., by purchasing equipment or by hiring researchers), the collaborations can also be used to attract and retain researchers, and, finally, there may be in-kind benefits resulting from the action of universities in R&D collaborations, such as an increase in the desire of companies to employ students and additional sponsored research [11].

Universities and industries have distinct objectives and face several conflicts and constraints [12]. Looking at the culture and interests of industries, we understand that they need to move fast and get results that can be translated quickly into economic advantages, as well as take ownership of those results, while the university culture emphasizes the long term and open discussion of results, leveraging more and more research projects and sticking their results in the public domain [13]. Also, the culture of open science, which refers to the free exchange and dissemination of new ideas, may be degraded by UIC, just as researchers at the university may also have restrictions on the dissemination of results obtained, a situation that might perjure the service provided at the university to students and other researchers [10]. Other challenges of UIC programs are the high levels of uncertainty, significant pressure in terms of creativity and innovation, individually oriented workers, and project members residing in different workplaces [14, 15]. All these issues have been widely addressed in recent years, along with the need to overcome the existing obstacles and with the increasing pressures of the globalization process [9]. In short, the existing contrast between Universities and Industry can increase the value of collaboration but is also a source of obstacles [12]. In the absence of appropriate coordination, conflicts of interest may be damaging the innovation of several sectors [13].

Given the effects of these collaborations, it is crucial to ensure successful management of UIC to realize the advantages on both sides and also to society [12]. Analysis of UIC has shown that many of these R&D projects, in spite of being highly promoted, have failed [16]. Chin et al. [16] indicate the main reasons present in the literature for these failures: different levels of commitment between partners; non-existence of a trusting relationship; different motivations and objectives; unclear requirements; and poor planning and progress monitoring. The literature on project management for research collaborations includes numerous prescriptions, suggesting that successful collaborations are management-intensive [13], i.e., the failure factors of collaborative university-industry R&D projects can be mitigated through the implementation of project management practices [6, 16, 17]. However, the unique nature of collaborative projects and the common conflicts among stakeholders make many conventional project management practices ineffective [6]. A practice is defined as a specific type of activity that may employ one or more techniques and tools to carry out a project management process [18]. In this regard,

Barnes et al. [2] developed a best practice model for managing university-industry R&D collaborations that provided important additional insights into the dynamics of collaborative R&D projects, particularly that a strong collaboration is based on strong personal as well as organizational relationships. Also, Fernandes et al. [19] emphasize the alignment of program goals and outcomes with stakeholder expectations in their integrated program and project management approach.

2.2 Stakeholder Management

The definition of stakeholder most widely used and recognized by researchers is authored by Freeman [20] who defines the concept as any group or individual who can affect or be affected by the achievement of the company's objectives. The definition of stakeholder includes all individuals, groups, and entities related to the project, regardless of whether they are internal or external to the organization [21]. The PMBOK [5] suggests four processes for conducting stakeholder management: identify stakeholders, plan stakeholder engagement, manage stakeholder engagement, and monitor stakeholder engagement.

Identify stakeholders is the process of regularly identifying project stakeholders and analyzing and documenting relevant information about their interests, involvement, influence, interdependencies, and potential impact on project success [5]. Jepsen and Eskerod [22] note that stakeholders need to be characterized concerning their required contributions, the expectations and benefits they have regarding rewards for delivering their contributions, and their power about the project. Although the importance that should be given to all stakeholders is extolled, in practice it is not always feasible to meet all expectations and deliver all benefits, and there is a need to pay more attention to certain specific groups to the detriment of others. Thus, the project manager should consider the most important stakeholders, ensuring that there is the necessary collection of information about their expectations [22]. In this field, the PMBOK [5] suggests mapping the stakeholders to categorize them. Mitchell, Agle, and Wood [23] allow the identification of the various types of existing stakeholders based on unique or combined attributes like power, legitimacy, and urgency. In turn, Bourne [24] developed the stakeholder circle, a methodology that allows you to identify and prioritize key stakeholders, develop an engagement strategy and an appropriate communication plan, to ensure that the needs and expectations of these stakeholders are understood and managed. This is a methodology that maps the key stakeholders according to their ability to influence the success of the project [24]. Eskerod and Larsen [25] argue that to understand the motivations and behaviors of project stakeholders, it is necessary to conduct an analysis that includes the stakeholders' past experiences and, more importantly, their perceptions and interpretations of the past, as well as their expectations for the future. Pouloudi and Whitley [26] advocate a set of principles that characterize stakeholder behavior. They are stakeholders depend on the specific context and time; stakeholders cannot be seen in isolation; the position of each stakeholder may change over time; and, finally, feasible options may differ from stakeholders' wishes. The authors advocate the use of these principles to understand organizational and inter-organizational reality, explain past circumstances, and use the findings to realistically plan future activity [26]. In short Pouloudi and Whitley [26] argue that within a specific context, the process of stakeholder identification and analysis must be iterative, adopting a long-term perspective in exploring who are the stakeholders and what are their views. The risk of considering all these aspects to obtain richer insights about the stakeholders is that the project representatives who should interact with the stakeholders become paralyzed due to excess of information [25]. Therefore, those responsible for stakeholder management must seek a balance between having as much information about stakeholders and not be overloaded of information [25]. The main output of the stakeholder identification process is the stakeholder register, a document that includes stakeholder identification, assessment, and classification information [5]. This information forms a basis for decisions, objectives, and plans for the project [22].

Plan stakeholder engagement is the process of developing methods to involve project stakeholders based on their needs, expectations, interests, and potential impact on the project [5]. The PMBOK [5] suggests that the project manager assesses stakeholder engagement through the assessment matrix. Bourne [24] proposes defining strategies for all stakeholders or stakeholder groups, with special attention to prioritized stakeholders. All information should be considered to develop a communication strategy (communication plan), which will manage the expectations and perceptions of stakeholders [24]. Finally, the frequency and regularity of communication varies according to the interest and level of support of each stakeholder, but also with the phases of the program [11].

Manage stakeholder engagement involves the process of working and communicating with stakeholders to meet their needs and expectations, address issues, and promote appropriate stakeholder involvement [5]. Stakeholder engagement management can also be defined as the various communication practices, processes, and actions that an organization (or project team) must undertake to engage stakeholders to ensure their commitment or reduce their indifference or hostility [27].

In turn, *monitor stakeholder engagement* is the process of monitoring project stakeholder relationships and adapting strategies to engage them by modifying engagement plans and strategies [5]. Hermann [28] defined several phases for a stakeholder management model where he included the continuous monitoring phase and the measurement phase and suggests defining performance indicators (KPIs).

Considering that collaborative R&D projects are an evolution of non-collaborative projects (traditional projects), driven by recent market demands imposed to organizations, it is necessary to adapt project management practices to the reality of the collaborative context. Urbinati et al. [7] summarized the main differences in stakeholder management in traditional and open or collaborative R&D projects. Table 1 shows the differences in stakeholder management in traditional and collaborative R&D projects, as well as the main requirements of stakeholder management in collaborative R&D projects. The literature on stakeholder management practices specifically in the collaborative context may still be scarce. Elias et al. [29] propose a stakeholder analysis methodology that assists stakeholder management in collaborative projects. There are eight steps to follow, the first seven are based on Freeman [20] and the last step is based on Mitchell et al. [23]. The steps are developing a stakeholder map of the project; preparing a chart of specific stakeholders; identifying the stakes of stakeholders; preparing a power versus stake grid; conducting a process level stakeholder analysis; conducting a transactional level stakeholder analysis; determining the stakeholder management capability of the R&D project; and analyzing the dynamics of stakeholders.

Table 1. Stakeholder management in traditional vs collaborative R&D projects (adapted from Urbinati et al. [7]).

	Project stakeholder management in traditional R&D projects	Project stakeholder management in collaborative R&D projects	Requirements for stakeholder management in collaborative R&D projects
Control and coordination mechanisms over the project	Control is exercised by a single firm and coordination is formalized and proceduralized: coordinated interaction among stakeholders	Control is spread among the different stakeholders involved and there is informal coordination: active participation of all the partners in the project	Assign responsibilities among the different stakeholders involved in the project
Conflict management	Conflicts are often managed with ordinary conflict management practices	Conflicts appear to be more frequent due to project openness and the higher heterogeneity degree of the stakeholders	Adapt common conflict management practices to the collaborative context or create new ones
Engagement of stakeholders	Stakeholders are engaged through clearly defined objectives and agreements	Stakeholder engagement is seen as a dynamic and continuous process that builds on on-going relationships	Involve stakeholders continuously in each phase of the project
Maintaining stakeholder relationships	A few stakeholders continue the collaboration after the closure of the project	Most of the stakeholders continue the collaboration after the closure of the project	Work on maintaining relationships with stakeholders over time, reinforcing the long-term view of collaborations
Stakeholder satisfaction	Stakeholder satisfaction is measured according to performance metrics: time, cost, quality	Stakeholder satisfaction is sensed through continuous project scoping to forecast the project success	Consider stakeholder satisfaction as a key objective of the project; understand stakeholder satisfaction and their subjective perceptions
Managerial commitment	A set of policies is designed to maximize stakeholder engagement in the project	Managerial commitment affects the propensity of stakeholders to disclose information to the others	Coordinate stakeholders and directly and positively influence their behavior; build relationships with stakeholders through engagement

Looking at the research of Fernandes et al. [30] that identifies key initiatives in the different phases of the program management lifecycle, it is notable that although these initiatives arise in an investigation into the general context of managing UIC, many of these initiatives are strictly related to stakeholder management, namely the identification of expected benefits and performance indicators (KPIs) and the regular management of project risks and problems.

3 Research Methodology

This is a qualitative research and is based on the interpretive paradigm with an exploratory orientation. The research strategy adopted is a longitudinal case study. The case study allows answering central questions of a research project, namely the "how" and "why" of certain events in a real-life context [31]. This research aims to identify the practices employed in stakeholder management and combine them with those proposed by other scholars to develop a methodology for stakeholder management for inter-organizational collaborations. In a first phase, a literature review was conducted to collect, analyze, and understand relevant information on UIC programs and stakeholder management, followed by data collection and analysis of the Bosch-UMinho collaboration through document analysis and participatory observation, thus enabling the identification of the practices carried out to manage its stakeholders. The governance model is included in the most relevant documents analyzed, as well as several documents that supported the management of the program and its constituent projects (e.g., project charters, benefits register, and lessons learned register). Observations were conducted at work routines, celebrations, workshops, and meetings at all organizational levels, as well as informal gatherings during the stakeholder management activities. Observation is commonly criticized for a potential lack of reliability [32]. However, coupled with other qualitative methods, observation is a crucial holistic research method, enabling researchers to gain a better understanding of the insider's perspective [33]. Finally, the methodology for stakeholder management in UIC is proposed.

4 Case Study

In 2013, due to the demand for increasingly advanced technologies, Bosch Car Multimedia, in Portugal, and University of Minho started what would become the largest innovation collaboration in Portugal and one of the largest collaborations between industry and university in Europe [34]. The first phase was called HMIExcel. After the success achieved with this R&D program, the collaboration moved forward in 2015 with a new program, named INNOVATIVE Car HMI, which also presented excellent results in June 2018 [34]. In July of the same year, the collaboration moved on to the 3rd and current phase, with three R&D programs in parallel: Sensible Car, Easy Ride, and Factory of the Future [33]. Since 2013, the beginning of the collaboration, more than 170 million euros have been invested in creating solutions for mobility and connected industry, creating critical knowledge that actively contributes to increasing the competitiveness of Portugal in the global market.

The collaboration has established a collaborative governance model which aims to provide roles and responsibilities, processes, and tools to ensure the governance of programs and its constituent projects developed within the collaboration for R&D [35]. The program coordination includes four people: two program directors and two program managers, each representing University of Minho and Bosch [35]. Looking specifically at their roles and responsibilities for the stakeholder approach, the governance model describes the main roles and responsibilities regarding stakeholders and communication, the phase of the program lifecycle they fall into, and the respective responsible person (body or member). These roles and responsibilities represent pragmatically the practices carried out in the Bosch-UMinho collaboration to perform stakeholder management. As a result of the participative observation and document analysis of the governance model and other management documents, it was possible to identify, in Table 2, the existence of practices to identify, plan, manage, and monitor the engagement of program stakeholders.

Table 2. Stakeholder management practices in Bosch-UMinho collaboration programs.

Process	Sub-activity
Identify stakeholders	Identify stakeholders Identify stakeholders' needs Identify stakeholders' expectations
Plan stakeholder engagement	Plan actions to ensure stakeholder engagement Plan communications with stakeholders
Manage stakeholder engagement	Align expectations among stakeholders Gain general stakeholder acceptance of program objectives Ensure the program stakeholders involvement and continuous commitment Promote the dissemination and sharing of the program's results Promote synergies between the projects of the program Coordinate the participation of subcontracted external entities Implement the program's communication plan Carry out knowledge talks and innovation workshops on the specific or transversal themes of the projects in the program Ensure effective communication between the various projects' stakeholders of the program Carry out other types of events, e.g., welcome sessions for the new human resources of the collaboration and technical and financial audits
Monitor stakeholder engagement	Define performance indicators (KPIs) Analyze stakeholder dynamics

5 Methodology for Stakeholder Management in UIC Programs

The proposed methodology is based on stakeholder management processes, practices, and tools collected from the literature as well as on the processes, practices, and tools used in the Bosch-UMinho collaboration. We highlight two central aspects to the success of stakeholder management in UIC. Firstly, the importance of the existence of a Program and Project Management Office (PgPMO) or similar structure, since this PgPMO team presents great relevance in actively supporting the program coordination and project teams, which in turn are responsible for stakeholder management [30, 35]. Secondly, the importance of defining the roles and responsibilities of the different stakeholders, commonly established in a governance model. The methodology comprises four main activities, based on the PMBOK stakeholder management processes [5]: identify stakeholders, plan stakeholder engagement, manage stakeholder engagement, and monitor stakeholder engagement. Each activity is broken down into sub-activities (practices), as shown in Table 3. The table discloses the research method used to reach them within the case study analysis, as well as the literature references where they were identified. As detailed in the next subsections, the activities proposed are linked to the program management lifecycle of Fernandes et al. [19], which is divided into four phases: program preparation, program initiation, delivery of program benefits, and program closure. This link is made considering the level of effort spent on different activities of stakeholders management, during the program management lifecycle, similarly to Figure 8 in Fernandes et al. [11], which intends to portray this abstract concept.

5.1 Identify Stakeholders

The first activity *'identify stakeholders'* (A1) aims to collect relevant information about the UIC program's stakeholders, information that will allow to efficiently manage their participation in the program. This is a step that should be performed at program preparation and program initiation, but it is not limited to these two stages due to the dynamic nature of the stakeholders. The information should be reviewed and updated periodically or whenever there are changes in the UIC program [5, 26]. It is up to the program coordination to define the most appropriate update periodicity considering the specificity of the program it coordinates. Therefore, the sub-activity *'Collect stakeholder identification information'* (A11) should be carried out dynamically, i.e., starting with an initially limited core of stakeholders and then each stakeholder identifies other stakeholders (snowball technique) [26, 36, 37]. Questionnaires, focus groups and workshops can also be conducted to collect stakeholders' information [5, 24, 30, 35, 37] , as well as to *'identify the contributions needed from each stakeholder'* (A12), to *'identify stakeholders' expectations'* (A13), to *'collect information on stakeholders' previous experiences'* (A14), to *'analyze interdependencies/relationships between stakeholders'* (A15) and to *'analyze the level of stakeholders' engagement'* (A16). Given the need to pay more attention to certain stakeholders, it is necessary to *'assess and classify the stakeholders'* (A17). There are several classification tools such as the power/interest matrix [35], the classification of stakeholder types by Mitchell et al. [23], that deems on the attributes power, legitimacy and urgency or stakeholder circle by Bourne [24]. This classification allows to *'prioritize stakeholders'* (A18). Finally, and as a result of all the previous sub-activities, it is essential to *'document all information in the stakeholder register'* (A19) [5], to establish a basis for planning stakeholder engagement.

Table 3. Stakeholder management methodology for UIC programs.

Activity	ID	Sub-activity	References	Research Methods
A1: Identify stakeholders	A11	Collect stakeholder identification information.	[5], [24], [29], [35]	Observation / Document Analysis
	A12	Identify the contributions needed from each stakeholder.	[5], [22], [24], [29]	-
	A13	Identify stakeholders' expectations.	[7], [22], [24], [26], [29], [30]	Observation / Document Analysis
	A14	Collect information on stakeholders' previous experiences.	[25], [26]	-
	A15	Analyze interdependencies/relationships between stakeholders.	[5], [26]	-
	A16	Analyze the level of stakeholders' engagement.	[5], [29]	-
	A17	Assess and classify stakeholders.	[5], [22], [29]	-
	A18	Prioritize stakeholders.	[5], [24], [29]	-
	A19	Document all information in the stakeholder register.	[5], [29]	-
A2: Plan stakeholder engagement	A21	Establish the desired level of stakeholders' engagement.	[5]	-
	A22	Develop strategies and actions to secure stakeholder engagement.	[5], [7], [24]	Observation/ Document Analysis
	A23	Create the communication plan.	[5], [24]	Document Analysis
A3: Manage stakeholder engagement	A31	Apply strategies and actions to ensure stakeholder engagement.	[5], [7]	Observation / Document Analysis
	A32	Implement the communication plan.	[5], [27]	Observation / Document Analysis
	A33	Address, anticipate and respond to risks and concerns related to stakeholder management.	[5], [30]	Observation / Document Analysis
A4: Monitor stakeholder engagement	A41	Define KPIs.	[28], [30]	Document Analysis
	A42	Monitor the effectiveness of stakeholder engagement strategies and actions.	[5], [29]	-
	A43	Monitor the effectiveness of communication.	[5]	Document Analysis
	A44	Analyze changes in stakeholder dynamics.	[26], [29]	Observation
	A45	Adapt strategies and actions to ensure stakeholder engagement.	[5], [7]	Observation / Document Analysis
	A46	Monitor stakeholders' management capability.	[29]	-

5.2 Plan Stakeholder Engagement

'Plan stakeholder engagement' (A2) occurs during the program initiation phase and aims to develop approaches to engage stakeholders based on the information gathered in the previous activity. The level of stakeholder engagement should be analyzed using the stakeholder engagement assessment matrix suggested in PMBOK [5] and *'establish the desired level of stakeholders' engagement'* (A21) to ensure the program success. The analysis of engagement (current and desired) allows to develop appropriate approaches to engage program stakeholders based on their needs, expectations, interests, and potential impact on the program. Literature suggests holding workshops [24] to *'develop strategies and actions to secure stakeholder engagement'* (A22) which results in the stakeholder engagement plan. Additionally, to better manage stakeholders' expectations and perceptions it is necessary to *'create the communication plan'* (A23) attending to the needs of each group of stakeholders. Within the scope of communication, it is essential to provide communication platforms to facilitate information sharing among stakeholders [30, 35].

5.3 Manage Stakeholder Engagement

'Manage stakeholder engagement' (A3) starts during program preparation and program initiation phases, however, it is during the delivery of program benefits phase that there is a higher proportion of effort in this activity. *'Manage stakeholder engagement'* is supported by the stakeholder engagement plan resultant from A22. It is during this activity that what has been planned is put into practice by *'apply strategies and actions to ensure stakeholder engagement'* (A31), *'implement the communication plan'* (A32), and *'address, anticipate and respond to risks and concerns related to stakeholder management'* (A33). In Bosch-UMinho collaboration, several initiatives fostered the involvement and commitment of program stakeholders, namely the implementation of mechanisms for recognition and motivation, meetings, events for sharing results, welcome sessions for new human resources of the UIC program, innovation meetings, knowledge talks, and workshop, as recognized in literature [5, 30, 35, 37].

5.4 Monitor Stakeholders Engagement

'Monitor stakeholder engagement' (A4) as *'manage stakeholder engagement'* (A3) is based on stakeholder engagement plan resultant from A22. This activity aims to analyze whether all activities of stakeholder engagement management are having the intended effect, and therefore *'monitor of stakeholder engagement'* takes place simultaneously within *'manage stakeholder engagement'*. To *'monitor the effectiveness of stakeholder engagement strategies and actions'* (A42) it is essential to *'define KPIs'* (A41) to create a way to measure the aspects to monitor stakeholders, many of them related to subjective aspects of human action, as well as to *'monitor the effectiveness of communication'* (A43). The KPIs can analyze aspects such as participation in meetings events, and even stakeholder satisfaction within the project and the program as a whole [35]. It is also recommended in these mentioned sub-activities the evaluation of the level of engagement of stakeholders through the stakeholder engagement assessment matrix suggested

in the PMBOK [5]. However, there is a need to collect the most accurate information possible regarding the level of stakeholder engagement. Since engagement is a subjective aspect of human action, perceptions for monitoring it can vary from person to person, so it is interesting to compare the perspective of the UIC program coordination with the stakeholder's perspective, throughout a stakeholders' self-assessment of their level of engagement. Another essential aspect of stakeholder engagement monitoring is to *'analyze changes in stakeholder dynamics'* (A44) since these changes influence the outcome of stakeholder engagement planning and management of stakeholders' engagement activities. In short, both the results of KPIs and changes in stakeholder dynamics lead to the need to *'adapt strategies and actions to ensure stakeholder engagement'* (A45). It is also suggested to *'monitor stakeholders' management capability'* (A46) using a capability index. In Bosch-UMinho collaboration the monitoring of stakeholders' engagement is made essentially by surveys every year, assessing various aspects such as whether stakeholders are satisfied with the communication - *'monitor the effectiveness of communication'* (A43).

6 Discussion and Conclusions

The research reported in this paper has both practical and theoretical contributions. This research builds knowledge in the area of stakeholder management by adopting a practical process view for developing a methodology focused on inter-organizational collaboration programs, in the form of university-industry collaborations, which brings together unique challenges such as different organizational cultures and often competing stakeholders' expectations and needs. Therefore, to answer the research question: *How can organizations effectively manage stakeholders of large UIC programs?* It is proposed a methodology that summarizes the management of stakeholders into four main interactive activities: *'identify stakeholders'*, *'plan stakeholder engagement'*, *'manage stakeholder engagement'*, and *'monitor stakeholder engagement'* that is mainly grounded on the stakeholder management processes from PMBOK [4]. Given the complexity of governance of these types of cross-organizational programs, the proposed methodology identifies a systematic set of sub-activities, i.e., stakeholders' management practices, that must be performed, as well as a set of tools and techniques to support the execution of these activities and sub-activities. The methodology differs from others by encouraging the use of multiple practices, tools, and techniques, and by introducing the participation of the stakeholder itself in its engagement assessment. In this way, it makes it possible to combine the opinion of the program coordination with the opinion of the stakeholder. This proposal aims to collect information that is as real as possible, since in stakeholder management many subjective aspects of human action are evaluated. The methodology also stands out for the implementation of a stakeholder management capacity index.

In future studies a focus group could be developed on participants in UIC programs, to confirm and expand these research findings. Participant opinions on the proposed methodology for managing stakeholders and new practices could be collected. Finally, it could be explored the suitability of this methodology in other inter-organizational contexts.

Acknowledgments. This work is supported by European Structural and Investment Funds, through the COMPETE 2020 [Project nº 39479; Funding Reference: POCI-01-0247-FEDER-39479] and by national funds through FCT – Project UIDB/00285/2020.

References

1. Fernandes, G., Pinto, E.B., Araújo, M., Machado, R.J.: The roles of a programme and project management office to support collaborative university–industry R&D. Total Qual. Manag. Bus. Excell. **31**, 583–608 (2020)
2. Barnes, T.A., Pashby, I.R., Gibbons, A.M.: Managing collaborative R&D projects development of a practical management tool. Int. J. Proj. Manag. **24**, 395–404 (2006)
3. Fernandes, G., O'Sullivan, D., Pinto, E.B., Araújo, M., Machado, R.J.: Value of project management in university–industry R&D collaborations. Int. J. Manag. Proj. Bus. **13**, 819–843 (2020)
4. International Standish Group: The Chaos Manifesto (2013)
5. PMI: A Guide to the project management body of knowledge (PMBOK Guide). Project Management Institute, Newtown Square (2017)
6. Barnes, T.A., Pashby, I.R., Gibbons, A.M.: Effective university - Industry interaction: a multi-case evaluation of collaborative R&D projects. Eur. Manag. J. **20**(3), 272–285 (2002)
7. Urbinati, A., Landoni, P., Cococcioni, F., De Giudici, L.: Stakeholder management in open innovation projects: a multiple case study analysis. Eur. J. Innov. Manag. Vol. ahead-of-print No. ahead-of-print (2020). https://doi.org/10.1108/EJIM-03-2020-0076
8. Rowlinson, S., Cheung, Y.K.F.: Stakeholder management through empowerment: modelling project success. Constr. Manag. Econ. **26**(6), 611–623 (2008)
9. Scandura, A.: University – industry collaboration and firms' R&D effort. Res. Policy **45**, 1907 (2016)
10. Poyago-Theotoky, J., Beath, J.S., Siegel, D.: Universities and fundamental research: reflections on the growth of university-industry partnership. Oxford Rev. Econ. Policy **18**, 10–21 (2002)
11. Fernandes, G., O'Sullivan, D.: Benefits management in university-industry collaboration programs. Int. J. Proj. Manag. **39**, 71–84 (2021)
12. Rybnicek, R., Königsgruber, R.: What makes industry–university collaboration succeed? A systematic review of the literature. J. Bus. Econ. **89**(2), 221–250 (2018). https://doi.org/10.1007/s11573-018-0916-6
13. European Commission: Study on Network Analysis of the 7 th Framework Programme Participation Final Report (2015)
14. vom Brocke, J., Lippe, S.: Managing collaborative research projects: a synthesis of project management literature and directives for future research. Int. J. Proj. Manag. **33**, 1022–1039 (2015)
15. König, B., Diehl, K., Tscherning, K., Helming, K.: A framework for structuring interdisciplinary research management. Res. Policy **42**, 261 (2013)
16. Chin, C.M.M., Yap, E.H., Spowage, A.C.: Project Management Methodology for University-Industry Collaborative Projects (2011)
17. Fernandes, G., Machado, R.J., Pinto, E.B., Araujo, M., Pontes, A.: A quantitative study to assess a program and project management approach for collaborative university-industry R&D funded contracts. In: Proceedings of 22th ICE/IEEE Conference, pp. 1–9. Institute of Electrical and Electronics Engineers Inc., Trondheim Norway (2016)
18. Tereso, A., Ribeiro, P., Fernandes, G., Loureiro, I., Ferreira, M.: Project management practices in private organizations. Proj. Manag. J. **50**, 6–22 (2019)

19. Fernandes, G., Pinto, E.B., Machado, R.J., Araújo, M., Pontes, A.: A program and project management approach for collaborative university-industry R&D funded contracts. In: Procedia Computer Science, pp. 1065–1074. Elsevier B.V. (2015)
20. Freeman, R.E.: Strategic Management: A Stakeholder Approach. Pitman, Boston (1984)
21. Eskerod, P.: Stakeholders. In: Cambridge Handbook of Organizational Project Management, pp. 172–185. Cambridge University Press (2017)
22. Jepsen, A.L., Eskerod, P.: Stakeholder analysis in projects: challenges in using current guidelines in the real world. Int. J. Proj. Manag. **27**, 375 (2009)
23. Mitchell, R.K., Agle, B.R., Wood, D.J.: Toward a theory of stakeholder identification and salience: defining the principle of who and what really counts. Acad. Manag. Rev. **22**, 853–886 (1997)
24. Bourne, L.: Project Relationship Management and the Stakeholder Circle (2005)
25. Eskerod, P., Larsen, T.: Advancing project stakeholder analysis by the concept 'shadows of the context.' Int. J. Proj. Manag. **36**, 161–169 (2018)
26. Pouloudi, A., Whitley, E.A.: Stakeholder identification in inter-organizational systems: gaining insights for drug use management systems. Eur. J. Inf. Syst. **6**, 1–14 (1997)
27. Bourne, L.: Targeted communication: the key to effective stakeholder engagement. Procedia Soc. Behav. Sci. **226**, 431–438 (2016)
28. Hermann, S.: Corporate Sustainability Branding. Deutscher, Wiesbaden (2005)
29. Elias, A.A., Cavana, R.Y., Jackson, L.S.: Stakeholder analysis for R&D project manage-ment. R D Manag. **32**, 301–310 (2002)
30. Fernandes, G., Amaral, A., Peixoto, J., Pinto, E.B., Araújo, M., Machado, R.J.: Key initiatives to successfully manage collaborative university-industry R&D: IC-HMI case study. In: Procedia Computer Science, pp. 414–423. Elsevier B.V. (2019)
31. Yin, R.K.: Case Study Research: Design and Methods (Applied Social Research Methods). SAGE Publications Inc., London (2013)
32. Saunders, M.N.K., Lewis, P., Thornhill, A.: Research Methods for Business Students. Pearson, Canada (2019)
33. Baker, L.M.: Observation: a complex research method. Libr. Trends. **55**, 171 (2006)
34. Universidade do Minho - Escola de Engenharia: Sensible Car. https://www.eng.uminho.pt/pt/investigareinovar/projetoscomempresas/Paginas/projetosensiblecar.aspx. Accessed 15 Mar 2021
35. Fernandes, A.G.G., et al.: I&DT em Colaboração Universidade-Indústria: Modelo de Governação, Portugal (2019)
36. Freeman, R.E., Harrison, J.S., Wicks, A.C.: Managing for stakeholders: Survival, reputation, and success (2007)
37. Yang, J., Shen, P.Q., Bourne, L., Ho, C.M.F., Xue, X.: A typology of operational approaches for stakeholder analysis and engagement. Constr. Manag. Econ. **29**, 145–162 (2011)

A Two-Stage Method to Solve Location-Routing Problems Based on Sectorization

Aydin Teymourifar[1(✉)], Ana Maria Rodrigues[1,2], José Soeiro Ferreira[1,4], Cristina Lopes[2], Cristina Oliveira[2], and Valeria Romanciuc[2,3]

[1] INESC TEC, Porto, Portugal
{aydin.teymourifar,ana.m.rodrigues,jsf}@inesctec.pt
[2] CEOS.PP/ISCAP/P.PORTO, Porto, Portugal
{cristinalopes,cteles}@iscap.ipp.pt, valeria@dps.uminho.pt
[3] Departamento de Produção e Sistemas, Universidade do Minho, Braga, Portugal
[4] Faculdade de Engenharia, Universidade do Porto, Porto, Portugal

Abstract. This paper deals with multi-objective location-routing problems involving distribution centres and a set of customers. It proposes a new two-stage solution method that comprehends the concept of sectorization. Distribution centres are opened, and the corresponding opening cost is calculated. A subset of customers is assigned to each of them and, in this way, sectors are formed. The objective functions in assigning customers to distribution centres are the total deviation in demands of sectors and the total deviation in total distance of customers from centroid of sectors, which must be minimized. Afterward, a route is determined for each sector to meet the demands of customers. At this stage, the objective function is the total distance on the routes in the sectors, that must be minimized. Benchmarks are defined for the problem and the results acquired with the two-stage method are compared to those obtained with NSGA-II. It is observed that NSGA-II can achieve many non-dominated solutions.

Keywords: Location-routing problems · Sectorization · Multi-objective optimization · Two-stage method · NSGA-II · Pareto optimality

1 Introduction

One of the most important problems in supply chain management is the location routing problem (LRP), which is about the placement of facility location and also routing between the customer and facilities. Several methods have been developed to solve LRP, which is an NP-hard problem [1–3]. LRP is generally related to geographical aspects, therefore sectorization can be useful to solve it. Sectorization Problems (SPs), tackles with dividing a large political territory or districts of sales, airspace, municipality, healthcare, electric power, emergency service, internet networking, police patrol, public transportation network, social facilities, collection and transportation of solid waste in municipalities, etc., into smaller regions [4–8]. SPs are usually multi-objective and the equilibrium of load, distance, client, contiguity, and compactness are the criteria that are usually considered in sectorization [9].

J. Machado et al. (Eds.): icieng 2021, LNME, pp. 148–159, 2022.
https://doi.org/10.1007/978-3-030-78170-5_14

In this work, we deal with a multi-objective location-routing problem (MO-LRP), in which there are potential distribution centres (DCs) and also customers. Considering some objective functions, sectorization is done such that a subset of the DCs is selected to be opened and then a subset of customers is assigned to each of the chosen DC. Thus, each DC and its assigned customers form a sector. A route is defined for each sector to meet the demands of customers. We propose a novel two-stage solution method, and apply it to solve benchmarks of the problem. Benchmarks are also solved with the non-dominated sorted genetic algorithm (NSGA-II) algorithm, which is a well-known and efficient multi-objective evolutionary algorithm [10]. The results of the methods are compared and the advantages and disadvantages of the methods are discussed.

There are studies in the literature that solve this problem on the basis of sectorization approach. Barreto et al. [11] integrate some hierarchical and non-hierarchical clustering techniques into a sequential heuristic algorithm to solve this problem. Martinho et al. [12] propose a method consisting of pre-sectorization, sectorization, routing, and multi-criteria evaluation phases to deal with multi-criteria and large dimensions of the capacitated location-routing problem. Different from these studies, we propose a new two-stage method on the basis of sectorization to solve MO-LRP, which is a hybrid method that includes both heuristic searches and a mixed-integer non-linear optimization model. To solve the problem based on the model, a solver of MATLAB's optimization toolbox is used. Another contribution of the article is the adaptation of the classical NSGA-II on the basis of sectorization to solve MO-LRP. We compare and interpret the results of the two applied methods.

The other chapters of the study are such that in Sect. 2 the problem and the proposed solution model are described. The experimental results are presented and discussed in Sect. 3. Conclusions about the study are provided in Sect. 4, as well as several possibilities for future work.

2 Problem Description and Solution Method

The MO-LRP tackled in this paper involves a set of potential DCs and customers in different geographical locations with known demands. A subset of the DCs is selected to be opened and the corresponding opening cost is calculated. Each customer must be assigned to a single DC, in such a way that the assignments are balanced and compact. That is accomplished by the objectives of minimizing the total deviation in demands of sectors and the total deviation in total distance of customers from centroid of sectors. The objective function of the routing stage is to minimize total distances on the routes in sectors. It is assumed that all customers are connected [12–14].

To solve this problem, we propose a two-stage method. In the first stage, different subsets of DCs, with at least one DC, are randomly generated. Both the number of DCs and selection of DCs to open are decided randomly. For each subset, the following steps are done: the first objective function, i.e. the cost of opening DCs, is calculated. Customers are assigned to the opened DCs to minimize the second and third objective functions, which are the total deviation in demands of sectors and the total deviation in total distance of customers from centroid of sectors. The number of sectors is equal to the opened DCs.

In the second stage, a route is defined for each sector. Each route starts from a DC and visits once all customers in the related sector and returns to the same DC. The objective function at this stage is to minimize the total of the routes in the sectors and for this aim, a travelling salesman problem (TSP) is solved for each sector. Thus, objective functions are acquired for each subset of DCs. This process is then repeated for other DCs subsets and the results are compared based on Pareto optimality.

To better understand the problem and the solution method, we present an illustrative example. In Fig. 1(a), squares represent potential DCs and circles are customers. As seen in Fig. 1(b), in the first stage, a subset of the DCs is selected to open and the corresponding cost is calculated. Here, the selection is made randomly. One of the opened DCs is in blue and the other in green. Customers are assigned to DCs to minimize the total deviation in demands of sectors and the total deviation in total distance of customers from centroid of sectors. Customers assigned to each DC are in the same color. As shown in Fig. 1(c), in the second stage, a route is defined for each sector, minimizing the total distance on them. Thus, all of the objective functions are determined for the opened DCs.

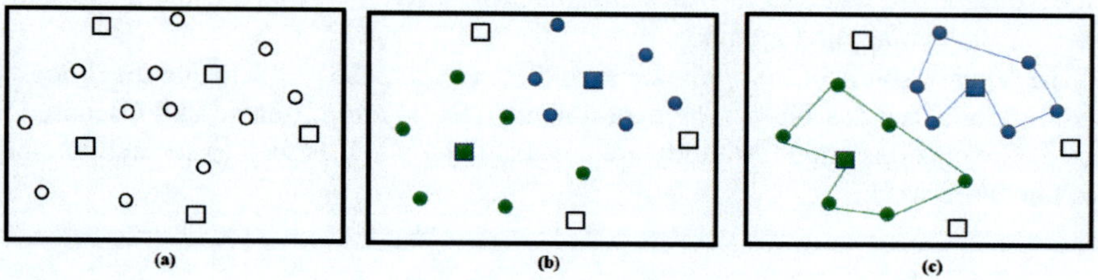

Fig. 1. An illustrative example

In the first stage, different subsets of DCs are generated and the process outlined in Fig. 1 is repeated for all of them, and the results are compared on the basis of the Pareto optimality concept. Different methods can be used to select DCs subsets. In this study, we choose them randomly.

In the next two subsections, the solution method is explained mathematically. Some of the used terminology and notations are summarized in Table 1.

Table 1. Used notations

Objective functions:	
f_1	Total cost to open DCs
f_2	The total deviation in demands of sectors
f_3	Total deviation in total distance of customers from centroid of sectors
f_4	Total distance on the routes in sectors
Sets and indexes:	
$i \in \bar{I} = \{1,...,I\}$	Index of customers
$l \in \bar{L} = \{1,...,L\}$	Index of DCs
$m \in \bar{M} = \{1,...,M\}$	Index of sectors
Parameters:	
CO_l	Opening cost of DC l
DS_{ij}	Distance between customer i and customer j
DE_i	Demand of customer i
VC	Capacity of each vehicle
Variables:	
DE^m	Total demand of customers in sector m
CE^m	Total distance between the centroid and customers in sector m
n	Number of opened DCs
Binary decision variables:	
x_l	Decision variable about if DC l opened or not
y_i^m	Decision variable about if customer i belongs to sector m or not
z_{ij}^m	Decision variable about if the path from customer i to customer j is on the route in sector m or not

2.1 First Stage of the Solution Method

At this stage of the problem, to open, different subsets of DCs are searched. For each subset of DCs, as defined in Eq. 1, f_1, is calculated.

$$f_1 = \sum_{l=1}^{L} CO_l \times x_l \tag{1}$$

where

$$x_l = \begin{cases} 1, & \text{if } DC\, l\, is\, opened \\ 0, & \text{otherwise.} \end{cases} \tag{2}$$

The number of opened DCs is obtained as in Eq. 3.

$$n = \sum_{l=1}^{L} x_l \tag{3}$$

A subset of customers is assigned to the opened DCs. Each DC and its assigned customers form a sector. The sectors must be balanced in terms of demand and distance. The objective functions for assigning customers to DCs are: the total deviation in demands of sectors and total deviation in total distance of customers from centroid of sectors, that must be minimized, defined as in Eqs. 4 and 6. The number of sectors formed in this way is equal to the number of open DCs.

$$Min\, f_2 = \sum_{m=1}^{M} |DE^m - \bar{DE}| \tag{4}$$

where $DE^m = \sum_{i=1}^{I} DE_i \times y_i^m$ and $\bar{DE} = \frac{\sum_{m=1}^{M} DE^m}{n}$ and

$$y_i^m = \begin{cases} 1, & \text{if } customer\, i\, is\, in\, sector\, m \\ 0, & \text{otherwise.} \end{cases} \tag{5}$$

$$Min\, f_3 = \sum_{m=1}^{M} |CE^m - \bar{CE}| \tag{6}$$

where $\bar{CE} = \frac{\sum_{m=1}^{M} CE^m}{n}$.

The coordinates of the centroid of each sector are calculated as the average of the coordinates of the customers in the sector.

Equation 7, imposes that each customer is only assigned to one sector.

$$\sum_{m=1}^{M} y_i^m = 1, \forall i \in \bar{I} \tag{7}$$

Each vehicle is allocated to a single sector. It is assumed that the fleets are homogeneous, i.e. the capacity of the vehicles is the same. Therefore, there is no need for a decision variable for assigning the vehicles. However, the total demand of customers in each sector must be less than or equal to the capacity of each vehicle, which is imposed as in Constraint 8.

$$DE^m \leq VC, \forall m \in \bar{M} \tag{8}$$

It should be noted that at this stage multiple subsets of DCs are searched, and each of them has corresponding f_1, f_2, and f_3. A subset of DCs that may give a good result in terms of one the objective functions may cause bad results for others, which is not desirable based on the Pareto optimality concept.

2.2 Second Stage of the Solution Method

At this stage, a route is determined for each sector. Starting from and returning to a DC, all customers of a sector are visited once on the route. The objective function of this stage, f_4, as defined in Eq. 9, is the total distance of routes in sectors, which must be minimized.

$$Min\, f_4 = \sum_{i=1}^{I} \sum_{j=1}^{I} DS_{ij} \times z_{ij}^m \tag{9}$$

where

$$z_{ij}^{m} = \begin{cases} 1, & \text{if } \textit{the path from customer i to customer j is on the route in sector m} \\ 0, & \text{otherwise.} \end{cases} \tag{10}$$

After achieving f_1, f_2, f_3 and f_4 for each DCs subset, a comparison is made between the results of different DCs subsets in terms of Pareto optimality.

2.3 Implementation of the Solution Method

As outlined in Fig. 2, the general steps of the implemented solution method can be summarized as: in the first stage of the method, 20 subsets of DCs are created randomly. For each subset of DCs, f_1 is calculated and sectorization is performed according to f_2 and f_3 objective functions. Here, sectorization means assigning customers to the opened DCs to minimize f_2 and f_3. In the second stage, a TSP is solved for each formed sector to minimize f_4. In this way, for each subset of DCs selected randomly, relevant f_1, f_2, f_3 and f_4 are achieved. The results obtained for different subsets are compared in terms of Pareto optimality.

Data: Problem
Result: Pareto optimal solutions
Generate 20 subsets of DCs randomly;
(Selected DCs and also the number of DCs in the subset are chosen randomly.);
foreach *Subset of DCs* **do**
    1. stage | **Calculate** f_1;
        | **Assign** customers of the opened DCs to minimize f_2 and f_3;
        | *(Each DC and its assigned customers form a sector.)*;
    2. stage | **foreach** *Sector* **do**
            | **Find** a route to minimize f_4;
            | *(A TSP is solved for each sector.)*;
        | **end**
end
Compare results achieved for all generated subsets of DCs based on the Pareto optimality concept;

Fig. 2. General steps of the two-stage method

We use the *intlinprog* function in both stages of the model to solve the problem, which is a binary integer linear programming solver in the optimization toolbox of MATLAB. The objective functions defined as in Eqs. 4 and 6 are non-linear, and since the *intlinprog* function of MATLAB requires linear objective functions, they are linearized. It is also possible to solve non-linear models in MATLAB but in this case, the nonlinear part of the objective function must be added as a constraint. The reason is that MATLAB, currently, does not have a solver for non-linear objective functions. To linearize the objective functions, as defined in Eq. 11, non-negative variables A^m and B^m, $\forall m \in \bar{M}$, are added to objective function f_2.

$$Min\, f_2 = \sum_{m=1}^{M} A^m + \sum_{m=1}^{M} B^m \tag{11}$$

Furthermore, the following m constraints are added to the model:

$$DE^m - \bar{DE} - A^m + B^m = 0, \forall m \in \bar{M} \tag{12}$$

$$A^m\, and\, B^m \geq 0, \forall m \in \bar{M} \tag{13}$$

In a similar way and using non-negative variables C^m and D^m, $\forall m \in \bar{M}$, objective function f_3 is linearized.

$$Min\, f_3 = \sum_{m=1}^{M} C^m + \sum_{m=1}^{M} D^m \tag{14}$$

Furthermore, the following m constraints are added to the model:

$$CP^m - \bar{CP} - C^m + D^m = 0, \forall m \in \bar{M} \tag{15}$$

$$C^m\, and\, D^m \geq 0, \forall m \in \bar{M} \tag{16}$$

2.4 NSGA-II

In addition to the two-stage method, the problem is solved using NSGA-II, which is a fast, elitist and efficient multi-objective evolutionary algorithm proposed by Deb et al. in 2002 [10]. We implement this algorithm on the basis of sectorization, whose stages are summarized in Fig. 3. As seen, the stages of this algorithm are similar to the two-step method.

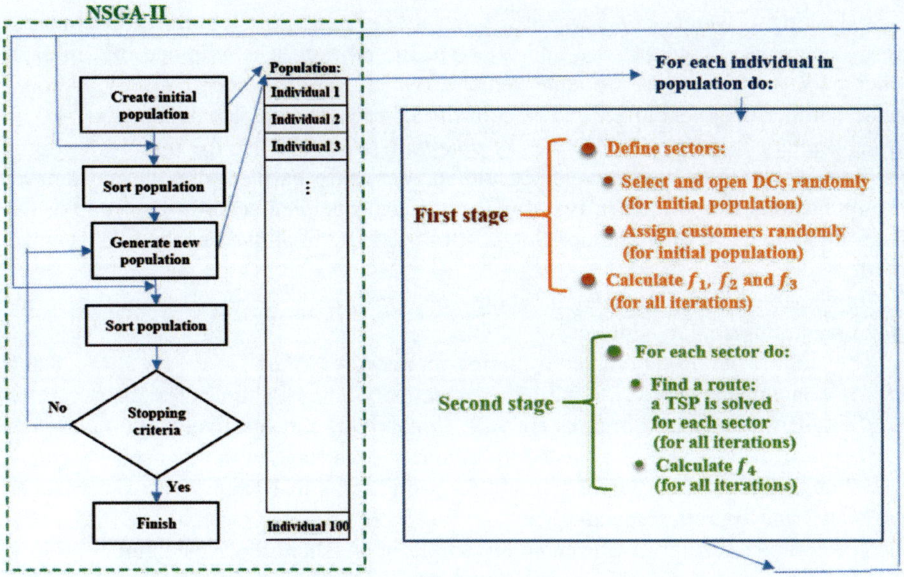

Fig. 3. Stages of NSGA-II to solve the problem

In the algorithm, for each individual in the initial population, a subset of potential DCs is opened, randomly. Each customer is assigned to the nearest opened DC. If the number of customers assigned to a DC is zero, then it is assumed as not opened. Afterward, using the nearest neighbor heuristic, a TSP is solved. A route is defined starting at a DC, visiting the nearest customer at each stage, repeating this process until all customers are visited, and returning to the DC. During other iterations, new individuals are derived using crossover and mutation operators. A one-point crossover is used to search for better customer assignments. Applying the mutation operator, an open DC is randomly selected to close and its customers are assigned to another DC, which is randomly selected to open. If all DCs are open, one of them is closed randomly and its customers are added to another DC, which is also chosen randomly. The objective functions of the new individuals formed in this process may change, in which case they must be recalculated. The parent and offspring populations are selected and merged with each other. Using non-dominated sorting, Pareto fronts are formed and then the new population is chosen using the selection operator. The general steps of the implemented NSGA-II can also be found in [15].

3 Experimental Results

The two-stage method and the NSGA-II algorithm, outlined in Figs. 2 and 3, were implemented in MATLAB on an Intel Core i7 processor, 1.8 GHz with 16 GB of RAM. The parameters used in NSGA-II are: *population size = 200*, *number of iterations = 500*, *crossover rate = 0.6* and *mutation rate = 0.1*.

Indicating as *Number of customers* $\times$ *Number of potential DCs*, three benchmarks are generated as 12×5, 15×8 and 20×10. In addition, it is assumed that in each benchmark the number of available vehicles, which have the same capacity, is equal to the number of potential DCs. The capacity of each vehicle is calculated as: $1.3 \times$ (*Total demand of customers/Number of potential DCs*). We use the discrete uniform distributions as $U(10, 100)$ to create customers' demands. Furthermore, the coordinates of both customers and DCs in two dimensions are generated according to the normal distribution as $N(50, 10)$. The opening cost of each DC in all the benchmarks is equal to 10.

Implemented codes and details about solutions can be accessed from the corresponding author's email address.

The number of non-dominated solutions obtained by NSGA-II for 12×5, 15×8 and 20×10 benchmarks are 60, 78 and 125, respectively. The CPU times for 500 iterations of NSGA-II for these benchmarks are 789, 876 and 948 s, respectively. The maximum running time of the solver is adjusted to 30 min in each stage of the two-stage method, even though finding the optimal solution for problems with 10 and 15 customers takes less than 2 and 10 min, respectively.

The obtained results are presented in Table 2, in which, *n* shows the number of open DCs. The two-stage method achieves only one solution for each subset of DCs, but NSGA-II finds multiple solutions. For each benchmark, results are presented for only two subsets of DCs. In addition, only two solutions of NSGA-II are presented for each subset of DCs. Results of the same DCs subset are given for each *n*. Therefore, it can be said that Table 2 contains only some example solutions, as there are similar states for other solutions. The selection of the presented solutions is done randomly.

Table 2. Acquired results for the benchmarks

Benchmark	NSGA-II					Two-Stage method				
	n	f_1	f_2	f_3	f_4	*n*	f_1	f_2	f_3	f_4
12 × **5**	5	50	77.60	44.23	24.13	5	50	5.20	9.41	42.10
	5	50	115.20	23.72	20.83					
	4	40	118	25.13	30.91	4	40	14	1.28	40.34
	4	40	236	4.06	45.10					
15 × **8**	8	80	161	22.70	7	8	80	57	7.61	29.66
	8	80	149.25	25.66	16.61					
	7	70	365.14	13.09	21.10	7	70	28.86	7.50	31.67
	7	70	167.14	23.75	13.03					
20 × **10**	9	90	439.77	39.90	11.05	9	90	24.66	10.99	30.17
	9	90	316.22	48.33	11.25					
	8	80	247	9.53	33.29	8	80	11.00	5.75	40.87
	8	80	160	12.04	35.21					

As seen in Table 2, considering Pareto optimality the solutions found by the methods for each benchmark do not dominate the other. The solutions obtained by the two-stage method are significantly better than the ones found by NSGA-II in terms of the total deviation in demands of sectors and the total deviation in total distance of customers from centroid, however, NSGA-II achieves better results in terms of the total distance on the routes in sectors. The two-stage method finds optimal solutions for the sectorization stage, but the best routes of them are worse than the non-dominated solutions found by NSGA-II. For each subset of DCs, NSGA-II achieves multiple non-dominated solutions, while the two-stage method offers only one. Although the two-stage method achieves good results, when the number of customers is more than 20, no optimal result was found within 30 min, which can be considered as a long time to solve a problem in this size.

4 Conclusion and Future Work

In this paper, MO-LRP is considered, which comprises four objective functions to be minimized as: minimizing the cost of opening DCs, the total deviation in demands of sectors and the total deviation in total distance of customers from centroid of sectors, and total distance on the routes in sectors.

Different from studies in the literature we propose a new two-stage solution method based on the sectorization. In the first stage of the solution method, different subsets of the potential DCs are searched. Also, a subset of customers is assigned to each of them and in this way, customers are divided into sectors. In the second stage, the problem of routing is solved for each sector of the first stage.

We also solve the problem with NSGA-II. As another novelty, we implement NSGA-II on the basis of sectorization, such that the problem is divided into sectors and then a route is defined for each the sectors. The two-stage method provides one output for each subset of DCs, while NSGA-II achieves multiple non-dominated results, which can be considered as an advantage by decision-makers. The results obtained by the two-stage method are better in terms of minimizing the total deviation in demands of sectors and the total deviation in total distance of customers from centroid of sectors, while NSGA-II finds better results according to total distance on the routes in sectors.

Although the operators used in NSGA-II are simple, the achieved solutions seem competitive. In future studies, it is planned to use more efficient operators and heuristics in NSGA-II, especially for routing.

The solution method is designed based on sectorization, meaning that customers are geographically divided into subsets. This procedure can be especially useful to solve large-scale problems. The benchmarks in this study are small in size and we plan to suggest new methods to solve larger benchmarks. Even though the two-stage method may not be successful in solving large problems in terms of computation time, it can be used as a tool to evaluate the performance of the method planned for the future to solve large-scale benchmarks.

The parameters used in the methods affect their performance. For example, in the first stage of the two-step method, 20 subsets of the DCs are randomly selected. This is just a practical choice and different numbers can be used instead of 20. Also, the used parameters in NSGA-II affect its performance. In future works, the effect of the selected parameters on the results will be investigated.

Also, it is planned to use the penalty or barrier technique to deal with MO-LRP. There are studies in the literature based on this approach to solve multi-objective problems [16], in which new objective function is defined including the initial objective function and penalization for constraints violation, dealing with feasibility and optimality together.

Acknowledgements. This work is financed by the ERDF - European Regional Development Fund through the Operational Programme for Competitiveness and Internationalisation - COMPETE 2020 Programme and by National Funds through the Portuguese funding agency, FCT - Fundação para a Ciência e a Tecnologia within project POCI-01-0145-FEDER-031671.

The authors would like to thank the editor and the anonymous referees for their valuable comments which helped to significantly improve the manuscript.

References

1. Karaoglan, I., Altiparmak, F.: A memetic algorithm for the capacitated location-routing problem with mixed backhauls. Comput. Oper. Res. **55**, 200–216 (2015)
2. Pekel, E., Kara, S.S.: Solving fuzzy capacitated location routing problem using hybrid variable neighborhood search and evolutionary local search. Appl. Soft Comput. **83**(10566), 5 (2019)
3. Oudouar, F., Lazaar, M., El. Miloud, Z.: A novel approach based on heuristics and a neural network to solve a capacitated location routing problem. Simul. Model. Pract. Theory **100**(10206), 4 (2020)
4. Camacho-Collados, M., Liberatore, F., Angulo, J.M.: A multi-criteria police districting problem for the efficient and effective design of patrol sector. Eur. J. Oper. Res. **246**(2), 674–684 (2015)
5. Filipiak, K.A., Abdel-Malek, L., Hsieh, H.N., Meegoda, J.N.: Optimization of municipal solid waste collection system: case study. Pract. Period. Hazardous Toxic Radioactive Waste Manag. **13**(3), 210–216 (2009)
6. Ghiani, G., Lagana, D., Manni, E., Musmanno, R., Vigo, D.: Operations research in solid waste management: a survey of strategic and tactical issues. Comput. Oper. Res. **44**, 22–32 (2014)
7. McLeod, F., Cherrett, T.: Quantifying the transport impacts of domestic waste collection strategies. Waste Manag. **28**(11), 2271–2278 (2008)
8. Mourão, M.C., Nunes, A.C., Prins, C.: Heuristic methods for the sectoring arc routing problem. Eur. J. Oper. Res. **196**(3), 856–868 (2009)
9. Teymourifar, A., Rodrigues, A.M., Ferreira, J.S.: A comparison between simultaneous and hierarchical approaches to solve a multi-objective location-routing problem. In: Graphs and Combinatorial Optimization: From Theory to Applications. CTW2020 Proceedings, vol. 251 (2020)
10. Deb, K., Pratap, A., Agarwal, S., Meyarivan, T.A.M.T.: A fast and elitist multiobjective genetic algorithm: NSGA-II. IEEE Trans. Evol. Comput. **6**(2), 182–197 (2002)
11. Barreto, S., Ferreira, C., Paixão, J., Santos, B.S.: Using clustering analysis in a capacitated location-routing problem. Eur. J. Oper. Res. **179**(3), 968–977 (2007)

12. Martinho, A., Alves, E., Rodrigues, A.M., Ferreira, J.S.: Multicriteria location-routing problems with sectorization. In: Congress of APDIO the Portuguese Operational Research Society, pp. 215–234 (2017)
13. Rodrigues, A.M., Ferreira, J.S.: Sectors and routes in solid waste collection. Oper. Res. **353–375** (2015)
14. Rodrigues, A.M., Soeiro Ferreira, J.: Waste collection routing—limited multiple landfills and heterogeneous fleet. Networks **65**(2), 155–165 (2015)
15. Teymourifar, A., Ozturk, G., Bahadir, O.: A comparison between two modified NSGA-II algorithms for solving the multi-objective flexible job shop scheduling problem. Univ. J. Appl. Math. **6**(3), 79–93 (2018)
16. Matias, J., et al.: Adaptive penalty and barrier function based on fuzzy logic. Expert Syst. Appl. **42**(19), 6777–6783 (2015)

Model of Academic Teachers Communication Competencies Management

Małgorzata Spychała[1](✉), Emanuel Said[2], and Agata Branowska[3]

[1] Poznan University of Technology, J. Rychlewskiego 2, 60-965 Poznan, Poland
malgorzata.spychala@put.poznan.pl
[2] University of Malta, MSD 2080 Msida, Malta
[3] Poznan University of Technology, J. Rychlewskiego 2, 60-965 Poznan, Poland

Abstract. While several attributes have been used to describe a teacher's profile, content knowledge and communication skills remain outstanding [1]. Oral communication, the ability to prepare and run interesting presentations, transferring the knowledge and skills in a clear and comprehensive manner is one of the most important competencies that an academic teacher should acquire. In this study, we propose a model of communication competencies management for academic teachers to identify competencies gaps and help make decisions related with training initiatives and development activities. In our study the communication profile of an academic teacher, which can be used to assess verbal and nonverbal skills is examined involving 92 doctoral students who would be future academic teachers at Poznan University of Technology (PUT). To build a profile a questionnaire was created. We use data exploratory steps and undertake data reduction procedures to enable us simplify observations into three key dimensions: non-verbal communication competences, verbal clarity competence and verbal richness competence. We examine these factors for the moderating effect of the source of evaluation (students assessment versus expert assessment) and find that whereas assessments tend to be consistent across student and expert evaluators, student evaluators tend to be more generous in their judgement than expert evaluators

Keywords: Model of competency management · Communication competencies · Competency profile of an academic teacher

1 Introduction

One of the most important task of academic teachers is passing the knowledge to students and taking care od high quality of didactic process. To reach those goals, the academic teacher need not only to academic up to date knowledge, but also social competencies [2, 3], which enable them to share information in effective way and build positive relationships with students. For teachers-to-be, one of the most important professional competences is the communicative competence [4–6].

The article is about communication competency level of doctoral students at Poznan University of Technology, who are going to be academic teachers in the near future. The paper focuses on the oral communication competence, the ability to share the knowledge

J. Machado et al. (Eds.): icieng 2021, LNME, pp. 160–173, 2022.
https://doi.org/10.1007/978-3-030-78170-5_15

with students, and give an effective presentation. The aim of this paper is to develop a model of communication competencies management for academic teachers. The model shows the basic steps of communication competence management.

To reach those goal the research on 92 students from five different groups was conducted. During the research the communication competency level of students during public speech was examined. To evaluate students a questionnaire to measure verbal and nonverbal aspects of communication was created. Each student gave a short presentation, after which he/she was evaluated by two persons: another student and an expert. Skills related with verbal and nonverbal communication were examined.

The empirical part of the study focused on following research questions:

1. What is the communication competency level of PhD students (future teachers) at Poznan University of Technology? What is a communication competency profile of a future academic teachers?
2. Are there differences in the level of communication competency between the students of different departments?
3. What are students communication competencies gaps?
4. Is there a difference in communication competency evaluation of an expert and students? In which areas the evaluation differs most?
5. Is there a correlation (and what is the strength of correlation) between any communication skill? Is there a correlation between verbal and nonverbal communication skill?

To answer those research questions, various statistical methods were used. In the article the profile of a communication competency necessary for public speaking was created. The profile consist of verbal and nonverbal behaviours, which a teacher needs to demonstrate while giving the presentations. The profile, including needed indicators can be used by universities to evaluate teachers communication competencies. Developing a profile is a first step in the presented model of communication competence management. The model can be used for identifying competencies gaps, making decisions related with training initiatives and development activities. The application of this model may thus improve the quality of teaching at different universities.

2 Communication Competencies of an Academic Teacher

The first step to build the communication competencies model is to analyze the job description of an academic teacher [7]. Universities publish different job descriptions, showing tasks, goals, responsibilities and requirements of academic teachers [8]. As an example, a job description at PUT is presented. At the university teachers are responsible for:

"- conducting didactic classes and developing educational offer (tasks related with this category include for instance designing, preparing and developing teaching materials; providing high quality of teaching, assessing and supporting students' learning process),

- conducting scientific researches (tasks related with this category include e.g. writing and publishing the articles, monographs etc. based on the conducted researches; receiving a patents and utility models),

- participating in organizational work of the university (tasks related with this category include: promoting the university and the unit e.g. participating in scientific or research and development projects; membership in the international bodies of the organizations; participating in the organization of the conferences)."

Basing on responsibilities and tasks which a person is to perform on particular job position, a set of competencies creating a profile is being designed. Competencies are aptitude, skills and knowledge used to effectively perform tasks in given work conditions [4, 9, 10]. Scholars distinguished competencies of the academic teachers and emphasize the following as most important: professional competencies, educational competencies, science and research competencies, publication competence, lifelong learning competencies, information and communication technologies (ICT) competencies, social-cultural competencies, motivational competence, and communication competencies [4, 11, 12]. Communication competencies of teachers are equally necessary for teachers' success as methodical skills [1, 13]. Communicative competence is a prerequisite for the didactic process, so it is essential for teachers to acquire a oral/public communication competencies [14, 15].

Teachers convey the information through the use of verbal and nonverbal communication. Poor speech delivery is a barrier to information transfer. Teachers should work on the verbal aspects of their presentations, make a structure of their presentation with clear purpose and present the content in such a way, that it is adapted to the characteristics of an audience [16]. Academic teachers content should present in a way that can be easily understood by the public. Speakers should stick to the main thread of the speech, avoid building overly long sentences and enrich the presentation with examples, anecdotes and jokes [17]. Equally, speakers should avoid use of notes during their speech. Nonverbal communication embraces all body language communication, and also includes clothing, environmental factors and the manner in which we use time [18]. Nonverbal communication includes ways of talking (e.g., pauses, stress on words), posture (e.g., slouching), facial expression, appearance, closeness, eye movements, head movements, sounds, hand movements, etc. [1, 19, 20].

3 Development of a Communication Skills Management Model for Academic Teachers

Our review of earlier literature suggests that new models and tools for competence management are required [21–26]. Competence management focuses specifically on identifying and locating competencies. Managing competencies involves administering several systems, processes and initiatives to ensure that the organization and its members are collectively utilizing their competencies to achieve organizational operational and strategic goals [27–31].

A profile of communication competencies is thus a very important competence management tool. It is a valid, observable, and measurable list of the knowledge, skills, and attributes that include the key behaviours required for excellent execution of a particular role [21, 32, 33]. The profile is organized into tiers of competencies and includes descriptions of the activities and behaviours associated with each competency [34, 35]. Analyzing the communication competences of the academic teacher, a profile of communication competencies necessary for public speaking was developed (Fig. 1). In the given example, behaviours regarding verbal and non-verbal communication have been specified. We use a rating from 0 to 5, where 0 means no communication skills, whereas 5 indicates a very high level of a given competence.

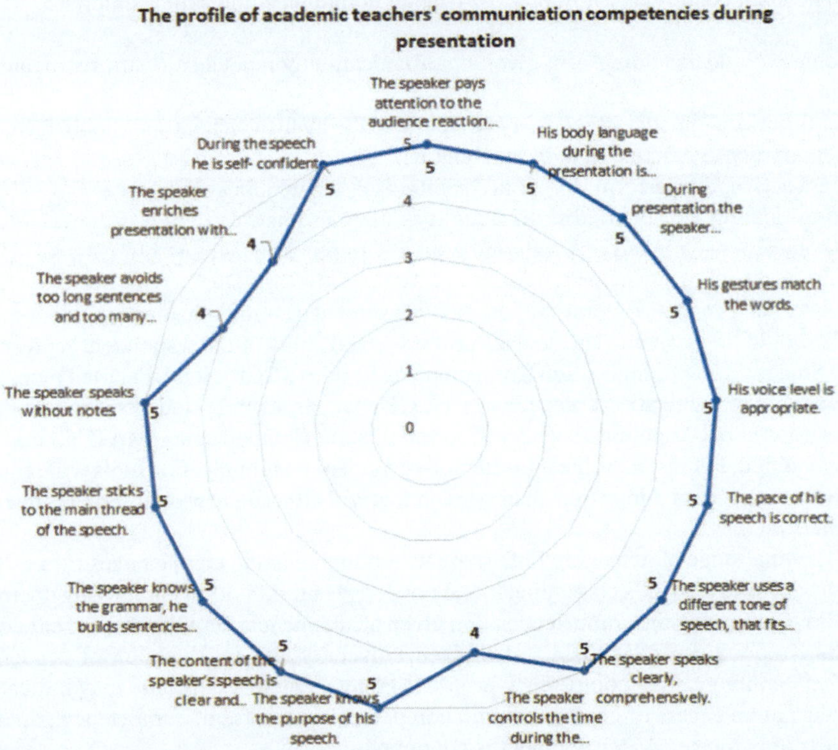

Fig. 1. Academic teachers profile of communication competences during presentation (Source: Authors)

The chart shows that the communication requirements of teachers were set at high and very high level. When managing the communication competences of academic teachers, it is necessary to recognize their abilities at the beginning of Doctorial studies.

Each university should have Matrix of Academic Teachers' Communication Competencies (MATCC) – a tool containing requirements towards academic teachers. MATCC involves the whole behaviours/indicators related with communication competency, not only those related with public speaking. The profile of communication competencies (Fig. 1) is a part of university requirements towards academic teachers.

Identification of the doctoral students' social skills is therefore a necessary stage. By examining the communication competencies of a given doctoral student at the beginning of the studies, and then comparing them with the MATCC, the competence gaps of each PhD student will be determined. Basing on this information, a plan for developing and improving competence gaps will be set up. Then the PhD student should improve his communication skills practically by completing tasks.

The main conditions for improving teacher communication skills comprise:

- cognitive understanding of a given communication competence (learn its meaning, purpose);
- acquiring and practising certain behavioural skills by the teacher (one should have the opportunity to practice it, to learn about it);
- feedback on the performance of tasks with usage of certain competence;
- consolidating the acquired skill, using it as often as possible until it becomes a fully automated, natural activity, fully integrated with other behaviours [36–38].

After each year of doctoral studies, the PhD student's communication knowledge and skills should be assessed. The actual knowledge and skills of the respondent represents PhD Student X' Communication Competencies Profile (SCCP). It is a detailed record of a person's communication competences. SCCP changes, as PhD students improve their skills. Matrix of Academic Teachers' Communication Competencies (MATCC) may be also modified. For instance, the introduction of modern communication tools will require the use of other communication channels and this will affect on new requirements toward teachers.

The final stage of managing PhD students' communication competences is to examine the doctoral student's communication knowledge and skills after completing doctoral studies. Comparing the competences of a given academic teacher with the current competency requirements (MATCC) at a given university, the final version of academic teacher communication competency profile is being created. The profile may indicate a shortage or an excess of communication competencies. In case of competency gap, the process of competency development is continued.

Communication competence management allows the organization to identify knowledge and skills gaps that may help to assess and refine the need for teachers and professional training initiatives. This information can be used for the mapping of professional development of the university members [39]. Through the development of an effective teachers' communication competence management system, the quality of teaching can be significantly improved [5, 40, 41].

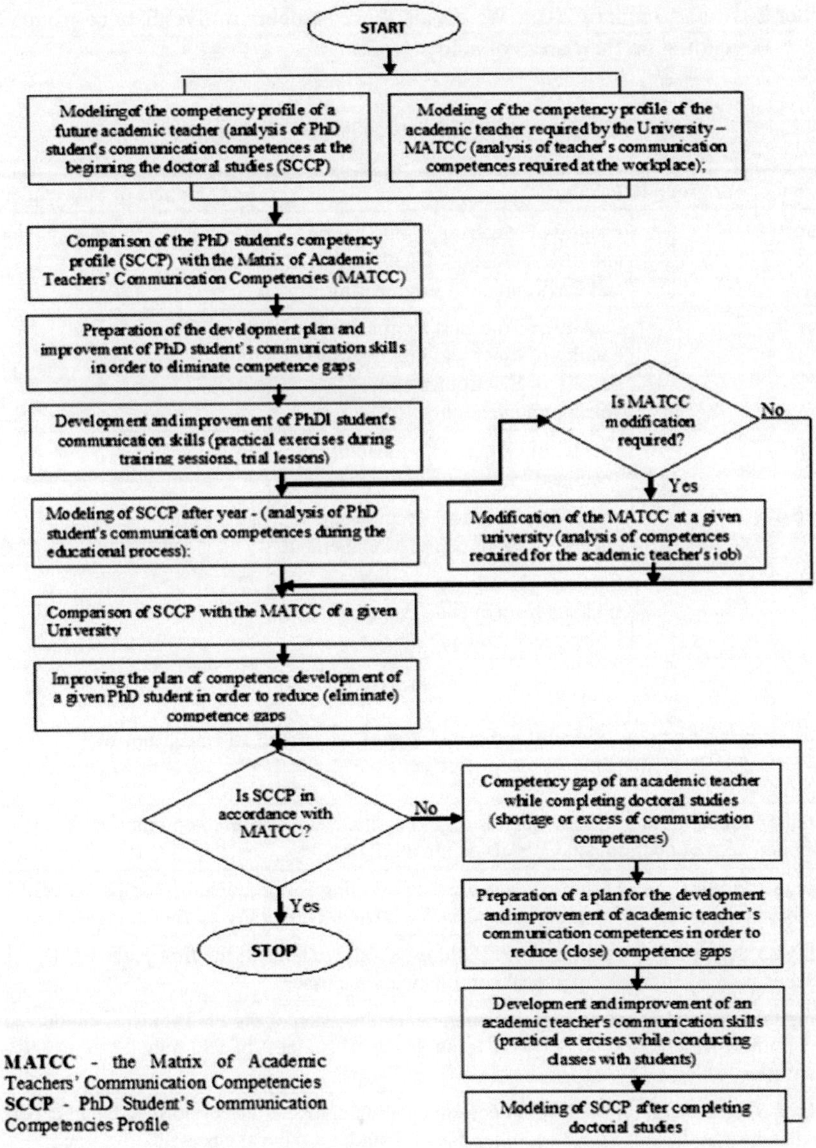

Fig. 2. Model of communication competences management of academic teacher (Source: Authors).

4 Methodology of Research

We study the communications competencies of doctoral studies by administering a survey among 92 first-year doctoral students at Poznań University of Technology between

October 2018 and January 2019. We divide these students in five distinct groups (see Table 1) depending on their areas of study.

Table 1. Student groups and data collection (Source: Authors).

The number of group	Faculty	The numbers of PhD students
Group 1	Faculty of Mechanical Engineering and Management + Faculty of Civil and Environmental Engineering	19
Group 2	Faculty of Transport Engineering + Faculty of Electrical Engineering + Faculty of Electronics and Telecommunications	14
Group 3	Faculty of Computing + Faculty of Technical Physics	23
Group 4	Faculty of Chemical Technology + Faculty of Engineering Management	16
Group 5	Interdisciplinary group - PhD students from outside Poznan University of Technology	22

We divided our study in seven distinct stages as set out in Fig. 3 below.

Stage 1 Preparation of questionnaires regarding the examination of communication competences of doctoral students/academic teachers
Stage 2 Preparation of a competence profile regarding communication competences of academic teachers during presentations at the Poznan University of Technology
Stage 3 Selection of the group of PhD students - all students of the first year of PhD studies who attended interpersonal communication classes
Stage 4 Conducting research of competence levels - each of the PhD students conducted a 5-minute presentation on the given topic. Doctoral students had 30 minutes to prepare their speech.
Stage 5 Assessment of the level of communication competences performed by an expert and auditor at the same time - another doctoral student. After the presentation, each doctoral student was assessed by an expert and auditor. Questionnaires prepared in stage I (two-source assessment) were used for the assessment
Stage 6 Comparison of competence requirements developed in stage 2 with average assessments of competence levels regarding communication made by an expert and PhD student
Stage 7 Statistical analysis of test results - linear regression **...**

Fig. 3. Main stages of PhD student's communication competencies research (Source: Authors).

It can be assumed that when conducting communication competence assessments, the method of assessing the competent activity of a PhD student in the workplace was used, as his task wasn't different from the tasks he performs during the classes. This method is one of the most valuable techniques for measuring professional competences, because it relies on observation and assessment of behaviours that are 'real' indicators of any given competence. In this case, doctoral students were evaluated by an observing expert and another observing doctoral student during the students' presentation. Two-source assessment, i.e. the use of information from listeners - doctoral students and an expert in the process of measuring communication competences provides more objective results than in the case of self-assessment.

5 Results

We analyse our data by following three phases of statistical analysis. First, we explore the items' correlation (Pearson Correlation coefficient, $p < 0.05$) by treating the data as two datasets – expert scores and student scores. Wheas all non-verbal competences items correlated with each other in the expert scores data set, we note only one item that could not correlate with the rest in the student scores data set ("*The speaker controls the time during the presentation*"). In a similar way, whereas all verbal competences items correlated with each other in the expert scores data set, one item would not correlate with the rest in the student scores data set ("*The speaker enriches his/her presentation with jokes, anecdotes and examples*").

Second, we looked closer at the items' variation and simplify the data set using exploratory factor analysis. During this stage, we aggregate the expert and student scores data sets in one data set by joining all responses under common variables but distinguishing the source of evaluation by creating a new dummy variable (source: 1 = student, 2 = expert). An initial correlation matrix (Pearson Correlation coefficient, $p < 0.05$) showed that all non-verbal competence items correlated with each other, as was the case of verbal competence items. In the case of the non-verbal competence items, we observed that the data was robust enough for subsequent data reduction through factor analysis (KMO measure of sampling adequacy = 0.905; Bartlett's Test of Sphericity, $p < 0.05$). Indeed, our principal component extraction offered a unidimensional solution that explains a cumulative variance of 52.5%. As for the verbal competence items, our data was also robust enough for further data reduction through factor analysis (KMO measure of sampling adequacy = 0.839; Bartlett's Test of Sphericity, $p < 0.05$). However, our principal component extraction of verbal competence items suggested a two-factor solution that explains a cumulative variance of 63.6%. We extracted two dimensions following Varimax rotation and Kaiser normalization that lead to a solution with greater contrast. The first dimension "clarity" resulted from the loading of six items (see Table 2), explaining 43% of the total variance across the data. The second dimension, "richness" resulted from the loading of two items and explained 20.5% of the total variance across the data.

Table 2. Factor loadings of verbal competences item set showing two dimension loadings (Source: Authors).

Rotated Component Matrix[a]	Component	
	1	2
The speaker knows the purpose of his/her speech	**.806**	.045
The content of the speaker's speech is clear and adapted to the auditorium	**.826**	.228
The speaker knows the grammar, he/she builds sentences correctly	**.573**	.399
The speaker sticks to the main thread of the speech	**.827**	.195
The speaker speaks without notes	**.735**	.186
The speaker avoids too long sentences and too many words in his/her speech	**.678**	.196
The speaker enriches his/her presentation with jokes, anecdotes, examples	.055	**.890**
During the speech he/she is self-confident (knowledge about the topic of the presentation and self-awareness)	.317	**.725**

Extraction Method: Principal Component Analysis
Rotation Method: Varimax with Kaiser Normalization[a].
a. Rotation converged in 3 iterations

To ensure repeatability of the study, instead of creating regression scores directly from our factor analysis, we created sum scores for each dimension by adding up the relevant items into one score. In this way, we created three new variables: Non-verbal competence score (NVCS) that involved tine summing up of the nine non-verbal item scores on a response-by-response basis, Verbal clarity score (VCS) that involved the summing up of the first six from the eight verbal competence items, and Verbal richness score (VRS) that involved the summing up of the last two items from the eight verbal competence items. We explore these new factors and find that all three factors correlate with each other at a 95% confidence interval ($p < 0.05$) (Table 3).

Table 3. Non-verbal and verbal competence scores correlation (Source: Authors).

		Non-verbal competence score	Verbal clarity score	Verbal richness score
Non-verbal competence scores	Pearson Correlation	1	0.473**	0.653**
	Sig. (2-tailed)		0.000	0.000
Verbal clarity score	Pearson Correlation		1	0.435**
	Sig. (2-tailed)			0.000

To assess any differences across the student groups (groups I through V, see Table 1) we conducted post-hoc (Tukey's) tests for significance of differences between groups and find no real difference in means except for between Group I and Group II in the case of verbal clarity. In a similar way we used post-hoc (Tukey's) tests to detect an significant difference in these factors across the two types of sources of evaluation (student evaluation vs expert evaluation) and find significant differences in means across the two verbal competence factors. A further independent samples t-test confirms our observations above as well as that that NVCS and VCS had unequal variance across the two sources of evaluation (Table 4).

Table 4. Independent samples t-test for significant differences in factor means and variance across evaluation source (Source: Authors).

		Levene's test for equality of variances		t-test for equality of means						
		F	Sig.	t	df	Sig. (2-tailed)	Mean difference	Std error difference	95% confidence interval of the difference	
									Lower	Upper
Non-verbal competence score	Equal variances assumed	7.174	**.008**	−1.397	184	.164	−1.301	.931	−3.138	.536
	Equal variances not assumed			−1.397	176.033	.164	−1.301	.931	−3.139	.536
Verbal clarity score	Equal variances assumed	5.224	**.023**	−5.624	184	**.000**	−2.484	.442	−3.355	−1.612
	Equal variances not assumed			−5.624	183.307	**.000**	−2.484	.442	−3.355	−1.612
Verbal richness score	Equal variances assumed	.290	.591	−2.050	184	**.042**	−.581	.283	−1.139	−.022
	Equal variances not assumed			−2.050	183.449	**.042**	−.581	.283	−1.139	−.022

The third stage of our analysis involved an analysis of Non-Verbal Competence Score (NVCS) as a predictor of Verbal Clarity Score (VCS) and Verbal Richness Score (VRS) moderated by the source of assessment (AT). For this step, we used Model 1 as a method developed for SPSS [42], and find that the correlation between NVCS and VCS is different across the two types of assessment (see Fig. 4). In similar way, also suggests that the correlation between VRS and VCS is different across the two types of assessment. However, we find no significant difference in correlation between NVCS and VRS across the two types of assessment.

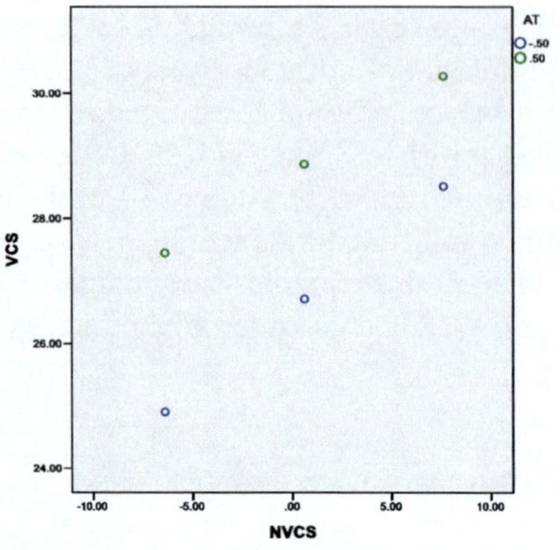

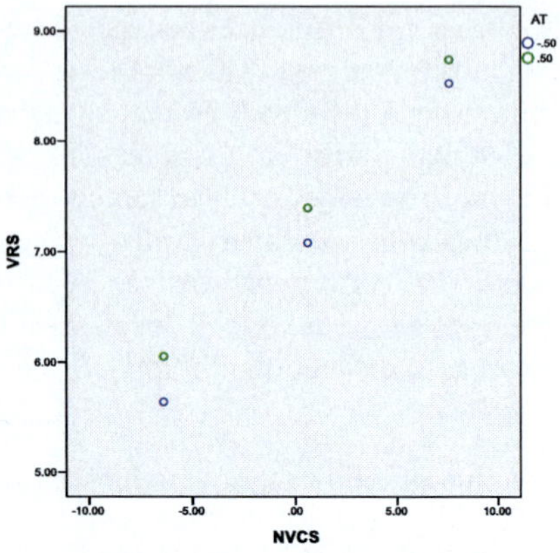

NVCS as predictor of VCS

NVCS as predictor of VRS

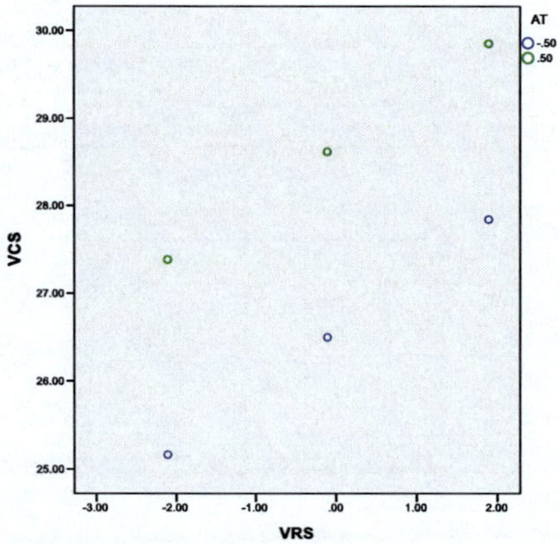

VRS as predictor of VCS

Note:
AT = -0.5 is PhD student assessment
AT = 0.5 is expert assessment

Fig. 4. Moderator analysis of NVCS, VCS and VRS as cross-predictors across evaluation source (Source: Authors).

6 Summary and Conclusions

Our findings suggest important implications on two areas. From a conceptual perspective, we find that our instrument that measures verbal and non-verbal competencies offers a reliable and consistent approach in measuring academic teacher performance in teaching. We observe that although student appraisers tend to be more generous in their judgement over their expert appraiser counterparts, assessments tend to be consistent across a diversity of competency levels. From a methodological perspective, we note that the proposed instrument has an opportunity to be used across universities in measuring the performance of academics in class – whether novice or expert, helping simplify existing student appraisal approaches while encouraging improved student response. Lack of student feedback on academics' classroom approaches is a frequent problem in many universities, and we hope that this instrument may help, at least in part, lead to better human resource performance management. Equally, our proposed instrument may track individuals' competency development during training – an especially important element as universities engage doctoral candidates in classroom work intent on helping candidates prepare themselves for future classroom roles.

Our proposed method is however still novel and warrants additional research. We acknowledge that this study relies on a rather small sample and consider this as a mere pilot approach that should be the starting point for further research. For instance, one important dimension that this study does not consider is culture, particularly context and explicitness [43, 44]. We recognize that context/explicitness has an important influence on styles of communication and associated effectiveness in delivering a message. Would our proposed method be equally applicable in countries and audience with different levels of context?

References

1. Okoli, A.C., Okpara, M.: Relating communication competence to teaching effectiveness: implication for teacher education. J. Educ. **8**(3), 150–154 (2017)
2. Papieska, J.: Children's and teachers' social competence in the context of teacher education, pp. 1–11. University of Leuven, Belgium (2012)
3. Kasikova, H., Valenta, J.: Teacher educational and social skills curriculum. Rocznik Lubuski **42**(2A), 289–299 (2016)
4. Blaškováa, M., Blaškoa, R., Kucharþíkováa, A.: Competences and competence model of university teachers. Procedia Soc. Behav. Sci. **159**, 457–467 (2014)
5. Zlatić, L., Bjekic, D., Marinkovic, S., Bojovic, M.: Development of teacher communication competence. Procedia Soc. Behav. Sci. **116**, 606–610 (2014)
6. Szafrański, M., Goliński, M., Graczyk-Kucharska, M., Spychała, M.: Cooperation of education and enterprises in improving professional competences-analysis of needs. In: Hamrol, A., Grabowska, M., Maletic, D., Woll, R. (eds.) Manufacturing 2019. LNME, pp. 155–168. Springer, Cham (2019). https://doi.org/10.1007/978-3-030-17269-5_11
7. Wilton, N.: An Introduction to Human Resource Management. SAGE Publications Ltd., Thousand Oaks, California, London (2016)
8. Statut Politechniki Poznańskiej uchwalony przez Senat Akademicki Politechniki Poznańskiej Uchwała Nr 175/2016-2020z dnia 10 lipca roku (2019). https://bip.put.poznan.pl/sites/default/files/attachments/uchwala_nr_175-_2019_-_zalacznik_statut_0.pdf. Accessed 29 Jan 2020
9. Pocztowski, A.: Zarządzanie zasobami ludzkimi. PWE, Warszawa (2007)
10. Armstrong, M.: Zarządzanie zasobami ludzkimi. Wolters Kluwer, Warszawa (2011)
11. Selvi, K.: Teachers' competencies. Cultura. Int. J. Philos. Cult. Axiol. **7**(1), 168–172 (2010)
12. Bhargava, A., Pathy, M.: Perception of student teachers about teaching competencies. Am. Int. J. Contemp. Res. **1**(1), 77–81 (2011)
13. Bakic-Tomic, L., Dvorski, J., Kirinic, A.: Elements of teacher communication comeptence: an examination of skills and knowledge to communicate. Int. J. Res. Educ. Sci. **1**(2), 157–166 (2015)

14. Greculescu, A., Todorescu, L., Popescu-Mitroi, M.M., Fekete, A.C.: Oral communication competence and higher technical engineering. Procedia Soc. Behav. Sci. **128**, 169–174 (2014)
15. Čižmešija, A., Diković, M., Domović, V., et al.: Handbook for teaching competence enhancement in higher education, Ministry of Science and Education, Zagreb (2018). https://mzo.gov.hr/UserDocsImages/dokumenti/Obrazovanje/VisokoObrazovanje/RazvojVisokogObrazovanja/EMPHASIS%20ON%20DEVELOPING%20AND%20UPGRADING%20OF%20COMPETENCES%20FOR%20TEACHING%20(EDUCA-T).pdf
16. Fierro, C., Brent, A.: Principles of Public Speaking. Lumen Learning USA. https://s3.amazonaws.com/lumenlearning/success/Master+PDFs/Principles-of-Public-Speaking_1-2-17.pdf
17. Wahyuni, A.: The power of verbal and nonverbal communication in learning. In: 1st International Conference on Intellectuals' Global Responsibility (ICIGR 2017) pp. 80–83. Atlantis Press, Netherlands (2018)
18. Eunson, B.: Communicating in the 21st century, 4th edn. John Wiley and Sons Australia Ltd., Milton (2016)
19. Mandal, F.: Nonverbal communication in humans. J. Hum. Behav. Soc. Environ. **24**, 417–421 (2014)
20. Bambaeeroo, F., Shokrpour, N.: The impact of the teachers' non-verbal communication on success in teaching. J. Adv. Med. Educ. Prof. **5**(2), 51–59 (2017)
21. Campion, M.A., Fink, A.A., Ruggeberg, B.J., Carr, L., Phillips, G.M., Odman, R.B.: Doing competencies well: best practices in competency modeling. Pers. Psychol. **64**, 225–262 (2011)
22. Miranda, S., Orciuoli, F., Loia, V., Sampson, D.: An ontology-based model for competence management. Data Knowl. Eng. **107**, 51–66 (2017)
23. Różewski, P., Małachowski, B.: Approach to competence modelling for enterprise knowledge management. In: Proceedings of the 14th IFAC Symposium on Information Control Problems in Manufacturing, Romania, pp. 1159–1164 (2012)
24. Sanchez, R.: Understanding competence-based management: identifying and managing five modes of competence. J. Bus. Res. **57**(5), 518–532 (2004)
25. Sampson D., Fytros, D.: Competence models in technology-enhanced competence-based learning. In: Adelsberger, H.H., Kinshuk, Pawlowski J.M., Sampson, D. (eds.) Handbook on Information Technologies for Education and Training. INFOSYS, pp. 155–177 Springer-Verlag, Heidelberg (2008). https://doi.org/10.1007/978-3-540-74155-8_9
26. Lindvall, M., Rus, I., Sinha, S.S.: Technology support for knowledge management. In: Henninger, S., Maurer, F. (eds.) LSO 2002. LNCS, vol. 2640, pp. 94–103. Springer, Heidelberg (2003). https://doi.org/10.1007/978-3-540-40052-3_9
27. Filipowicz, G.: Zarządzanie kompetencjami zawodowymi. PWE, Warszawa (2004)
28. Hong, J., Ståhle, P.: The coevolution of knowledge and competence management. Int. J. Manag. Concepts Philos. **1**(2), 129–145 (2005)
29. Hustad, E., Munkvold, B.E.: IT-supported competence management: a case study at Ericsson. Inf. Syst. Manag. **22**(2), 78–88 (2005)
30. Nordhaug, O.: Human Capital in Organizations: Competence, Training, and Learning. Scandinavian University Press, Oslo (1993)
31. Sadłowska-Wrzesińska J., Gabryelewicz I., Krupa P.: The use of IT tools for the analysis and evaluation of psychomotor efficiency of employees. In: MATEC Web of Conferences, vol. 94, p. 06017 (2017)
32. Spychała, M., Szafrański, M., Graczyk-Kucharska, M., Goliński, M.: The method of designing reference models of workstations. In: Marimon, F., Mas-Machuca, M. Berbegal-Mirabent, J., Bastida, R. (eds.) Proceedings of the 18th European Conference on Knowledge Management ECKM 2017, pp. 930–939. Academic Conferences and Publishing International Limited, Barcelona (2017)

33. Spychała, M., Goliński, M., Szafrański, M., Graczyk-Kucharska, M.: Competency models as modern tools in the recruitment process of employees. In: Sargiacomo, M. (ed.) Proceedings of the 10th European Conference on Intangibles and Intellectual Capital ECIIC 2019, pp. 282–291. Academic Conferences and Publishing International Limited, Chieti-Pescara, Italy (2019)
34. Chouhan, V.S., Srivastava, S.: Understanding competencies and competency modeling - A literature survey. IOSR J. Bus. Manag. **16**(1), 14–22 (2014)
35. Graczyk-Kucharska, M., Szafranski, M., Golinski, M., Spychala, M., Borsekova, K.: Model of competency management in the network of production enterprises in Industry 4.0—Assumptions. In: Hamrol, A., Ciszak, O., Legutko, S., Jurczyk, M. (eds.) Advances in Manufacturing. LNME, pp. 195–204. Springer, Cham (2018). https://doi.org/10.1007/978-3-319-68619-6_19
36. Dickson, D., Hargie, O.: Skilled Interpersonal Communication: Research, Theory and Practice. Routledge, London (2004)
37. Smółka, P.: Kompetencje społeczne. Metody pomiaru i doskonalenia umiejętności interpersonalnych. Oficyna Wolters Kluwer Business, Kraków (2008)
38. Joanna, M., Marek, G., Władysław, M.: The concept of the qualitology and grey system theory application in marketing information quality cognition and assessment. CEJOR **28**(2), 817–840 (2019). https://doi.org/10.1007/s10100-019-00635-y
39. Graczyk-Kucharska, M., Özmen, A., Szafrański, M., Weber, G.W., Goliński, M., Spychała, M.: Knowledge accelerator by transversal competences and multivariate adaptive regression splines. CEJOR **28**(2), 645–669 (2019). https://doi.org/10.1007/s10100-019-00636-x
40. Szafrański, M.: Acceleration of educating as an external factor supporting preventive and improving actions in businesses. In: Ahram, T., Karwowski, W., Schmorrow, D. (eds.) Procedia Manufacturing; 6th International Conference on Applied Human Factors and Ergonomics and the Affiliated Conferences. AHFE 2015, vol. 3, pp. 4948–4955. Elsevier BV. Netherlands (2015)
41. Zorn, T.E., Taylor, J.R.: Knowledge management and/as organizational communication. In: Tourish, D., Hargie, O. (eds.) Key Issues in Organizational Communication. Routledge, London (2003)
42. Hayes, A.F.: Introduction to Mediation, Moderation, and Conditional Process Analysis: A Regression-Based Approach. The Guilford Press, New York (2013)
43. Hall, E.T.: Beyond Culture. Anchor Books/Doubleday, New York (1976)
44. Hall, E.T.: Context and meaning. In: Samovar, L.A., Porter, R.E. (eds.) Intercultural Communication: A reader, vol. 9. Wadsworth, Belmont (2000)

Check for updates

Product Design and Development Strategies: Construction of Regional Innovation Systems, for New Achievements Brought by the Academy for Industry

M. J. Félix[1], G. Santos[1], R. Simoes[1,2], and C. Surrador[1](✉)

[1] Design School, Polytechnic Institute of Cávado and Ave, Barcelos, Portugal
[2] Institute for Polymers and Composites (IPC), University of Minho, Guimarães, Portugal

Abstract. In reference to the context of the theme, it is important to refer that the labour market is becoming more demanding, being up to the educational institutions to prepare their students, who in many cases develop projects without appreciating the industry's true difficulties and challenges.

The main objective of this present work will be to demonstrate that it is possible to build regional innovation systems, through the product design development of strategies that bring academy and industry closer together. Starting by preparing future design professionals with specific competences, trained also in knowledge in practice and experience-based learning.

Thus, in the methodology, a theoretical argumentation will be performed, supported on bibliographic analysis, where the construction of regional innovation systems will be studied, through the product design and development strategies, the reality of R&D activities in Portugal, and the profile of ceramic industry companies.

The result will be the development and application of a teaching model based on product design and development strategies, supported by experience-based-learning applied to the students of the Polytechnic Institute of Cávado and Ave (in Barcelos, Portugal) in partnership with a local ceramic industry.

In conclusion, it is expected that the project contributes to the construction of regional innovation systems, to new achievements brought by the academy for industry, which will culminate in product design and development strategies, performed in partnership with industry and accordingly to the expectations of all involved in this process.

Keywords: Regional innovation systems · Sustainability resources · *Experience-based* learning · Knowledge in practice · Ceramic industry

1 Introduction

The connection between academy and industry is essential in the training of professionals capable of adapting to several professional settings. With the growing technological

J. Machado et al. (Eds.): icieng 2021, LNME, pp. 174–184, 2022.
https://doi.org/10.1007/978-3-030-78170-5_16

innovation and globalization, several similar products emerge in the market, being fundamental to invest in the differentiation, which will allow consumers to distinguish between competing offers, being Design of crucial importance.

The role of institutions must be to prepare future professionals, with the objective to contribute for the development of the industry, which on the other hand, in many cases, industry is not prepared to understand the role of Design.

Preserving what a country has best, could be a way to face competition and the designer has a crucial role, thus it should take advantage of tangible value (properties, characteristics and material processing), and intangible values (cultural values, symbolic values) elaborating an analysis and an evaluation, that, consequently, could be applied in the context of the product.

During this research, it will be highlighted the way the bridge between the academy and industry could be done in a way to improve their students' training, based on experience-based learning [1], and direct contact with the realities of the professional world, being the ceramics the approached area because it is one of the oldest transformed materials, and a strong national resource, and because regional innovation systems [2] is a priority.

For Portugal, the ceramic industry is an activity with centuries of history, since the Arabic influences of the occupation period, going through the contacts with the east, till the consolidation and affirmation of their own productive system that nowadays exports all over the world. With the emergence of the theme, it also appears the cooperation with a faience production company, which was available to cooperate in this project, reinforcing the idea that this union is good for the students and for the company in a reinforcement of knowledge in practice.

This connection between design and regional innovation systems [3] is vital for the preservation and revitalization of obsolete techniques and knowledge, as well as for the product design and development strategies [4, 5], filled with memories and culture, but that fulfill the needs of the today's society.

In teaching design, the best way to stimulate systemic thinking is to expose students to industry, to the realities of production processes, and it is crucial to make a leap forward in how this teaching is carried out and what methodologies are addressed.

In conclusion our intention is to comprehend if it is possible to build regional innovation systems, through the product design and development of strategies that bring academy and industry closer together.

Starting by preparing future design professionals with specific competences, trained also in knowledge in practice and experience-based learning [6].

2 Literature Review

As it is acknowledged by a study supported by APICER[1], Portugal 2020[2], Compete 2020[3] and by the European Regional Development Fund [7], in terms of strategic positioning, each country is very dependent on factors like the perceived quality, the brand awareness, the industry identity, which may be more traditional or more innovative, the price of production factors and their own geography.

Regarding Portugal, the ceramic industry has a dynamic position, given that depending on the sub-sector, its competitive advantage lays more on the cost or on the differentiation. It is equally shown that the development of high quality products and value, added, as well as the concern with sustainability resources issues, has influenced the ceramics growth [8], and has helped in the recognition of the Portuguese image.

As the Portuguese ceramic industry differentiated itself from the international markets through the price factor, and competition ended up by responding to that strategy with low-cost prices, it became essential to change the strategic plan, investing this way in the differentiation of products, and in adapting them to new consumption patterns.

Being one of the strategic options, to a productive process level, the investment in research and development (R&D), which allows to develop new products and new and more efficient manufacturing ways, making possible the reduction of production costs without increasing the differentiation of offer to the final client.

One of the ways to reach that maturity in management, development and innovation could be the introduction and proper management of design skills in the business environment [9].

In 2013, Portugal spent on R&D, the equivalent to 1.36% of gross domestic product (GDP), less than the European Union (UE) average. One of the objectives of the Europe 2020 Strategy is to increase investment (public and private) in R&D and innovation to 3% of UE, GDP [10]. Ageing and climate will also have an impact on R&D agendas, future Research and Development (R&D) [11, 12] agendas will be strongly influenced by societies ageing, climate change, health challenges and the growth of everything that's digital, the Organization for Economic Cooperation and Development (OECD) considers [13].

According to APICER [7], the country has designed ceramics worldwide, recognizing the importance of commercial value and the value of sustainability, developing projects of greater durability and suitability in terms of use.

With the advance of technologies and the demands of new markets, it is, therefore increasingly desirable that the academic training of designers should no longer be essential- theoretical and should integrate practical components in various areas, promoting the participation of students in the business processes of product design and development strategies, through protocols that may be developed and that aim to enrich both parties.

[1] APICER—Portuguese Ceramics Association.

[2] Portugal 2020- Partnership Agreement adopted by Portugal and the European Commission in which are defined the principles of programming that establish the economic, social and territorial development policy to promote Portugal between 2014 and 2020.

[3] Compete 2020 - Managing Authority of the Competitivity and Internationalization Operational Program-Structure in the State's central Administration.

Contributing to the training of future design professionals with specific skills, also trained in knowledge in practice [14–16] and experience-based learning [17–19], may be the answer to the new challenges expected from the industry.

There are three strategies to achieve competitive advantage in companies: low cost, differentiation and focus [20]. Within innovation the main elements are products, materials, services, and organizational structures or processes, and companies generally begin this search for innovation when changes in competition, demand and technologies appear.

The percentage of active population in the R&D sector is slightly lower than the European average, with universities and public institutes standing out for their importance.

A policy of innovation and technology diffusion at regional level [21] has to be inherently horizontal, involving regional, industrial and economic policy, education and vocational training and science and technology policy.

New product design and development strategies will allow collaboration with regional innovation systems, finding ways to keep products from losing their character and origin, but also allowing the creation of contemporary products within the regional tradition, as well as to consider sustainable resources [22] and business models for enterprises that foresee the application of the circular economy [23] model.

In the Barcelos region, ceramic production is very high, the economic and social context of this municipality is closely linked to this production, due to the existence of a large community of potters. The production of products is a way of transmitting the identity and recognition of the region.

The ceramics sector is becoming more and more distinguished internationally, with the growing investment in design and innovation. The Portuguese industry in this sector has increased the quality of its products and invested in differentiation and competitive prices. Being characterized by tradition and handcrafted manufacturing, along with technological evolution, this industry is in 14th place in the world export ranking, employing around 16,704 workers in a total of 175 companies [24].

3 Methodology

The methodology used has evolved with the exploration of the problem, being the main philosophical influence in the work, the belief that all knowledge is socially constructed [25]. This study is exploratory, which according to Collis & Hussey [26], points to a conducted study when there are few researches on the theme, providing a bigger familiarity with the subject in analysis turning it more explicit for later research.

As far as the process is concerned, it will be used a mixed approach, being a methodology through which the investigator collects, analyses and mixes (integrates or relates) qualitative and quantitative data in one study or in many phases of the same research program Johnson, R., Onwuegbuzie A. & Turner L. [27], with the intention to produce/create data based on perspectives of different analysis. So, this study will focus on three distinctive approaches, being that in a first phase, the focus will be on the contextualization of the theme, which will englobe the collection of dissertations, thesis, scientific articles, books and websites, and where it will be presented a brief description of the essential aspects, so that then an interpretive reading of the analyzed bibliography can be performed, as a way of research of the existent relation between the authors' ideas and the researches problems outlined for this work.

In a second phase, data collection and analysis techniques will be used to understand a pattern that can be applied in different contexts.

To achieve the objective of this work, it will be necessary to analyze a varied universe of differentiated research in a type of companies, those that have shown interest in introducing innovation through their products and selecting one. Thus, it will be the categories of content analysis that were considered relevant in this research work, which should be investigated in each statement, later forming the argument related to the construction of regional innovation systems.

The practical part of the project relies yet on an analysis of practical cases of objects and products development in the area ceramic products, exposing and analyzing in detail one specific study case of a new product developed. The study case in analysis, apart from the collection and bibliographical review, through which it studied and systematized about the role of Design and its importance in the development of new products as the focus on industrial production, also considered the Design-Based-Research (DBR) research methodology based on practice, which allows a review and a perception of case studies through more democratic research practices and recognizing technology as a system beyond its tools [28], researchers can increase their impact on educational practice.

4 Results

Once the need has been identified and starting with the need of preparing design professionals with specific skills, also trained in knowledge in practical and experience-based learning, a teaching model supported by the objective of bringing academy and industry closer together will be analyzed, demonstrating that it is possible to contribute to the implementation of regional innovation systems through product design and development strategies. The area addressed is the industry of ceramics because it is a strong national resource and because regional innovation systems is a priority.

A teaching model based on experience-based learning was developed in the area of product design, which was carried out in the Design School of the Polytechnic Institute of Cávado and Ave, during the Designer's Work Methodology Curricular Unit in 2018, a project that consisted in working with a ceramic industry in the region. The work had as its title, Science Fiction, technological advances, sustainability and design.

For the study of this teaching model, the briefing, the performed working methods, the executed development process, and the final results were analyzed. The project briefing consisted in the identification of a ceramic industry in the Barcelos region, north of Portugal. A proposal was presented to the students, which after a careful analysis of the company, the type of products produced, the manufacturing processes, the customers, among other factors, stimulated new solutions that fitted in the regional innovation systems. This solution could be a product within the area of recycled or recyclable materials.

The project should reflect creative and innovative responses to all consumers' demands and needs that can be marketed and produced industrially, demonstrating the connection between design and project planning and development, and the union of areas such as technology, engineering and materials.

Throughout the project development, six crucial phases take place, the first three being more theoretical, and the following ones presenting a passage to the practical part, both described and presented in a document.

Once the theoretical phase is over and already with the collaboration of the company, it is initiated the process of plaster molds [29] in order to be filled with ceramic slip, mixture of clay and water used for moulding the products (Fig. 1).

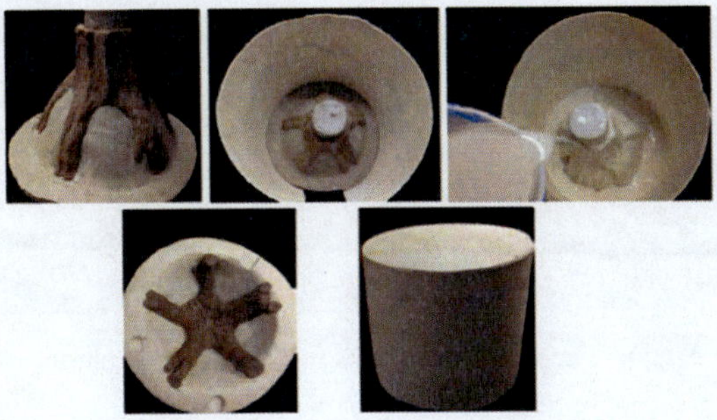

Fig. 1. Process of plaster moulds.

The obtained products result well in a formal way, presenting a white and opaque shade (Fig. 2), however, they do not pass to the final phase of finishing.

Fig. 2. Result of molded ceramic products

The analysis of this teaching model matched the needs of the company and there is already a demand from other companies to implement the same strategy. In total 43 new innovative proposals were developed by the students, which allowed the company to think through a partnership with the institution, present new collections and open the way to international fairs.

4.1 Case Study – Product Design and Development Strategies

Once the teaching model was presented, there was the decision to test it in a totally industrial environment, in a more technical proposal that aims to innovate more than just for the shape, but also technically.

The ceramic industry that hosted the project was the same one that supports the testing of the teaching model. For the production of the products in faience the company contemplates several processes, the paste arrives at the factory in the shape of discs, is transformed, by mixing it with silicate and water, in proper tanks (Fig. 3), controlling the density and weight.

The plaster moulds are made, in most cases two pieces are used, then moving to the creation of a three-dimensional model in clay, a first mould is produced covering completely half of the object and creating fittings (called male-female).

Fig. 3. The initial processes for mould preparation.

After a few minutes of drying (about 30 min), the box is removed and the used material is completely removed, so that no traces remain that could alter the shapes, leaving the other half uncovered, and then the previous procedure is repeated for the development of the second part of the mould.

Secondly, for the production of the product, the different parts of the mould are re-attached using tight rubber straps, and then the liquid faience paste is poured through a drying period of about one hour and a half.

In the end, the excess liquid paste is poured and then the mould is opened before the piece is completely dry, so that some finishes can be done, such as cutting excess material, or making details, such as gluing or holes. After a complete drying period, the pieces undergo the first firing of ceramics, which lasts for about seven hours, at a temperature of approximately 1100 °C. After that, the pieces are glazed and only then they pass to the final firing which takes less than one hour at a temperature of approximately 1070 °C, which serves as a bond between the glass and the raw material.

In the analysis of the opportunities for improvement, it is sought to understand what type of product can be developed, so that it brings advantages for the company, and therefore an analysis of the competing market is made. This way, it is possible to reach some conclusions, particularly that the production of parts often brings costs that are not overcome by the products' trade, namely, costs of materials such as faience pulp and plaster, electricity from the oven, among others.

How to innovate is always the question, and in this partnership, it was presented the possible use of a finish that gives insect repellent properties to the product. Thus, besides the project development, it was also indispensable to know what the market offers in terms of repellent products, because it is essential to know what is already being marketed, and how a new production can stand out from the others to face the competition.

The repellent the company had access to, consists of IR 3535, it was developed by the Smart Innovation Nanotechnology company located in Barcelos [30] and it is dedicated to research and development, working with various industries and using nanotechnology. This repellent called Repel Mosquito is biocompatible and keeps mosquitoes away without killing them. It can be applied to various materials such as paints, varnishes, tents, school, military and police clothes, home textiles, wipes, among other examples. It has a four-year duration, and it can keep insects away up to a radius of forty centimeters.

It can also prevent diseases caused by mosquito bites such as malaria and yellow fever [31]. In the present case, this product is applied to the pieces as a varnish. The repellent substance is in the liquid state and is previously mixed in this solution.

The mixture is made only at the time of applying the varnish or paint. This substance has a characteristic odor and transparent appearance. The company has also performed some repellency tests, which prove the effects, being its basis confidential.

4.2 Ceramic Products with Insect Repellent Properties

After the analysis of several concepts, some ideas to combine functionalities and forms of each one are established, in order to reach a more promising idea, analyzing some details such as fittings and forms.

This product consists of two main parts, the ceramic vase and a metal structure, the third part, is optional. It can be used with the vase and will be used in accordance to the user's choice.

The main vase can be used in conjunction with the metal frame with the optional part, or alone, with all parts fitting together. The metal structure can be used to guide plants, such as vines, helping their grow, and can be used repeatedly on itself. The set is a versatile product (Fig. 4), which the user can assemble according to his needs and preferences.

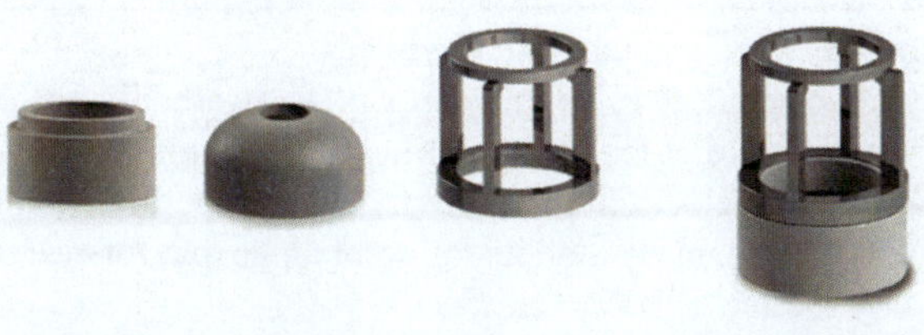

Fig. 4. Ceramic product set with insect repellent properties.

The prototype of the ceramic (Fig. 5) products was developed according to the previously analyzed process. In the end the piece is painted and only then the varnish with repellent properties is applied. For the application, 50 ml of the repellent substance for each liter of varnish, is introduced. It is mixed and ready for use as soon as a homogeneous result is obtained, with two applications.

After the production of the parts, a small test was done with the part where the housing with the repellent substance was applied, as a way to check if the repellent effect really works or not. For the test, it was used a faience vase with repellent varnish;

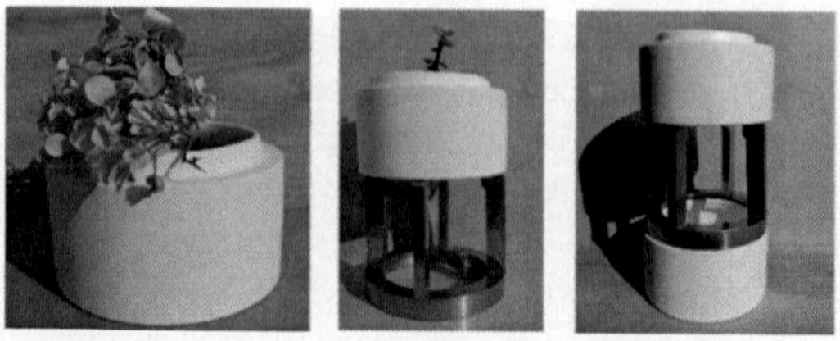

Fig. 5. Ceramic product set with insect repellent properties, prototype.

a white sheet with measurement mark; a transparent plastic box to retain the insects (flies; bees; flying insects, ants). The test was conducted in a transparent container in order to observe the behaviour of the insects.

A ceramic piece is placed under a white sheet where the repellent finish has been previously applied and marks with measurements of 10 in 10 cm are outlined. Along each mark an item of food (fruit and jam) was placed. The test was performed at room temperature of 24° and lasted 20 min. After an analysis during the test process, it was observed that the insects didn't approach the surface with repellent. Most of the insects remained on the walls of the plastic box between 10 and 20 cm from the vessel with repellent. No insect mortality rate was verified during the test.

5 Conclusion

This research arises from the need for cooperation between academy and industry, enhancing bridges that benefit the students' learning, who need to be prepared to face possible future scenarios in the labour market.

Starting this process, through a collection of information focused on the ceramic industry, product design and development and the need for regional innovation systems in Portugal, it was possible to apply a model based on experience, as a way to promote a closer relationship between academy and industry.

It is necessary to continue to raise the companies' awareness in the ceramic sector, for the adoption of practices that lead to a qualification of their social and environmental performance, and an improvement in productivity and competitivity, but also to consider sustainable resources and business models for companies that foresee the application of the circular economy model.

From the partnership between the ceramic company *Irmãos Ferreira*, from Barcelos in the north of Portugal, it was possible to demonstrate the application of the teaching model brought by the academy to the process of project development and get to know the results obtained. Therefore, we concluded that there is still a lack in the academy concerning the practical part, in the way that the students developed the products within the university, not having all the necessary means for the execution of the products.

The teaching model that empowers the execution of projects in partnership with companies is fruitful for both parts. This model is characterized by establishing some fundamental stages, such as getting to know the company, its objectives and vision, and production methods, the competition it faces in the market, customers and their needs, among others.

In this particular case, it was proposed to the company to innovate and to distance itself from the competition by applying a finish to their ceramic products in which a substance with an insect repellent effect is incorporated.

Upon this results the possibility of creating a product that improves people's quality of life has arised. The impregnated repellent is called Mosquito Repell, and is characterized for being biocompatible, preventing diseases caused by mosquito bites that are highly deadly, and is composed of IR 3535, keeping mosquitoes away without eliminating them, which is good for the normal functioning of ecosystems. This finish has a forty-centimeters radius of action, and a four-year duration.

One of the goals of this research is the development of a design project in partnership with a company, in order to explore, analyze and develop teaching models based on experience, and that aims for regional innovation through design strategies.

This way, it is possible to say that this purpose was achieved. On the side of the industry, it was verified the knowledge of new strategic approaches for the innovation and creation of value to the products, as well as the valorization by the competences of product design.

By the end of this development, the benefits of this cooperation between the academy and the industry are clear, for both.

Acknowledgments. Portuguese Foundation for Science and Technology (FCT).

References

1. Cameron, B.H.: Experience-based learning. In: Encyclopedia of Information Technology Curriculum Integration (2011)
2. Cooke, P.: Regional innovation systems, clusters, and the knowledge economy. Ind. Corp. Chang. **10**(4), 945–974 (2001). https://doi.org/10.1093/icc/10.4.945
3. Asheim, B.T., Gertler, M.S.: The geography of innovation: regional innovation systems. In: The Oxford Handbook of Innovation (2009)
4. Richey, R.C., Klein, J.D.: Design and Development Research Methods, Strategies and Issues (2007)
5. Cross, N.: Expertise in design: an overview. In: Design Studies (2004)
6. Andresen, L., Boud, D., Cohen, R.: Experience-based learning: contemporary issues. In: Understanding Adult Education Training, 2nd edn., pp. 225–239 (2000)
7. PwC: Capacitação da indústria da Cerâmica Portuguesa Um cluster, uma estratégia, mercados prioritários (2016)
8. Félix, M.J., Duarte, V.: Design and development of a sustainable lunch box, which aims to contribute to a better quality of life. Int. J. Qual. Res. **12**, 869–884 (2018). https://doi.org/10.18421/IJQR12.04-06
9. Best, K., Kootstra, G., Murphy, D.: Design management and business in Europe: a closer look. Des. Manag. Rev. **21**, 26–35 (2010)
10. Cappelletti, M.: Investimento das empresas da UE em investigação e desenvolvimento ficou abaixo da média mundial em 2013. https://observador.pt/2014/12/04/investimento-das-empresas-da-ue-em-investigacao-e-desenvolvimento-ficou-abaixo-da-media-mundial-em-2013/
11. Bremser, W.G., Barsky, N.P.: Utilizing the balanced scorecard for R&D performance measurement. R&D Manag. **34**, 229–238 (2004)

12. Hall, B., Hayashi, F.: Research and Development as an Investment. National Bureau of Economic Research (1989)
13. Garrido, N.: OCDE destaca impacto do envelhcimento e do clima nas agendas de I&D
14. Sanchez, R.: "Tacit Knowledge" versus "Explicit Knowledge" Approaches to Knowledge Management Practice. Department of Industrial Economics and Strategy, Copenhagen Business School (2004)
15. Dick, W., Carey, L., Carey, J.O.: The systematic design of instruction (2005)
16. Friedman, K.: Creating design knowledge: from research into practice. In: IDATER 2000 Conference (2000)
17. Andresen, L., Boud, D., Cohen, R.: Experience-based learning. Underst. Adult Educ. Train. **2**, 225–239 (2000)
18. Pruett, M., Thomas, H.: Experience-based learning in innovation and production. R&d Manag. **38**, 141–153 (2008)
19. Becker, J.J., Wits, W.: An experience-based approach to teaching product design. In: Bohemia, E., Eger, A., Eggink, W., Kovacevic, A., Parkinson, B., Wits, W. (eds.) Proceedings of the 16th International Conference on Engineering and Product Design Education, pp. 688–693. Design Society (2014)
20. Porter, M.E.: The Competitive Strategy: Techniques for Analyzing Industries and Competitors. Simon & Schuster, New York, United States (2004)
21. Doloreux, D., Parto, S.: Regional innovation systems: current discourse and unresolved issues. Technol. Soc. **27**, 133–153 (2005). https://doi.org/10.1016/j.techsoc.2005.01.002
22. Bhamra, T., Lofthouse, V.: Design for Sustainability: A Practical Approach. Routledge, Milton Park (2016)
23. Bocken, N.M.P., de Pauw, I., Bakker, C., van der Grinten, B.: Product design and business model strategies for a circular economy. J. Ind. Prod. Eng. **33**, 308–320 (2016). https://doi.org/10.1080/21681015.2016.1172124
24. Aicep Portugal Global: Exportações de cerâmica e cristalaria com forte crescimento. http://www.portugalglobal.pt/PT/RevistaPortugalglobal/2017/Documents/Portugalglobal_n98.pdf
25. Félix, M.J., Santos, G., Barroso, A., Silva, P.: The transformation of wasted space in urban vertical gardens with the contribution of design to improving the quality of life. Int. J. Qual. Res. **12**, 803–822 (2018). https://doi.org/10.18421/IJQR12.04-02
26. Collis, J., Hussey, R.: Business Research: A practical guide for undergraduate and postgraduate students (1942)
27. Johnson, R.B., Onwuegbuzie, A.J.: Toward a definition of mixed methods research. J. Mix. Methods Res. **1**(2), 112–133 (2007). https://doi.org/10.1177/1558689806298224
28. Amiel, T., Reeves, T.C.: Design-based research and educational technology: rethinking technology and the research agenda. Educ. Technol. Soc. **11**, 29–40 (2008)
29. Nagelberg, S.: Batch Manufacturing for Ceramics. Models and Molds, from Process to Product. Lulu.com (2014)
30. Smart Inovation: Repel Mosquito. http://www.smartinovation.com/index.php/produtos/produtos-repel-mosquito/
31. Mufamadi, S.: Si Repel Mosquito: the World's most effective and Non-Toxic Mosquito-Repellent using Smart Inovation Nanotechnology. http://www.nunnovation.com/2016/02/02/si-repel-mosquito-the-worlds-most-effective-and-non-toxic-mosquito-repellent-using-smart-inovation-nanotechnology/

An Integer Programming Approach to Sectorization with Compactness and Equilibrium Constraints

Valeria Romanciuc[1,2], Cristina Lopes[1(✉)], Aydin Teymourifar[3], Ana Maria Rodrigues[1,3], José Soeiro Ferreira[3,4], Cristina Oliveira[1], and Elif Göksu Öztürk[3]

[1] CEOS.PP/ISCAP/P.PORTO, Porto, Portugal
{cristinalopes,cteles}@iscap.ipp.pt
[2] Departamento de Produção e Sistemas, Universidade do Minho, Braga, Portugal
valeria@dps.uminho.pt
[3] INESC TEC, Porto, Portugal
{aydin.teymourifar,ana.m.rodrigues,jose.soeiro.ferreira,
elif.ozturk}@inesctec.pt
[4] Faculdade de Engenharia, Universidade do Porto, Porto, Portugal

Abstract. The process of sectorization aims at dividing a dataset into smaller sectors according to certain criteria, such as equilibrium and compactness. Sectorization problems appear in several different contexts, such as political districting, sales territory design, healthcare districting problems and waste collection, to name a few. Solution methods vary from application to application, either being exact, heuristics or a combination of both. In this paper, we propose two quadratic integer programming models to obtain a sectorization: one with compactness as the main criterion and equilibrium constraints, and the other considering equilibrium as the objective and compactness bounded in the constraints. These two models are also compared to ascertain the relationship between the criteria.

Keywords: Sectorization · Multi-criteria · Integer programming

1 Introduction

The problem of sectorization consists of dividing a set of points (which can represent clients, inhabitants, students, etc., depending on the specific context) into smaller subsets called *sectors*, respecting specific criteria. This procedure is often used to break down a large problem and divide it into smaller sub-problems [1], enabling a more organized and simplified approach to a complex scenario.

In sectorization, the goal is that the resulting sectors are as identical as possible concerning certain features, which vary with the applications. For instance, when designing sales territories for a company, balance in terms of the allocated workload (e.g. number of clients) between the salespersons should be assured.

J. Machado et al. (Eds.): icieng 2021, LNME, pp. 185–196, 2022.
https://doi.org/10.1007/978-3-030-78170-5_17

However, if the problem concerns the waste collection, the main feature to consider would be the (geographical) size of each sector in order to avoid unbalanced routes within each sector.

Despite the variety in criteria, there are more common features that appear in different contexts, such as *equilibrium* (or *balance*) and *compactness*.

Equilibrium is concerned with an equal division of the elements in the dataset among the sectors according to an activity measure (number of clients, workload, travel times, etc.). In political districting, a balanced sectorization would have approximately the same number of voters in each sector. In the case of sales territories, the regions of responsibility in which the market area is subdivided should be balanced in sales potentials [2]. In municipal waste collection, a similarity in the quantities of collected waste in each sector is desired [1]. Therefore, the balance of the number of points in each sector is the equilibrium criterion used in this work.

Compactness is related to the geographical regularity desired for the shapes of the sectors, meaning that sectors without pointy ends or irregular shapes are preferred, since a rounded shape will ensure the points inside it are as close as possible to each other [3]. However, since a rigorous definition of compactness does not exist [2], it is measured differently, according to the context of the problem and the solution method being applied. For instance, in sales territory design, compact sectors mean that no customer is too far away from the facility it is assigned to [4]. In this paper, we measure compactness using the distance between every two points in a sector.

Sectorization problems (SP) have been discussed widely in the existing literature, with their applications ranging from transportation mobility districting to the design of districts in healthcare [5], electrical supply [6], and foresting [7]. In commercial applications, the definition of the sectors has a great impact on the company costs and therefore needs to be optimized.

Solving these problems is a difficult task, and there is no one-size-fits-all method that can be applied in every context. On the contrary, the methods used are generally very dependent on the application, as the literature illustrates. Approaches mainly follow Location-Allocation, Facility Location or Set Partitioning [2]. The approaches described range from exact ones, using mathematical programming [8], to general and/or dedicated heuristics [5,9], or taking advantage of the strengths of both, known as hybrid methods [10].

With such vast and relevant applications, SP's study is of great interest to society and science. As such, we propose two models for sectorization where compactness and equilibrium are taken into account since these are the criteria that most SP strive to satisfy. This study aims to create sufficiently generalized models, which can be used in the most diverse scenarios. In the first model, compactness is in the objective function and equilibrium is controlled in the set of constraints, and in the second, the equilibrium criterion is considered for the objective function and compactness is bounded in the set of constraints. Furthermore, we solve both these models and evaluate the obtained results relative to the mentioned criteria.

The paper is organized as follows. Section 2 contains a brief literature review on sectorization applications. In Sect. 3, the used criteria are defined and the integer programming models are presented. Section 4 contains a discussion concerning computational tests for each model, and the comparison between the two models concerning the different criteria. Finally, Sect. 5 summarizes the conclusions of the presented results and provides possible directions for future work.

2 Relevant Literature

This section includes a selection of relevant publications regarding SP. The most cited publications in the surveyed literature on sectorization concern the Political Districting Problem [11] and Sales Territory Design [2,4,9], which can be used as benchmark to compare the trend in terms of extending the scope of sectorization problems to new applications (airspace, energy, harvesting) and methods.

Regarding policy applications, sectorization is of importance in designing political districts, assigning neighborhoods to schools, water supply and electrical power distribution [6].

Health management is also an important application field, where sectorization may be used either in locating health care services or in the transportation of patients, as well as for optimizing the service performance of health organizations. It can also be useful at a higher health management level: Farughi et al. [5] divide the health system of the South Khorasan Province in Iran into 10 health districts, so that each district can oversee the performance of its health organizations.

Commercial applications in logistics operations such as sales and services territory design usually integrate sectorization with vehicle routing problems [9]. Environmental problems that require sectorization concern the management of Municipal Waste Collection [12], street cleaning and maintenance operation zones, public transportation, forest management and harvesting.

Airspace sectorization is fundamental in an Air Traffic Control system [13]. Airspace is usually divided into several sectors, each assigned to a team of controllers that are responsible for the air traffic flow and capacity management of that sector.

Lin et al. refer that good districts with balanced resource allocation are the basis of good tactical and operational performance, improving delivery efficiency and effectiveness [8].

As for solution methods, in healthcare sectorization problems, the most used exact method is mixed-integer linear programming (MILP) and the vehicle routing problem is considered by many authors (VRP). The focus was mainly attributed to a wide range of criteria and other dimensions [8]. The main criteria follow balance and compactness, which dominate over the others (contiguity, boundaries, capacity, accessibility). Furthermore, the importance of patient admission, supplier selection and staffing were heavily studied [14].

The existing literature highlights the socioeconomic composition (neighborhood racial sorting or racial desegregation) and capacity as the two main problem characteristics in school districting problems. Bouzart [15] assigns students to schools' socioeconomic compositions, while considering the total travel distance, in a MILP model applied to data from Greenville, South Carolina. In Caro et al. [10], an optimization model together with the geographic information system is used to solve the school districting problem using real-life examples from the city of Philadelphia.

3 Sectorization Models

This section is concerned with the multi-criteria SP where compactness and equilibrium are taken into account. In order to understand the relation between these two criteria, we propose two quadratic integer programming models (inspired by [2]), where, alternately, one of the two criteria is used in the objective function and the other in the constraints.

In most sectorization problems, the concept of "sector center" is used, which represents the point of origin and/or distribution of activities to the remaining points in the sector. For instance, in problems involving distribution centers, there are fixed locations out of which the centers must be selected. Depending on the approach, these points can either be calculated through a method of choice or be an input of the model. In the scope of this paper, we consider the centers to be parameters that are known beforehand, as is the case with most applications.

Let us then start by considering Model A, where compactness is used for the objective function.

The following parameters are used in this model: a set S of n points, which are to be assigned to a predefined number of k sectors; d_{iw}, the Euclidean distance between every two points i and w in S; $Q = \lfloor \frac{n}{k} \rfloor$, a target number of points per sector; c_j, the center of sector j; and $0 \leq \tau_e \leq 1$, the tolerance for the equilibrium criteria. For some combinations of n and k, it was necessary to choose values of τ_e in this interval that ensure feasibility of the model.

The decision variables are defined as:

$$x_{ij} = \begin{cases} 1, & \text{if point } i \text{ is assigned to sector } j \\ 0, & \text{otherwise} \end{cases} \qquad i = 1, \ldots, n,\ j = 1, \ldots, k.$$

The definition adopted for compactness is that of [2], measuring the total distance of the points to the center of the sector they are assigned to:

$$\sum_{j=1}^{k} \sum_{i=1}^{n} d_{ic_j} x_{ij}. \tag{1}$$

The balance between sectors is determined by comparing the sector size with the target number of points per sector Q. Thus, Model A is defined as:

$$\text{Min} \sum_{j=1}^{k}\sum_{i=1}^{n} d_{ic_j} x_{ij} \tag{2}$$

$$\text{Subject to} \sum_{j=1}^{k} x_{ij} = 1, \quad \forall i = 1, \ldots, n \tag{3}$$

$$\sum_{i=1}^{n} x_{ij} \geq 1, \quad \forall j = 1, \ldots, k \tag{4}$$

$$Q(1-\tau_e) \leq \sum_{i=1}^{n} x_{ij} \leq Q(1+\tau_e), \quad \forall j = 1, \ldots, k \tag{5}$$

$$x_{ij} \in \{0,1\}, \quad \forall i = 1, \ldots, n,\ j = 1, \ldots, k. \tag{6}$$

The objective function (2) minimizes compactness. Constraints (3) ensure that each point must be assigned to one and only one sector. Constraints (4) ensure that each sector must have at least one point assigned to it. Constraints (5) limit the number of points in each sector (sector size) up to a given tolerance τ_e, guaranteeing that there is some equilibrium in the sectors' size. Constraints (6) define the domain of the decision variables.

Regarding Model B, the equilibrium function is defined according to [1], which measures the variance of the number of points in each sector:

$$\sum_{j=1}^{k} \frac{(q_j - \bar{q})^2}{k-1} \tag{7}$$

where $q_j = \sum_{i=1}^{n} x_{ij}$ are the quantities assigned to each sector j and $\bar{q} = \frac{\sum_{j=1}^{k} q_j}{k}$ is the average sector size, with $j = 1, \ldots, k$. The definition of compactness is maintained.

The decision variables remain the same. To the previous parameters, we also add $D_j = \sum_{i=1}^{n} d_{ic_j}$, the total distance from all points to the center of each sector $j = 1, \ldots, k$, and $0 \leq \tau_{com} \leq 1$, the tolerance for compactness. Model B is described below.

$$\text{Min} \sum_{j=1}^{k} \frac{(q_j - \bar{q})^2}{k-1} \tag{8}$$

$$\text{Subject to} \sum_{j=1}^{k} x_{ij} = 1, \quad \forall i = 1, \ldots, n \tag{9}$$

$$\sum_{i=1}^{n} x_{ij} \geq 1, \quad \forall j = 1, \ldots, k \tag{10}$$

$$\sum_{i=1}^{n} d_{ic_j} x_{ij} \leq \frac{D_j}{k}(1-\tau_{com}), \quad \forall j = 1, \ldots, k \tag{11}$$

$$x_{ij} \in \{0,1\}, \quad \forall i = 1, \ldots, n,\ j = 1, \ldots, k. \tag{12}$$

The objective function (8) addresses equilibrium by minimizing the variance of sector size. Constraints (11) limit the total distance within each sector up to a mean distance per sector with a given tolerance τ_{com}, so that there are no points in a sector that are too far from the center. Constraints (9), (10) and (12) are maintained from the previous model.

4 Computational Tests

Computational tests were conducted on 50 instances of different sizes (ranging from 30 to 1000 points) for 5, 10 and 15 sectors. The models in Sect. 3 were implemented in IBM ILOG CPLEX Optimization Studio 12.9 and all tests were performed on a machine with a Intel Core i5 CPU at 1.80 GHz and 8 GB RAM. Instances are generated randomly using the normal distribution following a two-step procedure to create unbiased and neutral data. Each instance consists of groups of nodes with different mean and variance. Moreover, instances contain the coordinates for each node and the adjacency matrix. The instances are available in [16].

In this section we will discuss the results obtained for two instances: one of size 438 and the other of size 1000. Although we only show the results for these two instances, the tests guarantee that they are representative of the behavior displayed by the remaining 48.

Table 1 and Fig. 1 display the results for an instance of 1000 points and 5 sectors. This table (and the remaining ones in this section) contains the compactness value and sector size (and the respective interval amplitude) for different values of the equilibrium tolerance τ_e, as well as the optimal value for compactness ("Comp*") when optimizing this criterion with no tolerance on the equilibrium. The figure visually represents the compactness function and the sector sizes' amplitude for a given equilibrium tolerance. The line for optimal compactness is used only as a visual reference for the compactness function, as it is independent of the equilibrium tolerance.

Table 1. Results of Model A for an instance of 1000 points and 5 sectors.

$n = 1000$, $k = 5$, Comp* $= 5222.27$, $Q = 200$			
τ_e	Compactness	Sector sizes	Amplitude
0.25	5222.27	[233 221 223 173 150]	83
0.2	5230.05	[233 221 213 173 160]	73
0.15	5245.72	[230 221 203 176 170]	60
0.1	5270.13	[220 220 197 183 180]	40
0.05	5312.12	[210 210 192 198 190]	20

We can observe a trade-off between the compactness function and equilibrium, since the higher the tolerance required for equilibrium (and, consequently, higher amplitude in sector size and less balanced distribution among the sectors), the better the value of compactness and vice-versa. Furthermore, a tolerance of 25% is enough for compactness to reach its optimal value of 5222.27. As for the equilibrium, for this tolerance, there is a big difference between the largest and smallest sectors, with 233 and 150 points, respectively (thus, an amplitude of 83). Table 2 and Fig. 2 show results for a smaller instance of 438 points and 5 sectors, where the same trade-off can be observed and similar conclusions are reached.

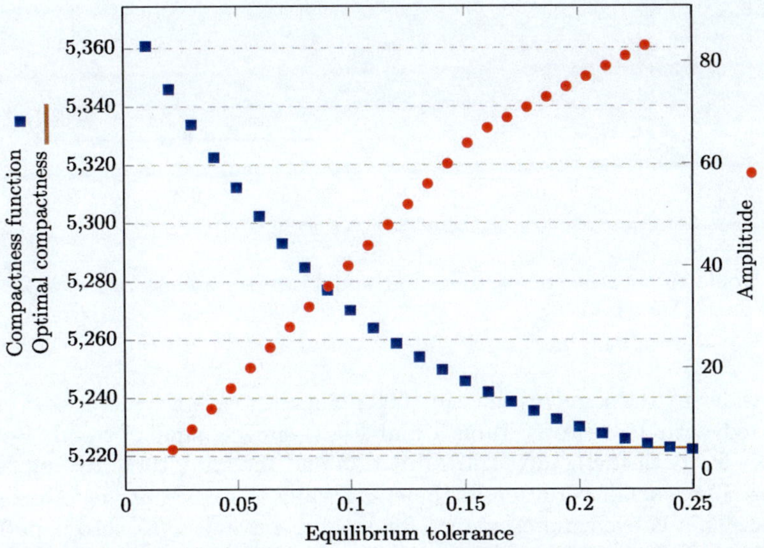

Fig. 1. Trade-off between compactness and equilibrium for an instance with 1000 points and 5 sectors (Model A).

Table 2. Results of Model A for an instance of 438 points and 15 sectors.

$n = 438$, $k = 5$, Comp* $= 8308.96$, $Q = 87$			
τ_e	Compactness	Sector sizes	Amplitude
0.25	8308.96	[69 98 69 98 104]	35
0.2	8309.44	[70 98 70 97 103]	33
0.15	8326.27	[74 99 74 91 100]	26
0.1	8383.31	[79 95 79 90 95]	16
0.05	8470.66	[83 91 83 90 91]	8

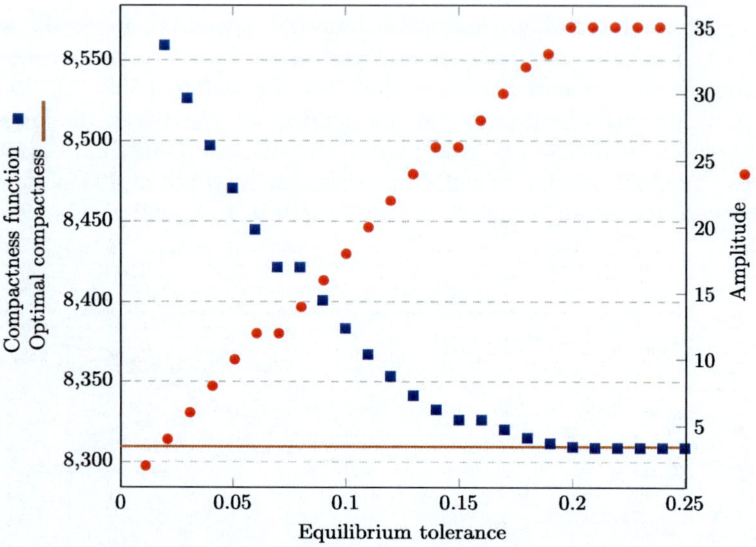

Fig. 2. Trade-off between compactness and equilibrium for an instance with 438 points and 5 sectors (Model A).

Considering the same instance of 1000 points but with the sectorization being conducted with 15 sectors, Table 3 and Fig. 3 show a similar trend. However, the plots for both the compactness function and the amplitude display "steps", showing that a small variation in the equilibrium tolerance has no effect on the solution. This is even more evident for smaller instances, as shown in Table 4 and Fig. 4, regarding an instance with 438 points divided into 15 sectors.

In both instances, we can also observe that with a higher number of sectors k, a higher equilibrium is achieved between the sectors, as the amplitude varies less and is smaller overall. That can be explained by the fact that there are fewer combinations for the assignment of points to sectors due to a large number of sectors.

Table 3. Results of Model A for an instance of 1000 points and 15 sectors.

$n = 1000$, $k = 15$, Comp* = 3108.66, $Q = 66$			
r_e	Compactness	Sector sizes	Amplitude
0.25	3325.06	[75 70 82 50 82 68 79 50 50 50 82 71 66 50 75]	32
0.2	3406.70	[76 70 79 53 79 70 69 53 53 53 79 73 68 53 72]	26
0.15	3545.92	[75 73 75 57 75 72 59 57 57 57 75 75 69 57 67]	18
0.1	3685.25	[72 72 72 60 72 72 60 60 60 60 72 72 72 60 64]	12
0.05	3883.89	[69 69 69 63 69 69 64 63 63 63 69 69 69 63 69]	6

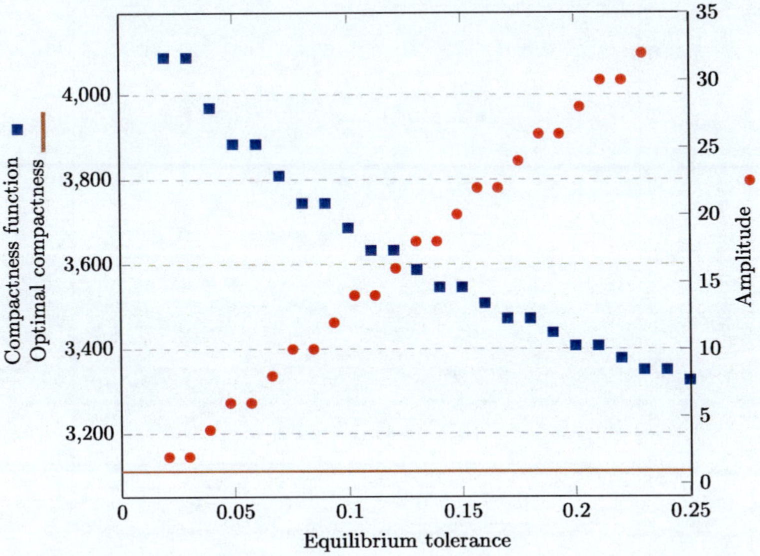

Fig. 3. Trade-off between compactness and equilibrium for an instance with 1000 points and 15 sectors (Model A).

Table 4. Results of Model A for an instance of 438 points and 15 sectors.

$n = 438$, $k = 15$, Comp* $= 4794.91$, $Q = 29$			
τ_e	Compactness	Sector sizes	Amplitude
0.25	5002.41	[22 32 29 26 22 28 30 36 36 28 36 22 22 33 36]	14
0.2	5151.42	[24 34 29 26 24 28 32 34 34 28 34 24 24 29 34]	10
0.15	5242.30	[25 33 30 25 25 29 33 33 33 29 33 25 25 27 33]	8
0.1	5477.38	[27 31 30 27 27 29 31 31 31 31 31 27 27 27 31]	4
0.05	5614.65	[28 30 30 28 28 29 30 30 30 30 30 28 28 29 30]	2

Given the various solutions obtained for different tolerance values, in a real-life scenario the best solution would be chosen by the decision makers, according to their preferences and the nature of the problem at hand.

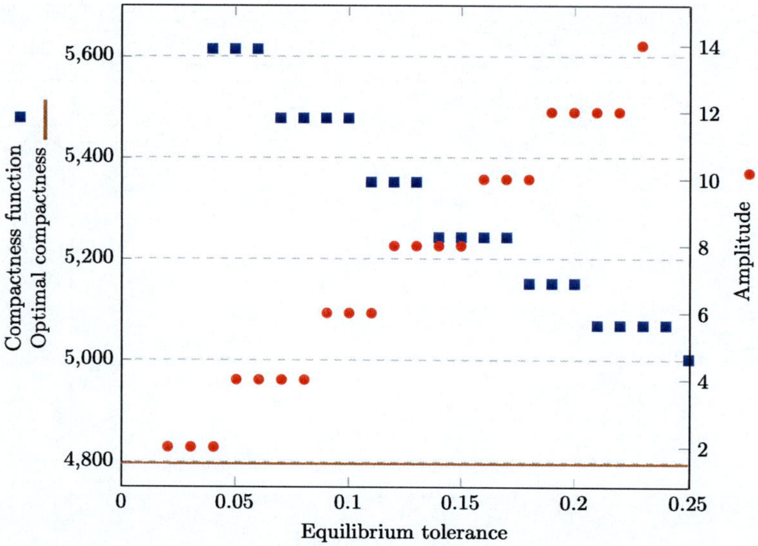

Fig. 4. Trade-off between compactness and equilibrium for an instance with 438 points and 15 sectors (Model A).

As for Model B, where the equilibrium function is optimized instead, Table 5 summarizes the results for the two previously tested instances of size 438 and 1000 points, considering 5 and 15 sectors, showing instance size, number of sectors, equilibrium function, compactness function, optimal compactness value, unbounded compactness value, ideal target value Q, sector sizes and amplitude. In the table, we observed that a variation in the compactness tolerance had no effect on the equilibrium function for the tested values ($0 \leq \tau_{com} \leq 0.25$).

Furthermore, the sectors are almost perfectly balanced in size, even though their compactness is much worse than the optimal possible value for that instance. However, by using Model B without any constraints for compactness (constraints (11)), the obtained results would be worse regarding compactness (represented in the column "Unbounded Comp." in Table 5).

Table 5. Results for Model B for two instances.

Size	k	Equi.	Compactness	Comp.*	Unbounded Comp.	Q	Sector sizes	Amp.
438	5	0.3	16744.28	8308.96	23411.88	87	[87 88 87 88 88]	1
438	15	0.171	18474.83	4794.91	26377.59	29	[30 29 ... 29 29]	1
1000	5	0	8487.52	5221.98	15853.74	200	[200 200 ... 200]	0
1000	15	0	11509.42	3108.66	17309.79	66	[67 67 ... 66 67]	1

5 Conclusions and Future Work

Sectorization problems arise in several contexts and are approached differently depending on the applications. Nonetheless, some similarities can be observed among them, namely the criteria according to which a sectorization should be conducted, such as equilibrium and compactness. One contribution of this paper is discussing the results of two generalized integer programming models that address those criteria due to their relevance. Model A optimized compactness with a bounded equilibrium, and Model B tackles the reverse situation.

Computational tests were conducted in CPLEX 12.9 on 50 instances (of sizes ranging from 30 to 1000 points, and for 5, 10 and 15 sectors). They allowed us to understand, for Model A, how did the compactness function behave for different values of the equilibrium tolerance, and, for Model B, the relation between the variance of sector sizes and the compactness tolerance. Moreover, for the first model we also considered the sizes of the obtained sectors and compared the overall amplitude (i.e. the difference between the largest and smallest sector sizes) in order to assess the quality of the sectorization in terms of equilibrium.

These tests allowed us to conclude that there is a trade-off between criteria: in the case of Model A, the higher the tolerance for equilibrium, the better the compactness of the sectorization; in the case of Model B, variations in the compactness tolerance did not have any effect on the equilibrium function, even though the values of compactness were better in comparison to Model B without constraints for compactness.

The inclusion of other criteria in these models (such as contiguity) and the tuning of tolerance parameters can be further explored. Furthermore, another distance metric could be used, such as the L_1 metric, or even the travelling time between two points. Additionally, different measures could be considered for the addressed criteria and compared to the obtained results.

Acknowledgments. This work is financed by the ERDF - European Regional Development Fund through the Operational Programme for Competitiveness and Internationalisation - COMPETE 2020 Programme and by National Funds through the Portuguese funding agency, FCT - Fundação para a Ciência e a Tecnologia within project POCI-01-0145-FEDER-031671.

References

1. Rodrigues, A.M., Ferreira, J.S.: Measures in sectorization problems. In: Povoa, A., de Miranda, J. (eds.) Studies in Big Data, pp. 203–211. Springer, Cham (2015). https://doi.org/10.1007/978-3-319-24154-8_24
2. Kalcsics, J., Nickel, S., Schröder, M.: Towards a unified territorial design approach — applications, algorithms and GIS integration. Top **13**(1), 1–56 (2005). https://doi.org/10.1007/bf02578982
3. Ríos-Mercado, R.Z., López-Pérez, J.F.: Commercial territory design planning with realignment and disjoint assignment requirements. Omega **41**(3), 525–535 (2013). https://doi.org/10.1016/j.omega.2012.08.002

4. Ríos-Mercado, R.Z., Fernández, E.: A reactive GRASP for a commercial territory design problem with multiple balancing requirements. Comput. Oper. Res. **36**(3), 755–776 (2009). https://doi.org/10.1016/j.cor.2007.10.024
5. Farughi, H., Mostafayi, S., Arkat, J.: Healthcare districting optimization using gray wolf optimizer and ant lion optimizer algorithms (case study: South Khorasan Healthcare System in Iran). J. Opt. Ind. Eng. **12**(1), 119–131 (2019). https://doi.org/10.22094/joie.2018.766.1489
6. Bergey, P.K., Ragsdale, C.T., Hoskote, M.: A decision support system for the electrical power districting problem. Decis. Support Syst. **36**(1), 1–17 (2003). https://doi.org/10.1016/S1344-6223(02)00033-0
7. Shan, Y., Bettinger, P., Cieszewski, C.J., Li, R.T.: Trends in spatial forest planning. Math. Comput. Forestry Nat. Resour. Sci. (MCFNS) **1**(2), 86–112 (2009)
8. Lin, M., Chin, K.S., Ma, L., Tsui, K.L.: A comprehensive multi-objective mixed integer nonlinear programming model for an integrated elderly care service districting problem. Ann. Oper. Res. (2018). https://doi.org/10.1007/s10479-018-3078-6
9. Lei, H., Laporte, G., Liu, Y., Zhang, T.: Dynamic design of sales territories. Comput. Oper. Res. **56**, 84–92 (2015). https://doi.org/10.1016/j.cor.2014.11.008
10. Caro, F., Shirabe, T., Guignard, M., Weintraub, A.: School redistricting: embedding GIS tools with integer programming. J. Oper. Res. Soc. **55**(8), 836–849 (2004). https://doi.org/10.1057/palgrave.jors.2601729
11. Ricca, F., Scozzari, A., Simeone, B.: Political Districting: from classical models to recent approaches. Ann. Oper. Res. **204**(1), 271–299 (2013). https://doi.org/10.1007/s10479-012-1267-2
12. Rodrigues, A.M., Ferreira, J.S.: Sectors and routes in solid waste collection. In: Almeida, J., Oliveira, J., Pinto, A. (eds.) Operational Research, pp. 353–375. Springer, Cham (2015). https://doi.org/10.1007/978-3-319-20328-7_20
13. Treimuth, T.: Dynamic optimization of airspace sector grouping. Ph.D. Thesis, Université de Toulouse, INPT (2018)
14. Emiliano, W., Telhada, J., do Sameiro Carvalho, M.: Home health care logistics planning: a review and framework. Procedia Manuf. **13**, 948–955 (2017). https://doi.org/10.1016/j.promfg.2017.09.165
15. Bouzarth, E.L., Forrester, R., Hutson, K.R., Reddoch, L.: Assigning students to schools to minimize both transportation costs and socioeconomic variation between schools. Socio-Econ. Plann. Sci. **64**, 1–8 (2018). https://doi.org/10.1016/j.seps.2017.09.001
16. Öztürk, E.G.: Normal distribution instances for sectorization (2019). https://drive.inesctec.pt/s/QfgAopoYXSCH9Ej

Supply Chains' Digitalization: Boosters and Barriers

Núbio Gomes Filho[1(✉)], Nazaré Rego[1,2], and João Claro[2,3]

[1] Escola de Economia e Gestão, Universidade do Minho, Braga, Portugal
id7657@alunos.uminho.pt
[2] INESC TEC, Porto, Portugal
[3] Faculdade de Engenharia, Universidade do Porto, Porto, Portugal

Abstract. Digitalization has spread across business and supply chains, becoming irreversible and affecting how companies run their businesses and fulfill their demand. This paper discusses the main aspects that propel and hinder digitalization in supply chains are. One could divide the boosters into two groups: the application of technological advances and circular boosters. On the other hand, the barriers are either sporadic or persistent. Despite the perceived barriers, if correctly applied, digitalization brings more benefits than problems to supply chains. Furthermore, recognizing this might help practitioners who are still reluctant about digitalization.

Keywords: Supply chain · Digitalization · Technological advances · Cyber-risk

1 Introduction

In the past, supply chains were defined by their physical flows, exchanging products, and information from one node to another, following a linear perspective [1,2]. From the two last decades on, we have witnessed the strengthening of digitalization among firms and, concurrently, among supply chains.

This process is not novel [2,3]; nevertheless, it became an irreversible trend after information gained its rightful importance for manufacturers and service providers who were seeking to increase their market share through customization. Digitalization represents the adoption of capabilities that improve organizations' competitiveness and involve all stakeholders [4–6]. This process is distinct from digitization (the replacement of processes or tools with digital analogs) [5,7].

According to Mussomeli et al. [1], traditional supply chains and digital supply chain networks are different.[1] Digital supply networks are digital supply chains where a dynamic environment with integrated networks exists. These integrated networks allow supply chains to disseminate information among stakeholders, facilitate automation, value addition, and insights [1].

[1] In [1, p. 06] there is an illustration that clarifies the difference among both concepts.

J. Machado et al. (Eds.): icieng 2021, LNME, pp. 197–208, 2022.
https://doi.org/10.1007/978-3-030-78170-5_18

Scrutinizing the supply chain's digitalization has become a priority, essentially due to increases in demands' volatility and the requirements supply chains face [8]. The customers try to maximize their value taking into account the trade-offs between price, delivery time, quality, uniqueness. Customers' access to the Internet and social media allows them to buy in physical stores next door or from sellers in a different part of the world.

Nevertheless, both processes (digitalization and digitization) are rooted in Information and Communication Technologies (ICT) because only through their implementation, business perspectives changed dramatically. ICT benefit firms because they foster information exchange [9–12], ease decision-making [11,13,14], and decrease the likelihood of production and logistics errors [15–17].

When digitalization and digitization are interconnected, a new phenomenon takes place: Industry 4.0. In this context, Industry 4.0[2] stands for a radical transformation [15,18] of industry processes, operations, and services [9], considering disruptive, connected, and intelligent technological solutions [14], which are combined with human resources[3] [4], creating smart factories [16] and integration [18]. Such a new paradigm modifies the foundations of how people and firms interact with each other, defining new priorities to researchers as they evaluate Industry 4.0 potential causes and effects under multiple scopes.

Although this transformation began in some specific sectors (automotive, technology, and biology industries [15]), it spread quickly to others. This spread happened because digitalization empowers firms, enabling the forecast of its products/services demand, resources' availability, and the identification of bottlenecks and process variability [7], and the reduction of mistakes [3].

Although some authors [4,9,14,20] did comprehensive studies in many digitalization dimensions, none has shed light onto the aspects that incentive the process, nor on the elements that may hinder it. Therefore, this paper describes the boosters and barriers of a digitalization process' in supply chains.

We divided the remainder of this work as follows: Sect. 2 will discuss aspects that induce digitalization in supply chains, Sect. 3 will provide elements that may hinder this process, and, finally, Sect. 4 makes some final remarks and proposes future research topics.

2 Boosting Digitalization

The digitalization booster represents every aspect that positively impacts the digitalization process in companies/supply chains, increasing their financial results and improving companies/supply chains' relations with stakeholders. These boosters split into two groups: application of technological advances and

[2] "Industry 4.0" has been used to represent the Fourth Industrial Revolution as did in [4,9,14,16,18], which is different from the idea of smart factories as in [11,19]. Additionally, the term 'Industry 4.0' rooted new ones, like Logistics 4.0 [15] or Supply Chain 4.0 [2].

[3] The participation of humans is not a consensus in this concept. For a different perspective, see [19].

circular boosters. The second group encompasses situations that simultaneously influence and benefit from digitalization.

The most common digitalization booster is the application of technological advances, divided by [14] as: *core* and *complementary.* According to Hahn [14], core technologies connect people, technical systems, and organizations producing an environment for smart relations, increasing autonomy, and well-informed decisions, such as Big Data Analytics, cloud computing, or artificial intelligence. Notwithstanding, complementary technologies aggregate tools to induce productivity, agility, and responsiveness, like the Internet of Things, blockchains, sensors, cyber-physical systems, 3D printing [21].

Companies make use of technological advances seeking to gain productivity in their processes in order to achieve scalability and flexibility [14], meaning that, in a broader sense, they are willing to increase their competitiveness [11]. However, Hahn [14] found a gap regarding the employment of technological advances' business impacts, especially considering their potential productivity outcomes, therefore:

- **Hypothesis 01**: Managers who employ technological advances in their production observe better financial results after doing it.
- **Hypothesis 02**: Application of technological advances increases companies' flexibility.

Even though applying such technologies is utterly necessary for boosting digitalization, other factors also contribute to the digitalization process within supply chains. The following boosters are defined as circular because they are simultaneously inductors and consequences of digitalization in a virtuous cycle, which means that they propel digitalization and benefit from it.

The first circular booster is the increase in market competitiveness. Considering that supply chains are focused on offering services or products that fit consumers' needs, creating value for them, digitalization enables mass customization [3,8,9,12,16,22], and redesigns production networks [19]. Through digitalization, supply chains expand their markets [21]. The second aspect of competitiveness improvement regards the simplification or the automation of processes (production, procurement, inventory, tracking, control),[4] reducing the financial costs of managing information, delivering to final consumers, resource planning [2,16,20,22,23], improving flexibility, accuracy, and efficiency [2].

As proposed by Alicke, Rexhausen, and Seyfert [2], Supply Chain 4.0 would impact firms by decreasing operational costs (by 30%), lost sales, and inventories (each by 75%). These values derived from the authors' "experience with numerous studies and quantitative calculations" [p. 08]; however, the authors do not present such studies database, and the calculations' methodology was not discussed in their article, therefore:

[4] For a more comprehensive improvement example, see [2, Exhibit 2].

- **Hypothesis 03**: The market pressures managers to increase digitalization within their companies.
- **Hypothesis 04**: Automatizing improves companies' financial results.

The second circular booster is transparency for stakeholders [10,20,24]. As supply chains become more explicit about their processes, stakeholders are assured about standards, equalizing information asymmetries [24,25]. A collaborative environment among supply chain nodes [9,26], and constant combat against data manipulation [23] creates transparency for consumers. Finally, as the information that flows through nodes can be tampered with by any agent, collaboration for continuously scrutinizing data is useful.

Despite realizing the importance of transparency as a critical success factor for supply chains,[5] Gani and Fernando [10] only discuss the transparency in frameworks for ensuring safer practices. Thus, in a broader perspective:

- **Hypothesis 05**: Transparency increases companies' responsiveness.

The third circular booster is traceability. Traceability contributes to eliminating fraud during transportation and logistics, which can benefit from authentication [27], and improve storage, avoiding losing track of materials/products in warehouses [28]. This booster is essential, especially in the food industry [8,27] because of concerns with quality and safety [3,23], or even specific characteristics, like kosher food. In the food industry, traceability also allows data-driven predictions regarding diseases or weather changes' impact on crops [3]. Other sectors, such as the clothing and pharmaceutical industry [24], also started benefiting from traceability. For example, the clothing sector enabled consumers to regard sustainability and labor practices. Traceability also allows the pharmaceutical industry to prevent counterfeit pharmaceuticals from entering the chain, improve the control over prescribed-drugs, and rapidly deal with manufacturing problems, for instance, recalling low-quality products.[6]

As recently discussed by Kittipanya-ngam and Tan [3], for traceability to turn into an asset in food supply chains, both inter-firm collaboration and government-to-government partnerships must occur, otherwise "no matter how advanced the technologies are to support the supply chain operations, such benefits will never happen." [p. 170]. Meaning that, for companies/supply chains to benefit from the results of quickly responding to problems or shifts in the market, they must have real-time awareness of their products/processes, then:

- **Hypothesis 06**: Traceability increases companies' resilience.

Despite circular boosters' apparent importance, there have been no studies dealing with digitalization impacts in supply chains. Regarding technological advances, Ehie and Ferreira's framework [21] was the only one dealing with

[5] These authors [10] use originally the term 'cyber supply chain', which is a "digitally integrated supply chain" [p. 5306].

[6] The examples are not comprehensive, but allow readers to grasp the booster's usefulness.

their effects on digitalization. Kittipanya-ngam and Tan's case-based conceptual framework [3] built guidelines for implementing these technological advances in food supply chains.

The conjunction of hypotheses 04 to 06 builds foundations for hypothesis 07. On the one hand, these elements (automatizing, transparency, and traceability), previously defined as inductors of digitalization, and their results (improvements in financial results, responsiveness, and resilience) foster firms' survivability. On the other hand, as survivability is an essential feature to firms' stakeholders, decision-makers' actions must account for it, creating grounds for digitalization as a strategy for reaching such a feature. This feedback relates to the circularity of boosters (market competitiveness, traceability, and transparency) and their relation with digitalization, therefore:

- **Hypothesis 07**: Circular boosters foster digitalization.

Figure 1 summarizes the concepts presented in this section. It describes the supply chains' positive impact side, which results from the application of technological advances and the impact of circular boosters. Only circular boosters contain a virtuous cycle for digitalization.

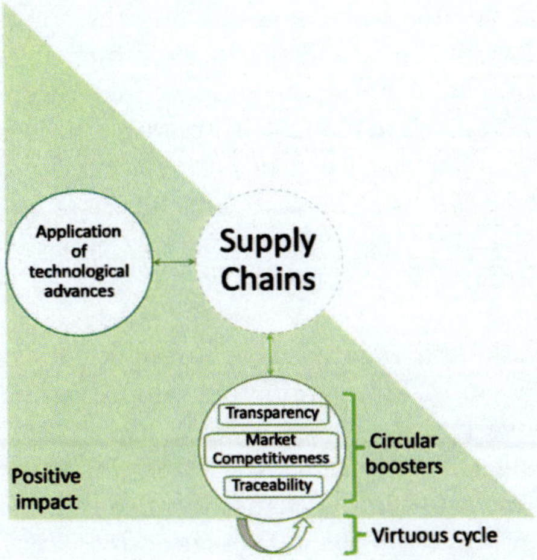

Fig. 1. Supply chains' digitalization boosters

3 Digitalization Barriers

Digitalization barriers are aspects that harm the digitalization process. These barriers represent everything that prevents firms/supply chains from achieving digitalization. Just as the boosters, two groups divide the barriers: persistent and sporadic. Managers' experience becomes a tool for dealing with their negative impacts in companies/supply chains, enabling the division of possible barriers into these two groups.

The digitalization process also comes with financial costs.[7] These costs might be related to the acquisition of technological advances or its development through Research & Development departments [6,29]. Knowledge transfer also creates financial costs through training and processes standardization, even if this transfer happens within companies or supply chain partners [12,20]. This knowledge sharing may create tensions costs amongst partners [3]. Financial costs are barriers because employing technological solutions or redesigning processes will continually demand high investments from industries despite the obtained know-how.

Moreover, decision-makers' common pitfall as they use technological advances or modify them in their businesses' processes is not to consider the full range of related costs [6,27]. Thus:

- **Hypothesis 08**: The costs of implementing technological advances are more restrictive than those associated with transferring technology.

Another potential barrier to the digitalization process relates to trust issues [5,29]. In cases where machine interaction replaces human contact, it becomes less likely that the partners know who is on the other side of negotiations [22]. Another situation that creates trust issues among the supply chain's stakeholders relates to the collaboration to engage in an industrial ecosystem [3,12,25], meaning that sometimes information is critical, and companies may not want to share it. Finally, on this barrier, it is also possible to lose confidence in partners if they suffer from cybersecurity issues [11,29]. As the manager's know-how increases, it becomes easier to deal with all the cases presented, which once built a wall of distrust issues among partners. This experience enables decision-makers to forecast potential scenarios that could obstruct the relationship with stakeholders.

Di Vaio and Varriale [22] pointed as a potential limitation of their study about the implementation of Port Communication Systems in sea-land supply chains the need to investigate further specific aspects of inter-organizational relationship systems, such as trust between players. This problem may extend the frontiers of this specific supply chain because trustful relations are a fundamental aspect of economic relations, and, the authors describe, trust as a "relevant inter-organizational characteristic(s)" [22, p. 229], then it is important to study if:

- **Hypothesis 09**: Digitalization decreases trust among partners.

A third barrier relates to technical requirements and standards to be fulfilled as supply chains add technological tools for reshaping their manufacturing processes, operations, and services [3,4,19,20]. A potential problem such requirements may bring regards supply chain members' capabilities [4], since adjusting capabilities may redesign the network, diminishing nodes, and increasing

[7] The costs exemplified here will not cover all possibilities. For other costs, see [27].

responsibilities in remaining ones. Added to that, full benefits from meeting such technical requirements/standards only come from horizontal integration among supply chain nodes [15,28]. This third barrier also represents a sporadic problem; once the gaining in managers' experience, facilitates reshaping or attending requirements.

The decision about digitalization, especially deriving from the employment of technological advances, builds a continuous condition of resources' allocation by decision-makers [10]. Despite these authors recognized this need, no study empirically assessed this engagement yet. Hence:

- **Hypothesis 10**: Process changes that emerged from applying technological advances take more engagement than other process changes.

Despite the barriers mentioned earlier, cyber-risk[8] is the last critical for digitalization. This barrier is crucial because digitalization creates technological dependency through technological application directly in production/processes or when employed to support production. Cyber-risk derives mainly from access points that enable information sharing between external supply chain partners, allowing exploitation of technical and human vulnerabilities [10,15,19]. Cyber-risk creates direct (e.g., production impediments, regulatory problems, and fees) and indirect (i.e., mistrust among partners, reduction of market value) costs [26,30]. Additionally, even after surpassing all previous barriers, cyber-risk remains a potential impediment due to its dynamics and unpredictability [30], meaning that, just like costs, cyber-risks are a persistent barrier.

Cyber-risk also poses in-company problems and inter-organizational ones, ergo "resilient supply chains can better prompt to react (to cyber-risks)." [30, p. 65]. Nevertheless, to become resilient, traditional supply chains must redesign their processes, focusing on increasing their resilience. Moreover, as presented by Siciliano and Gaudezi [30], "companies suffer from poor communication, overconfidence in technical investments and lack of integration and alignment of measures and procedures throughout the whole SC." [pp. 65–66], therefore:

- **Hypothesis 11**: Cyber-risks increase processes' changes at the firm level.
- **Hypothesis 12**: Managers are confident in firms' cybersecurity.
- **Hypothesis 13**: Managers are confident in the supply chain's cybersecurity.

Ergo, there are two types of barriers: sporadic (trust issues and technological requirements) and persistent (cyber-risks and costs). This division considers the managers' experience. If, on the one hand, experience builds a path to overcome sporadic barriers, persistent barriers, on the other hand, remain, and they are only manageable using this experience, thus:

- **Hypothesis 14**: Persistent barriers' impact on digitalization is higher than sporadic barriers' impact.

[8] We opt to aggregate IT and cyber risks under cyber-risk terminology. For those interested in their differences, see [30, p. 59].

Figure 2 condenses the negative impact side. There is a difference in each barrier's impact to supply chains, described by the arrow's continuity linking each barrier's group to supply chains. On one side, there is a dashed arrow linking sporadic barriers to supply chains. This arrow indicates that managers' experience would be enough to deal with these barriers' negative impact. Conversely, a straight-arrow links persistent barriers to supply chains, indicating that managers' experience can only reduce the flow but cannot block it.

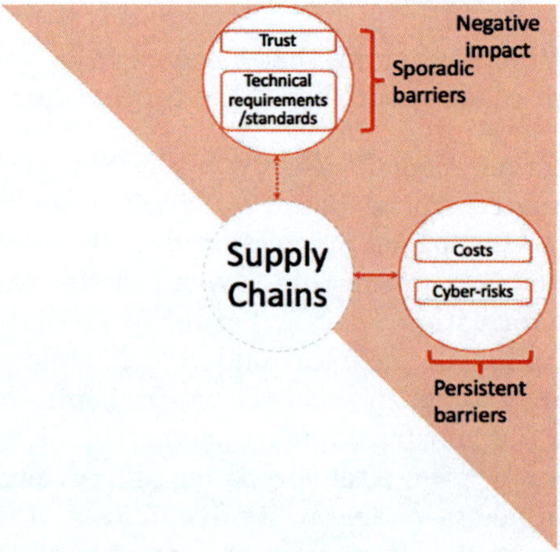

Fig. 2. Supply chains' digitalization barriers

4 Final Remarks

Industry 4.0 is a shift in paradigm that changed the way people and firms interact with each other. One of the primary sources of this shift is the digitalization that points to a new business direction. This paper described the aspects that interfere positively and negatively with digitalization in supply chains. Figure 3 sums up these concepts and condenses the crucial aspects of Figs. 1 and 2.

The hypotheses presented in Sects. 2 and 3 are the building-blocks for our framework. Collecting managers' perceptions from different sectors allow the study to assess these blocks. As we opt to work with companies spread throughout the economy, it may be possible to find sectoral specificities that can indicate the need for further focused studies. Furthermore, evaluating this set of hypotheses will help determine how boosters and barriers impact digitalization in companies and concurrently in supply chains. Figure 4 represents the diagram for a research framework created to observe managers' perspectives about the impacts of digitalization boosters and barriers.

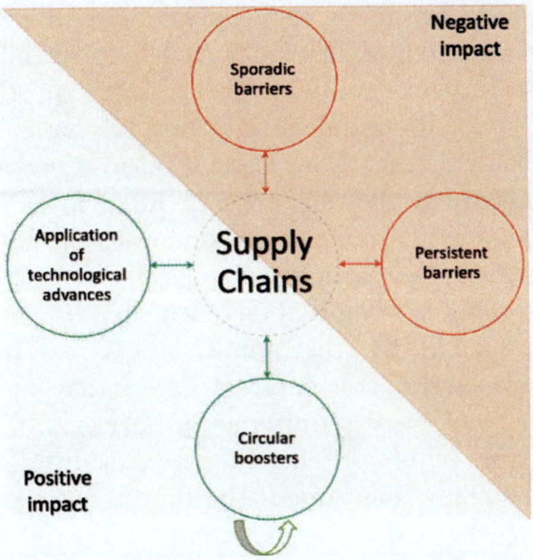

Fig. 3. Supply chains' digitalization: boosters and barriers

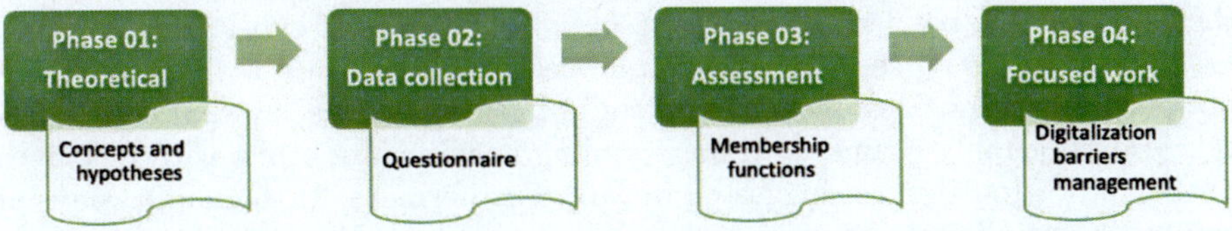

Fig. 4. Supply chains' digitalization research framework

The first phase defines concepts and hypotheses about digitalization boosters and barriers. The conceptualization derives from a literature review discussing the characteristics that divide boosters and barriers. After consolidating the hypotheses, a questionnaire collects managers' perceptions of digitalization's positive and negative impacts. The third phase measures the consensus through membership functions over digitalization's impacts on supply chains. Finally, the study focuses on finding ways to manage the negative aspects of digitalization.

Digitalization process creates a thin line between boosters and barriers, as perceived by [9], meaning that even though supply chains benefit from the process, managers must consider new ways to deal with potential problems. Such awareness tends to improve firms' management process[9] since it increases preparedness and flexibility through the analysis of what-if scenarios [19].

Alongside that, if correctly used, the benefits of passing through the digitalization process are higher than the costs for implementing it in supply chains [22]. Nevertheless, researchers need to focus on this quantification to solidify the

[9] For a strategic perspective of how inserting digitalization into supply chains, see [1, Figure 4].

idea. Besides that, not all products, especially food, can be cultivated anywhere [3], then digitalization plans must consider such impossibilities.

Digitalization also impacts the way integration[10] processes. Digitalization, for instance, creates space for omnichannel supply chains [15,16], representing an integration that flows further from horizontal and vertical integrations [18]. Using technological advances allows processes' integration in both directions; it also provides to supply chains' nodes real-time data on flows [2,31,32] that are useful for decision-making, increasing resilience.

Practitioners who want to begin their digitalization process must consider three steps [1]: 1) think big, 2) start small, and 3) act fast. The first step is related to the first circular booster (market competitiveness). The second step considers both the technological requirements barrier and the cyber-risks barrier. The third step tries to break with the conception of waiting for the right moment since, as previously mentioned, the digitalization trend is irreversible. Before operationalizing them, decision-makers should carefully plan this process. Otherwise, the companies or supply chains might suffer from massive setbacks.

Additionally, managers must be aware that there is no 'one-fits-all' solution; therefore, despite becoming more experienced, their experience should not blind them against the barriers' dynamics, especially in terms of cyber-risks.

Future studies on the digitalization process should concern with: 1) assessing how each group of boosters and barriers impacts managers' perception of digitalization in supply chains. 2) Understanding how humans relate with machines in their work routines, as suggested by [15]. Furthermore, 3) the implications of supply chains' digitalization on an omnichannel perspective, as in [16,20].

Acknowledgements. This research was partially funded by Fundação para a Ciência e a Tecnologia (FCT) under the Ph.D. Scholarship Reference: SFRH/BD/145941/2019.

References

1. Mussomeli, A., Gish, D., Laaper, S.: The rise of the digital supply network. Deloitte **45**(3), 20–21 (2015). http://ezproxy.lib.monash.edu.au/login?url=http://search.ebscohost.com/login.aspx?direct=true&db=bth&AN=110198124&site=ehost-live&scope=site
2. Alicke, K., Rexhausen, D., Seyfert, A.: Supply Chain 4.0 in consumer goods. Technical report, Exhibit 1, McKinsey (2016). https://www.mckinsey.com/industries/consumer-packaged-goods/our-insights/supply-chain-4-0-in-consumer-goods
3. Kittipanya-ngam, P., Tan, K.H.: A framework for food supply chain digitalization: lessons from Thailand. Prod. Plan. Control **31**(2–3), 158–172 (2019)
4. Queiroz, M.M., Pereira, S.C.F., Telles, R., Machado, M.C.: Industry 4.0 and digital supply chain capabilities: a framework for understanding digitalisation challenges and opportunities. Benchmarking (2019). https://doi.org/10.1108/BIJ-12-2018-0435
5. Seyedghorban, Z., Samson, D., Tahernejad, H.: Digitalization opportunities for the procurement function: pathways to maturity. Int. J. Oper. Prod. Manag. **40**(11), 1685–1693 (2020). https://doi.org/10.1108/IJOPM-04-2020-0214

[10] For a more comprehensive discussion about this topic, see: [18, p. 2952].

6. Hopkins, J.L.: An investigation into emerging industry 4.0 technologies as drivers of supply chain innovation in Australia. Comput. Ind. **125**, 103,323 (2021). https://doi.org/10.1016/j.compind.2020.103323
7. Holmström, J., Holweg, M., Lawson, B., Pil, F.K., Wagner, S.M.: The digitalization of operations and supply chain management: theoretical and methodological implications. J. Oper. Manag. **65**(8), 728–734 (2019). https://doi.org/10.1002/joom.1073
8. Demartini, M., Pinna, C., Tonelli, F., Terzi, S., Sansone, C., Testa, C.: Food industry digitalization: from challenges and trends to opportunities and solutions. IFAC-PapersOnLine **51**(11), 1371–1378 (2018)
9. Tjahjono, B., Esplugues, C., Ares, E., Pelaez, G.: What does Industry 4.0 mean to supply chain? Procedia Manuf. **13**, 1175–1182 (2017). https://doi.org/10.1016/j.promfg.2017.09.191
10. Gani, A.B.D., Fernando, Y.: Concept and practices of cyber supply chain in manufacturing context. In: Khosrow-Pour, M. (ed.) Encyclopedia of Information Science and Technology, 4th edn., vol. 460, pp. 5306–5316. IGI Global, Hershey (2018). https://doi.org/10.4018/978-1-5225-2255-3.ch460. http://services.igi-global.com/resolvedoi/resolve.aspx?doi=10.4018/978-1-5225-2255-3.ch460
11. Ivanov, D., Tsipoulanidis, A., Schönberger, J.: Digital supply chain, smart operations and Industry 4.0. In: Global Supply Chain and Operations Management, vol. 16, pp. 481–526. Springer, Cham (2019). https://doi.org/10.1007/978-3-319-94313-8_16
12. Zangiacomi, A., Pessot, E., Fornasiero, R., Bertetti, M., Sacco, M.: Moving towards digitalization: a multiple case study in manufacturing. Prod. Plan. Control **31**(2–3), 143–157 (2020)
13. Kamble, S., Gunasekaran, A., Arha, H.: Understanding the Blockchain technology adoption in supply chains-Indian context. Int. J. Prod. Res. **57**(7), 2009–2033 (2018)
14. Hahn, G.J.: Industry 4.0: a supply chain innovation perspective. Int. J. Prod. Res. **58**(5), 1425–1441 (2020). https://doi.org/10.1080/00207543.2019.1641642
15. Barreto, L., Amaral, A., Pereira, T.: Industry 4.0 implications in logistics: an overview. Procedia Manuf. **13**, 1245–1252 (2017). https://doi.org/10.1016/j.promfg.2017.09.045. https://linkinghub.elsevier.com/retrieve/pii/S2351978917306807
16. Szozda, N.: Industry 4.0 and its impact on the functioning of supply chains. Logforum **13**(4), 401–414 (2017). https://doi.org/10.17270/J.LOG.2017.4.2. http://www.logforum.net/volume13/issue4/abstract-2.html
17. Min, H.: Blockchain technology for enhancing supply chain resilience. Bus. Horiz. **62**(1), 35–45 (2019)
18. Xu, L.D., Xu, E.L., Li, L.: Industry 4.0: state of the art and future trends. Int. J. Prod. Res. **56**(8), 2941–2962 (2018). https://doi.org/10.1080/00207543.2018.1444806
19. Das, A., Gottlieb, S., Ivanov, D.: Managing disruptions and the ripple effect in digital supply chains: empirical case studies. In: Ivanov, D., Dolgui, A., Sokolov, B. (eds.) Handbook of Ripple Effects in the Supply Chain, chap. 13, vol. 276, pp. 261–285. Springer, Cham (2019). https://doi.org/10.1007/978-3-030-14302-2_13
20. Seyedghorban, Z., Tahernejad, H., Meriton, R., Graham, G.: Supply chain digitalization: past, present and future. Prod. Plan. Control **31**(2–3), 96–114 (2019)
21. Ehie, I., Ferreira, L.M.D.F.: Conceptual development of supply chain digitalization framework. IFAC-PapersOnLine **52**(13), 2338–2342 (2019)

22. Di. Vaio, A., Varriale, L.: Digitalization in the sea-land supply chain: experiences from Italy in rethinking the port operations within inter-organizational relationships. Prod. Plan. Control **31**(2–3), 220–232 (2019)
23. Yang, C.S.: Maritime shipping digitalization: blockchain-based technology applications, future improvements, and intention to use. Transp. Res. Part E Logist. Transp. Rev. **131**, 108–117 (2019). https://doi.org/10.1016/j.tre.2019.09.020
24. Peltoniemi, T., Suomi, R.: The impact of the digitalization to the medical supply chain. Inj. Prev. **22**(Suppl 2), A200.1–A200 (2016). https://doi.org/10.1136/injuryprev-2016-042156.555. http://injuryprevention.bmj.com/lookup/doi/10.1136/injuryprev-2016-042156.555
25. Ebinger, F., Omondi, B.: Leveraging digital approaches for transparency in sustainable supply chains: a conceptual paper. Sustainability **12**(15) (2020). https://doi.org/10.3390/su12156129
26. Kour, R., Karim, R., Thaduri, A.: Cybersecurity for railways – a maturity model. Proc. Inst. Mech. Eng. Part F J. Rail Rapid Transit. (2019). https://doi.org/10.1177/0954409719881849. http://journals.sagepub.com/doi/10.1177/0954409719881849
27. Reyna, A., Martín, C., Chen, J., Soler, E., Díaz, M.: On blockchain and its integration with IoT. Challenges and opportunities. Futur. Gener. Comput. Syst. **88**(2018), 173–190 (2018). https://doi.org/10.1016/j.future.2018.05.046
28. Fernández-Caramés, T., Blanco-Novoa, O., Suárez-Albela, M., Fraga-Lamas, P.: A UAV and blockchain-based system for Industry 4.0 inventory and traceability applications. In: Proceedings, vol. 4, no. 1, p. 26 (2018). https://doi.org/10.3390/ecsa-5-05758
29. Annosi, M.C., Brunetta, F., Bimbo, F., Kostoula, M.: Digitalization within food supply chains to prevent food waste. Drivers, barriers and collaboration practices. Ind. Mark. Manag. **93**, 208–220 (2021). https://doi.org/10.1016/j.indmarman.2021.01.005
30. Treku, D.N., Wiredu, G.O.: Information systems implementation and structural adaptation in government-business inter-organization. In: Lamboglia, R., Cardoni, A., Dameri, R., Mancini, D. (eds.) Network, Smart Open, vol. 4, pp. 21–41. Springer, Cham (2018). https://doi.org/10.1007/978-3-319-62636-9_2
31. Litster, J., Bogle, I.D.L.: Smart process manufacturing for formulated products. Engineering **5**(6), 1003–1009 (2019)
32. Merdin, D., Ersoz, F.: Evaluation of the applicability of Industry 4.0 processes in businesses and supply chain applications. In: 3rd International Symposium on Multidisciplinary Studies and Innovative Technologies, ISMSIT 2019 - Proceedings, pp. 1–10 (2019). https://doi.org/10.1109/ISMSIT.2019.8932830

Improving ELVs Components Locating in a Dismantling Company

Amanda Ebert Bobsin[1(✉)], Ricardo Augusto Cassel[1], Nuno O. Fernandes[2], and Paulo Torres[2]

[1] Production Engineering Department, College of Engineering, Universidade Federal do Rio Grande do Sul, Porto Alegre, Brazil
cassel@producao.ufrgs.br

[2] Escola Superior de Tecnologia, Instituto Politécnico de Castelo Branco, Castelo Branco, Portugal
{nogf,paulo.torres}@ipcb.pt

Abstract. The management of end-of-life vehicles (ELVs) is a key element for global sustainability. In this context, vehicles dismantling companies need to improve their production processes in order to achieve not only cost-effective, but also sustainable systems. Major challenges for these companies result from the enormous variety of components to be disassembled and stored, and from uncertainty at the ELVs' supply and at the components demand. In this paper, a study for the improvement of the processes associated with the locating and identification of stored components in the warehouse of a Portuguese ELVs dismantling company was developed, observing the company's current processes and the potential opportunities for improvement. From here, propositions were made to update the warehouse management system (WMS), warehouse layout, components location/ identification and its codification. This study is potentially useful for the case-study company as well as other businesses from the same sector.

Keywords: Process improvement · End-of-life vehicles · Warehouse management system

1 Introduction

The amount of consumer goods globally discarded has become one of the main challenges for sustainable development. In this regard, the management of end-of-life goods is a key-element for sustainability, which means answering to the negative impact they may have for the environment as pointed out by [1, 2, 3].

This is the case for the automotive industry. The automobile has become popular worldwide, transforming people's lifestyle and habits. There is an increasing number of vehicles circulating, and thus an increasing number of end-of-life vehicles (ELVs) [4]. This consequently makes the 3Rs (reuse, recovery and recycle) gain a more important place in public management (see e.g. [5, 6]). This is especially true because ELVs create great volumes of material residue, besides dangerous oils and fluids [2].

J. Machado et al. (Eds.): icieng 2021, LNME, pp. 209–219, 2022.
https://doi.org/10.1007/978-3-030-78170-5_19

In this context, European countries must assure the collection of residues from ELVs and their delivery to authorized treatment installations [7]. In Portugal the *Sociedade de Gestão de Veículos em Fim de Vida* (or Society for ELVs), Valorcar, is the institution responsible for promoting the correct management of waste related to the vehicle life cycle.

Dismantling companies have a particular manufacturing process: parts or residues from vehicles on the end-of-life are already a final product, differently from the manufacturing companies, which require the fabrication and assembly of a final product. Still, dismantling companies share similar processes to traditional manufacture, such as purchasing, storage, marketing, and distribution. They are also similar for the fact that, once they belong to a free-competition market, they need to aim the increase in productivity and production efficiency, as well as finding ways to address market demands in order to keep themselves competitive and profitable (see e.g. [8, 9]).

One of the strategies utilized to reduce cost of production and increase the competitive potential of a business is to improve the storage management [10]. The order picking is the activity that absorbs more time and labour in a warehouse and influences significantly the efficiency of the supply chain. A simple form of obtaining more efficiency in order picking involves the assignment of more proper places for the storage of items [11].

In the case of ELVs, to efficiently manage the operations for recovery of components is a challenging problem due to the uncertainty of quantity, age and quality of the returned vehicles [12]. Besides, the materials flow planning in a dismantling company is highly complex once it involves a high level of volatility on the demand for disassembled components, high variety of components for storing (with distinct sizes), differences in technical conditions for every component, and workers' subjectivity for taking decisions about choosing which components to reuse. Therefore, storage becomes one of the most important processes in a dismantling company [9].

This work aims to propose improvements in the locating and identification processes at the warehouse of ELVs components. This proposition was developed based on literature and on an already existing process in a car dismantling company.

The reuse of ELVs is a well-known subject in literature [13]. However, there are not works that consider problems in the area of dismantling companies storage installation, and every action taken to improve the competitive position of these companies is valued and desired [9]. Since there few studies on the subject and, to the best of knowledge, this does not address the problem of identifying/locating items in the warehouse of dismantling company, this work plays an important role as case study.

The remainder of the paper is as follows. Section 2 makes a concise literature review. Then, Sect. 3 presents the methodology used in the study, and Sect. 4 describes the applied work. Finally, Sect. 5 presents the concluding remarks, highlighting suggestions for further research work.

2 Background

Every product on the end of its life has a reverse logistics to be studied in order to decide its best way of treatment. Reverse logistics is the collection of goods used by consumers and their transportation to a suitable termination [14]. In this sense, ELVs draw attention to themselves because of the contradiction between fast increase in car sales and their low recovery rate (see e.g. [13, 15]). Knowing that vehicles have a considerable environmental impact on every stage of their lifecycle [16] and that the quick development of new technologies in automotive industry has shortened vehicles lifecycles, the disposal of ELVs exercises a strong pressure over the environment and human life [5].

Aiming for a sustainable development regarding the ELVs, reverse logistics management was adopted in several sectors and considered a highly highlighted option among the strategies suggested to solve these problems [5]. Besides environmental conservation, the application of strategies in reverse logistics services can also produce economic and social benefits [13, 17]. The economic aspect is presented by the maximization of profit - through recycling industry, for example [18]. If the system is profitable, it becomes more attractive to investors. The social dimension is related to the implementation of a recovery net. The creation of new jobs and the standardization of operations lead to human development [17].

Ene and Öztürk [14] and Phuc et al. [5], among others, investigated the optimization of vehicles reverse supply chain, which includes several stages for products recovery and residue management. These stages comprehend strategies such as recycling, reuse and remanufacture [19]. Recycling refers to the creation new products from the raw material from the original product [20]. To reuse is to use the same piece again for its original purpose, and remanufacturing is the processing of the main product of the component back into its original form [21]. The dismantling companies benefit especially from the sales of metals for recycling and of components for reuse. The low-value non-metallic residues, such as plastics, glass, and composites, are usually crushed and sent to a landfill site or for recycling (see e.g. [17, 22]).

For Xiao et al. [15], the features of the industry involved in the reverse supply chain are different for developed countries and developing ones. In developing countries, the illegal ELVs market is still very strong. This fact interferes in formal recovery and dismantling companies' performances, making them lose their competitiveness and bringing risks for traffic security, environmental protection and resources utilization. Therefore, the optimization of reverse logistics net efficiency is imminent, as well as the dismantling companies' necessity to improve their production processes and achieve sustainable systems for ELVs recovery [8].

3 Methodological Procedures

Gerhardt and Silveira [23] refer that applied research creates knowledge for practical application, intending to solve specific problems, involving local truths and interests. This study is of applied nature because it aims to apply process improvement theories into the reality of a certain model of company. According to Silva and Menezes [24] this research can be considered a qualitative approach, as it does not require the use of statistical methods and techniques, and researchers tend to inductively analyse their data. It can also be considered an exploratory research because it aims to provide greater intimacy with the problem, having as basis literature review, interviews with people and examples analysis [25]. The research strategy utilized in this paper is the application of

BPM (Business Process Management) concepts, composed by two steps: analysis and propositions, as suggested in literature.

The organization chosen for the application is an ELVs dismantling company, founded in 2008, located in the Industrial Zone of Castelo Branco, Portugal. The company operates with twelve employees and acts receiving and treating ELVs - which includes the removal and separation of materials for posterior reuse, recycle or suitable disposal. The main business area of the company is the commercialization of components extracted from the ELVs; but they also sell new components.

The work was executed in two stages: (i) Analysis of the current state (ii) Propositions for improvement. The analysis of the current state started with an interview with the company owner, so he could introduce the company's main goals and difficulties. In the local visits, an active observation of the macro process of the dismantling was made, since the removal from the company site until the storing of components. The aim was to understand how these processes interrelate. Later, the drawing of this macro process and inner processes was made, utilizing the software Bizagi Modeler and BPMN notation. In parallel, data and perceptions from the people involved in the activities were collected to help in the analysis making.

After understanding and visualizing these processes, it was possible to search for improvements [26]. The propositions made for improvements for this company were, therefore, based on the observation of their processes, on the theoretical framework presented here in this paper, alongside with the company's manager perceptions about their needs and challenges.

Dismantling companies can benefit from the application of Lean tools, even though these companies business models are different from those of production companies [27], and there may be some barriers for their implementation, especially regarding the possibility of reducing levels of storage, as pointed out by Hasibul et al. [8]. Thus, for the propositions, Lean concepts were also utilized to suggest improvements, aiming to reduce waste.

For analysing a process, ABPMP [28] suggests as a good practice to perform benchmarking - to learn from the best practices of similar organizations. The development of the propositions also utilized benchmarking, taking as reference another dismantling company in Portugal, appointed by Valorcar as the national reference for warehouse management; and a company from the textile sector, from the same region of the dismantling company, because it also had a large warehouse to manage.

At the end the proposals were critically analysed by the company's manager through a presentation - in which he also contributed with appointments. The results of the applied work are described in the next section.

4 Study

This section describes the current situation concerning of the company production processes. In addition, there are presented the difficulties of the process of locating components and propositions for its improvement.

4.1 Company Current Situation

The company deals with a wide range of reusable component types, which are dismantled and stored, with the uncertainty of demand (result of the durability of the components in vehicles in service, accidents, etc.) and with the uncertainty on the arrival of the ELV for dismantling. One strategy used by the company to deal with this kind of uncertainty is to keep in storage a high number of components. This however resulted in three main problems identified by the production manager: (i) high time required for order picking; (ii) lack of control of storage; and (iii) hard identification of components in the warehouse, because of the way they were stored.

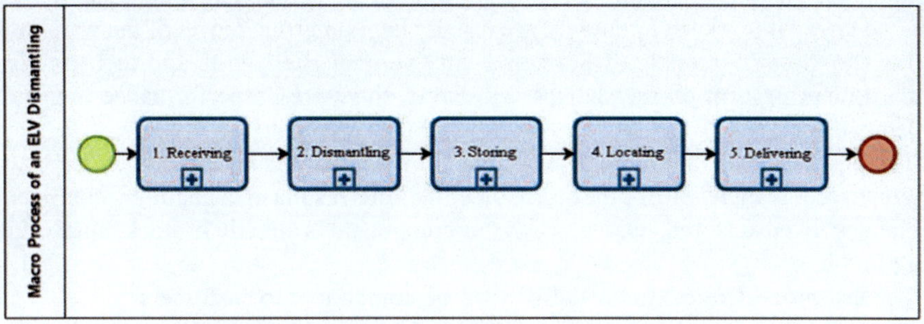

Fig. 1. Macro process of an ELV dismantling (Level 0).

The company follows the process activities imposed by Valorcar [29]. Figure 1 presents the mapping of level 0 (macro process) of the company. All data registration is based on a spreadsheet.

The first process, receiving the ELV, involves the tasks of requesting the necessary documents for cancelling the ownership and executing the cancellation, controlling destruction licenses, de-pollute the vehicle (removal of fluids and hazardous components) and storing it at the company's yard (open storage). All documentation tasks are performed by the administrative sector. In this step, it is not made a management register - as an inventory entry, for example. The handling of the vehicle is performed by internal logistics. They take the vehicle to the place where it is de-polluted and, later, the same employee allocates it back in the company's yard. The location of the vehicle in the yard is currently not documented.

The dismantling process includes the decision and order of which car to dismantle and which components to remove. Next, the vehicle is taken to the dismantling workstation where components are extracted. After dismantling, components are separated in two main types with economical value for the company: (i) components for reuse (e.g. engines); and (ii) components for recycling (e.g., bumpers). The decision concerning which vehicles and components are dismantled, depend on the manager's empirical knowledge. He decides what may or not be dismantled based on his empirical knowledge about potential demand, economic value of the components and (expected) quantities in stock.

The third process, storage, involves storing the compacted carcass and materials for recycling in the yard, besides the storing of the components which will remain in stock for sale. The focus here is in the storage of components. This process begins when an employee reaches the component in the dismantling workstation, uses a white marker and writes the name and model of the component, allocating it in a pre-established shelf. There is a person responsible for the warehouse, but there isn't a standard procedure for storing the components - workers allocate and withdraw the components from the warehouse without any record of these operations.

Regarding the pre-established locations in the shelves, a study about the warehouse layout was never done and the company utilizes a combination of open storage (yard) and dedicated storage (components warehouse). The components organization in the warehouse is done without considering criteria such inventory turnover rates or order piking costs. The only criteria occasionally used are the component weight and dimensions, so heavy items such as engines are usually kept on ground level of the warehouse shelfs. The storage is made of horizontal and vertical shelf units and there is not a standard identification on the shelves. Moreover, there are no performance indicators implemented.

The process of locating/identifying items in warehouse is based on the memory of two employees responsible for the task. Sometimes this results in dismantling component the client wants in a stored vehicle, when this component is already in stock, but workers didn't find it.

The last process refers to the delivery of the component to customers.

4.2 Improvement Propositions

After this initial analysis, it became clear that the process of locating/identifying components is critical and interferes directly with the delivery of value to the costumer, and the company has interest in improving it. Therefore, propositions are focused here.

Warehouse Management System. It is known that a warehouse management system must have tools for supporting the main steps and logistics of the entry process (or reception), production and exit (or expedition) [30]. When looking for possibilities of improvement for this process, it was clear that it was related to all the processes shown in Fig. 1. All could be facilitated if there were records of the components since the arrival of a new ELV at the company, until the delivery of reusable components or recycling materials. Thus, the first suggestion was the implementation of a warehouse management system, which should receive data through all the steps previously described. ABPMP [28] identifies principles that must be applied in a macro process in order to improve it. Among them, there is one about modelling: to invert the modelling course. This means to start the modelling by the deliverables that the process must generate and recognize, in this reverse path, the inputs, references, and the infrastructures associated with each transformation. This principle was utilized to understand how locating/identifying a component at the right moment in time affects every other process performed inside the company. In order to make the moment of the sale efficient - the implementation of a WMS to help locating components - all the other processes were remodelled.

Warehouse Layout. Throughout an analysis process it became clear that there were important improvements to be done in the warehouse layout, concerning the allocation of components to shelfs and identification of components. Although the warehouse is well located in the company site, between the dismantling site and the company store (selling point), what may favour the time spent in movements, its internal layout should be improved. This included developing a codification system for the warehouse' shelves in order to better identify the components location. The suggestion was also to keep the dedicated storage, with a fixed place for each product, and to adopt performance indicators such as the turnover storage, which means that the products are distributed by the storage area, according to the turnover - even though at first there would be scarce data to do it. The dedicated storage smooths the employees' adaption to the changes in the layout and it is useful for components with great variability of weight, decreasing the efforts with displacement; the turnover storage helps the company to also decrease the time spent by employees moving inside the warehouse [31]. The turnover information will be obtained by the WMS.

Components Codification. An alphanumeric coding was suggested to classify the components removed from the ELVs. It is composed by two initial digits, indicating the automobile brand, as it already is utilized by the company (e.g., Audi - AU). This is followed by three digits, which indicate the car model; The next two digits indicate the year of the model and the finally five digits (three for the component type and two for the components variations) indicate the component classification. An example of components variations is the window-winding mechanism there can have four variations for a same car model: right front, left front, right rear, and left rear.

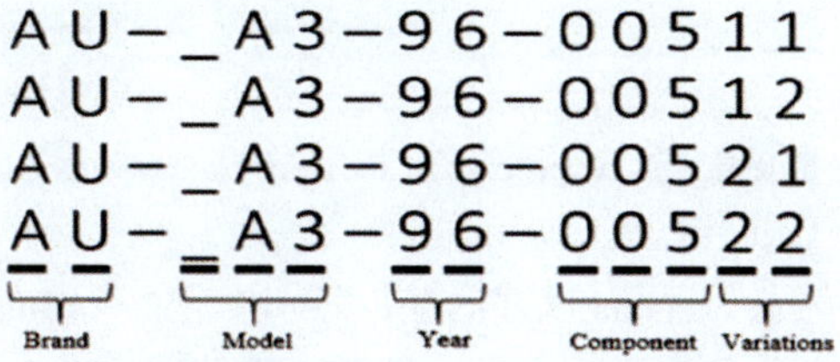

Fig. 2. Example of alphanumeric coding for ELVs components. Source: the authors.

Figure 2 shows an example of component coding. Acknowledging that code 005 refers to window-winding mechanism, the four codes mean that there are four variations for this type of component. This type of coding is useful even at the arrival of ELVs at the company, because the coding of brand, model, and year can be used. Some components may be used by more than one car brand or model (e.g. engine). This way, the component coding and its variations can be used in isolation, to locate this type of components. Searching in a data base for the last five digits, will allow all similar components can be found regardless of its brand and model. The storage coding can also be done by addressing. Addressing is a technique that facilitates locating the items. In the addressing system, the warehouse can be divided into places, streets (shelves), columns, and levels [32]. We adapted this system for the company in the following way:

- W_ = W0 (external warehouse); W1 (main warehouse); W2 (second floor warehouse); W3 (store storage); W4 (warehouse under construction).
- S_ = S1; S3; … S25 (Left side shelves – odd numbers); S2; S4; … S26 (Right side shelves – even numbers).
- C = A, B, C, D, E (corridor letter)
- L = 0 (ground); 1 (level 1); 2 (level 2); 3 (level 3); 4 (level 4); 5 (level 5).

The warehouse location may be implemented using bar codes or RFID to facilitate the register of new components in the system.

4.3 Remodelled Processes

After above improvement proposals, the receiving, dismantling, storing, locating, and delivering processes were remodelled. This subsection describes the changes each process suffered.

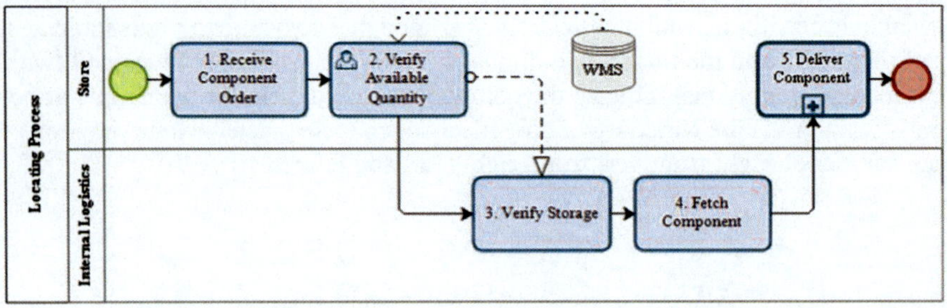

Fig. 3. Modeling of the process of locating components (Level 1).

The receiving process was changed to consider the tasks of inserting the ELV data (brand, model, and year), which of its components can be reused and where is in the company yard. This way, even though the ELV is not already dismantled, its components are already considered as part of company storage. The responsible for the external logistics can make this register with a mobile device or a tablet. Each vehicle can be coded as it is suggested here, as well as they can receive bar codes or RFID, to facilitate the communication with the system, decrease spent time, cost, and errors in the collection and insertion of data [33].

The dismantling process was provided with information concerning components demand and in-stock availability, etc. to support decision making. Demands for this type of business is hard to be determine, however, authors like Hao et al. [34] already suggested a way of do it. The information from administration to external logistics can be done via WMS. The storage process will substitute the white marker for labels with the component's codes. After dismantling each component must be registered in the system and updated with the new location inside the warehouse (replacing the first one given in the yard). With these changes, it is expected to decrease the customer's waiting time, which gives a better perception about the company's value.

In this way the locating/identification of components in the warehouse will not depend on employee's memory. All the locating can be done via the WMS (Fig. 3). The information flow between the company store and internal logistics can also be done in the WMS. At the delivering process, the sale must be registered in the same system so that the component is no longer accounted as available.

5 Conclusions and Future Research Work

Warehouse management in ELVs companies is not trivial, once the uncertainty of supply and demand for ELVs components creates a wide range of storage needs. This essential result from the large number of components and the large variability in size, weight, shape and volume of these.

In this paper, we presents some of the solutions developed for the problem of locating/identifying components at the components warehouse of Portuguese ELV dismantling company. It is acknowledged by the authors that the improvements discussed in here are not of instant implementation, given that data required by the electronic warehouse management systems. However, it constitutes a basis for the development of a strategic management of processes and a starting point for a culture of continuous improvement.

Future research should integrate demand forecast and financial issues in the warehouse management system, in order to allow for decision making considering the potential impact on the company's profit.

References

1. Cao, J., Chen, X., Zhang, X., Gao, Y., Zhang, X., Kumar, S.: Overview of remanufacturing industry in China: government policies, enterprise, and public awareness. J. Clean. Prod. **242**, 118450 (2020)
2. Kosacka, M., Golińska, P.: Assesment of sustainability in dismantling station-case study. Res. Log. Prod. **4**(2), 135–145 (2014)
3. Maia, A., Souza, G., Melquiades, J.A.R., Assunção, L., Arruda, F.: Logística reversa de veículos no fim de vida: a realidade internacional e os desafios do Brasil com vistas à sustentabilidade ambiental. Transport postgraduate program, Universidade de Brasília, Brazil (2009)
4. Tian, J., Chen, M.: Sustainable design for automotive products: dismantling and recycling of end-of-life vehicles. Waste Manage. **34**(2), 458–467 (2014)
5. Phuc, P.N.K., Vincent, F.Y., Tsao, Y.C.: Optimizing fuzzy reverse supply chain for end-of-life vehicles. Comput. Ind. Eng. **113**, 757–765 (2017)
6. Sakai, S.-I., et al.: An international comparative study of end-of-life vehicle (ELV) recycling systems. J. Mater. Cycles Waste Manage. **16**(1), 1–20 (2013). https://doi.org/10.1007/s10163-013-0173-2
7. Silva, J.R.B.D.: Tratamento de veículos em fim de vida: modelos de gestão internacionais e brasileiro. Masters dissertation, Universidade Federal de Santa Catarina, Brazil (2016)
8. Hasibul, I.M., Gustav, B., Malin, T.: Adoption of lean philosophy in car dismantling companies in Sweden-a case study. Procedia Manuf. **25**, 620–627 (2018)

9. Kudelska, I., Kosacka, M., Lewandowska, K.W.: How to support storage process in dismantling facility with IT solutions? – case study. Procedia Manuf. **11**, 703–710 (2017)
10. Braga, L.M., Pimenta, C.M., Vieira, J.G.V.: Gestão de armazenagem em um supermercado de pequeno porte. Revista P&D em Engenharia de Produção **8**, 57–77 (2008)
11. Zhang, R.Q., Wang, M., Pan, X.: New model of the storage location assignment problem considering demand correlation pattern. Comput. Ind. Eng. **129**, 210–219 (2019)
12. Ene, S., Öztürk, N.: Grey modelling based forecasting system for return flow of end-of-life vehicles. Technol. Forecast. Soc. Chang. **115**, 155–166 (2017)
13. Rosa, P., Terzi, S.: Improving end of life vehicle's management practices: an economic assessment through system dynamics. J. Clean. Prod. **184**, 520–536 (2018)
14. Ene, S., Öztürk, N.: Network modeling for reverse flows of end-of-life vehicles. Waste Manage. **38**, 284–296 (2015)
15. Xiao, Z., Sun, J., Shu, W., Wang, T.: Location-allocation problem of reverse logistics for end-of-life vehicles based on the measurement of carbon emissions. Comput. Ind. Eng. **127**, 169–181 (2019)
16. Vermeulen, I., Van Caneghem, J., Block, C., Baeyens, J., Vandecasteele, C.: Automotive shredder residue (ASR): reviewing its production from end-of-life vehicles (ELVs) and its recycling, energy or chemicals' valorisation. J. Hazard. Mater. **190**(1–3), 8–27 (2011)
17. Harraz, N.A., Galal, N.M.: Design of sustainable end-of-life vehicle recovery network in Egypt. Ain Shams Eng. J. **2**(3–4), 211–219 (2011)
18. Olugu, E.U., Wong, K.Y., Shaharoun, A.M.: Development of key performance measures for the automobile green supply chain. Resour. Conserv. Recycl. **55**(6), 567–579 (2011)
19. Go, T.F., Wahab, D.A., Rahman, M.A., Ramli, R., Azhari, C.H.: Disassemblability of end-of-life vehicle: a critical review of evaluation methods. J. Clean. Prod. **19**(13), 1536–1546 (2011)
20. Östlin, J., Sundin, E., Björkman, M.: Product life-cycle implications for remanufacturing strategies. J. Clean. Prod. **17**(11), 999–1009 (2009)
21. Gupta, S.M., Lambert, A.F.: Disassembly Modeling for Assembly, Maintenance, Reuse and Recycling. CRC Press, New York (2016)
22. Duval, D., MacLean, H.L.: The role of product information in automotive plastics recycling: a financial and life cycle assessment. J. Clean. Prod. **15**(11–12), 1158–1168 (2007)
23. Gerhardt, T.E., Silveira, D.T.: Métodos de pesquisa. Editora da UFRGS, Porto Alegre (2009)
24. Silva, E.L.D., Menezes, E.M.: Metodologia da pesquisa e elaboração de dissertação, 4th edn. UFSC, Florianópolis (2005)
25. Gil, A.C.: Como elaborar projetos de pesquisa. 4th edn. Atlas, São Paulo (2002)
26. Kim, S.H., Jang, K.J.: Designing performance analysis and IDEF0 for enterprise modelling in BPR. Int. J. Prod. Econ. **76**(2), 121–133 (2002)
27. Wiendahl, H.P., Seliger, G., Perlewitz, H., Bürkner, S.: A general approach to disassembly planning and control. Prod. Plann. Control **10**(8), 718–726 (1999)
28. ABPMP: Guia para gerenciamento de processos de negócio - corpo comum de conhecimento (BPM CBOK), versão 3.0, 1st edn., Brazil (2013)
29. Valorcar: Guia de desmantelamento de veículos em Fim de Vida. Lisboa, Portugal (2015). http://www.valorcar.pt/images/uploads/library/Guia_desmantelamento_VFV_3a_versao_1505988119.pdf
30. Assis, R.D., Sagawa, J.K.: Avaliação da implantação do Sistema de Gestão de Armazém em uma empresa multinacional do ramo de acionamentos. Gestão Produção **25**(2), 370–383 (2018)
31. Koster, R., Le-Duc, T., Roodbergen, K.J.: Design and control of warehouse order picking: a literature review. Eur. J. Oper. Res. **182**(2), 481–501 (2007)

32. Catto, M.F., Oliveira, K.F.R.: O endereçamento como ferramenta na armazenagem e estocagem. III Gepro Scientific Meeting. Fatec. Brazil (2013)
33. Groover, M.P.: Automação industrial e sistemas de manufatura. Pearson Education do Brasil, Brazil (2011)
34. Hao, H., Zhang, Q., Wang, Z., Zhang, J.: Forecasting the number of end-of-life vehicles using a hybrid model based on grey model and artificial neural network. J. Clean. Prod. **202**, 684–696 (2018)

The Impact of Women's Absenteeism on an Electric Components' Industry

José Dinis-Carvalho[1], Ana Lima[2], Ana Machado[2], Bárbara Santos[2], Daniela de Almeida[2](✉), Eva Correia[2], Maria Moniz[2], Rita Ferraz[2], Sara Machado[2], Susana Lage[2], and Tatiana Guimarães[2]

[1] Production and Systems Department, School of Engineering, University of Minho, Guimarães, Portugal
dinis@dps.uminho.pt

[2] Industrial Management and Engineering, School of Engineering, University of Minho, Guimarães, Portugal

Abstract. A common finding in absenteeism research is that women are, on average, more absent from work than men. Currently, there is a particular absence culture for women since gender inequality is still a problem in the Portuguese' workplace. As it is exposed in this article, there are several causes of absenteeism associated with gender that are qualitatively different for men and women. In order to face this problem, a study was developed on a real company producing electric components, although due to confidentiality it is not possible to name the company. By using lean principles and concepts and improving women's working conditions through different line operation mode, ergonomic changes and an incentive system, employers can reduce women's absenteeism but also have a significant impact on the companies' revenues. A Rabbit Chase operating mode was simulated and the results are promising in terms of reducing the impact of absenteeism in the production performance as well as having the potential of reducing the absenteeism itself.

Keywords: Absenteeism · Gender equality · Simulation · Lean production

1 Introduction

Absenteeism is frequently defined in dictionaries as a habitual pattern of absence from a duty or obligation without good reason. The part "without good reason" is not shared by every definition but even with or without a good reason, employee's absenteeism is a challenge that many companies must deal every day. Absenteeism definitely translates into a significant negative impact on companies' performance. Absenteeism is a cost for organizations since it frequently leads to labour productivity reduction, greater difficulty in meeting delivery dates, greater difficulty in planning the work, increase in stress at work, as well as other negative consequences. Moreover, in the long run, absenteeism can culminate in an employee getting fired or even quitting. Consequently, the recruitment, training and adaptation process of the new employee can have negative impacts on the

J. Machado et al. (Eds.): icieng 2021, LNME, pp. 220–233, 2022.
https://doi.org/10.1007/978-3-030-78170-5_20

normal operation of the company. Some interesting insights about cost of absenteeism is presented by [5].

In 2017 the absenteeism rate in Portugal was of 6.81% according to Portuguese institute of statistics [7] being sickness the main cause, representing 45,91% of absence. Sickness is also referred in other studies as being the main expression of absence behaviour [9]. Other common causes of absenteeism are family responsibilities, pregnancy and maternity leave, minor illnesses, acute medical conditions, injuries, burnout, stress among others. Going back to the Portuguese reality, on average in 2018 the Portuguese workers missed almost six days (5.9) of work per year, due to sickness [12] resulting in an estimated 2.200 million euros in costs for the national health system. It becomes quite clear that the absenteeism plays a major role in the economy and for that and many other reasons should be reduced.

Companies around the world face this problem with greater or lesser difficulty and with greater or lesser impact on their performance. This study takes place in a company where the managers are very concerned about the levels of absenteeism among their female workers. The managers are committed in finding ways of reducing this problem and took advantage of an interesting opportunity of having external help. This study was conducted by a team of industrial engineering and management students while carrying out a Project Based Learning program at the University of Minho in the north of Portugal. This one semester project took place in a local company, with around 300 employees, mainly dedicated to the assembly of electric components. One of the problems presented to the students to solve was high absenteeism rate among the female employees, which were around 90% of the workforce in the assembly cells. This absenteeism was having an extreme negative impact in the productivity and response capacity of the company.

2 Absenteeism and Gender

Since in this study the main problem was the female absenteeism it is important to understand the possible differences between female and male absenteeism. Since there is little literature on the subject and limited statistical data on absenteeism and its causes in Portugal, this topic turned out to be very relevant. In order to understand the context of the Portuguese population, it was collected statistical information from Pordata [14]. This is a certified statistic database with data related to Portugal.

The active Portuguese population by age group is presented in Table 1. The age groups with the higher number of people working were in 2018 the age groups from 35 to 44 years of age with a bit more than 1384 thousand workers followed by the age group from 45 to 54 years of age with a bit more than 1322 thousand workers. This is particularly important because the absenteeism can higher in these age groups because people are more likely to get ill and miss work.

Table 1. Active population: total and per age group [15]

Year	Age group (thousands of individuals)					
	Total	Under 25 years	25–34	35–44	45–54	55–64
2005	5461,4	541,3	1446,2	1367,0	1150,7	631,1
2010	5489,7	418,2	1325,4	1450,0	1273,1	699,3
2015	5195,2	369,5	1083,6	1443,3	1284,3	768,7
2016	5178,3	364,2	1054,8	1429,3	1295,6	795,6
2017	5219,4	371,3	1033,4	1407,5	1314,3	845,6
2018	5232,6	371,9	1016,4	1384,2	1322,5	881,2

Another factor that can be relevant is that although the average monthly basic wage of workers has been steadily increasing over the years, the difference between genders is maintained over the years, with men earning in average more than women. This fact that can be seen in Table 2 may lead to more absenteeism, as women can feel less motivated [17].

Table 2. Average monthly subsidy: total and per gender [16]

Year	Gender (euro - average)		
	Total	Masculine	Feminine
2005	764,7	832,5	672,0
2010	899,0	976,7	800,8
2015	913,9	990,1	825,0
2016	924,9	997,4	840,3
2017	943,0	1012,3	861,2

In Portugal, the gender salary gap tends to widen as the population gets older [17]. For workers under 30, the gender salary gap is € 5.6 per hour and this value continues to increase with age.

2.1 Causes of Absenteeism and Its Relationship with Gender

Despite all the progress made so far in promoting gender equality at work, when talking about differences in absenteeism values the reality is indeed disquieting. In general, there are stereotypes about women that create a culture of expectation and legitimacy around female absenteeism and lead to real absenteeism. In fact, it is often reported in the literature that absenteeism in women is higher than in men, and this is usually justified by several reasons. Studies show that women usually have a weaker immune system than men and ill more often. Others point maternity leave and childcare as absenteeism causes since women are usually associated with the role of caregiver [13]. Additionally, the authors point to bullying and stress as a cause of female absenteeism.

Women and absenteeism is a topic that has been discussed by politics since World War I and that has come up once more over World War II. Since then, absenteeism cultures have been settled according to gender, namely that absenteeism would be more acceptable for women than men. Several reasons have also been invoked to explain women's absenteeism culture such as less interesting jobs, poorer working conditions, more fragile psychological and physical health, family and childcare responsibilities. Finally, it can be argued that women are more absent because it is more socially acceptable for them to be [4].

Regarding the Sickness, and according to the same authors, women tend to call more sick days since they are the ones who suffer the most from physical symptoms such as insomnia, migraine headaches and depression. According to the Order of Psychologists, in Portugal, absenteeism has increased due to work-related stress and mental health problems. In addition to absenteeism, the consequences of poor mental health create numerous negative effects on businesses, such as reduced performance and productivity, low motivation and high employee turnover [2].

Regarding occupational diseases, in 2017 women had more occupational diseases than men and the age group that most suffered from occupational diseases was the one between 50 and 54 years old, as shown on Table 3 [1].

Table 3. Occupational diseases: total and per gender

Age group (years)	Total		
	Feminine	Masculine	Total
25–29	22	8	30
30–34	88	43	131
35–39	243	78	321
40–44	359	136	495
45–49	439	161	600
50–54	670	233	903
55–59	501	228	729
60–64	211	164	375
65–69	7	26	33
Total	2.554	1.087	3.641

Moreover, according to the Portuguese Health Ministry the main disease that is causing absenteeism amongst women is musculoskeletal injuries, specifically low back pain and neck pain [11], which are also the most common injuries work related.

Maternity Leave. When compared to men, women take more days off on maternity leave. In 2007, 92.6% of the women that had children took a maternity leave. Consequently, only 7.4% of the men took paternity leave. This means that women tend to be absent from work longer than men [3].

Assistance to Family Members. In most families the role of women is unequivocally linked to the role of caregiver, either because men refuse this responsibility, or because women prefer to be carers. On the other hand, for many couples, the choice of who should be absent from work to care for their family is a result of a rational economic choice since women earn on average 18.2% less than men [4]. The problem is that while maximizing couple income, women enter a cycle, where lower wages justify higher absenteeism which in turn justifies lower wages [10].

Moreover, about 18.1% of all women aged 15–64 regularly provide care for children under the age of 15 (in addition to their own children or spouse's children) and to sick, disabled or elderly people. In total, about 64.4% of caregivers were women [6].

2.2 The Costs of Absenteeism

The absence from work may be justifiable or unjustifiable. Absences motivated by the impossibility of working due to illness, accident, assistance to a child, grandson or household member are considered justified. Any other absence, such as maternity leave, is considered as unjustified [18].

Justified absences and those authorized by the employer are paid by the company. Concerning absences due to pregnancy, these are considered parental leave. In Portugal, this leave has a duration of 120, 150 or 180 consecutive days and is paid by Social Security [18].

In Portugal, direct and indirect costs of absenteeism are around 3.6% of Gross Domestic Product. Half a million Portuguese do not work daily for unjustified absences or casualties, representing about 3.7 billion euros per year [8].

This chapter emphasised the importance of studying absenteeism and the role it plays in companies, highlighting its high costs. In this context, it is crucial that employers properly manage their human resources and the absence of their employees, as this absence can cause a serious problem when it goes beyond normal limits.

3 Statistical Data Collected in the Company Under Study

First of all, to develop this study it is important to know this company population, their characteristics, in order to connect these aspects with the absenteeism involved in the company. The main features of this population that are going to be analysed are: age range, gender, level of education, type of worker and working time. All values shown on Table 4 and Table 5 were given by this company from 2018 data.

Table 4. Company's population: age range (%)

Age range	18–30	31–40	41–60
(%)	12%	8%	81%

Table 5. Company's population: gender (%)

Gender	Female	Male
(%)	82%	18%

By interpretation of these results, it is easily concluded that the majority of the company's workers are female (82%), have age comprehended between 41 to 60 years old (81%) and are working on the company for more than 20 years (66%). Also, concerning to the level of education of these workers, 34% of them have only done the 3rd cycle and almost half of the population has studied only until High School (48%).

Overall, the workers' population is mostly women and have been working on the company for a long time that reflects on the high number of people being on the oldest age group of 41 to 60 years old.

3.1 Causes of Absenteeism

According to the company data related to causes of absenteeism among their employees the numbers are presented in Table 6. The main cause of absenteeism, with 38%, is the professional off sick meaning that workers skipped work because of an illness that was developed while working. On the other hand, 33% missed work because they were sick, and it was not work related, while 14% did it because of family assistance.

Table 6. Company's population: causes of absenteeism (%)

Causes of absenteeism (%)	
Off sick	33%
Professional off sick	38%
Family assistance	14%
Others	15%

3.2 Absenteeism and Gender

Firstly, the relation between the causes referred above and the workers' gender was analysed. Through Table 7 it is evident the disparity between men and women absence, since this organization is mostly constituted by women, as shown above. Evidently, the company needs to analyse the causes of this absenteeism as well as figure out what is wrong and try to get around the problem.

Table 7. Company's population: causes of absenteeism by gender (%)

Gender	Off sick	Profissional off sick	Family assistance	Others
Female	91%	100%	96%	84%
Male	9%	0%	4%	16%
Total geral	100%	100%	100%	100%

This company's production processes require a very thorough manual operations. Therefore, women's professional off sick has the higher percentage in comparison to men that have none since man don't do this type of work in this company. Frequently, women are associated with being more precise and delicate. In result they are the ones hired to do this type of work.

The second big disparity is related to family assistance. This reflects on 96% of absence due to family assistance by women. Once again female workers have the highest percentage comparing to male workers. As mentioned before, in most families the role of women is unequivocally linked to the role of caregiver.

3.3 Absenteeism and Age Range

As shown above, the age group that predominates in the company is the oldest one (41–50 years old) and it is the same one that has the highest percentage of absenteeism, as it is possible to see by the results from Table 8.

Table 8. Company's population: causes of absenteeism by age range (%)

Age range	Off sick	Profissional off sick	Family assistance	Others
18–30 years	0%	0%	1%	5%
31–40 years	0%	0%	3%	11%
41–50 years	80%	64%	49%	72%
More than 50 years	20%	36%	47%	13%
Total	100%	100%	100%	100%

Furthermore, most of the workers of this company have been working on the company for a long time, which means they are doing thorough operations for a large period of time. As a result of this, 64% of the worker from 41 to 50 years old, have been absent due to professional sickness and 80% due to not Work-Related sickness, which can be caused by this being one of the oldest age groups (see Table 8 and Table 9).

These results are congruent with the ones shown on Table 9, related to the absence relation with working time, since the highest percentage of absenteeism is on the group that has been working on the organization for more than 20 years. In consequence, 90% of the workers have been absent due to professional diseases and 78% due to not Work-Related diseases. Once again, this happens because people who are working there for the longest time are probably on the oldest age group.

Table 9. Company's population: causes of absenteeism by working time

Working time	Off sick	Profissional off sick	Family assistance	Others
Less than 2 years	0%	0%	1%	7%
2–5 years	0%	0%	0%	2%
6–10 years	8%	0%	1%	6%
11–20 years	13%	10%	54%	25%
More than 20 years	78%	90%	44%	60%
Total	100%	100%	100%	100%

Overall, the group that has more impact on this company's absenteeism are the oldest women, with age comprehended between 41 and 50 and have been working on the company for more than 20 years.

4 Current Situation of the Assembly Process

The production area designated to this project is dedicated to the assembly of circuit breakers of 3 different types. The 3 types of products are assembled every day according to the master production schedule, but no line is assigned to a particular product.

There are 5 assembly lines with 5 workstations each (see Fig. 1) producing a mix of these three types of products. In the current condition the workers are allocated to fixed position in the assembly cell, changing to a different workstation every 4 h. These lines work in an 8-h shift per day. The shift does not correspond exactly to 8 h but to 457 min to be precise because of cleaning and a team meeting.

The absenteeism problem can be viewed here in the following two different perspectives: (1) How to reduce absenteeism and (2) how to reduce its impact in the production performance. The perspective that may be easier to solve is the second one and that was the reason why the team started to approach the problem. It was necessary to understand how workers' absenteeism was influencing the production performance. One question was raised: is the absenteeism just influencing the production output or also the productivity?

From the data collected for a quite long period resulted that when there are 5 workers in the line the production output is in average 232 units per shift. The team create a

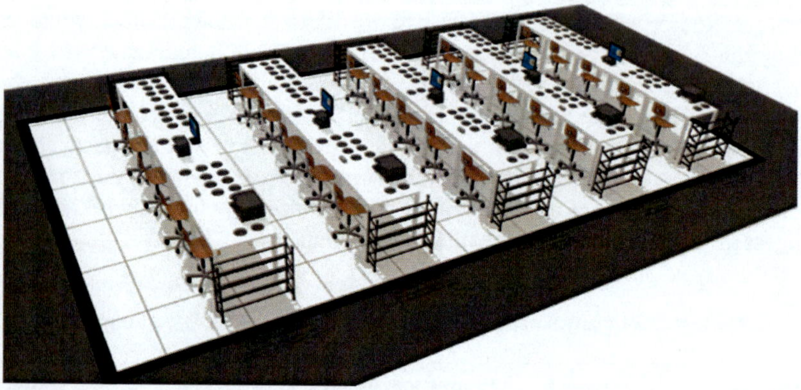

Fig. 1. Representation of the existing layout.

simulation model using SIMIO software representing the reality as much reliable as possible (see Fig. 2).

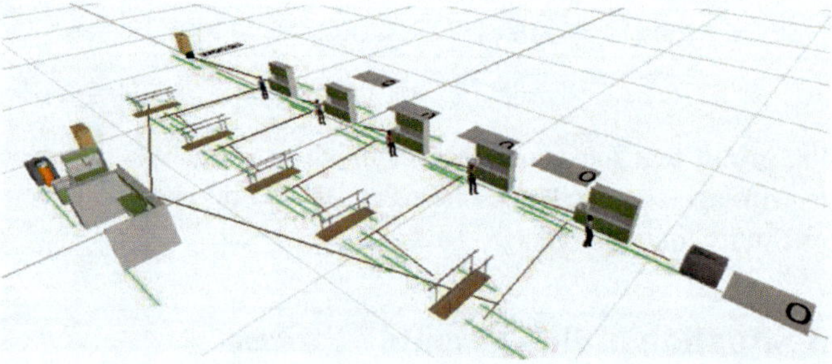

Fig. 2. Simulation model for one line.

The results obtained from the simulation model in terms of production output during one shift with 5 workers is quite similar to the average obtained from the production records. The real production is 232 units per shift in average and 233 units per shift was obtained from the simulation model (see Table 10). The workers' utilization rate obtained was very high (97.24%) and with the Work-In-Process was of 14 products in average.

When running the same simulation model with one worker missing and using the same kind of solution in terms of workers organization as the real solution the results in performance are greatly reduce as expressed in Table 10. This is caused by the highly unbalanced line resulting from the lack of one worker. With one worker missing the line produces only 123 devices per shift with productivity of only 30.75 units/worker.shift and with a Work-In-Process of 124 products. The productivity drops dramatically 34%.

By analysing the table above, it is possible to realize that when one worker is missing and another one from the same line must work on two different jobs, the line produces less 110 products, which means a decrease of 47,21%.

Table 10. Number of devices produced with one worker missing and without absenteeism.

Number of worker	Production output (units/shift)	Productivity (units/worker.shift)
4 workers	123	30.75
5 workers	233	46.6

When comparing the Work-In-Process from the different scenarios, the conclusion is that this number is almost 9 times bigger when one worker is missing. The reason behind this amount of WIP is that because of the unbalance nature of the line with only 4 workers the amount of WIP is necessary to keep all four workers busy all the time.

5 Suggestions to Decrease Absenteeism Rate

From the analysis of the previous results, it is possible to evaluate what needs to be modified in order to reduce the impact of the absenteeism in the production performance as well as reducing the impact of the existing production approach in the absenteeism. The assumption is that if the repetition of the same operations decreases then the risk of developing musculoskeletal disorders also decreases. With this purpose, one of the next steps to meets project goals is the attempt to implement a different operating mode for the line, including more job rotation, especially some kind of Rabbit Chase operating mode.

5.1 Implementation of Rabbit Chase

The Rabbi Chase operating mode has not only the advantage of being easy to set up and manage, but it also leads to low inventories and fast lead times. By using Rabbit Chase, it is expected that the lack of one or more workers will have almost no impact on productivity and production costs. Furthermore, as each operator is responsible for producing a product from the start until the end, which is not the current case, they have more responsibility. This leads to better quality of the products.

In order to prove the benefits of this operating mode that were stated before, SIMIO software was used once again. Using the same simulation run time, the scenarios of having 4 or 5 workers were tested, as it was performed before. A comparison between the existing operating mode with the Rabbit Chase operating mode in terms of production output and productivity is presented in Table 11.

The line produces less 15 products when changing the operating mode to Rabbit Chase with 5 workers. This reduction in productivity (from 46.6 to 43.6 units/worker.shift) was expected since this operating mode leads to waiting times when the number of workstations is the same as the number of workers and job stations are not perfectly balanced. On the other hand, when one worker is missing, the production output from the line only reduces 38 products, contrary to the fixed positions' scenario, where there is a negative production difference of 110 products. The productivity when one worker is missing is even higher than when all workers are present due to the reason

Table 11. Production output and productivity in the different scenarios.

	Production output		Productivity	
	Fixed position	Rabbit chase	Fixed position	Rabbit chase
4 workers	123	180	30.75	45
5 workers	233	218	46.6	43.6

explained before. This means that the impact of an employee missing work is much lower with Rabbit Chase. Due to the fact that there is an extremely high level of absenteeism in the company, this is the best option for them. Based on these results, since workers are missing every day, this Rabbit Chase option results better if all 5 lines work with 4 people than the existing solution where some lines work with 5 workers and one line with less than 5. The WIP comparison presented in Table 12 shows the benefits of the proposed operating mode.

Table 12. Total of work-in-process in the different scenarios (units).

	Fixed position	Rabbit chase
4 workers	124	4
5 workers	14	4

Besides the production differences, with Rabbit Chase there is a huge decrease in the Work-In-Process, as it can be seen in the table above. This is especially notorious with 4 workers, with a difference of about 3100%. This means that by adopting this operating mode, there is going to be less waste.

To conclude, this operating mode also reduces the utilization rate of workers (see Table 13). This may lead to less stress and, consequently, less absenteeism. Since in the Rabbit Chase mode each worker must perform all assembly and testing operations this leads to less risk for the development of musculoskeletal issues and less absenteeism in the long run. Moreover, in this new operating mode workers are standing and moving a bit from workstation to workstation instead of being sitting on a chair for 4 h straight.

Table 13. Workers' utilization rate in the different scenarios.

	Fixed position	Rabbit chase
4 workers	98,75%	93,57%
5 workers	97,24%	89,70%

5.2 Other Suggestions

In addition to the previous solution, there are several other important options, such as:

Rewards and Recognition. In the company under study there is a Post-It system where employees can make suggestions to improve their jobs. Every week, there is a team that analyses these proposals and checks if they are good changes to the workers. When the suggestion is applied, the company rewards the worker who suggested it, financially or otherwise. Although there is no study on the direct relation between this reward system and a less absenteeism rate or better performance, this system of recognizing the workers may be an option to overcome this problem.

Improve Supervisory Quality. The assembly line under study has a line manager who supervises and assists workers in their tasks while providing time for mutual communication. Still, it is always important that other managers do it regularly as well.

Promotion Expectations. By making employee effort result in a reward, including career advancement, is important. However, unlike what is explained above in the Rewards and Recognition point, this possibility is unknown in the company. Still, it is considered important that this can happen by increasing workers' motivation and interest in work.

Improve Workplace Conditions. In addition to improving the line layout, the aim is to improve workstation conditions at an ergonomic level. Among the most highlighted problems, after some observations and interviews with employees, it was found that the work chairs are not padded and little adjustable to the person, the foot rests are not adjustable in all workstations and the small dimensions in the stations and low shelves limits the operators' movements. In terms of raw material, it has small dimensions, which results in a high visual effort due to the exhaustive activities which results in detailed tasks and requires operators to adopt critical positions. In this sense, after some ergonomic evaluations, it is intended to improve the positions adopted by alternating sitting and standing, modifying benches and enlarge their workspace.

These may be possible solutions to decrease absenteeism in the company, always remembering that satisfied workers are more productive. However, a broader and more comprehensive reflection of the problem is essential in order to change absentee behaviours that are often not only related to workers' satisfaction at work. This improves not only the company's performance and its cooperative image, but also the productive costs that it loses.

6 Conclusion

Women's absenteeism is definitely higher than men's, therefore it is important to understand the causes of this absenteeism and its social context. It was proved that women continue to be the ones who take parental leave and provide care to family members, for this reason they are the main caregivers. Moreover, women still earn less than men. Consequently, in a situation where one spouse must stay at home, the woman is eventually chosen.

Since the company under study is mostly composed by women and they are the ones who are more absent, their reasons are not controllable. As a result of that, the company must be the one trying to adjust their productive system to that problem: women's absenteeism. So, in order to reduce the organization's absenteeism, it was suggested the adoption of a new operating mode that allows increasing the operators' versatility and turnover. It was also shown the impact that the absence of an operator causes if she misses work, as it can be seen in the simulation model presented.

Besides this, one of the major causes of absenteeism in the company is occupational disease. Therefore, it is important to use ergonomic studies and suggest improvements in this area to combat this cause. Finally, the adoption of methodologies and techniques that increase workers' motivation and the use of salaries incentives will be fundamental to decrease companies' absenteeism rate.

Acknowledgments. This work has been supported by FCT – Fundação para a Ciência e Tecnologia within the R&D Units Project Scope: UIDB/00319/2020.

References

1. ACT: ACT - Autoridade para as Condições do Trabalho: Relatório 2017, Lisboa (2018)
2. Antunes, S.: OPP envia carta às mil maiores empresas em Portugal | Ordem dos Psicólogos (2013). https://www.ordemdospsicologos.pt/pt/noticia/1065
3. CITE: A igualdade de mulheres e homens no trabalho e no emprego em portugal - Políticas e circunstâncias (2010)
4. GEP: Análises e Notas Técnicas. http://www.gep.mtsss.gov.pt/documents/10182/80545/Igualdade_salarial_8_11_2019.pdf/9961bdba-252a-4c80-94d6-8df7e965fdb1?fbclid=IwAR0gtmJ6D6Btn3tn5seMn7zX4lUxyUQQ0chiLKfik5oe7J4minEilF5LzPk
5. Huczynsk, A., Fitzpatrick, M.: End of the mystery - calculating the true cost of employee absence. Empl. Relations **11**(6), 12–15 (1989). https://doi.org/10.1108/01425458910134021
6. INE: Estatísticas no Feminino: Ser Mulher em Portugal. Instituto Nacional de Estatistica (2012)
7. Instituto Nacional de Estatística: Balanço Social (2017)
8. Lopes, R.: Absentismo pode duplicar os custos das empresas – JN (2004). https://www.jn.pt/arquivo/2004/absentismo-pode-duplicar-os-custos-das-empresas-463334.html
9. Macdonald, E.B., Asanati, K.: Absence from work. Contemp. Soc. Sci. **11**, 1 (2016). https://doi.org/10.1080/21582041.2016.1246749
10. Maximiano, S.: Expresso | Ciclos viciosos do absentismo feminino. https://expresso.pt/blogues/bloguet_economia/blogue_econ_aurora_teixeira/2017-09-15-Ciclos-viciosos-do-absentismo-feminino
11. da Saúde, M.: Retrato da Saúde. Serviço Nacional de Saúde, Portugal (2018)
12. Observador: Estudo. Os portugueses faltaram quase seis dias ao trabalho por doença, o que resultou num prejuízo de 2,2 mil milhões de euros – Observador. https://observador.pt/2019/04/02/estudo-os-portugueses-faltaram-quase-seis-dias-ao-trabalho-por-doenca-o-que-resultou-num-prejuizo-de-22-mil-milhoes-de-euros/
13. Patton, E., Johns, G.: Women's absenteeism in the popular press: evidence for a gender-specific absence culture. Hum. Relations **60**, 11 (2007). https://doi.org/10.1177/0018726707084301
14. Pordata: PORDATA - Estatísticas, gráficos e indicadores de Municípios, Portugal e Europa. https://www.pordata.pt/

15. Pordata: PORDATA - População activa: total e por grupo etário. https://www.pordata.pt/Portugal/População+activa+total+e+por+grupo+etário+-29
16. Pordata: PORDATA - Remuneração base média mensal dos trabalhadores por conta de outrem: total e por sexo. https://www.pordata.pt/Portugal/Remuneração+base+média+mensal+dos+trabalhadores+por+conta+de+outrem+total+e+por+sexo-367
17. Torres, A.: Igualdade de género ao longo da vida: Portugal no contexto Europeu. Fundação Francisco Manuel dos Santos (2018)
18. Decreto Lei n.º 7/2009 de 12 de fevereiro. Codigo do Trabalho, Portugal (2009)

Time Series Analysis for Anomaly Detection of Water Consumption: A Case Study

Marta Santos[1], Ana Borges[1], Davide Carneiro[1], and Flora Ferreira[2](✉)

[1] CIICESI, ESTG, Politécnico do Porto, Rua do Curral, Casa do Curral, Margaride, 4610-156 Felgueiras, Portugal
{8150180,aib,dcarneiro}@estg.ipp.pt

[2] Center of Mathematics, University of Minho, Campus de Azurém, 4800-058 Guimarães, Portugal
fjferreira@math.uminho.pt

Abstract. Water loss is one of the factors that most affect a concessionaire's financial sustainability. Early detection of any anomaly in water consumption is very valuable. This article aims to carry out a preliminary study to detect change points in consumption associated with water meter malfunction. The dataset is composed of water consumption measurements of two different companies (a hotel and a hospital) located in the north of Portugal, obtained during a complete year. Different methods were implemented in order to study its effectiveness in the detection of change points in the time series related to a sharp decrease in water consumption. Results suggest that the Seasonal Decomposition of Time Series by Loess method (STL) and the combination of several breakpoint detection methods is a suitable approach to be implemented in a software system, in order to help the company in anomaly detection and in the decision-making process of substituting the water meters.

Keywords: Time-Series Decomposition · Breakpoint detections · Water losses · Water meters

1 Introduction

Water is an essential good for human life, as water resources are increasingly scarce, so it is necessary to use measures to bridge current and future gaps in water supply and demand [1].

Water loss could have multiple causes including metering errors, leakage, public usage such as firefighting, and theft [2]. Defective water meters do impair water supply management, but they also affect the financial sustainability of utilities, justifying the need for a detailed analysis of water consumption data by meter [3]. The apparent loss has a significant role in water loss because these represent 30–40% of total leakage [4].

J. Machado et al. (Eds.): icieng 2021, LNME, pp. 234–245, 2022.
https://doi.org/10.1007/978-3-030-78170-5_21

In the literature, we can find various techniques for loss detection in water meters [5], suggesting the creation of indicators based on age, pressure and volume water estimating the meter replacement period [6].

This is a preliminary study of water consumption time-series, in order to implement/construct an algorithm that will be able to determine structural changes associated with water meter malfunction. The long-term purpose is to implement the algorithm in order to help a company, Águas do Norte, in anomaly detection and in the decision-making process of substituting the water meters.

To test de adequacy of the methods proposed in this study we apply them in two case studies: a hotel and a hospital in the time-horizon between 01/12/18 and 30/11/19.

This article is divided into five parts: the first part summarizes the existing literature on losses in water meters and factors that may lead to measurement failures. The second part describes the case study and the dataset, followed by the methodology chapter description. In the fourth, the part we perform an analysis of the results, ending with a brief discussion and conclusion of the results obtained, where remarks on the future investigation are also described.

2 Literature Review

Due to the consumption of uninvoiced water, concessionaires may experience breaks in their financial sustainability. Recently, regarding a European study, Liemberger and Wyatt [7] reported losses of approximately 26.8 million m^3 per day, equivalent to 9.8 billion m^3 in losses in a year. Another result of this same study is that Europe occupies the 6th position in 13th positions with the biggest breaks on the sustainability of concessionaires.

According to the water balance of the International Water Association (IWA), are able to obtain the amount of water without a prescription (Non Revenue Water (NRW)) through the difference between the amount of water that is introduced into the system and the amount of water that is billed by companies to their customers [8]. Water without revenue not only corresponds to actual and apparent losses but also to authorized and uninvoiced consumption. All systems have water losses [9], and it is neither possible nor economically viable to reduce them in their entirety. The reduction, however small, has a positive impact on companies revenues and is also one of the reasons why they seek new methods to avoid them.

Water meters are devices used by utilities to measure and bill the consumption of water that has been made available and consumed by their customers. No water meter is capable of registering 100% of the volume that passes, but there are more accurate meters according to the type of flow that passes through it [10].

Real losses or physical losses correspond to the volume of water that is lost in the network and in the infrastructure of the management entities [9]. These losses are usually caused by leaks in the pipes, valves, and other components of the network [11]. When there are high quantities of water in the network, and they remain for a long period of time there is a greater likelihood of leaks and broken pipes or plumbings [8]. Real losses exist in greater quantities and are more difficult to detect by concessionaires.

Unbilled authorized consumption is the use of water but the concessionaire has no monetary return from it [8]. These are consumptions that, although sometimes unmeasured, are actually consumed by the suppliers themselves, or by established social commitments such as when used in fires by firefighters [9].

The apparent losses are not physical losses but financial losses [12]. It is extremely difficult to quantify them and despite being in a smaller volume it generates a greater economic impact for companies [13]. The International Water Association (IWA), according to the water balance describes the apparent losses as unauthorised consumption and meter inaccuracies. These are the type of losses that this empirical work relies on.

Measurement errors can be avoided if the management company performs adequate maintenance of water meters and stipulates periods of exchange for them [14], since as the meters get older the water volume that is not registered tends to increase [15]. Criminisi et al. [12] affirm that the wear and the lack of maintenance and calibration of the meters, make the meters obsolete, thus losing their efficiency. The authors also point out that the incorrect choice of the class and application of it can also lead to these inaccuracies. Meter errors not only affect the control of water systems but also the financial management of entities. According to [3] the water losses of the meters caused by inaccuracies can be minimized by analyzing their performance and investigating the causes of these same inaccuracies.

Several approaches may be found in the literature to deal with the problem of water demand prediction (see Benítez et al. [16], for an exhaustive description of the latest works on that thematic). However, only a few dedicate their interest in explicitly exploring and finding changes in water consumption. In that context, we highlight the study of Hester and Larson [17] where they successfully employ breakpoint and decomposition analyses to explore changes in water use for three North Carolina municipalities between 1990 and 2014.

3 Case Study and Dataset Description

The company Águas do Norte, in the year 2018, had losses of around 2,700,000 m^3 of non-invoiced water, corresponding to about 3.67% of the value of the entire system [18]. With such a high volume of losses, the company seeks strategies to decide the best moment to replace the water meters. There is a need to seek and adopt strategies to make decisions to replace meters not only based on the maximum estimated volume, or the useful life, because it does not define which mechanism is working out of the ordinary.

In order to respond to this need, this study aims to develop an algorithm that considers different ways of detecting discrepancies in consumption, integrating later with a Business Intelligence platform where alerts will be given when anomalies in consumption are verified. This tool will give important support to the company in managing and the decision-making of replacing or repairing the water meters.

To demonstrate the adequacy of the methods proposed in this study, we apply them to two company datasets: a hospital water consumption time-series and a hotel water consumption time-series with a time horizon of 01/12/18 until 30/11/19. Both the hotel and the hospital are located in the north of Portugal. The collected data is broken down into periods of 15 in 15 min, hour to hour or daily observation. In the present study, the daily data corresponding a 2 water meters will be analysed.

4 Methodology

The methodology adopted focuses on the application of a set of statistical instruments that aim to validate for the subsequent implementation of an effective algorithm for change point detection on water consumption time-series.

Detecting the structural breaks in the trend of the series, related to a sharp decrease in water consumption, allows us to detect a possible failure in the water meter and alert the company. Note that, a malfunction in the water meter will be reflected in a lower than expected registration of the water consumption. Our analysis goes through three main phases:

Step 1 – Exploratory analysis, in order to detect consumption patterns, variations, and seasonality in the series under study;
Step 2 – Time-series decomposition, to extract the trend component associated with the water meter performance;
Step 3 – Time-series structural change detection, to detect critical moments on the water meter performance.

These steps will allow, in the end, the historical interpretation of data in order to decide if the water meter should be replaced.

4.1 Step 1 - Exploratory Analysis

During the last decades, the L-moments, introduced by Hosking [19], have gained great popularity and have been widely used in many hydrological applications, such as in the analysis and statistical characterization of residential water demand data [20]. In the present study, we examine three statistics based on the first three L-moments λ_i: mean value (=L-mean = λ_1), L-variation, $\tau_2 = \lambda_2/\lambda_1$, and L-skewness, $\tau_3 = \lambda_3/\lambda_2$. The mean value expresses the central tendency of the dataset. L-variation is analogous to the conventional coefficient of variation (ratio of the standard deviation to the mean value) and quantify the variability of the data values. For positive variables such as water flow measures, L-variation takes values in the range $[0, 1]$. L-skewness is a dimensionless measure of the asymmetry of the random variable analogous to the conventional skewness coefficient. L-skewness takes values in the range $[-1, 1]$ where L-skewness equal to zero means that the distribution is symmetric while the distribution is right or left for positive and negative values, respectively. The L-variation and L-skewness coefficients have the advantage to be more robust, to outliers in the data, than conventional measures, coefficient of variation, and skewness, which means that they suffer less from the effects of sampling variability [19].

4.2 Step 2 - Time Series Decomposition

When analyzing the time series of daily water consumption (in m^3) we need to be able to decompose the series eliminating seasonality and the non-explained variability (also known as remainder). Thus, eliminating measurement error due to noise (that varies depending on the flow rate at which the meter is working) and seasonality effects.

For that purpose, we consider in the proposed algorithm three different ways of decomposing a time series, depending on its nature (additive or multiplicative). The general representation of a time-series decomposition approach is given as:

$$Y_t = f(s_t, h_t, \epsilon_t)$$

where Y_t is the time series value (actual data) at period t, s_t is the seasonal component (or index) at period t, h_t is the trend-cycle component at period t, and ϵ_t is the irregular component at period t. In this particular analysis, water meter performance is directly related to the trend component, h_t. The additive decomposition is, then, given by: $Y_t = s_t + h_t + \epsilon_t$, and the multiplicative decomposition by: $Y_t = s_t \times h_t \times \epsilon_t$.

Seasonal trend decomposition based on Loess (Locally Estimated Scatterplot Smoothing), also known as STL decomposition, has been the preferred procedure for additive seasonal decomposition, in water related time-series decomposition (e.g. Hester and Larson [17] and Benítez et al. [16]). The smoothed Loess based in adjustment the polynomial regression, weighted for an observation time when the weights decrease with a distance nearest neighbor [21].

The package *stats* includes the function *decompose* that implements mechanisms to remove seasonality, trend, and error using moving averages, where they are calculated through the series sequential samples. This function allows the two types of decomposition, additive and multiplicative. To simplify, the function determines the trend through moving average, the seasonality is calculated through the mean in all periods for each time-space, centralizing it. The error is determined by removing the trend and seasonal values from the original series.

As we want a flexible algorithm to accommodate any kind of time-series we opt to implement the three decompositions described above and iterate each subsequent step (2–4) to the three decompositions and compare the final results.

4.3 Step 3 - Breakpoint Detection

As Hester and Larson [17] explains, structural change methods may be used to estimate a long-term trend, identifying periods of statistically significant change. They are, indeed, effective because these methods are specifically intended for exploratory cases where a regression parameter may have changed, but at an unknown point.

To clarify, in an ordered sequence of data $y_{1:n} = (y_1, \ldots, y_n)$ we say that a change point has occurred if there is a moment, $\tau \in (1, \ldots, n-1)$ such that the statistical properties of $(y_1, \ldots, y_\tau)$ and $(y_{\tau+1}, \ldots, y_n)$ are somehow different.

The detection of the change point can be put as a hypotheses test, with the null hypothesis being that there is no change point the alternative hypothesis is that there is a point of change.

There are several packages in R, with functions implemented able to detect breakpoints in a time-series, whether it is structural changes in the mean, variance, or the general distribution.

The *strucchange* package tests structural changes in linear regression models. In this package, we will use the *breakpoint* method that will allow us to calculate the breakpoints in the time series, given the number of breaks. The function will return the number of ideal points [22].

The *changepoint* package provides an algorithm that allows multiple detections of the change points, the package is composed of three methods: the *neighborhood segment*, the *binary segmentation*, and the *PELT (Pruned Exact Linear Time)*. In the study we will test the *binary segmentation* and the *PELT* procedure, testing the changes on the mean and variance of the time-series [23], not use a neighborhood segment because it's necessary attributes a maximum number of change points, limited the size of segmentation [24].

The *Binary Segmentation (BS)* is the most used method, regarding the changepoint analysis. Scott and Knott [25] applied the initial algorithm in search of binary segmentation. As Eckley et al. [26] simplifies, this method initiates by applying a single changepoint test statistic to the entire data if a changepoint is identified the data is split into two at the changepoint location. The procedure is then repeated on the two new data sets, before and after the change. If changepoints are identified in either of the new data sets, they are split further. The process continues until no changepoints are detected in any partition of the data.

The method *PELT* implements an algorithm proposed in [29], that minimizes the expression given by

$$\sum\nolimits_{i=1}^{m+1} [C(y_{(\tau_{i-1}+1):\tau_i}] + \beta f(m),$$

where C is a cost function for a segment and $\beta f(m)$ is a penalty to guard against overfitting. The minimization is done by using a dynamic programming technique to obtain the optimal segmentation for $m+1$ changepoints reusing the information that was calculated for m changepoints.

The package *Cpm* provides several functions to allow parametric and nonparametric distribution-free change detection in the mean, variance, or general distribution of a time-series, which constitutes an advantage since it relaxes parametric assumptions on the data under analysis.

In analysis will use only the nonparametric methods because the algorithm tests the time series without needing to know its nature. In particular, we made use of the function *processStream* used to detect multiple change points in a sequence of observations. As the Ross et al. [27] explain, the observations are processed in order, starting with the first, and a decision is made after each observation if changes occur at the point. In this empirical analysis, we address different tests statistics such as: *Mann-Whitney* test statistic, to detect location shifts in a stream with a (possibly unknown) non-Gaussian distribution; *Mood* test statistic, to detect scale shifts in a stream with a (possibly unknown) non-Gaussian distribution; *Lepage* test statistics to detect the location and/or shifts in a stream with a (possibly unknown) non-Gaussian distribution; *Kolmogorov-Smirnov* test statistic, to detect arbitrary changes in a stream with a (possibly unknown) non-Gaussian distribution; and *Cramer-von-Mises* test statistic, to detect arbitrary changes in a stream with a (possibly unknown) non-Gaussian distribution [27].

5 Results

5.1 Exploratory Analysis

Figure 1 displays the mean value, L-variation and L-skewness coefficients by month-to-month, day-to-day, and hour-to-hour for the hotel and hospital. The mean value in the hospital exhibits small monthly changes with slightly higher values between August and October. In contrast, large variation among months is observed in the hotel with a significant increase in the mean value (Fig. 1(a)) during the summer months. Regarding L-variation and L-skewness coefficients (Fig. 1(b), (c)), a small fluctuation in the L-variation coefficient is noted from month-to-month in both companies, while higher fluctuations are noted in the L-skewness coefficient mainly for the hospital during the summer months. Despite the high fluctuation, all L-skewness values are positive (i.e., right-skewed data).

From day-to-day, both companies show different mean values on the weekend days comparing to the mean values on the weekdays (Fig. 1(d)). However, while the hotel shows the higher mean values in the weekend days, the mean values in the weekend days in the hospital are the smallest ones. Regarding the two shape statistics (Fig. 1(e), (f)) a small fluctuation is noted in both. L-variation values remain almost invariant at around 0.4 in the hotel and at around 0.1 in the hospital. Most L-skewness values are concentrated in a narrow interval which ranges from 0.1 to 0.4 for both companies.

From hour-to-hour, while the hotel shows a typical daily consumption pattern, i.e., low flows during night hours, a sharp morning and evening/night peaks and a moderate flow during the day [28], (Fig. 1(g)). Observing the pattern of L-variation and L-skewness coefficients (Fig. 1(h), (i)) the main differentiation is noted between the daily ad night hours. For both shape statistics, a sharp night peak is observed in the hotel. In the hospital, the L-variation values remain almost invariant and a day peak is observed for the L-skewness.

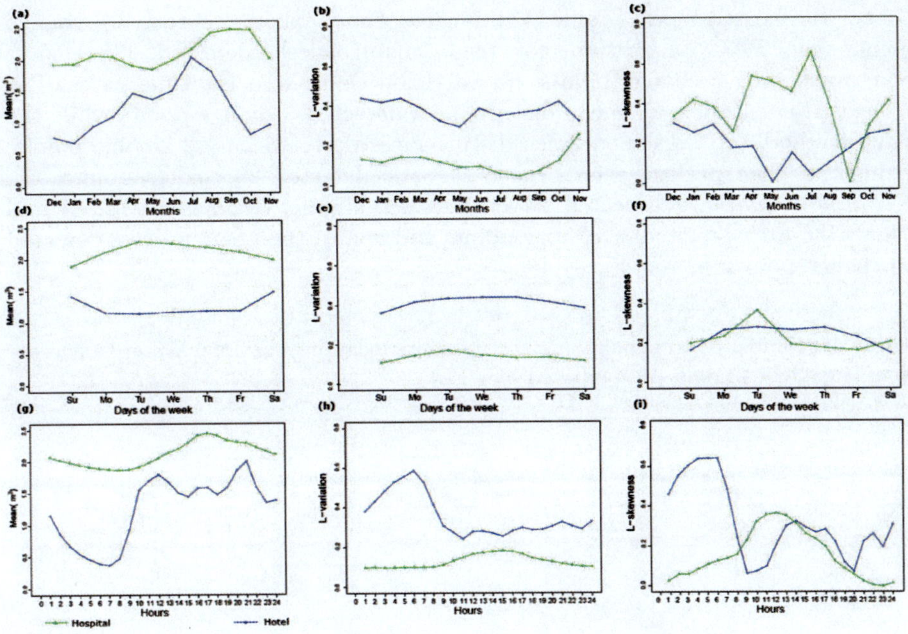

Fig. 1. Variation per month (Figures a, b and c), per day (Figures d, e and f), and per hour (Figures g, h and i) of the Mean, L-variation, and L-skewness values.

5.2 Time Series Decomposition

Additive decomposition is used when the variation in seasonality is constant throughout the series, whereas multiplicative decomposition is used when seasonality increases throughout the series. We note that the additive and multiplicative decomposition of both the hotel and the hospital, the trend, seasonality, and observations are identical.

5.3 Breakpoint Detection

Results of hotel water consumption time-series (Table 1) indicate that the package *Strucchange* has detected 3 change points. The *PELT* algorithm applied to changes on the series mean was the one that detected a higher number of change points (215). However, the same algorithm applied to the detection of changes in variance detected only 4 change points. The *Binary Segmentation* for detection on variance disturbances detected 3 change points, while detecting 5 change points on mean changes of the series. Using the *Cpm* package, the *Lepage* statistic detected 29 change points, the *Kolmogorov-Sminorv* detecting 26, the *Cramer-von-Mises* and *Mann-Whitney* detected 28, and, finally, the *Mood* process detected 17 change points.

For the hospital case (Table 1), the *Strucchange* package detected 4 change points, the *PELT* algorithm, for mean disturbances, detected 235 change points and only 2 change points, for variation changes in the time-series. The *Binary Segmentation* for mean disturbances detected 5 change points while the same method applied for variance disturbances detected only 2 change points. Using the *Cpm* package, the *Lepage* statistic detected 29 change points, the *Kolmogorov-Sminorv* detected 24 change points, the *Cramer-von-Mises* and *Mann-Whitney* detected 26 change points and finally, the *Mood* process detected 16 change points.

Table 1. Number of breakpoints detected and its locations, for hotel Water Consumption Time-Series Trend (decomposed by STL)

Method	Water meter 1			Water meter 2		
	Breakpoints detected	Location of breakpoint	Corresponding dates	Breakpoints detected	Location of breakpoint	Corresponding dates
Strucchange	3	135;236; 299	14/04/19; 24/07/19; 25/09/19	4	107;161; 225;279;	17/03/19; 10/05/19; 13/07/19; 05/09/19
PELT (change in mean)	215	——	——	235	——	——
BinSeg (Change in mean)	5	135;236; 297;318; 338;	14/04/19; 24/07/19; 23/09/19; 14/10/19; 03/11/19;	5	4;108; 162;231; 269;	04/12/18; 18/03/19; 11/05/19; 19/07/19; 26/08/19
PELT (change in variance)	4	133;236; 298;339	12/04/19; 24/07/19; 24/09/19; 04/11/19	2	252; 268	09/08/19; 25/08/19
BinSeg (Change in variance)	3	133;237; 297	12/04/19; 25/07/19; 23/09/19	2	228; 276	16/07/19; 02/09/19
processStream (Lepage statistic)	19	——	——	29	——	——
processStream (Kolmogorov-Smirnov)	26	——	——	24	——	——
processStream (Cramer-von-Mises)	28	——	——	26	——	——
processStream (Mann-Whitney)	28	——	——	26	——	——
processStream (Mood)	17	——	——	16	——	——

Analyzing Fig. 2, we can see that some methods detect the same change points, for both time-series under study, the method implemented by *strucchanges*, the *binary segmentation* for mean changes, and the *PELT* algorithm for variation changes, detect several proximal structural breaks between them.

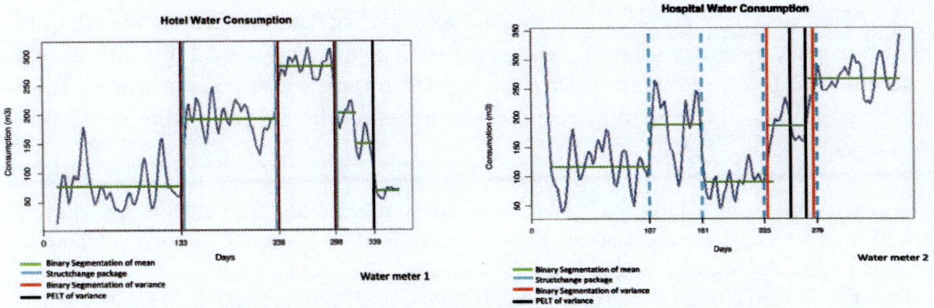

Fig. 2. Comparison of the different methods applied

6 Discussion and Conclusions

This empirical study consists in the application of methods to detect change points, i.e., breakpoints, in the water consumption time-series of a water company located in the north of Portugal. For that, we considered a case study concerning a hotel and a hospital data.

The analysis of seasonality shows a different water consumption pattern in the two understudy companies. While the hotel shows a pattern similar to residential consumption [20,28,30]; with higher consumptions during summer months (Fig. 1(a)), weekend days (Fig. 1(d)), and daily hours (Fig. 1(g)), the hospital shows a different pattern with higher consumptions in the last months of the summer and in the autumn months (Fig. 1(a)), weekdays (Fig. 1(d)) and small variation between daily and night hours consumptions (Fig. 1(g)).

The time-series analysis shows that there is a weekly seasonality and that, implementing the additive and multiplicative decomposition, we encounter no differences in the time-series trend between both procedures.

With the breakpoint detection methods used, we can understand that the *PELT* algorithm for detecting changes on the mean is not very effective since it detects more structural breaks, compromising the algorithm efficiency, augmenting the probability of false positives detection.

For a better prevision power of the algorithm, it will be used a combination of several methods for detection incorporated in software R. The alarm will only be given when the structural break detected has a negative slope and is above an error that the company considers critical. Also, the methods implemented in the *cpm* package may not be effective for our purpose since there was the detection of several points that could generate a great number of false positives and take the concessionaire making bad decisions, what we want to avoid.

Although the results obtained in the study are promising and allowed us to detect change points, critical indicators for the substitution of the water meters, it is necessary to continue to deepen and test new methods in order to obtain an improved algorithm to accurately detect the change points, not indicating false positives and lead the concessionaire into taking a bad decision, which we need to prevent.

Further research aims to extend the analysis of time series to hourly periods and, also, to every 15 min. Also, to perform an analysis of the influence of external variables (e.g. temperature, precipitation) on water consumption exploring different machine learning techniques, for example LSTM (Long short-term memory).

Acknowledgement. This work has received funding from FEDER Funds through P2020 program and from National Funds through FCT - Fundação para a Ciência e a Tecnologia (Portuguese Foundation for Science and Technology) under the projects UID/GES/04728/2017, UIDB/00013/2020 and UIDP/00013/2020.

References

1. Ncube, M., Taigbenu, A.: Assessment of apparent losses due to meter inaccuracy-a comparative approach. Water SA **45**(2), 174–182 (2019)
2. El-Abbasy, M.S., Mosleh, F., Senouci, A., Zayed, T., Al-Derham, H.: Locating leaks in water mains using noise loggers. J. Infrastruct. Syst. **22**(3), 04016012 (2016)
3. Mutikanga, H.E., Sharma, S.K., Vairavamoorthy, K.: Investigating water meter performance in developing countries: a case study of Kampala. Uganda. Water SA **37**(4), 567–574 (2011)
4. Szilveszter, S., Beltran, R., Fuentes, A.: Performance analysis of the domestic water meter park in water supply network of Ibarra. Ecuador. Urban Water J. **14**(1), 85–96 (2017)
5. Fontanazza, C.M., Freni, G., La. Loggia, G., Notaro, V., Puleo, V.: A composite indicator for water meter replacement in an urban distribution network. Urban Water J. **9**(6), 419–428 (2012)
6. Puleo, V., Fontanazza, C., Notaro, V., De Marchis, M., La Loggia, G., Freni, G.: Definition of water meter substitution plans based on a composite indicator. Procedia Eng. **70**, 1369–1377 (2014)
7. Liemberger, R., Wyatt, A.: Quantifying the global non-revenue water problem. Water Supply **19**(3), 831–837 (2019)
8. AL-Washali, T., Sharma, S., Al-Nozaily, F., Haidera, M., Kennedy, M.: Monitoring nonrevenue water performance in intermittent supply. Water **11**(6), 1220 (2019)
9. Sardinha, J., Serranito, F., Donnelly, A., Marmelo, V., Saraiva, P., Dias, N., Rocha, V.: Controlo ativo de perdas de água. EPAL-Empresa Portuguesa das Águas Livres, Lisboa (2015)
10. Richards, G.L., Johnson, M.C., Barfuss, S.L.: Apparent losses caused by water meter inaccuracies at ultralow flows. J. Am. Water Works Ass. **102**(5), 123–132 (2010)
11. Arregui, F.J., Gavara, F.J., Soriano, J., Pastor-Jabaloyes, L.: Performance analysis of ageing single-jet water meters for measuring residential water consumption. Water **10**(5), 612 (2018)
12. Criminisi, A., Fontanazza, C.M., Freni, G., Loggia, G.L.: Evaluation of the apparent losses caused by water meter under-registration in intermittent water supply. Water Sci. Technol. **60**(9), 2373–2382 (2009)
13. Arregui, F., Soriano, J., Cabrera, E., Jr., Cobacho, R.: Nine steps towards a better water meter management. Water Sci. Technol. **65**(7), 1273–1280 (2012)
14. Arregui, F., Cabrera, E., Cobacho, R., Garcia-Serra, J.: Reducing apparent losses caused by meters inaccuracies. Water Pract. Technol. **1**(4) (2006)

15. Arregui, F., Cobacho, R., Cabrera Jr., E., Espert, V.: Graphical method to calculate the optimum replacement period for water meters. J. Water Resour. Plan. Manag. **137**(1), 143–146 (2011)
16. Benítez, R., Ortiz-Caraballo, C., Preciado, J.C., Conejero, J.M., Sánchez Figueroa, F., Rubio-Largo, A.: A short-term data based water consumption prediction approach. Energies **12**(12), 2359 (2019)
17. Hester, C.M., Larson, K.L.: Time-series analysis of water demands in three North Carolina cities. J. Water Resour. Plan. Manag. **142**(8), 05016005 (2016)
18. Silva, D.V., Sampaio, M.J., Milagres, C., Alves, V., Ferreira, F.: Flow4Link - the flow in the hand. In: 18th International Flow Measurement Conference (2019)
19. Hosking, J.R.: L-moments: analysis and estimation of distributions using linear combinations of order statistics. J. Roy. Stat. Soc. Ser. B (Methodol.) **52**(1), 105–124 (1990)
20. Kossieris, P., Makropoulos, C.: Exploring the statistical and distributional properties of residential water demand at fine time scales. Water **10**(10), 1481 (2018)
21. Dagum, E.B., Luati, A.: Global and local statistical properties of fixed-length nonparametric smoothers. Stat. Methods Appl. **11**(3), 313–333 (2002)
22. Zeileis, A., Leisch, F., Hornik, K., Kleiber, C., Hansen, B., Merkle, E.C., Zeileis, M.A.: Package 'strucchange'. R package version, 1–5 (2015)
23. Killick, R., Eckley, I.: Changepoint: an R package for changepoint analysis. J. Stat. Softw. **58**(3), 1–19 (2014)
24. Killick, R., Fearnhead, P., Eckley, I.A.: Optimal detection of changepoints with a linear computational cost. J. Am. Stat. Assoc. **107**(500), 1590–1598 (2012)
25. Scott, A.J., Knott, M.: A cluster analysis method for grouping means in the analysis of variance. Biometrics **30**, 507–512 (1974)
26. Eckley, I.A., Fearnhead, P., Killick, R.: Analysis of changepoint models. Bayesian Time Series Models **205–224** (2011)
27. Ross, G.J.: Parametric and nonparametric sequential change detection in R: the CPM package. J. Stat. Softw. **66**(3), 1–20 (2015)
28. Walski, T.M., Chase, D.V., Savic, D.A., Grayman, W., Beckwith, S., Koelle, E.: Advanced water distribution modeling and management (2003)
29. Jackson, B., Scargle, J.D., Barnes, D., Arabhi, S., Alt, A., Gioumousis, P., Gwin, E., Sangtrakulcharoen, P., Tan, L., Tsai, T.T.: An algorithm for optimal partitioning of data on an interval. IEEE Signal Process. Lett. **12**(2), 105–108 (2005)
30. Memon, F.A., Butler, D.: Water consumption trends and demand forecasting techniques. Water Demand Manag. **2006**, 1–26 (2006)

From Research to Numbers: Role of Universities in Social Innovation

Jorge Cunha[1](✉), Carla Ferreira[1], Madalena Araújo[1], and Paul Benneworth[2]

[1] ALGORITMI Research Center, University of Minho, Braga, Portugal
jscunha@dps.uminho.pt
[2] Western Norway University of Applied Sciences, Bergen, Norway

Abstract. Commonly, social innovation is defined as new ideas proposals to the needs of humans. However, there is a lack of a well-definition comprehensive leading the fragmentation of field research. On the other hand, the contribution of universities for social innovation development is still less investigated. In this sense, this study intends to explore interrelations between universities and social innovation in relation to different lines of investigation employed. For this, a content analysis was applied to results obtained by Cunha et al. [1]. Five categories of analysis were defined. Findings revealed that there are a number of studies that applied qualitative research to investigate practical examples of social innovation inside of universities and few empirical studies. A link among keywords social innovation, higher education, and social entrepreneurship was found. Furthermore, the analysis showed that there is not a leader country but a distribution across several countries, where Spain and United Kingdom stand out. Thus, this result suggests that it would be helpful to develop an instrument to measure academics' engagement with social innovation research and practice. In addition, this research contributes to current knowledge regarding the role of universities in social innovation model, providing new theoretical and practical insights of investigation.

Keywords: Social innovation · University · Content analysis

1 Introduction

Innovation is commonly thought in terms of economic impact driven by research, development and training in science and technology, where important social aspects are sometimes ignored. Social innovation emerges to overcome complex societal challenges in an effective, efficient, and sustainable way, where the value created accrues primarily to society as a whole [2]. In this context, European Union (EU) has been incorporating social innovation into major policy documents, such as EU 2020 Strategy for smart, sustainable and inclusive growth, and EU budget to improve the european economy and create employment [3]. The concept of social innovation is not recent. For example, Agostini *et al.* [4] Showed that this concept has been around since at least the second half of the 19th-century. However, a clear and concise definition of social innovation does not exist in the current literature: Mulgan *et al.* [5] defined social innovation as

J. Machado et al. (Eds.): icieng 2021, LNME, pp. 246–256, 2022.
https://doi.org/10.1007/978-3-030-78170-5_22

"new ideas that work in meeting social goals, suggesting a somewhat narrower definition: innovative activities and services that are motived by the goal of meeting a social need and that are predominantly developed and diffused though organizations whose primary purpose are social"; while benneworth and cunha [6] defined social innovation as "a socially innovative practice that delivers social just outcomes by developing novel solutions in border spanning learning communities thereby creating social value by promoting community development, hence forming wider collaborative networks, and challenging existing social institutions though this collaborative action". The lack of a commonly comprehensive definition leads to a fragmentation of the research field and to a low use of social innovation practices in a large number of activities, such as new products or services produced by either the third sector, or the private or public sectors [7].

Oeij et al. [8] demonstrate that to analyze and to study social innovation an approach about the complexity of social innovation process is needed. In this sense, a number of models were developed to describe the process of social innovation, for example: Benneworth and Cunha [6]; Oeij et al. [8]; Mulgan [9]; Westley et al. [10]; Neumeier [11]. For Mulgan [9], the social innovation process follows a logic specific of development, identifying four main stages that explain the social innovation process: 1) generating ideas by understanding needs and identifying potential solutions; 2) developing, prototyping and piloting ideas; 3) assessing then scaling up and diffusing the good ones; and 4) learning and evolving. On the other hand, Benneworth and Cunha [6] based their model in three elements: the progress of the innovation, the agency of the innovator, and the building up of the societal capacity. These authors proposed a model with seven stages interlinked: 1) idea generation; 2) creation of experimental space; 3) demonstrator; 4) decision to expand; 5) support coalition; 6) codification; and 7) diffusion. The literature related suggests that processes of social innovation have more similarities than differences compared to technological innovations. However, the social innovation differs of technological and commercial innovations in the following aspects: goals (public/social value versus economic value); stakeholders (socially deprived groups versus investors; social entrepreneurs versus capitalist entrepreneurs); and infrastructure [10, 11].

Little literature conceptualizes the role of the university on social innovation process, partly due to the recent emphasis of university in profit–making and commercialization activities. Universities seek to improve the impact of their research by financing in technology commercialization services but it fails to support a number of impacts of non-commercial research. Several authors argue that universities respond to pressures for change adapting a process whereby service provision is co-delivered with different stockholders [12, 13]. Given competitive situation of universities, McKelvey and Zaring [14] claim that universities should be viewed as using resources and reacting to pressures from the external environment and stakeholders, felt over time. Thus, universities can be defined as delivering various knowledge-intensive services (or outputs). In this context, there is need to help the connection among researchers and research users so that research support decisions about public policy and professional practice [14].

For an economist, knowledge transfer provided by the university is associated with public and private goods. From a social innovation viewpoint, the services provided by universities should be the ones contributing to improved social welfare through both direct and indirect effects [14]. Several studies emerge in this research field. For example, [15] and [16] propose a framework for support the embedding of social innovation education across the disciplines of higher education. Almeida et al. [17] applied triplex helix dynamic model to explore the diffusion and expansion of cooperative incubator in social entrepreneurship. Benneworth and Cunha [6] intend to understand how universities contribute to knowledge- based urban development (KBUD). For these authors, universities are important sources of knowledge for society, that is, should be prepared to support social innovation actions and address societal challenges of the 21 century. In this sense, they identified six potential roles of universities (knowledge provider, knowledge bridge, financer, landlord, adviser, and mentor) based in their provided inputs which could contribute to social innovation process (knowledge, material resources, "know-how" and "know-who").

The higher-education is one of key players in creating state-of-the-art knowledge. By transferring knowledge into social use, universities can generate a number of benefits to society. Thus, understanding social innovation in universities and the transfer of knowledge into social sector is essential. The main objective of the present study is to fill in the gap identified in the literature by applying content analysis taken into account the connection between social innovation and universities found in Cunha et al. [1] in relation to different lines of investigation employed. The research contributes to current knowledge regarding the role of universities in social innovation model, providing new theoretical and practical insights. Practical contributions include the opportunity of university managers and policymakers to reflect about this interrelatedness in the development of new more effective policies.

2 Methodology

Cunha *et al.* [1] applied a systematic literature review taking into account the connection between social innovation and universities following a transparent and explicit protocol. In order to determine the research profile, these authors submitted 42 papers to a bibliometric analysis. In this sense, the present study employs a content analyses to results obtained by Cunha *et al.* [1], especially for the more cited research. Ten studies were imported to Nvivo qualitative data analysis software (version 11) for content analysis. The purpose of content analysis is to organize and elicit meaning from the collected data and to draw realistic conclusions from it [18]. For Krippendorff [19] content analysis is "a research for making replicable and valid inferences from texts to the context of their use". The content analysis method is commonly articulated to results of systematic literature reviews in order to explore data that are highly organized and contextualized. The process of content analysis comprises several steps that may be revisited and repeated many times as emergent insights dictate. Four main stages have been identified in the related literature: the decontextualisation, the recontextualisation, the categorisation, and the compilation [18]. Following this procedure, the data was analysed repeatedly by the authors and coded into categories consistent with approaches applied in prior research [19]. Some categories that originated these nodes had been previously defined in the systematic literature reviews, while others emerged during the analysis. After the coding phase, frequencies of concepts were computed, exploring the relation between nodes. Following the literature, it is assumed that the importance that each category corresponds to the frequency it appears in studies [18].

3 Key Findings

3.1 Sample Description

Based on results found by Cunha et al. [1], the ten most cited papers were selected for the content analysis, allowing to contribute for a better understanding of the role of universities in fostering social innovation. Particularly, this criterion allows to explore the research tendency, type of research, and authors. Table 1 shows the main researches characteristics under evaluation. By analysis of Table 1, it is verified that the two studies more often cited in literature were published in 2009 in area of social science. This observation suggests that this research topic is recent as cited in Cunha et al. [1]. Phipps and Shapson [21] paper with title "*Knowledge mobilisation build local research collaborations for social innovation*" is the most cited study. Despite most studies are in the area of social science, it is verified also that this field of investigation is comprehensive to more one area of knowledge: Business, Management and Accounting.

Table 1. Characterization of ten studies under evaluation

Author	Periodicals	Subject area	Document type	Number of citations
Phipps and Shapson [20]	Evidence and Policy	Social science	Article	22
Raufflet [21]	Journal of Business Ethics	Social science	Article	8
Cippolla and Bartholo [22]	International Journal of Design	Business management and accounting	Article	8
Benneworth and Cunha [6]	European Journal of Innovation Management	Business management and accounting	Article	7
Rivers *et al.* [15]	Higher Education, Skills and Work-based Learning	Social science	Article	7
Almeida *et al.* [17]	International Journal of Technology and Globalisation	Social science	Article	6
Rivers *et al.* [16]	Higher Education, Skills and Work-based Learning	Social science	Article	6
Unceta, *et al.* [23]	Innovation	Business management and accounting	Article	6

(*continued*)

Table 1. (*continued*)

Author	Periodicals	Subject area	Document type	Number of citations
Reznickova and Zepeda [24]	Journal of Community and Applied Social Psychology	Social science	Article	5
Meckelvey and Zaring [14]	Industry and Innovation	Business management and accounting	Article	5

3.2 Content Analysis

Content analysis fell on four topics of analysis. For each topic, several categories and nodes are identified. A summary list of the most relevant nodes for each topic is presented in Table 2.

Table 2. List of the most relevant nodes for each topic explored

Topic	Category	Nodes (Classification)
Type of research	Application	Pure research
		Applied research
	Objectives	Exploratory
		Descriptive
		Explanatory research
	Methods of data collection and analysis	Quantitative
		Qualitative
	Type of work	Literature review
		Research paper
		Conceptual paper
Region of publication	Country	Brazil
		Canada
		Italy
		South Africa
		Sweden
		United Kingdom
		United States
		Spain

3.3 Type of research

To of explore the type research, each study was grouped according to application, objectives, information sought, and type of work. Regarding to category Application, the results suggest that 80% of studies under evaluation employs an Applied research while 20% use a Pure research. In this sense, the results seem to indicate that the main aim of most studies was to answer certain specific questions, closely associated with the solution of specific problems. For most studies, the purpose is related to the analysis of case studies about social innovation, where the findings cannot be usually generalized. Applicability of the new knowledge generated is related to the specific topic, particularly to strengths and weaknesses of examples of social innovation inside or taken by universities [15–17, 22]. On the other hand, these results seem to suggest that few studies approach social innovation from the point of view of new knowledge creation about the role of universities on social innovation process and on ways of measuring the impact of social entrepreneurship [6]. In the same line, emerge the results about category Objective. This category was divided in three classifications or nodes: descriptive research, explanatory research, and exploratory research. Figure 1 summarizes the relative importance of each classification. The results show that 60% of studies under analysis followed an exploratory research. According to literature, exploratory studies are a value source to understand new phenomenon; to seek new insights and to assess the problem from a new point of view [25]. The exploratory research intends clarifying concepts, gathering explanations, and formulating hypothesis about the topic under research. In this context, the representativeness of exploratory research corroborated by the literature reviewed show that social innovation is an inspiring concept, but yet little defined in literature. This diversity poses important challenges to the definition, analysis and measure of social innovation. Additionally, little literature conceptualizes the role of the university on social innovation process. Thus, descriptive and explanatory research was only found in 20% of studies each, suggesting that few investigations about the role of universities in social innovation exist whether focus on cause-effect relationship through development of theories or hypotheses to analyse the forces that caused a certain phenomenon to occur [25].

In order to explore the methods of data collection and analysis, two classification emerged: qualitative and quantitative. Commonly, quantitative term refers to the tradition since the 17th century [26]. Quantitative approaches involve the generation of data in quantitative form which can be subjected to quantitative analysis. The purpose of this approach is to create a data base where it is possible to infer characteristics of or relationships between the population [27]. On the other hand, qualitative approach comprises to subjective assessment of attitudes, opinions and behavior. This approach generates data either in non-quantitative form, but it is possible to explore several factors which motivate individuals to behave in a particular way or which may like or dislike a particular thing. Thus, qualitative approach is concerned with the quality or nature of human experiences and what these phenomena mean to participants [26, 27].

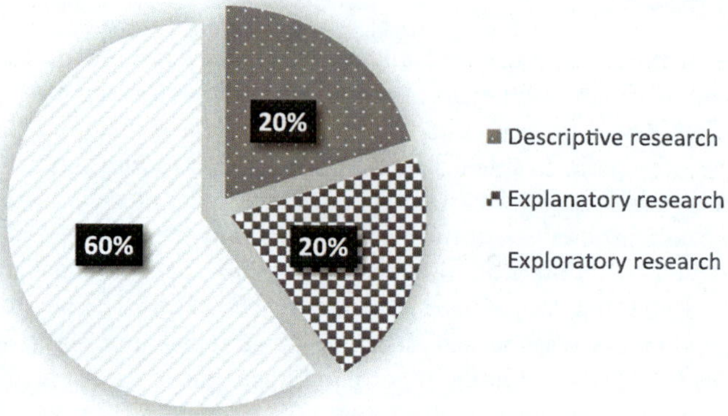

Fig. 1. Relative importance of each classification of category Objective.

The results obtained show that all studies reviewed applied a qualitative approach. In certain way, this result was expected given the main field of research of those studies: social sciences. The related literature describes qualitative research as naturalistic and interpretative, where their orientation derives from social sciences, especially of sociology and anthropology. Furthermore, the purpose of studies was exploring individuals' behaviour and attitudes in case studies related to social innovation.

Table 3 illustrates classification of studies by type of work. As outcomes, 80% of studies refer to research paper, while the remaining 20% comprise literature review and conceptual paper. These results seem to indicate that there is lack of studies to develop new concepts or to reinterpret existing ones, as reported by authors Benneworth and Cunha [6]. Following the previous results, it is found that several studies seek to interpret social innovation projects developed within universities, such as Cipolla and Bartholo [22] and Reznickova and Zepeda [24].

Table 3. Classification of studies by type of work

Classification (nodes)	Frequency relative (N = 10)
Literature review	10%
Research paper	80%
Conceptual paper	10%

Regarding keywords often employed in the sample studies, Fig. 2 demonstrates that of 28 keywords found, social innovation and higher education were the most cited, as expected by selection criteria. The keyword social entrepreneurship was the third more cited. Thus, these results seem to indicate there is a connection between social innovation, higher education and social entrepreneurship. This link to some extent can be explained by the transfer of knowledge between universities and society implied by academic entrepreneurship.

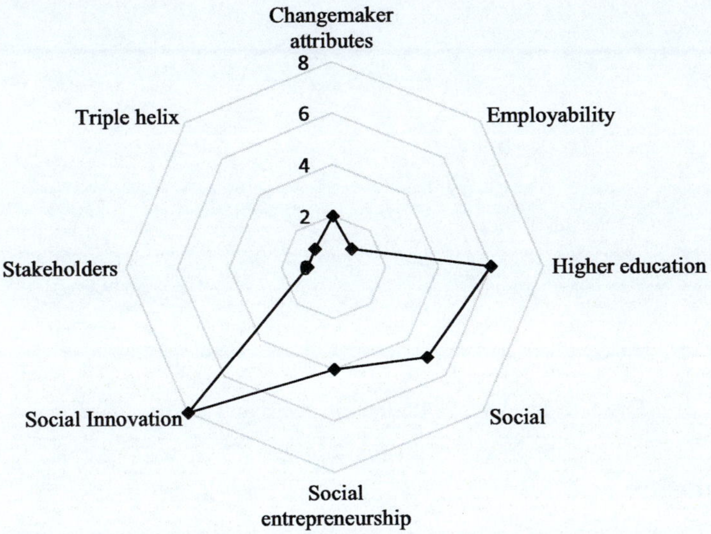

Fig. 2. Keywords often employed in the studies of sample (number of often).

3.4 Region of Publication

Currently, one of the core policies of European Union focuses on social innovation projects in terms of theory, methodology, policy areas, in order to bring strategic implications for policymakers, stakeholders and the broader public in a comprehensive way [28]. The three main objectives of these incentive policies are: promoting social innovation as a source of growth and jobs; sharing information about social innovation in Europe; and supporting innovative entrepreneurs and mobilising inventors and public organisations [28].

In this context, in order to investigate whether there is a leading country on the research topic of social innovation and universities, the papers were classified by country of publication. Figure 3 illustrates the number of publications by the eight countries found. The results show there is not a leader country but a distribution across several countries. Only Spain and United Kingdom stand out with two researches. In spite of European Union view of social innovation as a path of change there is no country or European region that focuses on the role of universities in social innovation.

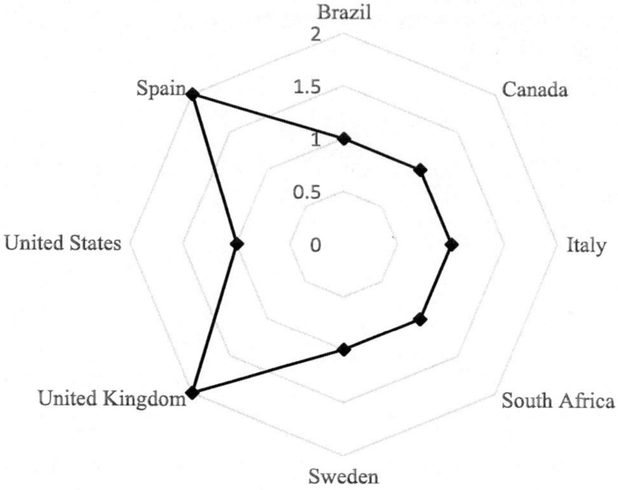

Fig. 3. Number of publications by the eight countries found

4 Conclusion

The aim of this study is to analyse the research profile that highlights the role of universities in social innovation. Considering the lack of studies in this subject and in accordance with social innovation concept and the core role of higher-education, a study was developed to understand how the research fits into the different lines of investigation defined. In this context, a content analysis was applied to results obtained by Cunha *et al.* [1].

Social innovation is an inspiring concept where the lack of a comprehensive definition leads to a fragmentation of the research field. Furthermore, universities play an important role stimulating technological change and innovation. However, their contribution for social innovation has been less investigated.

The findings of this study demonstrated that most studies employed an applied research approach, where the main purpose of study was exploratory. Regarding to methods of data collection and analysis the results showed that all studies applied qualitative methods to subjective assessment of attitudes, opinions and behaviour. For evaluation of work type, most studies referred to research paper, where they seek to interpret social innovation projects developed within universities campi. Social innovation, higher education, and social entrepreneurship were the most cited keywords. In relation to country of publication, the analysis showed that there is not a leader country but a distribution across several countries, where Spain and United Kingdom stand out with two researches.

To sum up, it can be concluded that the study of universities' role in the social innovation process is an emerging topic of research, although of paramount importance. This study highlights several researches on social innovation practical examples inside universities, where qualitative research is privileged. In this context, it is necessary to investigate in an empirical point of view. Particularly, it is critical to develop tools to measure the social innovation trends of academics, where university individuals are of utmost relevance in the process of social innovation. This approach will allow better understanding on higher education contribution for social innovation.

Acknowledgements. This work was financed by the Portuguese Foundation for Science and Technology, under Project PTDC/EGE-OGE/31635/2017.

References

1. Cunha, J., Ferreira, C., Araújo, M., Benneworth, P.: Role of universities in the social innovation model: contribution for a systematic literature review, Paper presented 8th International Conference on Social Responsibility, Ethics and Sustainable Busines, University of Minho, Braga, Portugal (2019)
2. Pol, E., Ville, S.: Social innovation: buzz word or enduring term? J. Socio-Econ. **38**(6), 878–885 (2009)
3. Hubert, A.: Empowering people, driving change: social innovation in the European Union (2010)
4. Agostini, M., Vieira, L., Tondolo, R., Tondolo, V.: An overview on social innovation research: guiding future studies. Braz. Bus. Rev. **14**(4), 1–13 (2017)
5. Mulgan, G.: Social Innovation: What it is, Why it Matters and How it Can Be Accelerated. Said Business School, Oxford (2007)
6. Benneworth, P., Cunha, J.: Universities' contributions to social innovation: reflections in theory & practice. Eur. J. Innov. Manag. **18**(4), 508–527 (2015)
7. Van der Have, R., Rubalcaba, L.: Social innovation research: an emerging area of innovation studies? Res. Policy **45**(9), 1923–1935 (2016)
8. Oeij, P., van der Torre, W., Vaas, F., Dhondt, S.: Understanding social innovation as an innovation process: applying the innovation journey model. J. Bus. Res. **101**, 243–254 (2019)
9. Mulgan, G.: The process of social innovation. Innov. Technol. Govern, Global. **1**(2), 145–162 (2006)
10. Westley, F., Patton, M., Zimmerman, B.: Getting to Maybe: How the World is Changed. Random House, Toron (2006)
11. Neumeier, S.: Why do social innovations in rural development matter and should they be considered more seriously in rural development research? – Proposal for a stronger focus on social innovations in rural development research. Sociol. Rural. **52**(1), 48–69 (2012)
12. Slaughter, S., Rhoades, G.: Academic Capitalism and the New Economy. The Johns Hopkins University Press, Baltimore (2004)
13. McKelvey, M., Holmén, M.: Learning to Compete in European Universities. Edward Elgar, Cheltenham (2009)
14. McKelvey, M., Zaring, O.: Co-delivery of social innovations: exploring the university's role in academic engagement with society. Ind. Innov. **25**(6), 594–611 (2018)
15. Rivers, A., Armellini, A., Nie, M.: Embedding social innovation and social impact across the disciplines: identifying 'Changemaker' attributes. Higher Educ. Skills Work Based Learn. **5**(3), 242–257 (2015)
16. Rivers, A., Armellini, A., Maxwell, R., Allen, S., and Durkin, C.: Social innovation education: towards a framework for learning design. Higher Educ. Skills Work-Based Learn. **5**(4), 383–400.5 (2015)
17. Almeida, M., Mello, J., Etzkowitz, H.: Social innovation in a developing country: invention and diffusion of the Brazilian cooperative incubator. Int. J. Technol. Glob. **6**(3), 206–224 (2012)

18. Bengtsson, M.: How to plan and perform a qualitative study using content analysis. NursingPlus Open **2**, 8–14 (2016)
19. Krippendorff, K.: Content Analysis: An Introduction to Its Methodology, 2nd edn. Sage Publications, Thousand Oaks (2004)
20. Phipps, D., Shapson, S.: Knowledge mobilisation builds local research collaborations for social innovation. Evidence Policy **5**(3), 211–228 (2009)
21. Raufflet, E.: Mobilizing business for post-secondary education: CIDA University, South Africa. J. Bus. Ethics **89**(2), 191–202 (2009)
22. Cipolla, C., Bartholo, R.: Empathy or inclusion: a dialogical approach to socially responsible design. Int. J. Des. **8**(2), 87–100 (2014)
23. Unceta, A., Castro-Spila, J., García Fronti, J.: Social innovation indicators. Eur. J. Soc. Sci. Res. **29**(2), 192–204 (2016)
24. Reznickova, A., Zepeda, L.: Can self-determination theory explain the self- perpetuation of social innovations? A case study of slow food at the University of Wisconsin—Madison. J. Community Appl. Soc. Psychol. **26**, 1–17 (2016)
25. Dinesh, J.: Research methodology. In: A Study of Buying Decision Process in Malls, Devi Ahilya Vishwavidyalaya (2016)
26. Draper, A.: The principles and application of qualitative research. Proc. Nutr. Soc. **63**(4), 641–646 (2004)
27. Kothari, C.: Research Methodology: Methods & Techniques, 2nd edn. New Age International (P) Ltd., New Delhi (2004)
28. European Commission: Social innovation: A Decade of Changes, Publications Office of the European Union (2014)

Modeling Waste Management and Boarded Personnel KAP

M. Filomena Teodoro[1,2](✉), José B. Rebelo[1], and Suzana Lampreia[1]

[1] CINAV, Center of Naval Research, Portuguese Naval Academy, Military Universitary Institute, 2810-001 Almada, Portugal
maria.alves.teodoro@marinha.pt

[2] CEMAT, Center for Computational and Stochastic Mathematics, Instituto Superior Técnico, Lisbon University, 1048-001 Lisbon, Portugal

Abstract. In this manuscript is extended from a previous study that carried out an analysis of the waste management system of the Portuguese Navy ships, restricted to the pollution contemplated in Annexes I, IV and V of Marpol 73/78, namely hydrocarbons, sewage and all types of garbage. Analyze the methods and types of storage and disposal of ship waste to ensure existing equipment, operating conditions and on-board waste management plans. However, it is not enough just to evaluate the materials and procedures. The knowledge and cooperation of the military on board the environment is also a decisive factor. A questionnaire was created and implemented for the garrison of some NRP warships with the permission of the commander of the Portuguese navy. Earlier we were able to use Exploratory Factor Analysis (EFA) to identify important queries to the latent variables that take into account waste management skills. Analysis of variance techniques (both univariate and multivariate cases) were used to obtain important independent variables that helped to explain the chosen EFA factors.

Keywords: Waste management · Navy and environment · Pollution of ships · Sea pollution · Literacy · Statistical analysis · ANOVA

1 Introduction

The illusion that the sea has the ability to absorb all pollution has long remained. Since the Navy has a very close relationship with the marine environment, it is a continuous concern to reduce the environmental impact of ships in their missions. How waste treatment is carried out on Portuguese Navy ships is a pertinent question. Implementing a waste management system (WMS) correctly on board helps to minimize the ecological footprint, which must comply with national and international regulations that cover the entire process from collection, separation and treatment types to discharges or storage of produced waste. Given that the Portuguese Navy has some aged ships in its fleet, the current task is to find out under what conditions the WPS is implemented, when possible This study takes interest on the garrison's status in terms of training and environmental awareness. It is important to note that ships are the main focus of this work, we have visited several ships and other ground support units and collected data in interview mode and applying questionnaires.

J. Machado et al. (Eds.): icieng 2021, LNME, pp. 257–270, 2022.
https://doi.org/10.1007/978-3-030-78170-5_23

This work started from [1], where the author resumed the specific regulations about the topic. In addition, in the same monograph, we implemented a questionnaire to assess knowledge, attitudes and practice (KAP) in waste management during boarding time for NRP ships. Using an incomplete sample of the embarked staff, a preliminary statistical analysis was performed on [2,3] to be used as a support for the current work. In this manuscript, we continued the statistical approach proposed in [4,5].

This article consists of an introduction, results and concluding remarks sections, a Sect. 2 where we present some motivations and references to the legal regulation of waste management in the marine environment. Section 3 has some details about the methodology. Empirical application can be found in Sect. 4.

2 Preliminaries

In accordance with the principles of sustainable development, the most important legal tool for dealing with marine environmental issues with regard to the management of ship-generated waste and cargo residues in EU ports is an important element to be studied. Receiving waste and cargo residues from ports is especially important for organization. In [7] the author describes the existing environmental tax system, the basis for calculating such taxes, several ports in Europe, namely Rotterdam, Antwerp, Klaipeda, Szczecin.

Another issue related with waste management is the life cycle of ships with age restrictions related to their operations. When the ship is no longer economically profitable, it is recycled or thrown away. Abandoned ships contain many hazardous materials, so special care must be taken to handle waste in accordance with national and internationally available rules. In this connection, dismantling and ship clothing, ship recycling facilities for recycling machinery and infrastructure must be well designed to comply with all regulations so as not to generate contaminated waste that is harmful to humans and the environment. The authors of [6] promoted a study to design an environmentally friendly or environmentally friendly ship recycling workshop, a pilot project that predicted the rise of old used merchant ships.

In [7], the authors placed particular emphasis on the legal provisions of international law that define how waste and exhaust gases are treated on ships (MARPOL; MARPOL Appendices 5 and 6). Due to the standardization and automation of the handling of shipping, shipping orders and documents are issued in accordance with classification societies in certain countries.

The Portuguese navy has been trying to become more and more green in recent years. The Navy's Admiral Chief of Staff has defined the security and health policy for the Navy's work and environment, where there are environmental guidelines on land and ships to which all military, military and civilian work and contribute for good waste management and environmental protection. In [8] the author gives a good description of this effort. This work has completed these studies in order to carry out a waste management analysis of Portuguese naval ships and restricts the study to the residues mentioned in Annexes I, IV and V of Marpol 73/78. For this analysis, a questionnaire with relevant questions about knowledge, practice and attitudes was completed and implemented on the crew of the NRP ship to gather the information necessary about the waste management.

3 Methodology

3.1 Analysis of Variance

Experimental Design One Factor. The purpose of these techniques boils down to comparing k treatments ($k \geq 2$). Suppose there are k groups of individuals chosen at random. Each group is subject to treatment, i, $i = 1,\ldots,k$. Each group does not necessarily have the same group of individuals. Consider n_i the number of individuals in group i. If in each group the number of individuals is equal, the design is denominated as balanced. When two independent ($k = 2$) random samples are available, t-tests can be established to compare means, when there are $k > 2$ independent samples there is no way to establish this test to proceed with their analysis. It is necessary to resort to a completely different technique known as analysis of variance. The data of k samples are generally presented as y_{ij}, the response of individual j in sample i.

Theoretical Model. Formal inference to compare means of different treatments implies the definition of probabilistic models. It is assumed that the relative data on the $i-th$ treatment have a normal distribution of mean μ_i and variance σ^2. If Y_{ij} is a random variable (rv) associated to the observed value y_{ij} the theoretical model can be represented by (1)

$$Y_{ij} = \mu_i + \varepsilon_{ij}, (j = 1,\ldots,n_i, i = 1,\ldots,k), \tag{1}$$

with ε_{ij} rv's independents and Gaussian

$$\varepsilon_{ij} \cap N(0,\sigma^2) \tag{2}$$

The model can be rewritten as (3)

$$Y_{ij} = \mu + \alpha_i + \varepsilon_{ij}, (i = 1,\ldots,k,\ j = 1,\ldots,n_i). \tag{3}$$

We are able to obtain confidence intervals at $(1-\alpha) \times 100\%$ for each μ_i. Since $\widehat{\mu}_i = \overline{y}_i$ and from (1) and (2) we can conclude

$$\overline{y}_i \cap N(\mu_i, \sigma^2) \tag{4}$$

following

$$\frac{\overline{y}_i - \mu_i}{\sigma/\sqrt{n_i}} \cap N(0,1) \tag{5}$$

or, if σ is unknown,

$$\frac{\overline{y}_i - \mu_i}{S/\sqrt{n_i}} \cap t_{[N-k]}. \tag{6}$$

The confidence intervals (CI) considering a confidence level $(1-\alpha) \times 100\%$ for each μ_i are given by (7)

$$\overline{y}_i - t_{[N-k;1-\alpha/2]} \frac{s}{\sqrt{n_i}} \leq \mu_i \leq \overline{y}_i + t_{[N-k;1-\alpha/2]} \frac{s}{\sqrt{n_i}}. \tag{7}$$

To investigate whether the treatments have identical means or not can be tested the hypothesis (8) using as statistical test the formula (6)

$$H_0 : \mu_1 = \cdots = \mu_k \text{ versus } H_1 : \text{ some } \mu_i \text{ are not equal.} \tag{8}$$

A rejection of H_0 means that there is experiential evidence that treatments differ from each other. In terms of the effect of treatment α_i, the above hypothesis can be described as

$$H_0 : \alpha_1 = \cdots = \alpha_k = 0 \text{ versus } H_1 : \text{ some } \alpha_i \text{ are not null.} \tag{9}$$

To perform the test defined by (9) it is necessary to calculate the total variability of $\widehat{\mu}_i = \bar{y}_i$ around $\widehat{\mu} = \bar{y}$ given by the sum of squares of treatment deviations (16)

$$SS_{TREAT} = \sum_{i=1}^{k} n_i \left(\bar{y}_i - \bar{y}\right)^2 . \tag{10}$$

A large SS_{TREAT} suggests that the treatments are distinct. It is necessary to take its average value, dividing by the degrees of freedom (11)

$$MS_{TREAT} = \frac{\sum_{i=1}^{k} n_i \left(\bar{y}_i - \bar{y}\right)^2}{k-1} \tag{11}$$

and compare with the variability of each observation within the sample, i.e., with the mean sum of squared errors defined as (12)

$$MSE = s^2 = \frac{\sum_{i=1}^{k} \sum_{j=1}^{n_i} \left(y_{ij} - \bar{y}_i\right)^2}{N-k} = \frac{SSE}{N-k}. \tag{12}$$

The F test statistic associated to test hypothesis (9) is given by (14)

$$F = \frac{\text{average sum of squares due to treatments}}{\text{mean sum of squares of residuals}} \tag{13}$$

$$= \frac{MS_{TREAT}}{MSE} \cap F(k-1, n-k). \tag{14}$$

The critical region of test (9), for significance level α is given by (26)

$$F_{|H_0} > F_{(k-1,n-k,1-\alpha)}. \tag{15}$$

The total variability SST (16) is measured by the squared mean of the deviations of each observation from the overall mean

$$SST = \sum_{i=1}^{k} \sum_{j=1}^{n_i} \left(y_{ij} - \bar{y}\right)^2 \tag{16}$$

and can be decomposed by the sum of two terms: the inter-group variability given by SS_{TREAT} and a variability within each group SSE, given by (17)

$$SST = SS_{TREAT} + SSE. \tag{17}$$

This relationship is generally presented in tabular form, the ANOVA table.

Contrast Analysis. There is often a need to infer about differences in expected values (e.g. $\mu_3 - \mu_5$). At other times, the parameter of interest is more general than the simple difference in means and can be seen as the difference in expected value 'blocks' (e.g. $\frac{1}{2}(\mu_3+\mu_5)-\frac{1}{3}(\mu_1+\mu_2+\mu_4)$. Such differences lead to the notion of contrast. Contrast is no more than the linear combination of $\mu_i' s$,

$$L = \sum_{i=1}^{k} c_i \mu_i \tag{18}$$

such that the sum of the coefficients c_i is null

$$\sum_{i=1}^{k} c_i = 0.$$

The sample estimator contrast L defined by é is given by (19)

$$\widehat{L} = \sum_{i=1}^{k} c_i \widehat{\mu}_i = \sum_{i=1}^{k} c_i \overline{Y}_i. \tag{19}$$

A Priori Comparision. From the fact of $\widehat{L}$ be Gaussian ??, we conclude that

$$T = \frac{\widehat{L} - \sum_{i=1}^{k} c_i \mu_i}{S\sqrt{\sum_{i=1}^{k} \frac{c_i^2}{n_i}}} \cap t_{[N-k]}. \tag{20}$$

A CI with a confidence level of $(1-\alpha)\times 100\%$ for the contrast L given by (3.1) can be deduced from (20) and it is given by

$$\widehat{l} \pm t_{[N-k,1-\alpha/2]} \times s\sqrt{\sum_{i=1}^{k} \frac{c_i^2}{n_i}} \text{ ou} \tag{21}$$

$$\sum_{i=1}^{k} c_i \bar{y}_i \pm t_{[N-k,1-\alpha/2]} \times s\sqrt{\sum_{i=1}^{k} \frac{c_i^2}{n_i}}. \tag{22}$$

A Posteriori Comparisions. When the hypothesis of equality of means of the different treatments is rejected, it is urgent to make more detailed multiple comparisons between the different averages, since there are at least a couple of distinct averages between them. You only know which or which of these pairs are. Thus, if the explanatory variable influences the dependent variable, it is necessary to ascertain which levels have actually expected different values. The Tukey method or Scheffé method are the most used. In the present manuscript we use the Scheffé method. Follows a brief summary about Scheffé method.

Scheffé Method

The Scheffé method [9] allows you to test any kind of contrast and not just simple differences.

A posteriori CI with a confidence level of $(1-\alpha)\times 100\%$ for the contrast L (see definition (3.1) by Scheffé é method is given by

$$\widehat{l} \pm \sqrt{(k-1)\,F_{[k-1,N-k,1-\alpha]}} \times s\sqrt{\sum_{i=1}^{k}\frac{c_i^2}{n_i}} \text{ ou} \tag{23}$$

$$\sum_{i=1}^{k} c_i\bar{y}_i \pm \sqrt{(k-1)\,F_{[k-1,N-k,1-\alpha]}} \times s\sqrt{\sum_{i=1}^{k}\frac{c_i^2}{n_i}}, \tag{24}$$

with $F_{[k-1,N-k,1-\alpha]}$ the quartile with order $1-\alpha$ from a rv with distribution F de Snedcor with $k-1$ e $N-k$ degrees of freedom.

To test the hypothesis (25)

$$H_0 : L = \sum_{i=1}^{k} c_i\mu_i = 0 \text{ versus } H_1 : L \neq 0 \tag{25}$$

considering a significance level α it is enough to verify if the zero value belongs to the CI (23) with a $(1-\alpha)\times 100\%$ confidence level for the contrast L or considering the critical region (26)

$$F_{|H_0} > (k-1) \times F_{[k-1,N-k,1-\alpha]} \tag{26}$$

with F the statistic test given by

$$F = \frac{\left(\sum_{i=1}^{k} c_i\overline{Y}_i\right)^2}{S^2 \times \sum_{i=1}^{k}\frac{c_i^2}{n_i}} \cap F(k-1, N-k). \tag{27}$$

Notice that the method of Scheffé [9] is just one of the methods that can be found in the literature to perform multiple comparisons, or in general, to test all possible contrasts at the same time, to see if at least one is significantly different from 0. Other possibilities can be considered, e.g. the Tukey method, a parametric approach to test all possible pairwise differences of means to determine if at least one difference is significantly different from 0, or a non parametric approach like Jonckheere-Terpstra test [10, 11] or Kruskal Wallis test [12, 13]. Kruskal Wallis test is a nonparametric alternative to the one-way ANOVA, a rank-based nonparametric test that determines if there are statistically significant differences between two or more groups of an independent variable on a continuous or ordinal dependent variable. Some nonparametric methods are described with some detail by the authors of [14–17].

3.2 MANOVA

Multivariate analysis of ANOVA (MANOVA) is an extension of univariate analysis of ANOVA (ANOVA). In ANOVA, we examined the statistical difference between the continuous dependent variable and the independent grouping variable. MANOVA expands this analysis with several consecutive dependent variables in mind and groups them into weighted linear combinations or complex variables. MANOVA compares whether the newly created combination is different from other groups or levels of independent variables. In this way, MANOVA primarily tests whether the independent grouping variables at the same time constitute statistically significant fluctuations in the dependent variable. MANOVA assumes that the observations are independent of each other, there is no standard for sampling, and the samples are completely random. The independent variable is categorical and the dependent variable is continuous or scaling variable. The dependent variables cannot be closely related. The authors of [18] suggest that the correlation should be at least $r = .90$. The data are multivariate gaussian with homogeneity of variance in distinct groups.

The Levene variance test is applied to evaluate the variance homogeneity between groups of independent variables. The Box M test evaluates the similarity of covariants between groups. This corresponds to multivariate variance homogeneity. The significance of this test is usually determined considering low significance levels due the fact that this test is generally considered very sensitive. The statistics η^2 shows the variance of the independent variable. Used as power size on the MANOVA model. Some posterior tests cam be applied to evaluate if there exists significant differences between the groups. The multivariate F test is basically the sum of the mean squares SS of the source variable divided by the mean error of the source variable MSE.

4 Empirical Application

4.1 The Questionnaire

After authorization by the Portuguese Surface Fleet Commander, the data collection was carried out through questionnaires and some successive visits to the ships, in which the person responsible for waste management was boarded on board each ship and the questionnaires were distributed to the military belonging to each garrison. After approval by the commander of the Portuguese Navy Surface Fleet, data collection was carried out through a questionnaire applied to the embarked personnel of each authorized ship. The questionnaire is divided into two parts, the first part includes socio-demographic variables and personal details (detailed in Fig. 1).

- $Q1_1$ - "'*Gender.*"'
- $Q1_2$ -"'*Age.*"'
- $Q1_3$ - "'*Grade.*"'
- $Q1_4$ - "'*Have you ever attended an environmental training course?*"'
- $Q1_{41}$ "'*If you answered "Yes" in the previous question, it was in Navy.*"'
- $Q1_5$ "'*Do you recycle at home?*"'

Fig. 1. Questionnaire questions. Questions about the socio-demographic information about each participant.

The second part of questionnaire is composed with questions that allow to evaluate KAP about waste management (detailed in Fig. 2).

- $Q2_1$ - "'*The environmental concern on board is always present in my daily life.*"'
- $Q2_2$ - "'*I consider good waste management practice on board ships important.*"'
- $Q2_3$ - "'*There are regular lectures on board on waste management .*"'
- $Q2_4$ "'*Sometimes I dump small waste into the sea.*"'
- $Q2_5$ - "'*I think there is a good waste management policy on board ships.*"'
- $Q2_6$ -"' *There are some types of waste that we can discharge into the sea.*"'
- $Q2_7$ - "' *The glass can be discharged into the sea, as it ends up in the bottom of the sea, having no interaction with the environment.*"'
- $Q2_8$ - "'*Paper and cardboard can be discharged at sea because they easily degrade.*"'
- $Q2_9$ "'*Proper packaging of waste contributes to the welfare, hygiene and safety of the trim.*"'
- $Q2_{10}$ - "'*Waste storage space is adequate.*"'
- $Q2_{11}$ - "'*The conditions of shipboard equipment allow for the treatment of different types of waste.*"'
- $Q2_{12}$ - "'*Even if conditions are not adequate, there is an effort and concern from the trim to minimize the environmental impact of the ship.*"'
- $Q2_{13}$ - "'*The educational offer of the Navy in the environment preservation is sufficient.*"'
- $Q2_{14}$ - "'*The Navy promotes, with its military staff, the preservation of the environment.*"'
- $Q2_{15}$ -"' *There has been an increase in people's awareness of environmental preservation .*"'
- $Q2_{16}$ - "' *I know the Navy Environmental Policy and I know where I can consult it.*"'
- $Q2_{17}$ - "'*I am aware of national and international regulations for reducing environmental impact.*"'
- $Q2_{18}$ - "'*Sometimes on board, environmentally harmful acts are performed due to lack of waste treatment conditions.*"'
- $Q2_{19}$ - "'*Feels that their role in minimizing waste generation on board is important for good waste management in the organization.*"'
- $Q2_{20}$ - "'*On board are used environmentally friendly consumables.*"'
- $Q3$ - "'*Has the waste generated on board ever compromised your well-being?.*"'
- $Q4$ - "'*elect from 1 to 2 factors that undermine the proper functioning of onboard waste management.*"'
- $Q5$ - "'*As the Navy is a military organization, do you consider your concern about the ecological footprint at sea important?*"'

Fig. 2. Questionnaire questions. Questions about the knowledge, Attitude and Practice of each participant about waste management.

The major part of questions from second part use of Likert scale with four levels from 1 to 4 (1 - Totally Disagree, 2 - Partially Disagree, 3 - Partially agree, 4 - Totally Agree; also some questions have a "yes" or "No" answer. Question related to identifying the factors that contribute to poor waste management has an open answer.

4.2 Sample Characterization

The collection of data was done taking into consideration that we are interested to know how waste treatment is carried out on Portuguese Navy ships, if there exists a waste treatment plan adapted to the characteristics and needs of each ship and which are the conditions, how is condition of the equipment on board for processing waste and if the garrisons are aware of the importance of minimizing the environmental impact of ships. Were performed fifteen interviews to on board Waste management Officers to understand the application of a WMS; there were 241 participants from garrisons that responded the questionnaires; the chief of the port support service and the head of financial management and supply section were also interviewed (Fig. 3). In this manuscript we have focused our attention in questionnaires data set.

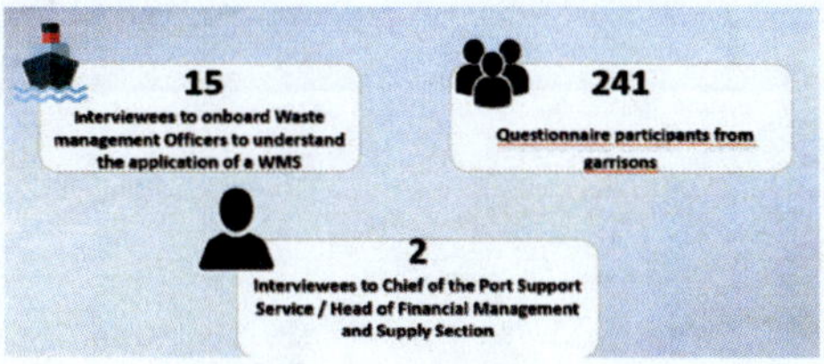

Fig. 3. Collection data.

The questionnaire answers are described as follows. In Fig. 4, the distribution of gender and age is summarized. The sample is composed by 64% of men and 36% of women. People at almost 46% are under 30, and between 38% between 30 and 40. The oldest participant is 51 years old. Information about the environmental issues training and recycling at the home can be found in Fig. 3. Approximately one third of the participants (36%) performed environmental education training. About 68% of participants recycle at home.

Fig. 4. Sample distribution per gender, per age, per environmental education training attendance, per use of recycling at home. Left: Gender (male/female); Center left: Age; Center right: environmental training (no/yes); Right: Recycling use (no/yes). Source [1]

4.3 Results and Final Remarks

In [4,5], the statistical analysis of the questionnaire was performed. The alpha-Cronbach coefficient was coherent to a good internal consistency, the We measure the homogeneity and internal consistence of questionnaire and respective validation.

A preliminary data analysis using the questionnaire data set and taking into account the non-quantitative nature of the relevant variables, we have computed some measures of association between the distinct questions: the non-parametric Spearmann correlation coefficient and Friedman's non-parametric test for samples [3,5]. Both tests were significant ($p - value < 0.001$) meaning that in general questions have not the same distribution Fig. 5.

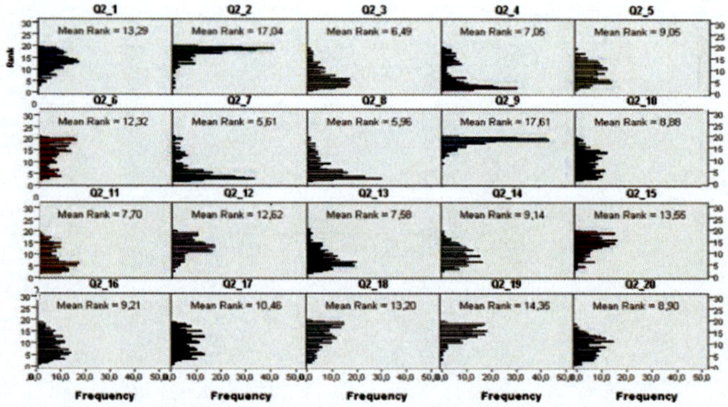

Fig. 5. Sample distribution of ranks $Q2_1$–$Q2_{20}$. Source [4].

The paired T-test, McNemar's test for frequencies comparison, Crochan's Q test for binary variables comparison. Also were performed the Friedman test (p-value < 0.001) and kendalls coefficient of concordance test ($p - value < 0.001$). All tests conduced to the same conclusion: the distributions of considered questions are distinct [2].

In [5] were also performed unilateral tests for median considering each question $Q2_1$–$Q2_{20}$, the Wilcoxon test and the signal test. We can say that more than 50% of participants have declared that there exists the daily environmental care, consider it an important procedure, also consider that some waste can be left in sea, the waste storage contributes to welfare, security and hygiene of staff. Also consider that the existent equipment to process waste is not enough. The staff declares to know the internal and external rules but claims that there is not a good offer of formation in environmental education. Also we could confirm that environmental awareness is increasing.

In Fig. 6 is displayed the composition of sample per rank and the distribution of the attendance of environmental training per rank. Clearly, it is not uniformely distributed. The correlation between the training course attendance and the rank is strongly significant ($p - value < 0.001$).

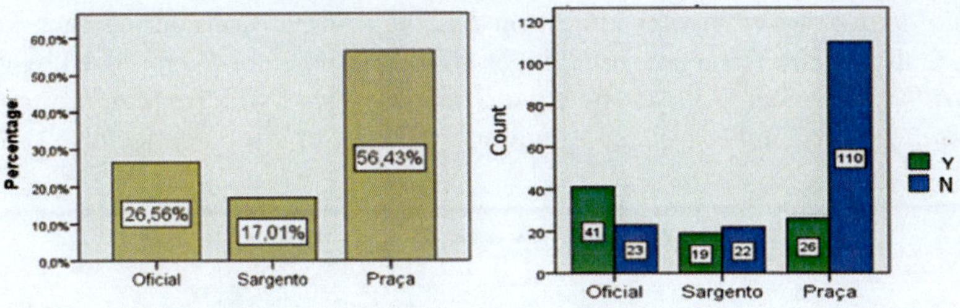

Fig. 6. Left: Rank percentage; Right: Distribution of environmental education attendance per rank. Source [1].

In [1] we also have applied an exploratory factorial analysis to reduce the dimensionality of the problem. Four factors were selected and was identified a meaning for each: F_1 combines variables that characterize *Awareness*, F_2 combines variables from *Hygiene and Safety*, F_3 combines variables from *Practice*. This new variables were introduced as independent variables in a ANOVA uni-dimensional model in [5]. The factor associated with *Hygiene and Safety* issues conduced a significant model ($p - value < 0.05$) when is related with the kind of ship.

In the present work, we could obtain a model with a global F test with a $p - value =$ 0.073, using a two dimensional anova (see Table 1 where the kind of ship and the first interaction between rank and the kind of ship were significant as covariates to explain the *Awareness* variable (the dominant factor $F1$ obtained in [4] by EFA).

Table 1. Analysis of variance. Dependent variable: *Awarness* (factor 1 obtained by EFA in [4]). Explanatory variables: Kind of ship, Rank (first order interaction with kind of ship).

Source	Type III sum of squares	df	Mean square	F	Sig.
Corrected model	29.297a	20	1.465	1.529	0.073
Intercept	0.800	1	0.800	0.835	0.362
Q1_3	1.916	2	0.958	1.000	0.370
Kind_ship	10.680	6	1.780	1.858	0.089
Q1_3 * Kind_ship	18.411	12	1.534	1.602	0.092
Error	210.703	220	0.958		
Total	240.000	241			
Corrected total	240.000	240			

The rank revealed significant differences in the first factor F_1 when we consider different levels of ranks in intersection with the kind of ship. The F test conduced to a $p - value = 0.09$. We can find such differences in the Fig. 7 where is displayed the difference (effect) of the global mean of the factor F_1 relatively to the mean per rank for each level of kind of ship. Notice that from Fig. 7, in general we can evidence that the *oficial* and *sargento* greater effects on $F1$, than *praça*. Considering each kind of ship level, the *lancha* is the one with bigger effect on the factor F_1, followed by *veleiro* and *patrulha oceânica*. Distinct effect padrons are achievd with *fragata*, *hidrográfico*, *lancha hidrográfica* and *corveta* contributing with smaller effects on factor $F1$.

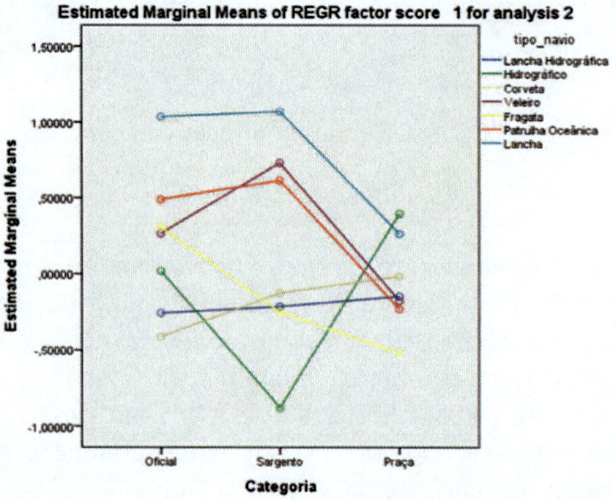

Fig. 7. Effects of the kind of ship in factor 1: difference between the mean estimates per each level with the global average.

In Table 2 we have displayed the Scheffe simultaneous 90% intervals for the mean difference of *Awarness* when are considered distinct ranks. The only significant interval for the mean difference of *Awarness* is obtained when we consider *oficial* and *oficial* ranks. The results appear to be adequate and in accordance with the fact that a higher rank corresponds to a major rate of an environmental training.

Table 2. Multiple Comparisons: Scheffe simultaneous 90% intervals for the mean difference of *Awarness* for several levels of ranks. The mean difference is significant at the 0.10 level.

(I) Categoria		Mean difference (I-J)	Std. error	Sig.	90% Confidence interval	
					Lower bound	Upper bound
Oficial	Sargento	0.0817538	0.19576567	0.917	0.3405607	0.5040684
	Praça	0.3202526*	0.14834720	0.100	0.0002312	0.6402739
Sargento	Oficial	−0.0817538	0.19576567	0.917	−0.5040684	0.3405607
	Praça	0.2384987	0.17436088	0.394	−0.1376405	0.6146379
Praça	Oficial	−0.3202526*	0.14834720	0.100	−0.6402739	−0.0002312
	Sargento	−0.2384987	0.17436088	0.394	−0.6146379	0.1376405

The results are compatible with the fact that smaller ships have no WMP, with few elements on board and with short time missions, the waste volume is reduced, does not represent a problem. With bigger ships, the majority have a WMP, some times without enough conditions to apply the plann, a minority without a WMP corresponds to older ships (like *corveta*) that were built in a time where environmental issues were not considered. Also were studied another kind of relations with the 3 factors obtained by EFA, including the computation of an indicator of good practices combining the information of some questions. The results appear to be promising but will be presented and discussed in a future article.

Acknowledgements. This work was supported by Portuguese funds through the *Center of Naval Research* (CINAV), Portuguese Naval Academy, Portugal and *The Portuguese Foundation for Science and Technology* (FCT), through the *Center for Computational and Stochastic Mathematics* (CEMAT), University of Lisbon, Portugal, project UID/Multi/04621/2019.

References

1. Rebelo, J.: Impacto Ambiental da Marinha Portuguesa. Análise e resolução da Gestão de Resíduos no mar. Master thesis, Escola Naval (2019)
2. Rebelo, J., Jerónimo, J., Teodoro, M., Lampreia, S.: Preliminary reflexion about waste management plan in NRP ships. In: Simos, T., et al. (eds.) Computational Methods in Science and Engineering, vol. 2186, pp. 183–188. AIP (2019)
3. Rebelo, J., Jerónimo, J., Teodoro, M., Lampreia, S.: Modeling the waste management in NRP ships. In: Entrepreneurial Ecosystems and Sustainability. Proceedings of Regional Helix 2019 (2019). ISBN 978-989-98447-7-3
4. Rebelo, J., Jerónimo, J., Teodoro, M., Lampreia, S.: Literacy about the waste management in boarded staff. In: Proceedings of EUROGEN 2019, Guimarães, 12nd–14th September (2019)
5. Teodoro, M., Rebelo, J., Lampreia, S.: Waste management and embarked staff. In: Gervasi, O., et al. (eds.) Computational Science and Its Applications – ICCSA 2020. LNCS, vol. 12251, pp. 402–503. Springer, Cham (2020). https://doi.org/10.1007/978-3-030-58808-3_36
6. Sunaryo, S., Pahalatua, D.: Green ship recycle yarn design. J. Nav. Archit. Mar. Eng. **12**(1), 15–20 (2015)
7. Deja, A.: Organization of the reception of ship-generated wastes and cargo residues illustrated by selected examples of European union seaports. Sci. J. **33**(105), 16–21 (2013)
8. Monteiro, J.: Poluição. Marítima; Normas e Controlo nos meios Navais das Forças Armadas. Monography, Instituto Universitário Militar (2016)
9. Scheffé, H.: The Analysis of Variance. Wiley, New York (1959)
10. Jonckheere, A.: A distribution-free k-sample test again ordered alternatives. Biometrika **41**, 133–145 (1954)
11. Terpstra, T.: The asymptotic normality and consistency of Kendall's test against trend, when ties are present in one ranking. Indag. Math. **14**(3), 327–333 (1952)
12. Bewick, V., Cheek, L., Ball, J.: Statistics review 9: analysis of variance. Crit. Care **7**, 451–459 (2004)
13. Bewick, V., Cheek, L., Ball, J.: Statistics review 10: further nonparametric method. Crit. Care **8**(3), 196–199 (2004)
14. Conover, W.: Practical Nonparametric Statistics, 3rd edn. Wiley, New York (1999)
15. Moore, D., McCabe, G.: Introduction to the Practice of Statistics, 4th edn. W H Freeman & Co., New York (2003)

16. Sprent, P., Smeeton, N.r.e.: Applied Nonparametric Statistical Methods. Chapman & Hall/CRC, London (2001)
17. Whitely, E., Ball, J.: Statistics review 6: nonparametric methods. Critical Care **6**, 509–513 (2006)
18. Tabachnick, B., Fidell, L.S.: Using Multivariate Statistics. Person Education, Boston (2012)

Mapping of Bioeconomic Development Centers: Trends and Opportunities in Research, Development and Innovation

Jayson Luis da Silva Ribeiro[1](✉), Anapatricia Morales Vilha[1,2], Carlos Antonio Medeiros Gambôa[1], Fábio Danilo Ferreira[1,2], and Débora Maria Rossi de Medeiros[1,2]

[1] Federal University of ABC, Santo Andre, SP 09210-580, Brazil
[2] Innovation Center, Federal University of ABC, Santo Andre, SP 09210-580, Brazil

Abstract. The models of production management and business organization have been constantly improving. Industrial conglomerates have emerged in the last decades, resulting in large centers of biotechnological and pharmaceutical knowledge. Initiatives related to the sharing of demands between companies, and more recently, sharing of research and development infrastructure, have transformed business models creating new opportunities for the biotechnology segment around the world. Following this trend, this article seeks to map large companies in the biotechnology sector, analyzing data from regulatory agencies, scientific articles, reports from regional development and innovation centers and other market information disclosed, Aiming to pointing out agglomerations in large hubs and initiating a discussion of the possibilities for the development of new technologies and the strengthening of the Brazilian biotechnology market. This article analyzed the main clusters in Europe and United States of America, listing some of its largest companies, market sectors, research and innovation projects. As a result, we concluded that these large conglomerates have been benefited from their location and commercial relationship. We believe that this research can serve as a basis for new studies related to models that can foster the development of new biotechnologies between companies and their innovation ecosystem. As a suggestion for new discussions, the continuity of studies related to the concepts of sharing economy is required, especially in the Brazilian context, seeking innovative solutions to factors that hinder the strengthening of the world biotechnological market, mainly those related to the costs of research, development and innovation (RD&I) of new technologies.

Keywords: Biotechnology market · RD&I · Innovation

1 The Biotechnology Market

During the convention on biological diversity of the United Nations Organization - UN (1992), the concept of biotechnology used worldwide was defined, establishing Biotechnology as the area of biology that uses biological systems, living organisms, or derivatives

J. Machado et al. (Eds.): icieng 2021, LNME, pp. 271–279, 2022.
https://doi.org/10.1007/978-3-030-78170-5_24

thereof, to develop or modify products and processes for technological and specific use. Since then, the biotechnology sector has been a promising bet for solving several complex problems in contemporary society, from the treatment of diseases, development of biomaterials, agricultural techniques, clean production and innovative solutions to fight climate change in a sustainable way.

Biotechnology has revolutionized markets and has the potential to improve various fields of knowledge through the development of new products and processes. Among some of the current research with great impact, we can mention the use of bioplastics, biofuels, enzymatic detergents, production of food in culture (meat for example, reducing the need to use land, natural resources and antibiotics), biopesticides, biofertilizers, cosmetics, clothing and so many other markets that can benefit from the use of new techniques and inputs. Following this global trend, Brazil biotechnological market can be improved by understanding international models of cooperation on research, development, innovation - RD&I of the global biotechnology market, strengthening the biotechnology industry in the region and assisting the study of future public sector incentive policies.

In this article, we researched some of the main biotechnological companies and hubs in the Europe and North America regions, listing the major technology centers by area of knowledge and business concentration. In parallel, some initiatives and projects to strengthen the biotechnological sector in the national level were analyzed, initiating a discussion on the future of biotechnology in Brazil. The data was obtained through reports from regulatory agencies, scientific articles, reports from regional development and innovation centers and other market information disclosed.

2 The International Market and Analysis of Data from the Biotechnology Segment

The world consumer market has undergone a change in its production models and product engineering. Countries like the United States, Spain and China observe the exponential growth of startups focused on research, development and biotechnological innovation (RD&I), advancing in several areas of knowledge in a collaborative and interdisciplinary way (Phillips 2019).

According to a report by the consultancy Global Market Insights (2019), the biotechnology market should surpass the mark of $ 729 billion dollars in the year 2025. Similarly, analyzing the year 2018, it was possible to evidence an exponential growth especially in the bioagriculture segment, moving the equivalent of $ 35 billion dollars in the year. Several fields of research and biotechnology hubs have emerged in recent years, in Europe, for example, the biotechnology segment is identified as one of the sectors that most employs and generates income. In the next paragraphs, we seek to map some of the major hubs in the European biotechnology segment. The list of companies analyzed in this research is shown in Table 1 – “Biotechnological research centers”.

Regarding to data analysis, cities with a large concentration of companies focused on biotechnological research, development and innovation were chosen for this research. Factors such as territorial location, development of new projects (presence of scientific

Table 1. Biotechnological research centers.

Region	Biotechnology companies
Amsterdam	UniQure, Arthrogen, Agendia, Kiadis
Barcelona	Ysios capital, Minoryx Therapeutics, Aelix Therapeutics, Leukos Biotech, Peptomyc
Berlin	Mologen, RenovaCare, ImmunoLogik e Solaga
London	Orchard Therapeutics, Nightstar, Autolus, Cell Medica, Motif Bio e ReViral
Basel	Actelion, Basilea, Polyphor, Allecra, Idorsia (spin-off da Actelion) e Santhera Pharmaceuticals
Oslo	Zelluna, Nordic Nanovector, Targovax, PCI Biotech e AlgiPharma
Munich	MorphoSys, Medigene, Wilex, 4SC e Oryx
Lyon	Adocia, Poxel e Alizé Pharma, Imaxio, Fab'entech, Enyo Pharma, Erytech, APCure e Orega Biotech
Vienna	Hookipa, AFFiRis, Themis, Nabriva e Apeiron
Oxford	Oxford Nanopore, Oxford Biomedica, OxSyBio e Oxitec

articles or patents) and local cooperation were evaluated. Concerning the main data collected, we would like to emphasize the cities described below, as follows.

Amsterdam is considered one of the major centers of European biotechnological knowledge, has several companies conducting research focusing on gene therapy, new models and techniques to improve compatibility in bone marrow transplantation, development of new procedures aimed at combating diseases such as arthritis and cancer breast cancer (reducing the need for chemotherapy that has more intense and harmful side effects to patients), among other important researches of worldwide relevance. Some of the its main companies: uniQure, Arthrogen, Agendia, Kiadis.

In Spain, the city of Barcelona is a reference as the largest national biotechnological center, with exponential growth in recent decades. The region has a large number of startups and research focused on the treatment of rare diseases such as HIV, cancer and leukemia. Among the main companies in the segment can be mentioned: Ysios capital, Minoryx Therapeutics, Aelix Therapeutics, Leukos Biotech, Peptomyc.

The city of Basel stands out as the headquarters of the company Actelion, one of the largest biotechnology companies in Europe. It also has other large companies, like Basilea, Polyphor, Allecra, Santhera Pharmaceuticals and Idorsia (spin-off from Actelion).

In Berlin, companies with a focus on the use of cyanobacteria for the production of biogas, treatment of cancer, HIV and methods to heal burns and wounds by stimulating the development of stem cells stand out. Most relevant companies: Mologen, RenovaCare, ImmunoLogik and Solaga.

The city of London has a structured ecosystem for supporting and developing companies (startup to the highest levels of maturity) and benefits from the proximity to institutions, financing agents and investors. The city has been the subject of major investments in biotechnology in recent decades and has companies such as Orchard Therapeutics, Nightstar, Autolus, Cell Medica, Motif Bio and ReViral.

Munich is known as a business hub, it has several universities and research institutes and because of that it has attracted many biotechnology startups in the last decades, among which we can mention: MorphoSys, Medigene, Wilex, 4SC and Oryx.

In the region of France, Lyon is the country's second largest city and is often cited as one of the cities with the best quality of life in Europe. It has more than 60 thousand workers allocated in the biotechnology segment and also has a cluster of companies focused on the treatment of diabetes (Adocia, Poxel and Alizé Pharma), infectious diseases (Imaxio, Fab'entech, Enyo Pharma, Erytech, APCure and Orega Biotech) among others.

The Oslo region stands out mainly in the development of research related to oncology and microbial infections, the city has important biotechnology companies such as Zelluna, Nordic Nanovector, Targovax, PCI Biotech and AlgiPharma.

The Oxford business conglomerate has companies working on DNA sequencing, gene therapy, regenerative medicine and methods to prevent Zika, malaria and dengue through mosquito reengineering. Among the main members of this cluster we can highlight Oxford Nanopore, Oxford Biomedica, OxSyBio and Oxitec. The city has large laboratories and universities that help to strengthen its research and applied scientific development conglomerate.

The capital Vienna has large companies working on vaccines, immuno-oncology and development of antibiotics for multi-resistant bacteria, among which we can mention Hookipa, AFFiRis, Themis, Nabriva and Apeiron.

In the United States, in view of the importance of the biotechnological segment and following this global trend, in 2016 the businessman and creator of Facebook Mark Zuckerberg and Priscilla Chan, highlighted the potential of the biotechnological and medical research market by making large investments in the creation of a Biohub in a collaborative and integrated with three leading teaching and research institutions: Berkeley, UCSF and Stanford.

Youtie and Shapira (2008) implemented studies on Georgia's innovation and technology hub, establishing the importance of this type of organizational arrangement for the development and economic strengthening of the region. In the same way, Baark and Sharif (2005) studied the transformation of a shopping center into an innovation hub in Hong Kong, strengthening the business ecosystem in this region of China. In both cases, the proximity and sharing of physical and technological resources resulted in an efficiency gain in the development of new technologies.

Analyzing the data from these global biotechnological centers, we can conclude that the organization of knowledge conglomerates around the world are present in cities that have a policy of fostering innovation. Through public policies and the provision of infrastructure and skilled labor, these cities were able to create large centers of biotechnological research and development, new clusters and Hubs were established on the last decade in this regions and changed the way the market relates to society. Understanding these models and mapping these determinants is a relevant factor for the development of this type of experience in the Brazilian market.

A relevant fact was to verify the common presence in these centers of a cooperative model between companies, government and society, one of the main pillars of the triple helice concept (for more information see (Bruneel et al. 2010; Etzkowitz et al. 2017; Ribeiro 2018).

Another interesting information is that all these centers have a knowledge generator university, with biotechnological research projects, and technical related degrees on the region. The interaction between this ecosystem generates a beneficial cycle in which the university is aligned with the needs of the local market and serves as a factor of regional development through the supply of qualified technical labor for the region. Likewise, the presence of laboratories and specialists proved to be an attractive factor for the establishment of these companies in the region.

Economic factors were also analyzed and considered as limiting factors to the development and creation of new companies in the analyzed biotechnological production centers. The operational cost, security, intellectual and legal protection were the limiting factors for the creation of these research centers.

Considering the results of the data analysis, as well as the concept of shared economy, we see an opportunity for the creation of large centers of shared biotechnological research, focusing on the interaction between companies, government and university, creating an environment conducive to strengthening the biotechnological segment and reducing operating costs, thus making the region more efficient and competitive in a globalized market.

The study of the large biotechnological conglomerates around the world mentioned above has a lot to contribute to the development of models of cooperation and promotion at the national level, creating new business, research and industrial development opportunities in our innovation ecosystem, especially for the small biotechnological companies and startups.

Through strategies aimed at product engineering and market alliances (horizontal and vertical cooperation), small biotechnology companies have shown an increase in representativeness in the world market in branches previously dominated exclusively by large pharmaceutical companies (Heitzmann 2019).

In more developed countries, the biotechnology clusters are located in regions with competence in the area, as they depend on efforts of interaction between actors, such as teaching and research institutions, incubators and technology parks, commercial and industrial associations, quality and standardization entities, risk investors, in addition to organization and resources capable of strengthening interaction, synergy, collaboration and competition (Suzigan et al. 2005).

Thus, the creation of knowledge-intensive clusters converges with the prospect of strengthening the critical basis for coagulation of the actors responsible for the technological performance of countries and regions, such as companies, teaching and research institutions, startups, financing agencies and venture capital, technology parks and government.

Similarly, business hubs are collaborative environments capable of stimulating the development of technologies through the sharing of tangible and intangible resources; with a focus on connectivity, these spaces bring together companies in order to create innovative products and services using the common coworking infrastructure in a shared and collaborative way. The city of St. Louis in the United States, for example, outlined strategies for creating its BIOHub and emerged as one of the fastest growing markets with a focus on life sciences (Marris 2008).

Nevertheless, John (2014) defines Hubs as business, research and development environments capable of providing adequate spaces for PD&I support and especially valuable networking for the promotion and maturation in micro and small companies (MSEs) and startups. With regard to clusters, John (2014) argues that:

> "... clusters can be described as a group of mutually dependent and related businesses grouped in the same locality or region in a city ... clusters can also be established virtually through networking networks ... hubs and clusters are closely interconnected and associated with the same phenomenon (p. 03) ".

Universities and research institutes are an important part of the existing cluster models. The relevance of the role of universities and institutes of science and technology (ICTs) in building sustainable creative economies is recognized as one of the main reasons for the development and rise of this type of practice. Ruiz (2010), analyzes in his research the role of universities and agents of science and technology in the structuring of an innovation hub in the UK organization Nesta, located in the United Kingdom, demonstrating the impact of these organizations in promoting the development of new technologies in the region.

In this sense, Chesbrough (2003) describes a model of open innovation and industrial cooperation in order to facilitate the interaction and transfer of knowledge and technologies. In his point of view, the researcher states that in order to obtain benefits from research and development of a product in competitive economies, it is necessary to exchange knowledge with other organizations, creating mutual gain relationships and reducing transaction costs.

Considering the results of this research, we have evidenced through the final data analysis that the main concepts related to the success in the experiences of this biotechnological centers are related to practices of open innovation, shared economy and triple helix aligned with economic, political and local infrastructure factors. The study of these factors has great potential to contribute to the Brazilian biotechnology research, development and innovation market.

3 The Brazilian Biotechnological Market: Considerations and Opportunities

According to a survey organized by the OECD (2013), Brazil occupies the 12th position in the number of biotechnology companies - the State of São Paulo occupying the 20th place - with a population of 314 firms, 152 of which are located in São Paulo (Fig. 1).

Considering the 27 federation units, 14 of them have technology-based companies. The State of São Paulo covers almost half of the companies in this sector (48%), Minas Gerais has a little more than a fifth of the companies (21.3%) and in Rio de Janeiro they have 10%.

Considering the field of public policies for biotechnological industrial development, the government of the state of São Paulo (Brazil), through its economic development

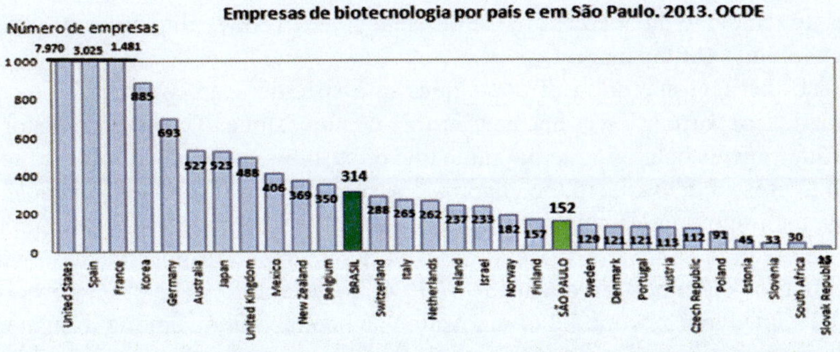

Fig. 1. Biotechnology companies by countries and in São Paulo. Source: OECD 2013, adapted from Freire (2014).

secretary, has encouraged studies of new models for sharing and promoting industrial production and innovation. Debates about Open innovation projects, relationships with startups and the creation of innovation hubs have been encouraged, especially those related to the triple Helix concepts, highlighting the role of state and federal institutes, universities and research centers.

The multidisciplinary nature of the biotechnology field, connecting large areas of knowledge, requires scientific research on interaction models capable of strengthening scientific research that allows investigating the results from a technological, economic and social point of view.

The variables involved in the process of innovation and development of technologies are not linear and suffer great interference according to the sector in which the company is inserted. While sectors that are intensive in scale direct a large part of their investments towards the maintenance of an already developed process, science-based sectors such as biotechnology companies invest a large part of their resources in the generation of knowledge and new technologies.

It is relevant to mention the research by Guiliani (2005), evaluating the organization of clusters and their capacity for evolution and consolidation. We can emphasize the fact that in the last decades, research, development and innovation models have been the object of study in the most diverse fields of knowledge. Research relating new techniques and analysis of complex systems began to modify the way that companies organize their production model and interact with the market. The growth of startups, competing directly with traditional large companies, has caused many entrepreneurs and researchers to rethink the existing PD&I models in their business, seeking the dynamism and culture of reengineering present in these new entrants in their business chain.

Likewise, following the concepts of the Sharing economy, large companies and markets around the world have devised strategies to foster innovation through the sharing of goods and services. Ribeiro (2018) found the main factors related to the mortality of basic research in micro and small companies (MSEs), showing the sharing of laboratory infrastructure as an alternative for solving the main factors related to the abandonment of promising projects in their field: lack of R&D equipment, laboratory infrastructure and specialized operational/technical knowledge. Thus, the principles of sharing economy are directly related to the increase in the efficiency rate in converting basic research into new technologies for the market.

In fact, the agglomeration of companies in a specific segment in order to obtain better market opportunities is not new in this century alone. Throughout history, the most diverse agents belonging to the same line of business have joined together seeking strength to present demands to the government, negotiate with suppliers and customers and even the use of tactics now considered unethical, such as controlling the supply of determined well to control its price in times of high demand (as was the case with the organization of oil exporting countries - OPEC, for example). The great news of this decade is related to a new model of innovation and collaboration among its agents, and not only to the sharing of information inherent to its market itself.

Innovation R&D hubs share their infrastructure and technical knowledge applied in the production and development of new products and present a good opportunity for new studies, thus identifying more efficient production models through alternatives for better use and use of their available physical resources.

This article analyzed the main clusters in Europe, listing its largest companies, market sectors, research and innovation projects. We infer that these large conglomerates have benefited from their location and commercial relationship. As a suggestion for new discussions, we verified the possibility of analyzing Brazilian biotechnological centers by regions, comparing the Brazilian's development centers experience (especially in the regions of São Paulo, Rio de Janeiro and Minas Gerais) with the formation of European clusters listed in this scientific article.

Through the analysis of the world organization of biotechnological clusters, their formation, structure, specificities and individual skills, we aim to contribute to the discussion of new models and possibilities of productive arrangement. We believe that the continuity of studies related to the concepts of sharing economy is relevant, especially in the Brazilian context, seeking innovative solutions to the factors that hinder the strengthening of the world biotechnological market, mainly those related to the costs of research, development and innovation (RD&I) of new technologies.

References

Baark, A., Sharif, N.: From trade hub to innovation hub: the role of Hong Kong's innovation system in linking China to global markets. Innovation: management, Policy Practice **8**(1), 193–209 (2006)

Bruneel, J.: The Search for Talent and Technology: Examining the Attitudes of EPSRC Industrial Collaborators towards Universities. Advanced Institute of Management Research, Londres (2010)

Chesbrough, H.: Open Innovation: The New Imperative for Creating and Profiting from Technology. Harvard Business School Press Books, Boston (2003)

Etzkowitz, H., Zhou, C.: Hélice Tríplice: inovação e empreendedorismo universidade-indústria-governo. Estud. av.**31**(90), 23–48 (2017)

Giuliani, E.: Cluster absorptive capacity: why do some clusters forge ahead and others lag behind? European Urban and regional studies (2005)

Global Market Insights: Biotechnology Market Size by Application Potential, Competitive Market Share & Forecast (2019)

Heitzmann, F.: Proposal for an innovating business model for supporting biotechnology companies, ecosystem and their founders. J. Commercial Biotechnol. **24**(3) (2019)
John, N.: Hubs, clusters and regions. British Council. Creative economy guide (2014)
Marris, E.: Nature Biotechnology. Nature Publishing, vol. 26, pp. 471–472 (2008)
Nações unidas: Convention sur la diversité biologique signée le 5 juin 1992 à Rio de Janeiro. Recueil des Traités des Nations Unies, vol. 1760, Numéro de l'enregistrement: 30619 (1992)
Phillips, T.: The top countries for Biotech firms and research. Disponível em https://www.thebalance.com/ranking-the-top-biotech-countries-3973287. Acesso em: 18 de julho de 2019
Ribeiro, J.L.S.: Avaliação da interrupção no desenvolvimento de novas tecnologias nas micro e pequenas empresas do setor químico do abc paulista com base no TRL. Anais - simpep. **1**, 13 (2018)
Ruiz, P.: NESTA's innovation hub: an analysis based on network theory. Bath University (2010)
Suzigan, W.: Padrões setoriais de inovação na indústria de transformação brasileira Estudos econômicos. São Paulo **44**(2), 277–321 (2005)
Youtie, J., Shapira, P.: Building an innovation hub: a case study of the transformation of university roles in regional technological and economic development. Res. Policy **37**(1), 1188–1204 (2008)

New Hybrid Algorithm for Supply Chain Optimization

Carlos Cevallos[1], Mario Peña[2], and Lorena Siguenza-Guzman[3](✉)

[1] Faculty of Engineering, Universidad de Cuenca, Cuenca, Ecuador
carlos.cevallos@ucuenca.edu.ec

[2] Research Department, Universidad de Cuenca, Cuenca, Ecuador
mario.penao@ucuenca.edu.ec

[3] Department of Computer Sciences, Faculty of Engineering, Universidad de Cuenca, Cuenca, Ecuador
lorena.siguenza@ucuenca.edu.ec

Abstract. Optimization is the process of obtaining the best solutions to specific problems. In the literature, those problems have been optimized through a plethora of algorithms. However, these algorithms have many advantages but also disadvantages. In this article, a New Hybrid Algorithm for Supply Chain Optimization, NHA-SCO has been proposed in order to improve the benefits of objective function convergence. For the analysis of the results, three assembly companies have been utilized as case studies. These companies present their supply chains, i.e., networks where products flow from their raw material to the final products delivered to clients. These supply chains must satisfy different objectives, such as maximize benefits and service level and minimize scrap. For the evaluation of results, NHA-SCO has been compared to other well-known optimization algorithms. In the presented case studies, the NHA-SCO algorithm performs faster, or it converges in fewer iterations, obtaining similar or even better results than the other algorithms tested.

Keywords: Optimization · Supply chain · Metaheuristics · Hybrid algorithm

1 Introduction

The technological wave has grown in the last years. These changes have greatly improved decision-making; however, they have also brought new problems that involve more variables and information, complicating and streamlining these processes. In this scenario, it is essential to find fast and optimal solutions. Optimization is the process of obtaining the best solutions (optimal state) to specific problems. In this sense, Global Optimization is a technique to solve any optimization problem to find an optimal state from a large set of elements or states [1]. In the literature, a plethora of algorithms has been developed to optimize different problems. However, these algorithms have many advantages but also disadvantages. Thus, new algorithms have been developed which combine different characteristics of other algorithms to optimize faster while getting better results [2].

J. Machado et al. (Eds.): icieng 2021, LNME, pp. 280–292, 2022.
https://doi.org/10.1007/978-3-030-78170-5_25

In the industry, companies try to improve their services by optimizing their Supply Chain (SC). [3] defined SC as an entity network where the raw material or product flows through it. This network includes distribution centers, markets, clients, transporters, industries, and providers. The study of SC is significant within organizations since it offers information to company stakeholders to make decisions and changes. These decisions and changes could enhance the behavior of an organization to improve its benefits and the quality provided to final users [4]. In this sense, global optimization allows optimizing a SC, but an optimizer must make the process; in this case, the chosen optimizer is an algorithm.

In [5], a hybrid algorithm was developed to optimize a SC. It consisted of combining different characteristics of several optimization algorithms to obtain a new algorithm that optimizes faster than others. The optimization algorithms used are known as metaheuristics. Metaheuristics are algorithms usually inspired by nature [6] that help to solve general problems utilizing objective functions [7–9]. This paper aims to describe an improvement of the previous algorithm proposed. Specifically, the changes made to enhance the algorithm are discussed, and also the reasons why these changes were made. To analyze the benefits of the new algorithm, three case studies were utilized. The first case study is the same as shown in [10], but variables were modified. The second case is similar to the previous; however, the SC is handled individually per product in each period. Finally, the last case study is the optimization of raw materials cutting through patterns. The rest of the paper is organized as follows. In Sect. 2, the case studies are presented. Section 3 briefly explains the optimization algorithms utilized. Section 4 describes the New Hybrid Algorithm for Supply Chain Optimization, NHA-SCO. Section 5 reports the results of the proposed algorithm. In Sect. 6, these results are discussed. Finally, the conclusions are included in the last part.

2 Case Studies

Three case studies were utilized in this work. The first case study is related to a company where televisions are assembled. The second and third case study belong to the same company in charge of assembling furniture. The names of the companies are not mentioned due to a confidentiality agreement signed with the companies.

2.1 Case Study I

The first case study corresponds to maximizing the benefit and service level in an assembly company. **Benefit:** Maximize the profit of the company, i.e., the resulting value of the sum of all distributed products multiplied by its sell price, subtracted by the amount of the costs of assembly, distribution, transportation, and storage (1). **Service Level:** Maximize the service level offered to clients (2).

$$f1 = \sum_z \sum_j \sum_k \sum_b (PD_{zjkb} * PV_{kb}) - \sum_z \sum_j \sum_b P_{zjb} * (CM_{zjb} + CE_{zjb}) - \sum_z \sum_j \sum_k \sum_b (PD_{zjkb} * CT_{zjkb}) - \sum_z \sum_j \sum_b (I_{zjb} * CI_{zjb}) \tag{1}$$

$$f2 = \frac{\sum_z \sum_j \sum_k \sum_b PD_{zjkb}}{\sum_z \sum_k \sum_b D_{zkb}} \tag{2}$$

where: b is the product, j is the plant, k is the market, z is the period. **Variables:** P_{zjb} is the total amount of product b assembled in plant j at period z, PD_{zjkb} is the quantity of product b sent from plant j to market k at period z, I_{zjb} is the final inventory of product b in plant j at period z. **Constants:** CM_{zjb} is the unit cost of raw material supplied to assemble b to plant j at period z, CE_{zjb} is the cost of assembly of product b in the plant j at period z, CI_{zjb} is the unit holding cost of product b in plant j at period z, PV_{kb} is the sale price of product b in market k, CT_{zjkb} is the cost of transporting one unit of product b from plant j to market k at period z, D_{zkb} is the demand for product b of customer k at period z, $I_{jb}0$ is the initial inventory level of product b in the plant j, pt_{jb} is the time required to produce product b in plant j, tt_{zj} is the time available to produce in plant j at period z, SS_{zjb} is the safety stock of product b in plant j at period z.

Subject to the following **constraints**. Constraint (3) ensures that the quantity of distributed products in a market at each instant is less than the total amount of produced products in every plant. Constraint (4) guarantees that the time available at each plant in each moment is higher than the time required to produce products. Constraint (5) ensures that the products distributed to the markets from the plants should be less than the demand at each instant. Equation (6) defines the inventory balance for the products in each plant at each moment. Constraint (7) ensures that the inventory of products at the end of each instant is higher than the safety stock in each plant. The quantity of products produced, stored, and distributed in a market in each plant at every moment is always positive and are guaranteed through Constraint (8).

$$\sum_j P_{zjb} \geq PD_{zjkb}; \quad \forall z, k, b \tag{3}$$

$$\sum_b pt_{jb} * P_{zjb} \leq tt_{zj}; \quad \forall z, j \tag{4}$$

$$\sum_j PD_{zjkb} \leq D_{zkb}; \quad \forall z, k, b \tag{5}$$

$$I_{zjb} = I_{(z-1)jb} + P_{zjb} - \sum_k PD_{zjkb}; \quad \forall z, j, b \tag{6}$$

$$I_{zjb} \geq SS_{zjb}; \quad \forall z, j, b \tag{7}$$

$$P_{zjb},\ PD_{zjkb},\ I_{zjb} \geq 0; \quad \forall z, j, k, b \tag{8}$$

2.2 Case Study II

The second case study is the maximization of the SC benefit. **Benefit:** Maximize the value of the benefit obtained subtracted by the sum of distribution, inventory, and transportation costs (9).

$$f3 = \sum_{k,b} PVP_{kb} * PV_{kb} - \sum_{b} PR_b * CP_b - I * \sum_{b} CP_b * Inv_b * T_b - \sum_{k,b} CT_b * PD_{kb} \quad (9)$$

where: b is the product, k is the market. **Variables:** PV_{kb} is the quantity of product b sold in market k, PR_b is the production of product b, PD_{kb} is the quantity of product b distributed to market k. **Constants**: PVP_{kb} is the selling price of product b in market k, CP_b is the unit production cost of product b, Inv_b is the inventory of product b, T_b is the average inventory time of product b, I is the defined interest rate, CT_b is the cost of transport of product b, D_{kb} is the demand in the market k of the product b.

Subject to the following **constraints**. Constraint (10) ensures that the products distributed to the markets should be higher than the demand. Constraint (11) guarantees that the production plus the inventory of the products meet at least the demand. The quantity of products produced, distributed, and sold in a market is always positive and are guaranteed through Constraint (12).

$$\sum_{k,b} PD_{kb} \geq \sum_{k,b} D_{kb}; \quad \forall\, k, b \quad (10)$$

$$PR_b + Inv_b \geq PD_{kb}; \quad \forall\, k, b \quad (11)$$

$$PR_b,\ \ PD_{kb},\ PV_{kb}, \geq 0; \quad \forall\, k, b \quad (12)$$

2.3 Case Study III

The last case study corresponds to the minimization of scrap produced by the cut of raw material. **Scrap:** Minimize scrap generated by plants (13).

$$f4 = \min\{cx\ : x \in \mathrm{Z}_{+}^{\eta}\} \quad (13)$$

where: c is the cut pattern, x is the cut frequency, i is the raw material, j is the pattern. **Variables:** aij are the produced cuts, xj are the frequency cuts applied. **Constants:** ej is the inventory available, di is the demand for cut pieces.

Subject to the following constraints. Constraint (14) guarantees that the demanded quantity of cuts pieces is satisfied by the sum of the produced cuts multiplied by the frequency of the cuts. Constraint (15) verifies that the frequency cuts applied to each pattern do not exceed the available inventory. Constraint (16) satisfies that the quantity of cut produced and the frequency cuts implemented are always positive.

$$\sum_{j=1}^{\eta} a_{ij} x_j \geq d_i; \quad \forall\, i, \mathrm{j} \quad (14)$$

$$\sum_{j=1}^{\eta} x_j \leq e_j; \quad \forall \mathrm{j} \quad (15)$$

$$a_{ij}, x_j, \geq 0; \quad \forall\, i, \mathrm{j} \quad (16)$$

3 Optimization Algorithms Utilized

The algorithms utilized were the following, Multi-objective Pareto archives evolutionary strategy (M-PAES), Multi-objective particle swarm optimization (MOPSO), Non-dominated Sort Genetic Algorithm 2 (NSGA-II) and NSGA-II K-means.

M-PAES is a memetic algorithm designed for multi-objective problem resolution. It uses the technique of local searching, specifically with the PAES (Pareto archives evolutionary strategy) procedure. The individuals used in each iteration combine and mute themselves to generate new individuals [11]. This algorithm takes time to be executed since iterations consists of many processes with several operations. MOPSO is a modification of the Particle Swarm Optimization (PSO) algorithm. It uses the Pareto frontier to determine the movement of the particles through the search space [12]. This change was made to compete with algorithms that work with multi-objective problems. NSGA-II is a genetic algorithm modified to work in multi-objective problems. It changes the sorting of individuals through a new method called non-dominated sorting [13]. uNSGA-II is a modification of NSGA-II that works with a few numbers of individuals. Therefore, the runtime is also reduced [14]. NSGA-II K-means is a modification of the NSGA-II algorithm. In this case, this algorithm tries to split the population into different clusters. Each cluster will pass for a NSGA-II process that takes less time for each group due to the number of members [15].

4 New Hybrid Algorithm Developed

4.1 Previous Hybrid Algorithm

[10] presented a hybrid algorithm which consisted of a combination of three algorithms: micro-algorithm (i.e., an optimization algorithm with a few number of individuals), NSGA-II, and MOPSO. NSGA-II was the basis algorithm. The parents' selection was made with MOPSO procedures. After each iteration, the population was reduced multiplying by a factor to decrease runtime and computer memory. As genetic algorithms, the crossover and mutation are based on probability, which relies on a randomly generated number. Mutation usually takes a random value for the entire space of the problem, but in this algorithm, a sub-space was created to find solutions faster. The previous hybrid algorithm was compared with the other two algorithms in only one case study. The case study was the minimization of costs of a SC and the maximization of the service level offered to clients. The algorithm was compared with NSGA-II and MOPSO algorithms. The results obtained showed that the hybrid algorithm was faster in runtime in comparison to the other algorithms [10].

4.2 Creation of Sub-spaces

The creation of sub-spaces was developed to improve the algorithm mutation. This consists of reducing the search space of individuals in each iteration. At the start of the problem, this sub-space would be like the same space, which is a width space. However, in the next generations, this should be reduced to have fewer options knowing that there are the best individuals. This reduction relies on a variable *xz*, which will be cut in some iterations. This is something similar to PSO, where the particle is limited to the max speed.

4.3 New Hybrid Algorithm for Supply Chain Optimization

This algorithm for SC optimization, which was named NHA-SCO, has substantially changed since many features have been replaced, and others eliminated. In the selection operation, the MOPSO process is not always perfect because on some occasions could not converge. Therefore, it was necessary to change this selection method to NSGA-II non-dominated sorting. In each generation, the crossover operation is executed. The crossover and the mutation are similar to the previous algorithm. The mutations of the individuals have more diverse values from the new sub-space generated by *xz*. After generating the new child, the population is re-evaluated through NSGA-II non-dominated sorting. Before finishing the generation, it is necessary to calculate the generation mod of a number. This number depends on the max number of iterations. If the result is zero; then, *xz* reduces its value, and the population is increased. Micro-algorithm made a twist, instead of reducing the population, it has increased. This change was made due to, at the start of the process, individuals have more variable values. Thus, it would be easier to get parents in a short set of individuals. However, at a large number of iterations, the individuals will have similar values. For that reason, it is necessary to have more options to choose from. It is essential to know that the algorithm starts with a few individuals, e.g., 10 or 20. These changes produce that NHA-SCO takes more iterations to converge, but the speed has been reduced incredibly. Figure 1 presents the resulting flowchart of the new hybrid algorithm NHA-SCO.

5 Results

5.1 Case Study I

Many runs were made to find a mean of runtime and the number of iterations of each algorithm. Some algorithms converged to better values than others. These values are presented as the SC benefit and service level. In Table 1, the results of running the algorithms are presented. The results obtained were the runtime and the number of iterations required to converge; and values of the benefit and service level offered.

Table 1. Comparison of optimization algorithms for the Case Study I

Algorithm	M-PAES	MOPSO	Previous hybrid	NHA-SCO
Runtime	1020 s	51.32 s	24.79 s	8.08 s
Number of iterations	8141	2848	3110	3050
SC benefit ($)	4314,239	4263,268	4359,815	4360,030
Service level	99%	98.6%	99.9%	99.9%

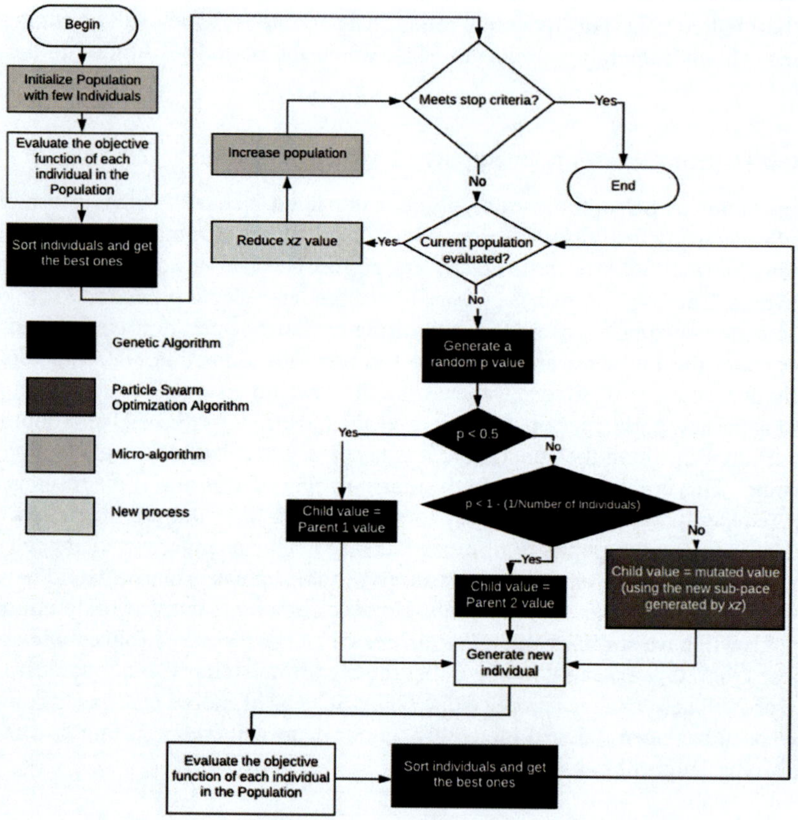

Fig. 1. Flowchart of the new hybrid algorithm NHA-SCO.

5.2 Case Study II

It is necessary to maximize the SC benefit. However, for this case, the SC is partially optimized. The running of the algorithm is for each product per period. There are 12 months; however, it is presented only the first two months of the year for illustration purposes. Table 2 and Table 3 show the expected benefit taken from [15], and the output values after running the four optimization algorithms.

Table 2. Results after running case study II with the optimization algorithms in January

Product	Expected value	uNSGA-II output value	NSGA-II K-means output value	Previous hybrid algorithm output value	NHA-SCO algorithm output value
PTC002B	1,258.25	1,084.21	1,258.25	1,258.25	1,258.25
PTC007B	1,027.91	801.91	1,027.91	1,027.91	1,027.91
PTC009D	156.73	148.65	148.65	156.73	156.73
PTC012B	801.87	676.73	801.02	801.87	801.87
PTC018B	775.21	533.49	775.21	775.21	757.46
PTC021B	593.47	509.77	593.47	593.47	593.47

Table 3. Results after running case study II with the optimization algorithms in February

Product	Expected value	uNSGA-II output value	NSGA-II K-means output value	Previous Hybrid algorithm output value	NHA-SCO algorithm output value
PTC001B	70.4	70.39	70.39	70.39	70.39
PTC002B	428.58	418.59	418.59	428.57	428.57
PTC007B	12.58	0	0	12.57	12.57
PTC008B	1,681.5	1,681.49	1,681.49	1,681.49	1,681.49
PTC009D	519.19	519.19	519.19	519.19	519.19
PTC010B	2.66	0	0	2.65	2.65
PTC012B	269.56	266.81	266.81	269.56	266.81
PTC018B	913.43	913.43	913.43	913.43	913.43
PTC021B	609.7	609.69	609.69	609.69	609.69

5.3 Case Study III

In this case, it is necessary to minimize the scrap generated by the plants after cutting the raw material. It would be running each product per month. As in Case Study II, Table 4, and Table 5 present the expected values and output values of each optimization algorithm.

Table 4. Results after running Case Study III with the optimization algorithms in January

Product	Expected value	uNSGA-II output value	NSGA-II K-means output value	Previous hybrid algorithm output value	NHA-SCO algorithm output value
PTC002B	1,258.25	1,084.21	1,258.25	1,258.25	1,258.25
PTC007B	1,027.91	801.91	1,027.91	1,027.91	1,027.91
PTC009D	156.73	148.65	148.65	156.73	156.73
PTC012B	801.87	676.73	801.02	801.87	801.87
PTC018B	775.21	533.49	775.21	775.21	757.46
PTC021B	593.47	509.77	593.47	593.47	593.47

Table 5. Results after running Case Study III with the optimization algorithms in February

Product	Expected value	uNSGA-II output value	NSGA-II K-means output value	Previous hybrid algorithm output value	NHA-SCO algorithm output value
PTC001B	70.4	70.39	70.39	70.39	70.39
PTC002B	428.58	418.59	418.59	428.58	428.58
PTC007B	12.58	0	0	12.577062	12.577062
PTC008B	1,681.5	1,681.5	1,681.5	1,681.5	1,681.5
PTC009D	519.19	519.19	519.19	519.19	519.19
PTC010B	2.66	0	0	2.655796	2.655796
PTC012B	269.56	266.81	266.81	269.56	266.81
PTC018B	913.43	913.43	913.43	913.43	913.43
PTC021B	609.7	609.69	609.69	609.69	609.69

6 Discussion

Several algorithms have been tested to compare the results. Each case study has its algorithms to execute. Findings represent important algorithms features, such as speed and convergence values, and the number of iterations that the algorithm takes to obtain the objective value. In the next subsections, the results obtained in each case study will be discussed.

6.1 Case Study I

Figure 2a, and b show the convergence of the benefit obtained and service level offered to clients, respectively. These figures present the next features:

- M-PAES takes more runtime, but it gets better solutions to the problem.
- MOPSO does not converge to good solutions, but it takes less time than M-PAES.
- Hybrid algorithms have faster runtime, but they use more iterations to converge than MOPSO.

NHA-SCO and the previous algorithm are better in convergence to the solution. However, the previous algorithm converges faster in the first iterations. Although this could seem a disadvantage, it is not, since the number of iterations is compensated with the speed of the algorithm. This is presented in Table 1, where the number of iterations to converge of the hybrid algorithms are similar, but the runtime of the new algorithm and the old are different. New hybrid algorithm runtime is 33,33% of the previous hybrid algorithm runtime. The number of iterations of both hybrid algorithms is more significant than MOPSO, but the service level convergence is better than MOPSO, as shown in Figure 2a, and b.

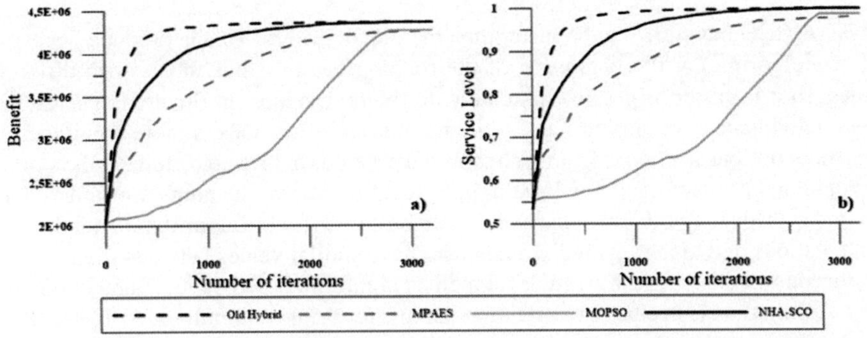

Fig. 2. **a)** Convergence of the benefit function. **b)** Convergence of the service level function.

6.2 Case Study II

Case Study II results, presented in Table 6, show that NHA-SCO takes more time in completing the running than the old algorithm. However, the new hybrid algorithm converges to the solution in fewer iterations than other algorithms. uNSGA-II takes the minimum runtime, but it has the worst number of iterations to converge. NSGA-II K-means takes more time but gets better results than uNSGA-II.

Table 6. Comparison of optimization algorithms for Case Study II.

	uNSGA-II	NSGA-II K-means	Previous hybrid	NHA-SCO
Runtime	0.33 s	59.54 s	2.31 s	7.32 s
Iterations	1013088	390410	148059	131695

6.3 Case Study III

In the last case study shown in Table 7, there is a change. NHA-SCO is faster than the previous algorithm; however, the number of iterations is the same as Case Study II. As the results presented during the last case study, uNSGA-II takes the minimum runtime and the worst number of iterations to converge. NSGA-II K-means takes more time but gets better results.

Table 7. Comparison of optimization algorithms for Case Study III.

	uNSGA-II	NSGA-II K-means	Previous hybrid	NHA-SCO
Runtime	0.66 s	104.4 s	51.74 s	18.7 s
Iterations	71180	40178	28018	26570

NHA-SCO has shown to be faster than the old algorithm. This is produced because the new algorithm at the beginning of the running has few individuals with different values, so it is easier to get new parents with the best values. In the next generations, when individuals have similar values, the number of individuals is more significant to have more options to choose from. Whereas the previous hybrid algorithm, at the start of the running, has many individuals; and, in the next iterations, the number of individuals is reduced. This is a problem because it would be harder for the algorithm to get the new best solutions due to the few individuals who have similar values between them.

Jamshidi, Fatemi, and Karimi developed an algorithm based on the Taguchi method [16]. This method is in charge of selecting parents for the next iterations. As NHA-SCO, the algorithm presented by the authors uses PSO methods for the selection. Therefore, the algorithm is as fast as the new hybrid algorithm. Although this last algorithm uses Micro-algorithms, thus, the population will decrease; therefore, the algorithm will be faster and will use less machine memory. Kuo and Han present three variations of algorithms [17]. These hybrid algorithms use the processes of the PSO and Genetic Algorithm. PSO is very slow in runtime; hence, the algorithms take more time to converge to the results. On the contrary, NHA-SCO uses the genetic process to increase speed to converge to the solution.

7 Conclusion

There are always failures in the systems; thus, there will always be space for improvements. A newly developed algorithm could still be improved. The new hybrid algorithm has found flaws in the theory of the previous algorithm. NHA-SCO has taken the old algorithm disadvantages to strengthen it and get better results and solutions to the problems related to SC. M-PAES algorithm has shown excellent results at the expense of having a high value of runtime. MOPSO has not converged to best solutions but optimizes faster than M-PAES. The proposed algorithm has demonstrated to be better in speed and convergence in comparison to MOPSO and M-PAES.

It is important to mention that, in this paper, three different case studies have been utilized to evaluate the algorithm. However, this not ensure that the algorithm would be better in other cases. The algorithm uses a parameter *xz* to change the size of *th* sub-space. Hence, this parameter must change according to the problem. In the future, this algorithm developed could be enhanced with new techniques of optimization.

Acknowledgments. This study is part of the research project "Modelo de optimizaci6n de costos en la Cadena de suministro en Empresas de ensamblaje," supported by the Research Department of the University of Cuenca (DIUC). The authors gratefully acknowledge the contributions and feedback provided by the IMAGINE Project team.

References

1. Yang, X.-S.: Computational Optimization, Methods and Algorithms (2011)
2. Xiong, F., Gong, P., Jin, P., Fan, J.F.: Supply chain scheduling optimization based on genetic particle swarm optimization algorithm. Clust. Comput. **22**(6), 14767–14775 (2018). https://doi.org/10.1007/s10586-018-2400-z
3. Lummus, R.R., Vokurka, R.J.: Defining supply chain management: a historical perspective and practical guidelines. Ind. Manag. Data Syst. **99**, 11–17 (1999). https://doi.org/10.1108/02635579910243851
4. Garcia, D.J., You, F.: Supply chain design and optimization: challenges and opportunities. Comput. Chem. Eng. **81**, 153–170 (2015). https://doi.org/10.1016/j.compchemeng.2015.03.015
5. Cevallos, C., Siguenza-Guzman, L., Peña, M.: A hybrid algorithm for supply chain optimization of assembly companies. In: 2019 IEEE Latin American Conference on Computational Intelligence (LA-CCI), Guayaquil, Ecuador, pp. 1–6 (2019)
6. Slawomit, K., Xin-She, Y.: Computational Optimization, Methods and Algorithms. Springer, Heidelberg (2011). https://doi.org/10.1007/978-3-642-20859-1
7. Blum, C., Roli, A.: Metaheuristics in combinatorial optimization: overview and conceptual comparison. ACM Comput. Surv. (CSUR). **35**, 268–308 (2003)
8. Gendreau, M., Potvin, J.-Y. (eds.): Handbook of Metaheuristics. ISORMS, vol. 272. Springer, Cham (2019). https://doi.org/10.1007/978-3-319-91086-4
9. Gonzalez, T.F.: Handbook of Approximation Algorithms and Metaheuristics: Contemporary and Emerging Applications. CRC Press Taylor & Francis Group, New York (2018)
10. Cevallos, C., Siguenza-Guzman, L., Peña, M.: A hybrid algorithm for supply chain optimization of assembly companies. Presented at the IEEE LA-CCI, Ecuador (2019)
11. Knowles, J., Corne, D.: M-PAES: a memetic algorithm for multiobjective optimization. In: Proceedings of the IEEE Conference on Evolutionary Computation, ICEC, 1 (2000). https://doi.org/10.1109/CEC.2000.870313
12. Coello Coello, C.A., Lechuga, M.S.: MOPSO: a proposal for multiple objective particle swarm optimization. In: Proceedings of the 2002 Congress on Evolutionary Computation, CEC 2002 (Cat. No. 02TH8600), vol. 2, pp. 1051–1056 (2002)
13. Deb, K., Pratap, A., Agarwal, S., Meyarivan, T.: A fast and elitist multiobjective genetic algorithm: NSGA-II. IEEE Trans. Evol. Comput. **6**, 182–197 (2002). https://doi.org/10.1109/4235.996017
14. Coello Coello Coello, C.A., Toscano Pulido, G.: A micro-genetic algorithm for multiobjective optimization. In: Zitzler, E., Thiele, L., Deb, K., Coello Coello, C.A., Corne, D. (eds.) EMO 2001. LNCS, vol. 1993, pp. 126–140. Springer, Heidelberg (2001). https://doi.org/10.1007/3-540-44719-9_9

15. Berrezueta, N.: Optimización de la cadena de suministro mediante uso de un algoritmo genético basado en clusterización (Working Paper). Presented at the (2020)
16. Jamshidi, R., Fatemi Ghomi, S.M.T., Karimi, B.: Multi-objective green supply chain optimization with a new hybrid memetic algorithm using the Taguchi method. Scientia Iranica **19**, 1876–1886 (2012). https://doi.org/10.1016/j.scient.2012.07.002
17. Kuo, R.J., Han, Y.S.: A hybrid of genetic algorithm and particle swarm optimization for solving bi-level linear programming problem - a case study on supply chain model. Appl. Math. Model. **35**, 3905–3917 (2011). https://doi.org/10.1016/j.apm.2011.02.008

Is Romania Ready for the Development of Smart Industry 4.0?

Gheorghe Gheorghe[1(✉)], Badea Sorin-Ionut[1], Iulian Ilie[1], and Despa Veronica[2]

[1] INCDMTM Bucharest, Pantelimon Road, 6-8, 2nd District, 021631 Bucharest, Romania
[2] Valahia Univerity of Targoviste, 130105 Targoviste, Romania

Abstract. The scientific work deals with Romania's level of readiness for the development of the Intelligent Industry (4.0), in relation to the national, European and international trends, with the identification of the strengths and weaknesses and the steps needed to be taken to reduce the gap compared to the developed countries. The scientific work will focus next on examples of contributions and elaborations of original concepts and constructions of intelligent systems, technological platforms and cobot networks that support the digital enterprise and the Intelligent Industry (4.0) in Romania.

Keywords: Smart Industry 4.0 · Process chain · Smart manufacturing · Mechatronic and cyber-mechatronic systems · Cobot technology platforms · Cobot networks

1 Introduction

1.1 International Context

At European level, a complex and ambitious program called Industry 4.0 has been elaborated, with the aim focused on a new industrial revolution understood more towards a much improved or modernized industry, by using and integrating cyberphysical and cyber-mixmechatronics systems in intelligent manufacturing.

The initiative of starting Industry 4.0, took Germany in 2011, through the so-called Industry 4.0 project, after which other countries in the EU and from around the world have created and adopted similar programs: Factory of the future (in France and Italy), Catapult (in the UK), respectively Smart Manufacturing in the USA, Made in China - 2005 (in China) or Innovation 2025 (in Japan), respectively the principles of the German program Industry 4.0, a program fully adopted by the EU, where integrated industrial products can interact with the production of equipment by transferring the information corresponding to the different stages of processors and development of an intelligent manufacturing environment having the capacity to communicate and make optimal decisions in a self-contained way.

The major economic and political challenge of allowing all industrial sectors to take advantage of digital innovation in products, manufacturing processes and business models.

J. Machado et al. (Eds.): icieng 2021, LNME, pp. 293–304, 2022.
https://doi.org/10.1007/978-3-030-78170-5_26

The European Industry 4.0 program has become of global importance, with the initiative of the World Economic Forum in Davos, to organize a debate on this topic, at the level of 2016.

Industry 4.0 is characterized by the automation, digitization and cybernetization of all components of the intelligent production processes and it is based, mainly on the completely new approach to the design and development of cyber-physical systems. Several works have been developed, in this context, dealing physical part and control part of cyber-physical systems and respective integration in the context of Industry 4.0 [1–7].

In order to reach this very ambitious goal, Europe has set itself to invest over 1.3 billion euros in the next 15 years.

The major political and economic challenge is to make all industrial sectors take full advantage of digital innovation in products, processes and business models.

Then, is Romania ready to align itself with the European countries that have already started national/European programs to meet this challenge and not to remain an outsider in this competition?

For this, one of the directions of strategic development of Romania must be oriented to Industry 4.0.

Knowing that in this period, the intelligent product options are chosen by the customer, purchased from the manufacturer and then realized with advanced processing systems such as: rapid prototyping technologies, cloud manufacturing, augmented reality, stochastic simulation, Internet of Things - IoT, Internet of people - IoP, data security, data processing, parts manufacturing in the cloud, processing by adding material, Big Data, autonomous robots, process simulation, vertical and horizontal systems integration. The intelligent technical and technological evolution takes place from Embedded Systems to the IoT (Internet of Things) and has been achieved through the Networks of Embedded Systems and Cyber-Physical Systems, and Cyber-MixMecatronics Systems.

For a better understanding of the difference between the process chain structures in a classic manufacturing system and a 4.0 Industry specific manufacturing system, these were presented in Fig. 1: In the classical system, the production process takes place in a well-defined manufacturing flow, between independent work cells, as seen in Fig. 1(a); In the new concept Industry 4.0, there is a flow of both products and data, integrated (see Fig. 1(b)), with the following characteristics: an integrated communication throughout the entire working cycle (1); a high degree of automation (2), increasing the number of highly qualified persons for the monitoring and management of the intelligent manufacturing flow (2); a high degree of communication between machines (machine to machine-M2M), respectively between machine and human (machine to human-M2H) (3); optimization of the entire process chain by using artificial intelligence programs in each structure of the technological chain [8].

1.2 National Context (in Romania)

In Romania, the field of Intelligent Industry is promoted very intensively by the multinational companies located in Romania: Siemens, Bosch, FESTO, Vodafone, etc., by national R&D institutes: NIRD-Mechatronics and Measurement Techniques - Bucharest, NIRD - Agricultural Machines - Bucharest, NIRD - Advanced Research - Bucharest,

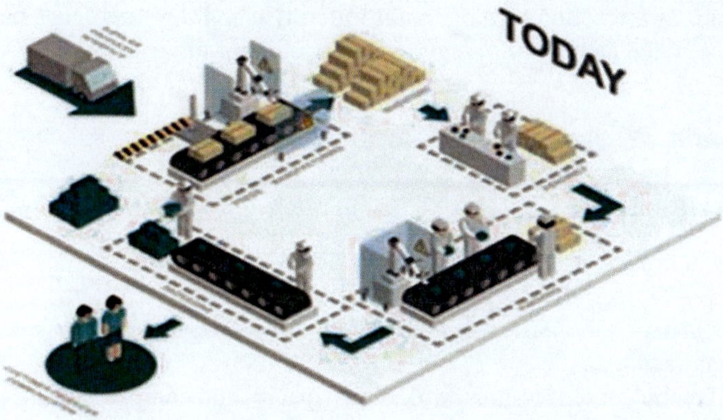

(a)

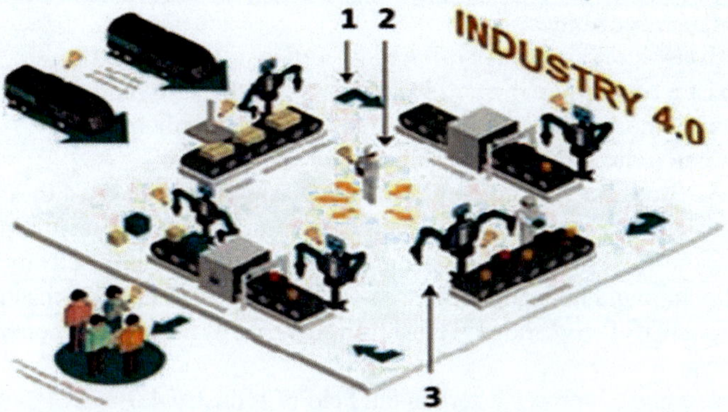

(b)

Fig. 1. The structure of process chains in a conventional manufacturing system (a), respectively one specific to Industry 4.0 (b)

NIRD - ICI - Bucharest, NIRD - IMT - Băneasa, etc., of technical universities: Politehnica University of Bucharest, Transilvania University of Brașov, Technical University of Cluj-Napoca, Politehnica University of Timișoara, Technical University of Iași, etc., of academies: Romanian Academy, Romanian Academy of Technical Sciences, etc. and more innovative SMEs from most industrial areas.

In Romania, the strategic discussions and directions regarding the Industry 4.0 Strategy were addressed, in the literature: Science and Technology, Financial Newspaper, Technique and Technology Review, etc. and in specialized conferences, in specialized magazines, technical and technological books and in sectoral project proposals in different ministries, as follows: International Mechatronics and Cyber-MixMechatronics Conference, International Journal of Mechatronics and Applied Mechanics, Mechatronics and Cyber-Mix Mechatronics 4.0 and Mechatronics and Cyber-Mix Mechatronics

Engineering for the Construction of the Digital Enterprise and the Intelligent Industry (4.0) and respectively the National Strategy for the Intelligent Industry (4.0).

2 Strenghts and Weaknesses for Romania

2.1 Strenghts for Romania

In Romania today, there are strengths for Industry 4.0:

- the existence of national R&D institutes, with top European infrastructure, on intelligent specialized fields;
- the existence of technical universities, with high capacity and competence infrastructure;
- the existence of developing industries, such as the smart car industry;
- the existence of research groups, from national institutes of RDI, technical universities and branch academies;
- the existence of companies offering expertise in Industry 4.0;
- the existence of some industries-suppliers, for the auto industry;
- the existence of excellent collaboration with German industry, French industry, American industry, etc., the promoters of Industry 4.0;
- the existence of a very active Romanian diaspora in Industry 4.0;
- the existence of nuclei and research centers in specialized technologies for Industry 4.0, in the field of "additive manufacturing", in the field of advanced simulation of intelligent manufacturing processes - multiscale modeling, stochastic modeling, etc.;
- the existence of workshops in the field of Cyber-Physical and Cyber-Mix Mechatronics Systems;
- the existence of some masters in the field of Industry 4.0;
- the existence of high-performance centers in important fields for Industry 4.0: (computer science, computers, robotics, automatic, mechatronics, cyber-mixing, etc.);
- the existence of a highly performing IT sector, with competences in the areas essential for Industry 4.0;
- the existence of a highly performing mechatronics and cyber-mechatronics sector, with competencies for Industry 4.0;
- the existence of a very efficient electronics and computers sector, with competencies for Industry 4.0;
- the existence of an artificial intelligence sector, data security, Big Data and communications networks, very performing, with competencies for Industry 4.0;
- the existence of a sector of intelligent systems, technological platforms, cobot platforms and networks of cobot systems and platforms, very performing, with competences for Industry 4.0;
- the existence of a highly qualified workforce, corresponding to Industry 4.0;
- etc.

2.2 Weaknesses for Romania

In Romania today, there are weaknesses for Industry 4.0:

- lack of a strategic program of the Romanian governments in the field of Industry 4.0;
- the lack of a Strategy for Industry 4.0 and for its integration in the Development and Outlook Strategies of Romania;
- lack of efficient and competent representation in international forums, where the Industry Agenda 4.0 is discussed;
- the lack of participation, with a few exceptions, of the Romanian researchers in the international conferences in which the problems of Industry 4.0 are debated;
- the lack of visibility of the Romanian researchers in the field of advanced processing technologies, with a few exceptions, in the relevant literature relevant to the international scientific community for Industry 4.0;
- lack of an organizational culture in Industry 4.0;
- lack of financial resources, at national level, for Industry 4.0;
- lack of financial resources, of Romanian companies in the Industry 4.0 field;
- lack of interest of Romanian banks in financing Industry 4.0;
- lack of specialists in organizing processes and productions related to Industry 4.0;
- lack of qualification of the labor force, related to Industry 4.0;
- lack of interdisciplinary and multidisciplinary specializations (computers - information - sensory - actuator - advanced technologies - intelligent materials - organization of intelligent production) specific to Industry 4.0;
- the poor quality of students, masters and even doctoral and post-doctoral students, in areas essential to Industry 4.0;
- etc.

By analyzing and inter correlating Romania's strengths and weaknesses, a first answer to the question from the title of the article can be given:

Romania is not yet ready to face the challenges of the fourth industrial revolution (Industry 4.0)!!

What can be done for Romania, to reduce the gap with developed countries in the EU and in the World, regarding Industry 4.0?

3 What Can Be Done for Romania, to Reduce the Gap with Developed Countries in the EU and in the World, Regarding Industry 4.0?

3.1 Romania's Program Plan to Reduce the Gap Regarding Industry 4.0

The Plan - Program of Romania, regarding Industry 4.0, will include:

1. Defining the "Industry 4.0 Agenda" of Romania, in the short, medium and long term;

2. Establishment of a "Register of Industry 4.0 Experts", with specialists - experts from Universities, INCDs, institutes of the Romanian Academy and the Academies of Branch, Industry, IT Companies, etc.
3. Elaboration of the Industry Strategy 4.0;
4. The integration of the Industry 4.0 Strategy and the Industry 4.0 Agenda in the Romanian Development Strategies for a long term;
5. Integration of the Industry 4.0 Strategy and the Industry 4.0 Agenda in the National Program for Research - Development and Innovation 2015–2020, PN III;
6. Accentuated promotion of the Industry 4.0 concept in the academic, research and business environment in Romania;
7. Introduction in the degree, master, doctoral and postdoctoral degree programs of some inter and multidisciplinary courses addressing Industry 4.0, to different university specializations (technology of machine building, machine tools and intelligent production systems, industrial engineering, mechatronics, robotics, instrumentation and data acquisition, communications networks and software, computers, information technology, automobiles, etc.);
8. Development of existing and promotional initiatives, through debates, round tables, workshops, workshops, etc. by the Romanian Academy, the Romanian Academy of Technical Sciences, INCDs by domains, the Romanian Chamber of Commerce and Industry, associations professionals and employers (APROMECA, ARIES, AGIR, ACAROM, MECHATREC, etc.) and others to promote the Industry 4.0 Concept;
9. The motion, on the part of the Romanian Government, to involve the Romanian companies, in the Industry Agenda and Strategy 4.0;
10. Increasing the interest of the Romanian banks in financing certain initiatives of the enterprises involved in the Industry 4.0 program;
11. Development and collaboration with different people from the diaspora in Industry 4.0 specific programs;
12. Promoting and integrating competent persons in technological platforms and European agencies, having as an area of interest Industry 4.0;
13. Increasing the efficiency of the communication between the representatives of Romania in the European/international forums, where the issues in the field of Industry 4.0 and those of the national bodies (financing agencies, INCDs, Universities, Companies, Associations, Employers, etc.) are debated.

3.2 Romania's Warning

At this moment it can be concluded that, all the initiatives started in Romania, do not make it possible to register Romania on the direction of achieving Industry 4.0 or to integrate it into the Industry 4.0 Agenda.

This, together with the late start to the alignment of the European countries for Industry 4.0, will lead to removing Romania from the map of the European Agenda for Industry 4.0, with many long-term negative repercussions.

That is why, for Romania, an active awareness and support campaign is needed, until it is not too late, for Romania to adapt to the European and global trends of digitalization of manufacturing in the digital enterprise and the intelligent industry (4.0).

4 Examples of Contributions and Developments of Some Original Concepts and Constructions of Smart Systems, Technological Platforms and Cobot Networks that Support the Digital Enterprise and the Smart Industry (4.0) in Romania

4.1 General Data

Industry 4.0 consists of the application of digital tools for profound transformation of business and operational models along the value chain. This digitization already allows companies in many sectors [9] to bring major and lasting improvements to the efficiency and flexibility of their operations [10].

The implementation of Industry 4.0, conditions for the success of its opportunities, not to make major mistakes, such as:

- lack of strategic direction;
- incorrect technological evaluation;
- lack of forecast and impact measurement;
- failure to comply with short-term results;
- failure to comply with the plan for operating and implementing the solutions.

For the success of this transformation in a dynamic and efficient way and to overcome the barriers that appear in the implementation of Industry 4.0, it is necessary to resort to a differentiated and original methodology, as well as to extensive consulting capabilities (strategy, business and technology), and to technological assets and products which allow the implementation of end-to-end digital solutions tailored to each client.

Industry 4.0 consists of the application of digital tools in order to initiate a profound transformation of operations and business models: in product design and design (agile prototyping); in manufacturing (increasing efficiency and flexibility); in the supply chain (improving accuracy and efficiency, from end to end) and in marketing (digitization of the point of sale and adapted multi-channel distribution).

4.2 Examples of Original Contributions and Development of Intelligent Systems Concepts and Constructions Supporting the Digital Enterprise and the Intelligent Industry in Romania

Intelligent System

Application in the industrial field of the electronic and electrotechnical industry (pick and place assembly and inspection platform), (Fig. 2).

Most pick and place applications can run autonomously using Universal Robots and Kuka collaborative robots and essentially contribute to increased process productivity and application flexibility. It takes superhuman abilities to repeat the same movements for many hours with the same precision. Therefore, the repeatability of ±0.1 mm of the bumpers is perfect for the automation of precision pick and place applications.

Due to their small size and constructive shape, collaborative robots can be easily integrated into small workspaces.

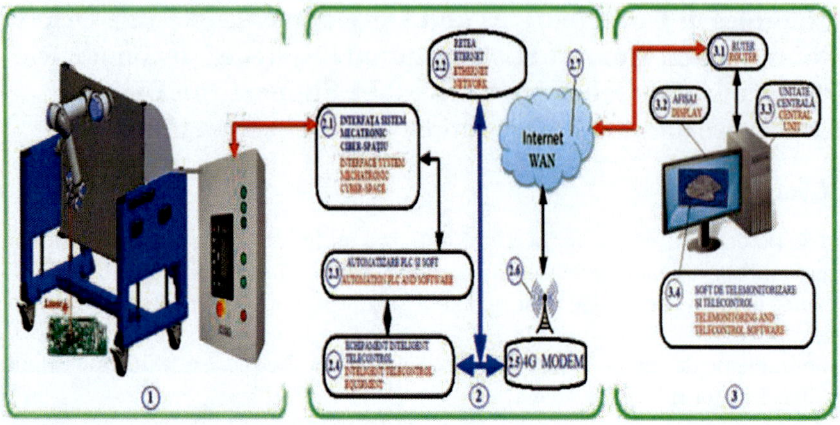

Fig. 2. Application in the industrial field of the electronic and electrotechnical industry (pick and place assembly and inspection platform)

Due to the easy programming and the short start-up time of Universal Robots, these are ideal for small volumes of production, they can be easily reprogrammed and relocated for different handling applications.

Adapting the collaborative robot to other handling applications is quick and easy, allowing you to automate almost any manual process, including for small series production.

Application in the industrial field of machine tools (Fig. 3).

Automating the loading and unloading of machine tools with collaborative robots.

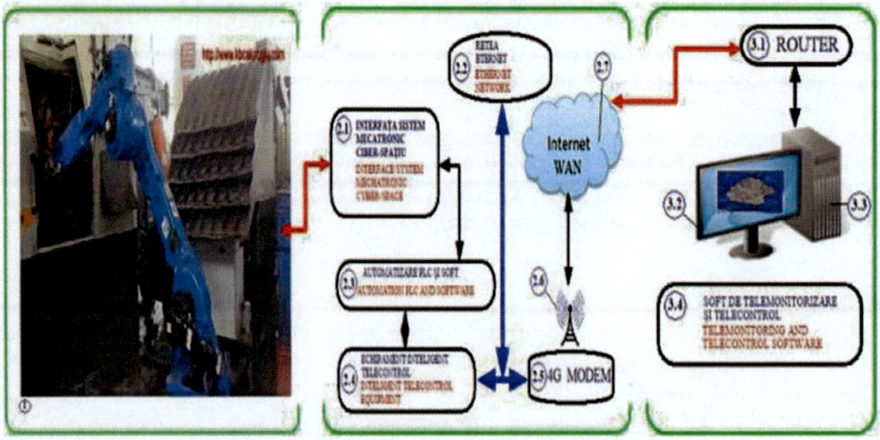

Fig. 3. Application in the industrial field of machine tools

It improves the operating speed and at the same time reduces the risk associated with accidents due to work near machine tools. If the collaborative robot comes into contact with the human operator, it stops the movement so that the operator is not injured. In addition, Universal Robots can be programmed to work at low speeds in certain workspaces where it can intersect with the human operator, or, using a laser scanner, the robot automatically reduces the speed when the operator is near the work area and resume the cycle at full speed as the operator exits the scanned area.

Universal Robots collaborative robots can be used in most CNC applications. Due to the high degree of flexibility, robots can be quickly adapted to other machine tools and equipped with other peripherals;

Developing applications for machine tool servicing (CNC) is done quickly and easily, allowing the automation of most repetitive and sometimes dangerous manual tasks for the human operator;

Transformation of the labor force by releasing people of repetitive or non-ergonomic work and reintegration into other processes that are less dangerous or that add value to the production;

The operating costs are reduced by eliminating the traditional costs with the necessary security systems when integrating industrial robots. The average time to recover the investment is 195 days;

Application in the field of auto - body/engine assembly line (Fig. 4 and Fig. 5).

Collaborative robots can reduce assembly time, increase productivity by increasing speed and at the same time improve production quality on auto assembly lines.

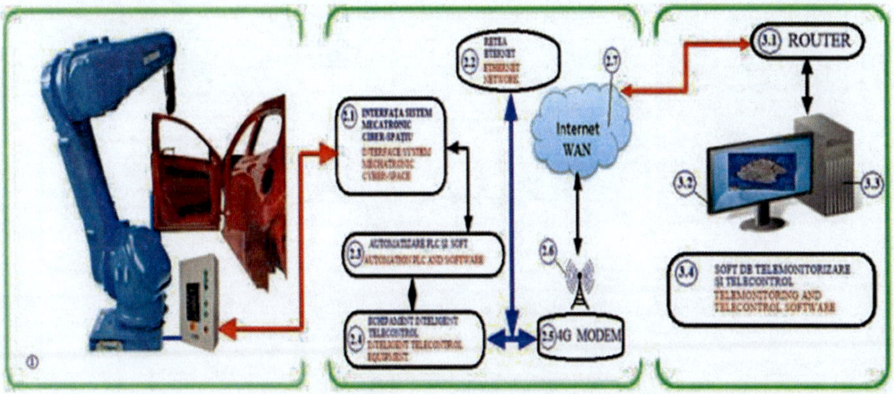

Fig. 4. Application in the field of auto - body

Increases the quality of the products while reducing the risk of injury associated with operating near heavy machinery. With proper equipment, with collaborative robots can be assembled parts of plastic, metal, wood or other materials. If during the assembly process when the human works in collaboration with the robot, physical contact occurs, the latter automatically stops so as not to injure the human operator.

Collaborative robots [11] can be used in most assembly processes, offering repeatability and speed to the human operator (Fig. 6); Collaborative robots can be quickly equipped with certified accessories so that adaptation to other assembly processes is quick and efficient;

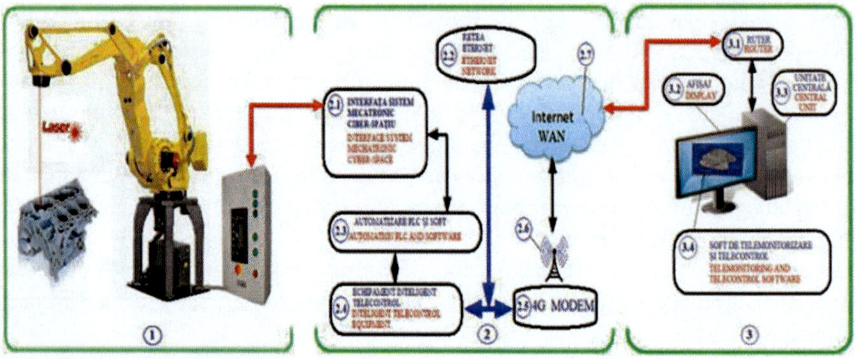

Fig. 5. Application in the field of auto - body/engine assembly line

Production capacities are expanded by increasing the quality, consistency and speed of production;

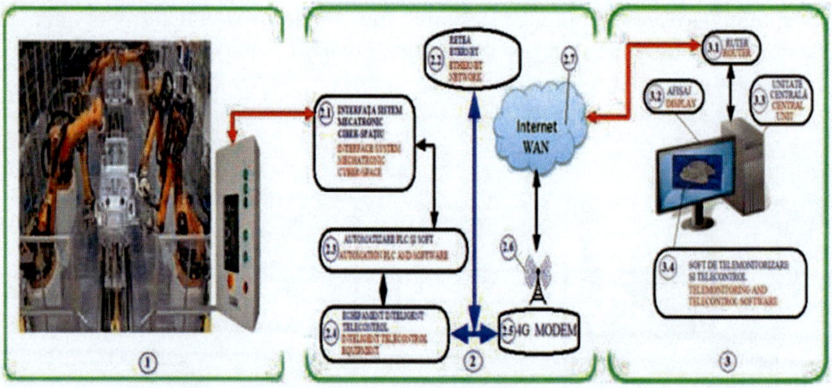

Fig. 6. Application in the field of collaborative robots

5 Conclusions

At the present moment, Romania is not ready for the development of the Intelligent Industry (4.0) as it has not started at the national level the initiation procedures of Industry 4.0, not to mention the development of Industry 4.0.

In an initial thought, the basic organizations responsible for promoting Industry 4.0, had to propose and elaborate the phases necessary to achieve Industry 4.0 assimilated at national level with the responsibility of involving the Government of Romania.

In this regard, various organizations in Romania, have elaborated some phases to start the initiation of Industry 4.0, without the centralized approval of the Government of Romania.

Thus, the strategy for industry 4.0, action plans to implement what was proposed and proposals to different ministries to start their implementation, either as sectoral projects or as projects of some organizations entitled to this new field, were elaborated on different levels the Intelligent Industry.

Some organizations, either national research development institutes, universities, academia and other organizations, have proposed to carry out, initial stages of creation and development of the Intelligent Industry, through research projects in Romania and the European research programs, dealing with new topics for this domain.

Depending on the specialized smart fields, carried out by the specialized organizations in Romania, they have gone on to develop intelligent products, technologies and services, which are part of what is meant by Industry 4.0 and the Digital Enterprise.

Moreover, the first steps in approaching Industry 4.0 are included in the scientific paper.

References

1. Rancalanza, E., Borg, J., Constantinescu, C.: A knowledge-based tool for designing cyber physical production systems. Comput. Ind. **84**, 39–58 (2017). https://doi.org/10.1016/j.compind.2016.08.001
2. Canadas, N., Machado, J., Soares, F., Barros, C., Varela, L.: Simulation of cyber physical systems behaviour using timed plant models. Mechatronics **54**, 175–185 (2018). https://doi.org/10.1016/j.mechatronics.2017.10.009
3. Gunes, V., Peter, S., Givargis, T., Vahid, F.: A survey on concepts, applications, and challenges in cyber-physical systems. KSII Trans. Internet Inf. Syst. **8**(12), 4242–4268 (2014). https://doi.org/10.3837/tiis.2014.12.001
4. Martins, L., Varela, M.L.R., Fernandes, N.O., Carmo-Silva, S., Machado, J.: Literature review on autonomous production control methods. Enterp. Inf. Syst. **14**(8), 1219–1231 (2020). https://doi.org/10.1080/17517575.2020.1731611
5. Kunz, G., Machado, J., Perondi, E., Vyatkin, V.: A formal methodology for accomplishing IEC 61850 real-time communication requirements. IEEE Trans. Ind. Electron. **64**(8), 6582–6590 (2017). https://doi.org/10.1109/TIE.2017.2682042. Article no. 7878522
6. Silva, M., Pereira, F., Soares, F., Leão, C.P., Machado, J., Carvalho, V.: An overview of industrial communication networks. In: Flores, P., Viadero, F. (eds.) New Trends in Mechanism and Machine Science. MMS, vol. 24, pp. 933–940. Springer, Cham (2015). https://doi.org/10.1007/978-3-319-09411-3_97
7. Kunz, G., Perondi, E., MacHado, J.: Modeling and simulating the controller behavior of an Automated People Mover using IEC 61850 communication requirements. In: IEEE International Conference on Industrial Informatics (INDIN), pp. 603–608 (2011). Article no. 6034947. https://doi.org/10.1109/INDIN.2011.6034947
8. Gheorghe, G.: Concepts and mechatronics and cyber-mixmechatronics constructions, integrated in COBOT type technology platform for intelligent Industry (4.0). In: Gheorghe, G.I. (ed.) Proceedings of the International Conference of Mechatronics and Cyber-MixMechatronics – 2019. LNNS, vol. 85, pp. 281–300. Springer, Cham (2020). https://doi.org/10.1007/978-3-030-26991-3_26
9. Costa, D., et al.: Performance evaluation of different mechanisms of production activity control in the context of Industry 4.0. In: Gheorghe, G.I. (ed.) ICOMECYME 2019. LNNS, vol. 85, pp. 82–103. Springer, Cham (2020). https://doi.org/10.1007/978-3-030-26991-3_9

10. Gheorghe, G., Bajeanru, V., Ilie, I.: Ingineria Mecatronică şi Cyber-MixMecatronică pentru Construcţia Intreprinderii Digitale şi Industriei Inteligente (4.0). CEFIN Pubishing House, Bucharest (2019)
11. Gheorghe, G.: Challenges and Research in the Innovation of Digital Enterprise and smart Industry (4.0). In: Proceedings of 2019 International Conference on Hydraulics and Pneumatics – HERVEX, Baile Govora, Romania, 13–15 November 2019 (2019). ISSN: 1454-8003

Sustainable Supply Chain Research and Key Enabling Technologies: A Systematic Literature Review and Future Research Implications

Akshay Patidar[1](✉), Monica Sharma[1,2], Rajeev Agrawal[2], Kuldip Singh Sangwan[3], Anbesh Jamwal[2], and Manuela Gonçalves[4]

[1] Department of Management Studies, Malaviya National Institute of Technology, J.L.N. Marg, Jaipur 302017, Rajasthan, India

[2] Department of Mechanical Engineering, Malaviya National Institute of Technology, J.L.N. Marg, Jaipur 302017, Rajasthan, India

[3] Department of Mechanical Engineering, Birla Institute of Technology and Science Pilani, Vidya Vihar, Pilani 333031, Rajasthan, India

[4] DMAT - Department of Mathematics, CMAT – Center of Mathematics, University of Minho, 4800-058 Guimarães, Portugal

Abstract. The concept of sustainability is becoming a buzzword these days. Incorporation of sustainability into Supply Chain Management is reviewing great attention from the wide range of industries in recent years. It is also an important study to review the impact of enabling technologies like Blockchain technology, Internet of Things (IoT), Big data, AR and VR-Immersive technologies, Artificial Intelligence on the economic, social and environmental aspects of Supply Chain performance. The importance of this study has attracted many academic researchers to publish their research in journals and proceedings, where discussion is about research progression both quantitatively and qualitatively. Following the progression, this article focuses on the review of key enabling technologies and sustainable supply chain providing insight into the development. Besides providing the insights of the progression this study also focuses on application areas of technologies along with its acceptance. Study presents a Systematic Literature review for showing the insights of key enabling technologies and sustainable supply chain along with the toppers of the research in terms of top authors, top journals, top research areas and top countries conducting this research thereby concluding the current status of the technology applications in sustainable supply chain and elaborating future research implications.

Keywords: Sustainable supply chain management · Triple bottom line · Top performers · Key enabling technologies

1 Introduction

The major challenge faced by a business is about the integration of its intermediaries and that too efficiently and effectively [1]. To capture the synergy of inter-functional and inter-organizational integration and coordination and to make strategic decisions, the concept

J. Machado et al. (Eds.): icieng 2021, LNME, pp. 305–319, 2022.
https://doi.org/10.1007/978-3-030-78170-5_27

of Supply Chain Management evolved. The concept, having its roots in logistics started when Father of Scientific Management Fredrick Taylor gave principles of Scientific Management and focused his early research on improving manual loading processes [2]. Gradually with the advent of time and resources and after globalization, the concept of Supply Chain Management evolved and developed [3]. Mentzer et al. (2002, p. 18) have described SCM as, "the systemic, strategic coordination of the traditional business functions and the tactics across these business functions within a particular company and across businesses within the supply chain, for the purposes of improving the long-term performance of the individual companies and the supply chain as a whole" [4]. Excessive Human Interference has always been a bane to the environment and led the concern about sustainability. With the growing phases of Developments, the need for Sustainable Development aroused. According to Brundtland Report Sustainable development can be defined as meeting the needs of the present generation without compromising the ability of future generations to meet their needs [5]. This led to an approach where the focus should be on environmental economic and social-oriented development. Some of the leading Researchers, Academicians also called it a Triple Bottom Line Approach (TBL) also [6]. Sustainability can be an end and means to achieve this end is sustainable Development. Sustainable development also emphasizes that the rate of consumption of resources should be equal to the rate of substitution of resources [7].

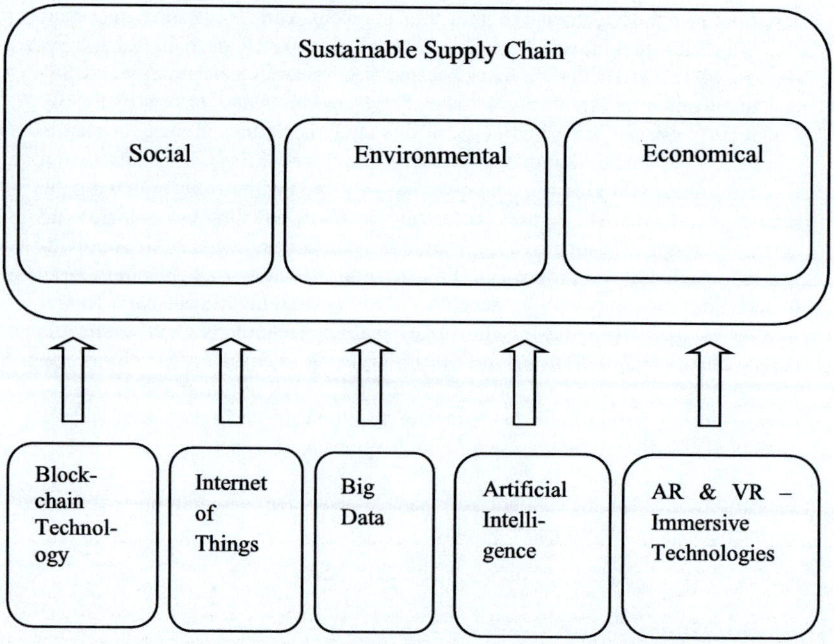

Fig. 1. Interaction between technologies and sustainable supply chain

Information technology is always considered as a tool to enhance efficiency and effectiveness. With a long-term focus use of technology also affects the optimized performance of the supply chain. The recent wave of technology advancements led dynamicity in them and led business areas to adapt them. One such area is supply chain management. To enhance the social, economic and environmental performance of Supply Chain technology played a vital role and the use of ICT helped businesses meet dynamicity in supply chains [8]. Figure 1 shows some of the technologies that have seen significant researches in reshaping business, society, and supply chain are Blockchain Technology, IoT-Internet of Things including cyber-physical systems, Big data including advanced analytics and distributed cloud, Augmented Reality and Virtual reality as immersive technologies for enhanced customer experience, Artificial Intelligence including Machine learning and deep learning, robotic process automation for increasing efficiency and efficacy in sustainable supply chain management.

2 Methodology

The paper focuses on structure Literature review commonly known as SLR. The study initiates with a collection of research articles related to the sustainable supply chain and the technologies enabling them. For solving the same purpose Web of Science Database was accessed using "((“sustain* Supply Chain”) AND (“IOT” OR “Internet of things” OR “BT” OR “Blockchain Technology” OR “Data Analytic*” OR “AR” OR “VR” OR “Augmented Reality” OR “Virtual Reality” OR “AI” OR “Artificial Intelligence” OR “ML” OR “Machine Learning” OR “Automation”))" as search string for basic search and in all fields. Using these strings total 43 research article was found and this database was further refined using exclusive criteria and inclusion criteria. Research articles from journals only were included. The time frame of all years was used for study purposes. Since all the articles found in the database were in the English language so no exclusion criteria for language were used. Analyzing this database concept of supply chain and sustainable supply was understood and key enabling technologies were identified.

The article is organized into 7 sections. Section 1 provides the introduction and builds the premises of the research problem. Section 2 elaborates on the methodology of review i.e. SLR which database is used and how and which research articles are used for analysis purposes. Further Sect. 3 enlightens the different supply chain which is in our focus of study and is more relevant to our research problem. Section 4 discusses the technologies which we are evaluating during our literature review. Five technologies are chosen based on the Gartner proposal of strategic technological trends for 2020 Sect. 5 focuses on the analysis and discussions of the articles selected for identifying the top articles, top countries, top areas, and top contributing authors in the research area followed by their citation analysis to get the insight of the research problem and to identify the further progression of the research area. Section 6 focuses on the findings and observations over the review done along with the future research implications. Finally, the study is concluded with the Sect. 7. The structured literature review is used to remove the biases in the study.

3 Sustainable Supply Chain

Sustainability is also gaining attention to supply chain management in both academic and industry practices. Companies across the globe are implanting sustainability initiatives in their supply chain because of the extensive pressure from the various stakeholders including government and customers [9]. Melnyk, Davis, Spekman, and Sandor (2010) states supply chain needs to be designed in such a way that it delivers sustainability as their outcomes [10].

Hence Sustainable supply chain can be defined as "Management of raw materials and services from suppliers to the manufacturer to customer and back with improvement of the social and environmental impacts explicitly considered". For achieving success creation of a sustainable supply chain is essential these days [11].

The supply chain sustainability concept revolves around socially responsible products and practices that not only preserve the planet's health and are good for people's dwelling, also good for building brand awareness along with minimizing environmental impact and improving profitability [12]. Emphasis needs to be given on sustainable changes in the supply chain that are long term in nature. (Goodland and Daly 1996) [13]. The idea of a sustainable supply chain covers supplier selection, procurements, product designing, manufacturing, distribution, and disposal [14]. By managing and seeking to improve environmental, economic and social performance, organizations act in their stakeholders and societal interest at large [11]. L'Oreal's vision statement says L'Oreal has Chosen to integrate the principles of Sustainable Development into its business model to build sustainable growth both responsible and united [15]. Hence, the true sustainability occurs at the intersection of all three dimensions—environmental, social and economic, and also involves those activities that companies comprehensively associated with environmental, social and economic goals in developing long-term strategic objectives [11]. To discuss this insight some of the key supply chains where ICT implications are very wide are identified and selected.

3.1 Classification of Supply Chain

Supply Chain exists in both Manufacturing and Service organizations and the complexity of these supply chains varies from product to product and service to service. Table 1 shows the classification of the supply chain and some of the examples of supply chains both in manufacturing as well as service sectors. Product variation, volume, and product robustness are some of the factors deciding the complexity of the supply chain [16].

Number of researches elucidate that visibility of the material which is being transported is of utmost importance to all the intermediaries involved in this supply chain [17]. Advancements in technologies led to identifying the practices that directly or indirectly affect the environmental, social and economic performance of the supply chain [8].

Table 1. Supply chain classified

Manufacturing		Service	
Food SC	Food Supply Chain is the complex and interconnected network of farmers, processing companies their distributors and then retails to consumers facilitating inputs from farmers and retailing output to consumers	Logistics SC	Logistics refers to the movement of goods from one location to another that starts when the supplier tries to move raw material and includes all kinds of goods movement until the customer gets the finished product. Entre flow can be called a Logistics supply chain
Pharmaceutical SC	Pharmaceutical SC is an amalgamation of key business processes to generate value to supply prescribed drugs, over the counter and generic medicines to the consumer	Healthcare SC	The hospital supply chain is a series of processes and integration of various departments to provide the best healthcare services to the patients right from the procurement of equipment to hiring skillful professionals

4 Technologies in SSCM

Technological advancements are leading most of the supply chain to adopt the latest technologies that help to increase the efficiency and effectiveness of the entire supply chain. Gartner [18], a leading research and advisory company publishes the latest technologies and their state of acceptance which they call it a plateau of productivity and many researchers clearly explicate the era of Industry 4.0. In the era of Industry 4.0 supply chain is also observing a paradigm shift towards industry 4.0. Industry 4.0 is all about transparency traceability, edge, and fog computing, better consumer experiences through augmented and virtual reality, cyber-physical systems or IoT devices, automation and artificial intelligence including machine learning and deep learning to meet the customer expectations [19].

4.1 Blockchain Technology

A blockchain is simply a time-stamped series of an immutable record of data that is maintained by the nodes of the network and not by a centralized authority [20]. It is a Distributed ledger technology whereby a transaction between any nodes of the network or new node entering the network is stored in the distributed ledger with the consensus of all the nodes [20].

The main benefits of using Blockchain Technology is the creation of transparency, traceability, and trust in the network [21]. Blockchain is said to be a disruptive technology as it has the capability of transforming the entire scenario [21]. The first use case of this technology is the Bitcoin by Satoshi Nakamoto in 2008 [22]. Blockchain technology has wide applications; one of such application is Supply Chain. Blockchain can provide information to any level of supply chain intact, based on this information flow efficiency of product flow can also be increased [23]. Using smart contracts, financial transactions in a supply chain can also be made efficient thereby making the entire supply chain efficient [23]. With its distinguished feature and commendable benefits Blockchain technology can help improving social performance, economic performance and environmental performance thereby enhancing the sustainability of the supply chain [23].

4.2 Internet of Things (IoT)

Internet of Things as its name implies, it is simply things interconnected over the internet and these things are devices or sensors interconnected. These devices have a basic function of monitoring, collecting and reporting or exchanging the data [24]. The measures that these devices cater are location, temperature, movement, handling, and other relevant environmental factors. In the area of the supply chain, these devices can be used to track and authenticate products and shipments at any time using RFID chips, smart devices, and mobile sensors [24].

The major benefits of using IoT in the supply chain can be it provides assurance related to goods being received, lost orders, inventory visibility, order tracking, and monitoring warehousing and storage conditions [25]. Also, they have some major acceptance challenges like the installation of these devices needs expertise and they are entirely dependent on the strength of internet connectivity for their communication purposes.

Using these devices can commensurately contribute towards the sustainability aspect of the supply chain by contributing environmental, social and economic performance enhancement.

4.3 Big Data

Data is the essence of Industry 4.0 Everyone in this era is talking about data. Data is used for strategy making, decision making, and many more decisions. Data is converted into big data in industry 4.0. Big data is simply a large set of structured and unstructured data leading to meaningful information by means of computing techniques. It's not the volume of data that matters, how fruitfully it is extracted and its insights are analyzed matters for effective decision making.

The integration of this big data with a number of computation techniques or high-powered analytics leads to better decision making [26]. It helps in determining the root cause of failures problems and defects. It also helps in determining consumer's buying habits, calculating risk portfolios.

Big data includes advanced analytics for predictive purposes and it also including computing at cloud whereby the user uses various algorithms to convert data in meaningful information at the cloud level itself. But in the era of Industry 4.0, data storage and computing are being decentralized. This decentralization of data led to decentralized computing techniques i.e. Edge computing and fog computing. Edge computing occurs at the cyber-physical system level or at IoT devices level and fog computing is at the application level to increase the efficiency of the computations.

Big data in the supply chain helps in supply chain planning for demand forecasting and evaluating supplier's performance to improve sourcing, increases delivery experience in terms of accuracy and speed and also helps in back ordering by reducing return costs and increasing the visibility.

Big data along with cloud computing can contribute valuably. For example, IoT devices along with AI can bring down asset downtime, weather data can help in optimizing product delivery, and social media data can reveal customer insights and their purchase habits [24].

4.4 AR and VR-Immersive Technologies

Immersive technologies are being considered as transforming technologies in the supply chain area [27]. Augmented reality and Virtual reality are the most preferred technologies in creating an augmented supply chain. Smart glasses, simulators or shop floor displays are some examples that increase the efficiency of supply chain partners.

Augmented reality is just an interactive real-world experience with the objects that reside in the real world. A worker working in a warehouse can use this kind of interaction by using smart glasses which have a bar code scanner built in it and can increase his efficiency of work whereas Virtual reality is a simulated experience with the real world. Augmented reality improves the order picking process as goods can be traced where they are placed and can be picked up quickly which belongs to the cart. It also helps in reducing job training in the case of seasonal workers.

Virtual reality can be used to simulate what-if scenarios in the supply chain. It helps in making delivery more efficient also it can help in superimposing more information for drivers. Packages can be added with scan-able images to provide information such as content and handling instructions.

Companies like IKEA and Amazon are trying to integrate these technologies into their supply chain for better customer experiences. Imagine if one person wants to buy furniture or any home décor product, they can use the camera of their smart phone and can use this technology to ensure where the product can be placed and how it looks like easing their decision of buying and bridging the gap of lack of personal touch in e-commerce.

4.5 Artificial Intelligence

Artificial Intelligence is a technology that is a simulation of human intelligence with the machines having traits similar to human beings lie learning and problem-solving. It enhances customer experience improved facility and production planning, Scheduling, etc.

Artificial Intelligence-powered by Machine learning and deep learning is the algorithms that help in decision making and can help in inventory and demand forecasting and optimization of supply chain decisions [28]. Machine learning also contributes to warehouse management by simplifying overstocking and understocking. Supplier selection and sourcing are other key areas of the supply chain where machine learning is contributing a lot with the help of intelligent algorithms by supplier assessments, audits, and their credit score thereby reducing costs and increasing efficiency of the supply chain. Demand supply mismatch is a big problem in the supply chain and that can be minimized using AI. AI-powered supply chain enhances customer experiences.

Major use cases of AI now a day are Chabot that uses machine learning and Natural language processing (NLP) to interpret the customer needs and wants [29]. AI-powered Chabot can be an effort to recreate humans.

Because of the predictive nature of AI and its contribution to enhancing and empowering the supply chain, a major focus of supply chain research is towards an AI-powered supply chain.

5 Analysis and Discussions

5.1 Journals and No of Articles

Figure 2 framed after analyzing a database of 43 research articles to identify journals for key technologies enabled sustainable supply chain and analysis clearly shows that Journal of Cleaner Production leads the list with 9 publications followed by Sustainability and Resource, Conservation and Recycling with 4 articles each which are approx. 44.4% of the leader. International Journal of Production Economics and International Journal of Production Research with 3 articles each holds the third position in the list followed by Industrial Management Data Systems, International Journal of Logistics of Management and Journal Manufacturing Technology Management having 2 articles each. The figure clearly shows rest of 14 research articles are published in 14 different journals implicating the infancy of the research.

5.2 Year and No of Articles

Figure 3 shows the distribution of the research status of our database by showing the number of publications in the year 2015 to 2020. From the beginning of the year, publications grew as 1, 8, 2, 12, 17, 3 number of research articles. 39.5% of total research was conducted in the year 2019 and the figure clearly shows the progression of the research. After getting insight research grew in the year 2016 but a sudden dip in the number of publications was seen in the year 2017. After that number of publications grew exponentially in the year 2018 and 2019. Since review is being conducted at the starting of the year 2020 so researcher got the lesser number of publications in the year 2020, but the trend explains that the number of publications will be higher at the end of the year 2020.

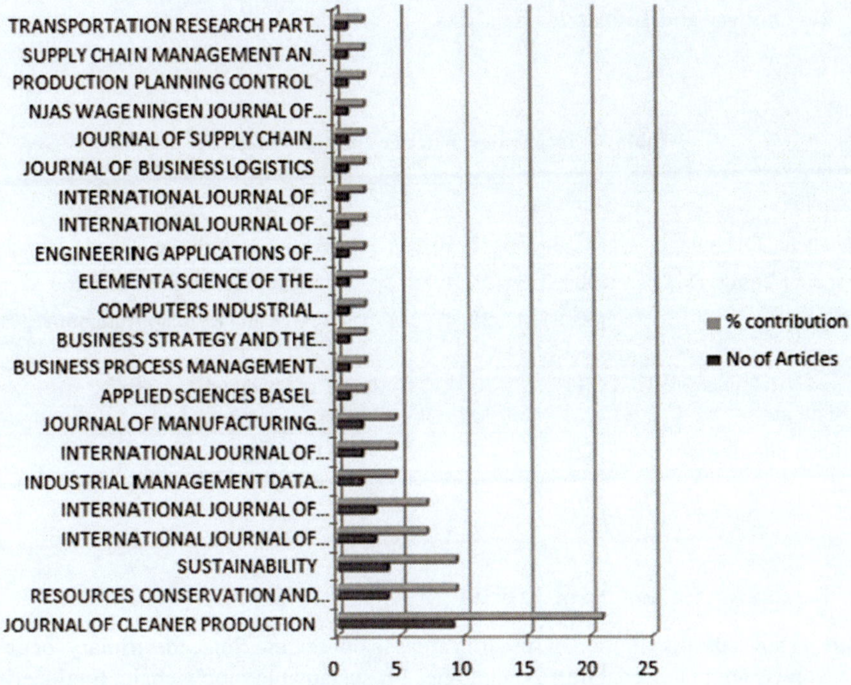

Fig. 2. Journals and their contribution

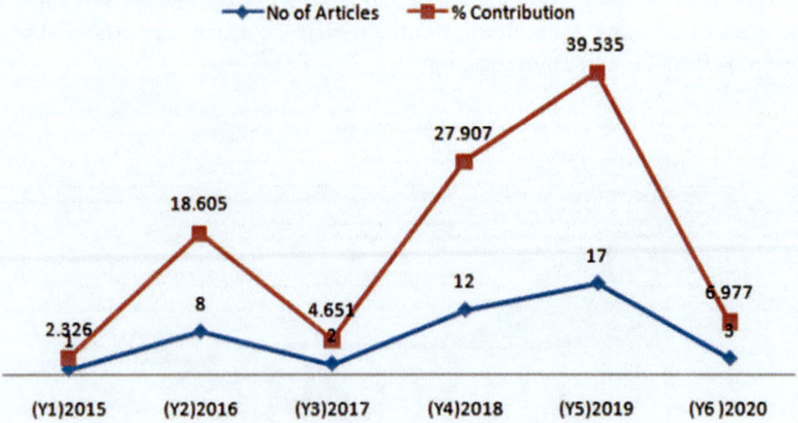

Fig. 3. Year-wise progression of publications

5.3 Technology and Industries Analysis

See Table 2.

Table 2. Technology, industry and study mapping

Technology	Authors	Industries
Blockchain Technology	[33, 35, 36, 43, 50, 51]	Textile, Refurbished Goods, Blood
Internet of Things (IoT)	[30, 33, 34, 38, 44, 51]	Healthcare, Automotive, Food
Big Data	[30, 31, 37, 38, 47, 51]	Textile and Mining, Automotive, Food
AR and VR-Immersive Technologies	[51]	Not Specified
Artificial Intelligence (AI)	[52]	Oil and Gas

5.4 Research Area and No of Articles

Figure 4 shows the distribution of research areas in our review. Since the primary focus of our review is about key enabling technologies for sustainable supply chain, Engineering leads the areas of research with 27 research articles and 62.8% researches. Environmental science is the second area since our review's secondary focus is on sustainability and is having 19 articles with 44.2% hold. These two main areas are followed by other areas like Business Economics, Agriculture, Computer Science, Chemistry, Material science, Meteorology, Physics, and Transportation.

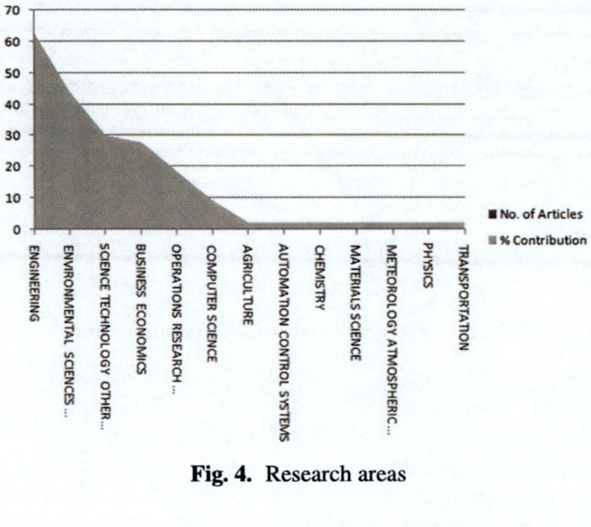

Fig. 4. Research areas

5.5 Countries and No of Articles

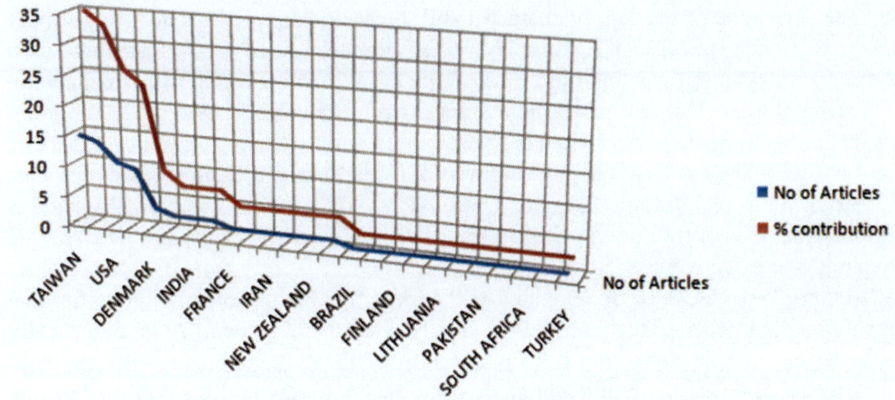

Fig. 5. Countries conducting research

Analyzing the database countries conducting research on sustainable supply chain were found and key enabling technologies as shown in Fig. 5 where Taiwan leads the list with 15 research articles followed by People R China having 14 and USA with 11 and England with 10 research articles respectively. These countries are followed by Denmark, Australia, India, Malaysia France, Germany and other countries with 4, 3, 3, 3, 2, 2 research articles respectively.

5.6 Citation Analysis

Table 3. Research articles and citations

Articles	[45]	[47]	[32]	[41]	[49]	[40]	[42]	[39]	[48]	[46]
Total citations	87	75	56	50	39	31	28	20	17	16
Average per year	17.4	12.5	11.2	10	7.8	7.75	9.33	4	8.5	5.33

Table 3 shows the top 10 studies of the database which are under review and their respective citations with the average citations per year. It is pretty evident that the topper of the list is an article related to the improvement of sustainable supply chain management using the DEMATEL approach with total citations of 87 and an average citation of 17.4 per year followed by various researches viz. Tseng et al. [47], Busse et al. [32], Lin et al. [41], Wu et al. [49], Lim et al. [40], Lin et al. [42], Kirchoff et al. [39], Wang et al. [48], Taleizadeh et al. [46] with 75, 56, 50, 39, 31, 28, 20, 17 and 16 citations respectively. This clearly shows the level of interest in the research area and people's contribution to the area.

6 Findings and Future Research Implications

Researchers have proposed many ways to increase the environmental, economic and social performance of the supply chain but still, there are several gaps affecting supply chain performance because of its complexity. This research clearly shows that these gaps are transparency in information flow, traceability of product movement and trust among the intermediaries. The research article shows that engineering is the most important research areas being discussed when the Sustainable Supply Chain comes into the frame of the study. Hence a review and current status of research considering new technologies like Blockchain Technology, IoT, Big Data, AR & VR, and Artificial Intelligence is proposed that can overcome the gaps in the sustainable supply chain by increasing its efficiency and effectiveness.

Further research can be done related to the identification of potential industries and the applicability of these technologies in supply chain management. Also, empirically research needs to be done about the integration of these technologies with the supply chain to evaluate the applicability of these technologies and their impact on the sustainable supply chain.

7 Conclusion

Supply Chain Management discipline aims at efficient and effective procurement, production, distribution of goods and enhancing coordination among intermediaries thereby emplaning on the effective flow of goods information and money. Frequent researches are being conducted to get the technological implications on sustainability of supply chain. With the up gradation of technology and its amalgamation with supply chain, sustainability aspects of supply chain should be enhanced in turn contributing towards society and better customer experiences.

References

1. Mair, J., Marti, I., Ventresca, M.J.: Building inclusive markets in rural Bangladesh: how intermediaries work institutional voids. Acad. Manag. J. **55**(4), 819–850 (2012)
2. Nilsson, F., Gammelgaard, B.: Moving beyond the systems approach in SCM and logistics research. Int. J. Phys. Distrib. Logist. Manag. **42**, 764–783 (2012)
3. Hobbs, J.E., Kerr, W.A., Smyth, S.J.: The perils of zero tolerance: technology management, supply chains and thwarted globalisation. Int. J. Technol. Glob. **7**(3), 203–216 (2014)
4. Mentzer, J.T., et al.: Defining supply chain management. J. Bus. Logist. **22**(2), 1–25 (2002)
5. Schubert, A., Láng, I.: The literature aftermath of the Brundtland report 'our common future'. A scientometric study based on citations in science and social science journals. Environ. Dev. Sustain. **7**(1), 1–8 (2005). https://doi.org/10.1007/s10668-003-0177-5
6. Slaper, T.F., Hall, T.J.: The triple bottom line: what is it and how does it work. Indiana Bus. Rev. **86**(1), 4–8 (2011)
7. Lorek, S., Spangenberg, J.H.: Sustainable consumption within a sustainable economy–beyond green growth and green economies. J. Clean. Prod. **63**, 33–44 (2014)
8. Melville, N.P.: Information systems innovation for environmental sustainability. MIS Q. **34**(1), 1–21 (2010)

9. Meixell, M.J., Luoma, P.: Stakeholder pressure in sustainable supply chain management: a systematic review. Int. J. Phys. Distrib. Logist. Manag. **45**(1/2), 69–89 (2015)
10. Melnyk, S.A., Davis, E.W., Spekman, R.E., Sandor, J.: Outcome-driven supply chains. MIT Sloan Manag. Rev. **51**(2), 33 (2010)
11. Putri, A.S., Hisjam, M.: Research development of sustainable supply chain management: a comparative study. In: AIP Conference Proceedings, vol. 2097, no. 1, p. 030102. AIP Publishing, April 2019
12. Kumar, S., Teichman, S., Timpernagel, T.: A green supply chain is a requirement for profitability. Int. J. Prod. Res. **50**(5), 1278–1296 (2012)
13. Goodland, R., Daly, H.: Environmental sustainability: universal and non-negotiable. Ecol. Appl. **6**(4), 1002–1017 (1996)
14. Hu, A.H., Hsu, C.W.: Critical factors for implementing green supply chain management practice: an empirical study of electrical and electronics industries in Taiwan. Manag. Res. Rev. **33**(6), 586–608 (2010)
15. Loreal. https://www.lorealusa.com/csr-commitments/sharing-beauty-with-all/our-vision. Accessed 10 Jan 2020
16. Yadav, S.R., Mishra, N., Kumar, V., Tiwari, M.K.: A framework for designing robust supply chains considering product development issues. Int. J. Prod. Res. **49**(20), 6065–6088 (2011)
17. Heaney, B.: Supply chain visibility. A critical strategy to optimize cost and service (2013)
18. Gartner. https://www.gartner.com/smarterwithgartner/5-trends-appear-on-the-gartner-hype-cycle-for-emerging-technologies-2019. Accessed 10 Jan 2020
19. Lasi, H., Fettke, P., Kemper, H.-G., Feld, T., Hoffmann, M.: Industry 4.0. Bus. Inf. Syst. Eng. **6**(4), 239–242 (2014). https://doi.org/10.1007/s12599-014-0334-4
20. Abeyratne, S.A., Monfared, R.P.: Blockchain ready manufacturing supply chain using distributed ledger (2016)
21. Niranjanamurthy, M., Nithya, B.N., Jagannatha, S.: Analysis of blockchain technology: pros, cons and SWOT. Cluster Comput. **22**(6), 14743–14757 (2018). https://doi.org/10.1007/s10586-018-2387-5
22. Nakamoto, S.: Bitcoin: a peer-to-peer electronic cash system. Manubot (2019)
23. Saberi, S., Kouhizadeh, M., Sarkis, J., Shen, L.: Blockchain technology and its relationships to sustainable supply chain management. Int. J. Prod. Res. **57**(7), 2117–2135 (2019)
24. Liu, T., Lu, D.: The application and development of IoT. In: 2012 International Symposium on Information Technologies in Medicine and Education, vol. 2, pp. 991–994. IEEE, August 2012
25. Ben-Daya, M., Hassini, E., Bahroun, Z.: Internet of Things and supply chain management: a literature review. Int. J. Prod. Res. **57**(15–16), 4719–4742 (2019)
26. Katal, A., Wazid, M., Goudar, R.H.: Big data: issues, challenges, tools and good practices. In: 2013 Sixth International Conference on Contemporary Computing (IC3), pp. 404–409. IEEE, August 2013
27. Reiners, T., Wood, L.C.: Immersive virtual environments to facilitate authentic education in Logistics and supply chain management. In: Learning Management Systems and Instructional Design: Best Practices in Online Education, pp. 323–343. IGI Global (2013)
28. Sarhani, M., El Afia, A.: Intelligent system based support vector regression for supply chain demand forecasting. In: 2014 Second World Conference on Complex Systems (WCCS), pp. 79–83. IEEE, November 2014
29. Raj, S.: Building Chatbots with Python: Using Natural Language Processing and Machine Learning. Apress, New York (2018)
30. Ardito, L., Petruzzelli, A.M., Panniello, U., Garavelli, A.C.: Towards Industry 4.0: mapping digital technologies for supply chain management-marketing integration. Bus. Process Manag. J. **25**(2), 323–346 (2019)

31. Bag, S., Wood, L.C., Xu, L., Dhamija, P., Kayikci, Y.: Big data analytics as an operational excellence approach to enhance sustainable supply chain performance. Resour. Conserv. Recycl. **153**, 104559 (2020)
32. Busse, C., Schleper, M.C., Niu, M., Wagner, S.M.: Supplier development for sustainability: contextual barriers in global supply chains. Int. J. Phys. Distrib. Logist. Manag. **46**(5), 442–468 (2016)
33. Cousins, P.D., Lawson, B., Petersen, K.J., Fugate, B.: Investigating green supply chain management practices and performance. Int. J. Oper. Prod. Manag. **39**(5), 767–786 (2019)
34. Daú, G., Scavarda, A., Scavarda, L.F., Portugal, V.J.T.: The healthcare sustainable supply chain 4.0: the circular economy transition conceptual framework with the corporate social responsibility mirror. Sustainability **11**(12), 3259 (2019)
35. Duan, Y., Aloysius, J.A.: Supply chain transparency and willingness-to-pay for refurbished products. Int. J. Logist. Manag. **30**(3), 797–820 (2019)
36. Eskandari-Khanghahi, M., Tavakkoli-Moghaddam, R., Taleizadeh, A.A., Amin, S.H.: Designing and optimizing a sustainable supply chain network for a blood platelet bank under uncertainty. Eng. Appl. Artif. Intell. **71**, 236–250 (2018)
37. Govindan, K., Cheng, T.E., Mishra, N., Shukla, N.: Big data analytics and application for logistics and supply chain management. Transp. Res. Part E Logist. Transp. Rev. **114**, 343–349 (2018)
38. Gružauskas, V., Baskutis, S., Navickas, V.: Minimizing the trade-off between sustainability and cost effective performance by using autonomous vehicles. J. Clean. Prod. **184**, 709–717 (2018)
39. Kirchoff, J.F., Omar, A., Fugate, B.S.: A behavioral theory of sustainable supply chain management decision making in non-exemplar firms. J. Supply Chain Manag. **52**(1), 41–65 (2016)
40. Lim, M.K., Tseng, M.L., Tan, K.H., Bui, T.D.: Knowledge management in sustainable supply chain management: Improving performance through an interpretive structural modelling approach. J. Clean. Prod. **162**, 806–816 (2017)
41. Lin, Y.H., Tseng, M.L.: Assessing the competitive priorities within sustainable supply chain management under uncertainty. J. Clean. Prod. **112**, 2133–2144 (2016)
42. Lin, K.P., Tseng, M.L., Pai, P.F.: Sustainable supply chain management using approximate fuzzy DEMATEL method. Resour. Conserv. Recycl. **128**, 134–142 (2018)
43. Liu, Y., Quan, B.T., Xu, Q., Forrest, J.Y.L.: Corporate social responsibility and decision analysis in a supply chain through government subsidy. J. Clean. Prod. **208**, 436–447 (2019)
44. Manavalan, E., Jayakrishna, K.: A review of Internet of Things (IoT) embedded sustainable supply chain for industry 4.0 requirements. Comput. Ind. Eng. **127**, 925–953 (2019)
45. Su, C.M., Horng, D.J., Tseng, M.L., Chiu, A.S., Wu, K.J., Chen, H.P.: Improving sustainable supply chain management using a novel hierarchical grey-DEMATEL approach. J. Clean. Prod. **134**, 469–481 (2016)
46. Taleizadeh, A.A., Soleymanfar, V.R., Govindan, K.: Sustainable economic production quantity models for inventory systems with shortage. J. Clean. Prod. **174**, 1011–1020 (2018)
47. Tseng, M.L., Tan, K.H., Geng, Y., Govindan, K.: Sustainable consumption and production in emerging markets. Int. J. Prod. Econ. **181**, 257–261 (2016)
48. Wang, C., Ghadimi, P., Lim, M.K., Tseng, M.L.: A literature review of sustainable consumption and production: a comparative analysis in developed and developing economies. J. Clean. Prod. **206**, 741–754 (2019)
49. Wu, K.J., Liao, C.J., Tseng, M., Chiu, K.K.S.: Multi-attribute approach to sustainable supply chain management under uncertainty. Ind. Manag. Data Syst. **116**(4), 777–800 (2016)
50. Wu, K.J., Chen, Q., Qi, Y., Jiang, X., Gao, S., Tseng, M.L.: Sustainable development performance for small and medium enterprises using a fuzzy synthetic Method-DEMATEL. Sustainability **11**(15), 4119 (2019)

51. Zhang, A., Zhong, R.Y., Farooque, M., Kang, K., Venkatesh, V.G.: Blockchain-based life cycle assessment: An implementation framework and system architecture. Resour. Conserv. Recycl. **152**, 104512 (2020)
52. Madsen, E.S., Mikkelsen, L.L.: The need for knowledge modification in technology change: a framework to consider changes in domain complexity, knowledge and productivity. Prod. Plann. Control **29**(2), 91–105 (2018)

Definition of a Decision Support Model for Calculating the Economic Production Lot Size

M. Teresa Pereira[1,2](✉), Hugo Magalhães[1], F. A. Ferreira[3], and Alcinda Barreiras[1]

[1] CIDEM, School of Engineering of Porto (ISEP), Polytechnic of Porto, 4200-072 Porto, Portugal
mtp@isep.ipp.pt

[2] INEGI - Instituto de Ciência e Inovação em Engenharia Mecânica e Engenharia Industrial, Rua Dr. Roberto Frias, 400, 4200-465 Porto, Portugal

[3] Applied Management Research Unit (UNIAG), School of Hospitality and Tourism of Polytechnic Institute of Porto (ESHT), Rua D. Sancho I, 981, 4480-876 Vila do Conde, Portugal

Abstract. This paper proposes a Decision Support System (DSS) for Calculating the Economic Production Lot Size. The focus is on the analysis of improvement of manufacturing productivity in the packaging industry, specifically in a metal packaging company. Through combining different SKUs (Stock Keeping Units) in the same production order, as opposed to the current single SKU orders production paradigm of, it is possible to increase the overall equipment effectiveness (OEE). This study identified the most critical OEE products within the company portfolio. One critical OEE was selected from within company's strategic fitting process for the assessment of all relevant parameters set for the manufacturing process so that no product variant was disregarded. Evaluation of the improvement achieved by the new production paradigm was possible through a DSS tool developed to combine different SKUs in due production order, taking into account real data from customers' orders. Following the tool development, it was possible to proceed with simulations for various review periods. As this study is an iterative process, the results obtained within each simulation enabled an improved definition for subsequent analysis focusing on the goal set - improving the OEE without increasing stock value. Finally, results were evaluated and the operational and economic benefits of combining different SKUs within the same production order were unequivocal, representing an OEE increase of 20% and a maximum decrease of 367,215€ in stock value.

Keywords: Overall Equipment Effectiveness · Decision Support System · Metal packaging

1 Introduction

With the up growing competitiveness in the retail-market, the need for companies to stand out from their competitors and to increase the flexibility and promptness of their reply to market is paramount. Thus, it is up to the suppliers to provide reduced quantities

J. Machado et al. (Eds.): icieng 2021, LNME, pp. 320–329, 2022.
https://doi.org/10.1007/978-3-030-78170-5_28

and comply with progressively shorter timeframes, hence decreasing their Minimum Order Quantity (MOQ) and their Lead Time (LT).

This paper describes the study developed for a worldwide metal packaging manufacturer. We sought to evaluate the impact on Overall Equipment Effectiveness (OEE) through the application of printing schemes combining different Stock Keeping Unit (SKU) to the metal sheets. As a result, we conducted an analysis of a critical product from the company's portfolio. This analysis sustained a decision on the MOQ and enabled the revision of the product's frame. Simultaneously we conducted and evaluation of the operational and economic benefits of using the new production paradigm.

The study started with an evaluation of the company's portfolio to identify primary critical products according to their production efficiency. As a result, one was selected taking into consideration the company's strategic fit. This stage was followed by the assessment of all the manufacturing parameters relevant to this study.

The following stage required the development of a tool for simulating the combination of different SKUs in production orders, taking into account real data from customers' orders. The optimization of the SKUs grouping into production orders was achieved by an algorithm developed in Visual Basic. Once the tool development phase was concluded, it was possible to proceed with the simulation of various review periods. This was an iterative recursive process that generated results that enabled an improved definition for the subsequent analysis.

The performance combination of SKU scheme analysis in this study for OEE increase without SQU increase is innovative, as no literature on this issue from reliable sources has been identified.

This paper is organized in 5 sections. The first - introduction - introduces the problem and details the objective. Section 2 presents theoretical concepts that support the developed DSS tool. Section 3 defines the DSS. Section 4 demonstrates where simulations are showed, and a Sect. 5 with the main conclusions and its impacts in OEE, productivity and sustainability.

2 Theoretical Concepts

The project team analysed the problem scope and the motivations for the development of the DSS.

Key theoretical concepts underlying this study are as follows:

Print layout: the prepress teams must modify the designs and/or artwork to be printed in the metal cans, according to the print technology specificities. One of the tasks is to organize artwork in a printing layout respecting manufacturing process' restrictions - sheet dimensions and margins required/blanks [1].

OEE is a broadly spread concept for the evaluation of manufacturing processes' efficiency. OEE equals a) the ratio between production time and time available for production; times b) the ration between real production time and possible standard production time; time c) the ration between viable items produced and the total amount of items produced – which is to say: (Availability)*(Performance)*(Quality). The main causes of efficiency loss are setups, micro stops, speed losses and defects [2]. Related with OEE another related concept is the Changeover.

Changeover, commonly called Setup, is the process including all necessary activities to switch products being made in the production line/machine. Changeover duration is defined by the time elapsed between the last a good product in the previous production order, at the standard line efficiency, and the first produced viable product, at standard line efficiency, of the following order [3]. Thus, the Kaizen or Continuous Improvement methodology can contribute to eliminating waste activities, bringing the focuses to added-value tasks and increasing its flexibility [4].

Kaizen is a continuous improvement process that involves the entire corporation equally, from top-management to shop-floor workers. It is a strategy that includes concepts, systems, and tools, to improve efficiency and is triggered by the focus on the client [5].

3 Methodology

This study consists of four major steps:

- Selection of the products in the Company's portfolio with the lowest productive efficiency from amongst these, one is set as the object of the study, considering the Company's strategic fit
- Collecting all relevant commercial and productive data on the selected product
- Developing a tool capable of simulating different scenarios and changing different variables, in order to achieve the most beneficial solutions
- Performing an iterative study

4 Study Development

4.1 Object of Study

Considering clients' annual packaging orders' data from 2018, we calculated the annual quantity of sheets needed, taking into account the printing scheme of each packaging. Afterwards, we calculated the proportion of each product in the global production and, lastly, the average printing order.

Of the top ten annual sheet consumption averages, two products appear to be possible objects of study: CY 99 × 118 - with an annual volume of about 580,000 sheets and 213 sheets average production per order; and TL 188 × 197 - with an annual volume of about 320,000 sheets and average series of 191 sheets per order.

Considering there is already a project for using digitally printed labels for the CY 99 × 118, the TL 188 × 197 was chosen as object of study.

4.2 Data Gathering

A proper definition of the relevant production parameters was required to ensure that the results achieved with this tool would be accurate.

As the client's orders are made into pallets of finished pails, the parameters were evaluated from the end to the beginning of the manufacturing process.

Packing Scheme
The packing scheme defines the palletising of products - quantity of pails, type of pallet and packing articles to be used. For the TL 188 × 197 four types of packing schemes are used. For this study we considered the packing scheme with a lower quantity of pails (504 pails), requiring lower MOQ (Minimum Order Quantity), thus representing less risk for clients.

Scrap Rate
From the beginning of the manufacturing process until the end of slitting, the scrap is 4.5%. Afterwards, in assembling, the rate is 3%.

Coating
There are different coatings according to clients' needs. In the company there are 11 possibilities of base coat and inside lacquer combination. As some of those coatings are similar and have small annual consumption, we selected three coating possibilities, which are used in 82% of the annually produced sheets.

Printing Layout
The TL 188 × 197 has a printing layout of eight pail bodies per sheet. Considering the stacking process of the slitters, the maximum number of different artwork able to be printed per sheet is four.

Units Conversion
According to the packing scheme and the printing layout, one pallet has 456 pails made with 57 sheets.

4.3 Tool Development

The tool being developed was to be used by the Planning Department of the company's Printing Plant, so it was decided to develop it in software with which they were familiar - Microsoft Excel.

The fields to be filled by company planners are: Minimum number of pallets that customers can order (MOQ); packing scheme to be used; and customers' orders, which are automatically rounded to quantities matching the packing scheme chosen (if the quantity is lower than MOQ, it will be increased to MOQ).

Then, the developed Visual Basic algorithm optimizes the orders to be printed considering the four different grouping possibilities (see Table 1).

The rationale behind the algorithm created is presented in Fig. 1.

4.4 Scenarios Simulation

Once the development phase of the tool was completed, we proceeded to simulate different scenarios to evaluate the impact of the paradigm shift from one SKU per sheet to a maximum of four SKU's per sheet. This study was carried out with real customer orders dated 2018. It started by evaluating a small revision period solution, aiming at a lower lead-time, and then perform several iterations according to the results obtained.

Table 1. SKU's grouping.

Possibility	Row1	Row2	Row3	Row4
1	SKU1	SKU2	SKU3	SKU4
2	SKU1	SKU1	SKU2	SKU3
3	SKU1	SKU1	SKU2	SKU2
4	SKU1	SKU1	SKU1	SKU2

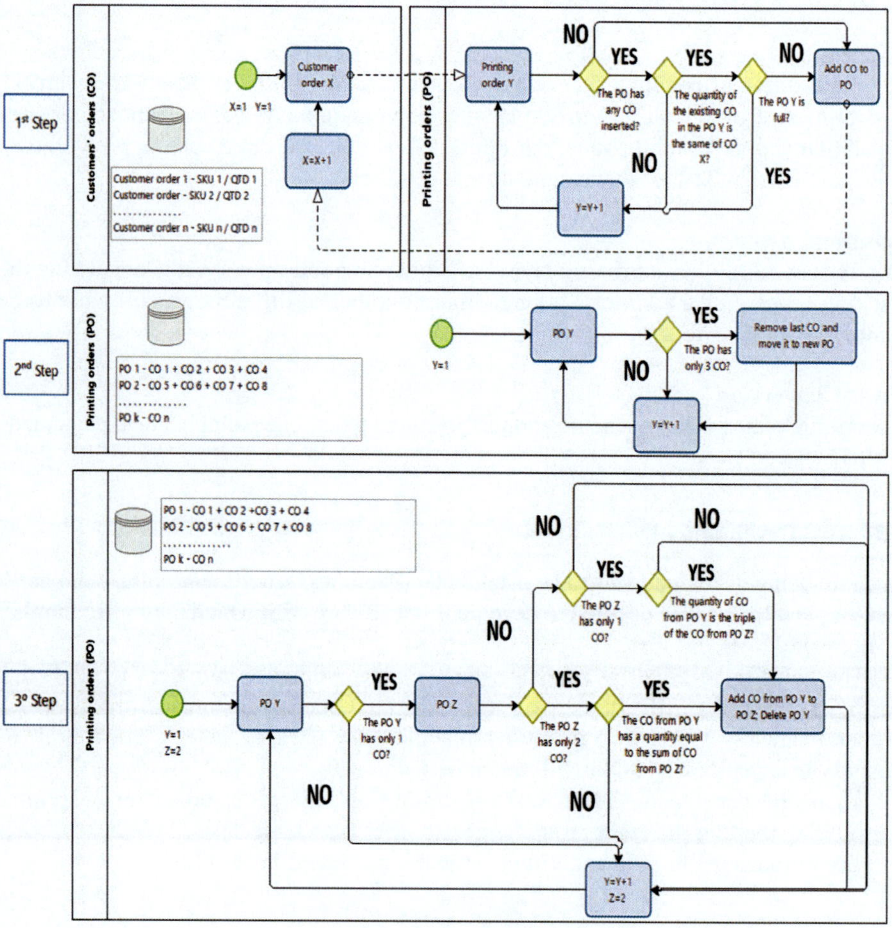

Fig. 1. Algorithm rationale.

Current Status

In 2018 the **average size of the company's production orders** was of 900 sheets for the product in the study. Clients must comply with the company's minimum order size of a 500 sheets MOQ. The item produced is fully billed and is made available to the clients for the period on one year – at whatever time rate – upon which it is converted to scrap.

This solution is not advantageous to the company in terms of stock value. For the customers this implies a commitment of 500 sheets that they may not need.

This study aims to increase the average production size with the usage of combined printing orders and improve the company's flexibility by lowering stock value. The current restriction of a minimum of 500 sheets was disregarded for the purpose of this study in order to properly evaluate the impact of the new production paradigm.

Weekly Review

Aiming at a lower lead-time and more flexibility the study started by simulating a review period of 1 week.

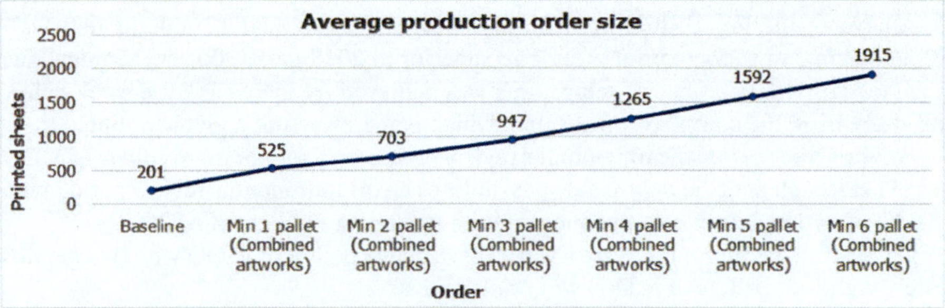

Fig. 2. Weekly review.

Figure 2 illustrates the benefits of using combined artworks instead of the current paradigm, increasing the average production order size from 201 sheets to 525. Although this might be quite a significant increase, it is still too small a value, considering that an offset machine can print up to 6.000 sheets per hour and has a setup time of one hour.

Customer orders' profiles were evaluated to assess if it would be reasonable to increase the MOQ to five or six pallets and consequently to increase the average order size. We concluded that 74% of the orders were for one or two pallets and 85% were for three pallets. Hence, it may be unrealistic to consider MOQ's of five or six pallets.

Yearly Review

As most of the orders are for less than three pallets, we analysed the annual consumption of the SKU's to determine whether the bigger part of the SKUs have a demand smaller than three pallets or if the customers make multiple small size orders.

The Table 2 presents the results of that analysis. The volume of sales for products with an annual volume of six pallets - 10% - is significant. For eight pallets, this value increases up to 16%. It is not the purpose of this study to determine which customers should be made to increase their annual volume to cope with MOQ, but rather to provide adequate information to company's decision-making process and managing bodies.

Table 2. Annual demand.

Quantity	Volume-SKUs with smaller demand
2	1%
4	5%
6	10%
8	16%

Sensitivity Analysis

Considering the results obtained with the previous simulations we decided to perform a sensitivity analysis to evaluate the behaviour of stock value and average production order size considering various review periods (Note: In 2018 the fewer sheets than the demand were printed because of the existing stock).

Analysing the Fig. 3, it is possible to conclude that six and 12 months reviews - despite having an average print order size superior to 2018 data (900 sheets/order) due to the lowering the number of setups, this was achieved by increasing the stock value, thus decreasing the company's flexibility, which is therefore not a viable option.

For the same reason, the four months review period with four or more pallets of MOQ was excluded. Hence, the only viable possibilities are a) four months review period with a MOQ of two pallets and b) two months with two or four pallets of MOQ.

Comparing the four months review period with two pallets of MOQ with two months review with four pallets of MOQ, we conclude that:

- Stock value is similar
- Average print order size is superior in the two months review
- Number of print runs are tantamount
- Final stock is larger in the two months review

The analysis carried out in conjunction with the company's management team allowed to conclude that:

- final stock is higher in two months review
- all other Key Performers Indicators (KPIs) were similar in all tested possibilities
- using a smaller review period enables agility increase

As a consequence, the decision was made to exclude the four months review period, working from then on only with the remaining as possibilities - two months review period for two or four pallets of MOQ.

Considering that the best results were obtained by the shorter review period, we decided to simulate a shorter review period. A two weeks review period was simulated for two and four pallets of MOQ, as presented in Fig. 4.

The two weeks review period with a MOQ of two pallets enabled a higher stock value reduction when compared with the two months period and with the current status.

		2 Months	4 Months	6 Months	1 Year	Current status (2018 data)
		Total	Total	Total	Total	
Minimum 2 pallets	Average stock (pallets /mont)	359	617	881	1,673	
	Average stock (sheets /month)	20,449	35,141	20,316	95,333	
	Final stock (pallets)	149	149	149	149	
	Stock value - TL (€/month)	180,404 €	310,018 €	442,901 €	841,046 €	
	Stock value baseline - 2018 (€/month)	---	---	---	---	367,315 €
	Average print order size	1143	1431	1786	2728	900
	Print runs/period	41	66	79	104	---
	Print runs/year	246	197	157	104	194
Minimum 4 pallets	Average stock (pallets /mont)	624	886	1,175	2,029	
	Average stock (sheets /month)	35,592	50,521	66,989	115,625	
	Final stock (pallets)	505	505	505	505	
	Stock value - TL (€/month)	313,999 €	445,708 €	590,995 €	1,020,067 €	
	Stock value baseline - 2018 (€/month)	---	---	---	---	367,315 €
	Average print order size	1600	2128	2421	3336	900
	Print runs/period	32	49	65	95	---
	Print runs/year	194	146	129	95	194
Minimum 6 pallets	Average stock (pallets /mont)	908	1,178	1,459	952	
	Average stock (sheets /month)	51,728	67,137	83,177	54,236	
	Final stock (pallets)	835	835	835	343	
	Stock value - TL (€/month)	456,352 €	592,294 €	733,809 €	478,478 €	
	Stock value baseline - 2018 (€/month)	---	---	---	---	367,315 €
	Average print order size	2182	2703	3077	3980	900
	Print runs/period	26	47	55	85	---
	Print runs/year	154	141	110	85	194
Minimum 8 pallets	Average stock (pallets /mont)	1,254	1,521	1,832	1,176	
	Average stock (sheets /month)	71,488	86,688	104,396	67,004	
	Final stock (pallets)	1,251	1,251	1,251	567	
	Stock value - TL (€/month)	630,680 €	764,778 €	921,002 €	591,121 €	
	Stock value baseline - 2018 (€/month)	---	---	---	---	367,315 €
	Average print order size	2705	3157	3603	4314	900
	Print runs/period	23	40	52	87	---
	Print runs/year	137	119	104	87	194

Fig. 3. Sensitivity analysis – summary

Nevertheless, this represents smaller print orders, that generate more setups and consequently lower OEE. Comparing with 2018 data, the decrease is low but when comparing with the two months review the difference is considerable.

For the two weeks review with four pallets of MOQ, the average print size is bigger than the one being currently used. However, the stock value almost triplicates when compared with the same review with two two pallets of MOQ. Nonetheless, both cases present better KPIs than that of the current status. Considering the obtained results it was decided not to proceed with further simulations.

		Current status (2018 data)	2 Months	2 Weeks
			Total	Total
Minimum 2 pallets	Average stock (pallets /mont)		359	173
	Average stock (sheets /month)		20,449	9,878
	Final stock (pallets)		149	138
	Stock value - TL (€/month)		180,404 €	87,149 €
	Stock value baseline - 2018 (€/month)	367,315 €	---	---
	Average print order size	900	1,143	821
Minimum 4 pallets	Average stock (pallets /mont)		624	470
	Average stock (sheets /month)		35,592	26,787 €
	Final stock (pallets)		505	470
	Stock value - TL (€/month)		313,999 €	236,317 €
	Stock value baseline - 2018 (€/month)	367,315 €	---	---
	Average print order size	900	1,600	1,254
	Print runs/period	---	32	9
	Print runs/year	194	194	240

Fig. 4. Sensitivity analysis - two weeks

5 Conclusions

The simulations carried out within this study highlight the improvements attainable through the paradigm shift from 1 SKU per production order to various combined SKUs per order. It was possible in each simulation to reach a considerable increase on average print order size, merely by the usage of various SKUs per order, which required fewer setups and therefore increased the effectiveness of OEE.

In parallel, it is possible to increase OEE by changing the review period or the MOQ. The review period was increased - despite improving OEE the SKU increased which in turn decreased flexibility. Thus, this study proceeded focusing solely on smaller review periods.

Considering the variable MOQ, there is a clear increase in the OEE when this variable is increased. Nonetheless, this change has to be accepted by the company customers. One of the trade-offs could be the removal of the current to the 500 sheets per annum commitment.

It was also possible to conclude that, for the purpose of this study, the review periods that better suit the company's needs are those of two weeks and two months.

Summarizing, with the implementation of this new productive paradigm, and taking into account only the review periods considered acceptable (two weeks and two months), it is possible to increase the average print order size from 900 to 1,600 sheets, which would represent an increase of about 20% to the OEE. It is also possible to lower the stock value from 367,215€ to virtually 0€ when the customers accept the MOQ restriction. Related to stock value decrease there is the storage space liberation that becomes available for other uses - an evident possibility being the increase of productive space. At the same time, the stock decrease prevents the sheets from becoming obsolete or suffering damage and having to be scrapped - environmentally unsustainable measure.

Acknowledgments. Fernanda A. Ferreira acknowledges the financial support of UNIAG, R&D unit funded by the FCT - Portuguese Foundation for the Development of Science and Technology, Ministry of Science, Technology and Higher Education, under the Project UIDB/04752/2020.

References

1. Johansson, K., Lundberg, P., Ryberg, R.: A Guide to Graphic Print Production. Wiley, Hoboken (2011)
2. Hedman, R., Subramaniyan, M., Almström, P.: Analysis of critical factors for automatic measurement of OEE. In: 49th CIRP Conference on Manufacturing Systems (CIRP-CMS 2016), Stuttgart (2016). Procedia CIRP **57**, 128–133
3. Henry, J.R.: Changeover of packaging lines. In: Yam, K.L. (ed.) The Wiley Encyclopedia of Packaging, pp. 254–261. Wiley, Hoboken (2009)
4. Singh, J., Singh, H.: Kaizen philosophy: a review of literature. ICFAI Univ. J. Oper. Manag. 51–72 (2009)
5. Imai, M.: Kaizen: The Key to Japan's Competitive Success. McGraw Hill, New York (1986)

Check for updates

Essay on the Contribution of Companies to Social Security Based on Added Value

Ana Borges[1], Conceição Nogueira[2], Fábio Duarte[1], Isabel Cristina Lopes[3](✉), Leonel Vicente[4], Manuel Cruz[5], and Carlos Poço[6]

[1] CIICESI, ESTG, Politécnico do Porto, 4610-156 Felgueiras, Portugal
{aib,fdd}@estg.ipp.pt

[2] CMAT (UMinho), ESTG, Instituto Politécnico de Leiria, 2411901 Leiria, Portugal
conceicao.veloso@ipleiria.pt

[3] CEOS.PP, ISCAP, Politécnico do Porto, 4465-004 S. Mamede de Infesta, Portugal
cristinalopes@iscap.ipp.pt

[4] Instituto Politécnico de Leiria, 2411901 Leiria, Portugal
leonel.vicente@ipleiria.pt

[5] LEMA, ISEP, Politécnico do Porto, 4249-015 Porto, Portugal
mbc@isep.ipp.pt

[6] POCO - Equipamentos Industriais, Leiria, Portugal
cp@poco.pt

Abstract. In the 155th European Study Group with Industry, the company POCO - Equipamentos Industriais, proposed the challenge of developing a mathematical model which allowed to calculate, in a transparent and clear way, the variable amount of fees that each company has to pay to the Portuguese Social Security (SS) in order to take into account its contribution to the added value of the different products produced. This paper presents a mathematical model built under the Modeling, Simulation and Optimization (MSO) paradigm, that is able to be parametrized against a different number of echelons concerning decreased/increased SS contribution of each Portuguese firm. For its implementation, we considered the empirical cumulative distribution function of the SS contributions with respect to four different indexes. The proposed indexes were tested with a sample of Portuguese companies. Several comparisons concerning different Key Performance Indexes were performed. Regarding the model created and the results of the different scenarios comparison, the challenge main questions were addressed. The model was developed in a way that may be applied to any indexes, being ready to simulate and compare different scenarios and parameterizations.

Keywords: Social security contribution · MSO · Gross value added · Key perfomance index

1 Introduction

Traditionally, one of the main functions of the public pension system is to protect people who lack the foresight to save for their own old age (myopia), from

J. Machado et al. (Eds.): icieng 2021, LNME, pp. 330–341, 2022.
https://doi.org/10.1007/978-3-030-78170-5_29

irrational saving behaviors. Several studies defend that a pay-as-you-go (PAYG) public pension system should correct the agents' myopia problem, supported by a positive relation between the optimal social security tax and the degree of myopia. However, [1] shows that the optimal social security tax should be lower when people are more myopic, given the discovery that the welfare cost of the social security tax actually increases with people's degree of myopia. According to [2], there seems to be a tendency in some countries to shift some legal burden of Social Security Contributions (SSC) from employers to employees. In fact, equal sharing of SSC was abolished in Germany in 2005, and countries, such as Germany, Denmark, and France, increased value added taxes in order to finance a decrease in SSC. In his study [2], Neumann utilizes the discontinuities induced by earnings caps for SSC in Germany to analyse the effect of SSC on gross labour earnings and finds that the burden of SSC is shared equally between employers and employees. Social security contributions have effects on labor cost, hours of work, and labor cost per hour. [3] examined a long panel dataset covering 35years of policy reforms in the United Kingdom to find that labor cost falls much more when average employer SSCs rates are reduced than when average employee SSCs rates are reduced, with most of this differential effect coming through reductions in hourly labor cost. The fairness of a social security system can be addressed in many perspectives. The fairness of the SSC is of interest, not only in the perspective of the worker, but also in the perspective of the employer. [4] studied the extent to which the employers shift the burden of compliance with social security obligations back to employees, in the form of lower wages. I their study, results from a fixed effects panel model using a sample of audited firms from Shanghai in 2002 and 2003 found that 18.9% of the compliance cost was shifted back to employees in the form of lower wages [4].

The challenge proposed at the 155th European Study Group with Industry was to develop a mathematical model which allows to calculate, in a transparent and clear way, the variable amount of fees that each company has to pay to the Portuguese Social Security in order to take into account its contribution to the added value of the different products produced. The model proposed shouldn't change the global contribution amount to the Social Security (SS) system, currently 23.75%× Global Salaries for the general companies. It should distinguish companies that add value to the products produced and contribute to improve the competitiveness and productivity of the Portuguese Companies.

Portugal is an economy mainly populated by a large number of micro and small enterprises, specialy from the secondary and tertiary sectors. According to the Portuguese National Institute of Statistics (INE), in 2017, there were 1,260,436 companies in Portugal, of which 68% were individual companies and 32% were corporations. Non-financial companies accounted for 93.4% of turnover and 88.9% of the Gross Value Added (GVA) in the Portuguese business sector. The 1,242,693 non-financial companies produced a turnover of 371,478 million euros and a GVA of 92,690 million euros in 2017, with staff expenses of 52,619 millions of euros. The purpose of this study is to obtain approximately the same sum of wages with a breakdown of the contribution of each company in a *fairer* way, taking into account the added value of the company to the Portuguese economy. This may be achieved by creating a set of 3 or 5 echelons, with different values of percent applicable to companies within each of these levels defined according to the value added to Portuguese Economy per capita. Encouraging to add value to products or services, contributes more to the Gross Domestic Product (GDP) and thus to economic growth. Usually, a greater added value is associated with more labour intervention. As so, the model should alleviate the contribution of enterprises by a percentage in proportion to the added value or more manpower. The alleviated percentage should be understood as a bonus to support the largest employers without ever lowering productivity and competitiveness. Companies generating more employment and creating greater added value are already the largest contributors in value, so they can/should have an incentive through a slight reduction of their contribution to the SS system. This may be understood as a prize to the best employers and an incentive to create jobs in order to add value. Also, in a situation of crisis due to the breakdown of orders, the company that creates the greatest value (and therefore a large employer) is currently being doubly penalized, because it keeps the costs of labour unchanged, while another one that adds less value has other means of control of costs. The idea underneath this challenge does not intend to incentive the creation of jobs without creating productivity, but rather the first objective is to add value and the second one is to improve productivity and therefore competitiveness. The amount of contributions is currently calculated by the application of a contributory rate to the gross remuneration due to the exercise of the professional activity (known as base of incidence). And additionally, by the application of a tax rate to conventional bases of incidence determined by reference to the value of the index of social support (IAS), in 2019, 435.76 EUR. The updating of the incidence base shall take effect on the first day of the month following the publication of the diploma that defined its value.

2 Empirical Design

In addition to the restrictions imposed on the model, described in the previous section, it should be able to be parametrized against different number of echelons with respect to decreased/increased SS contribution. Although the initial request was to simulate the results with 3 or 5 echelons, we generalized it to any odd number. The only constraint that must be take into account is that the echelons should be symmetric with respect to the regular contribution, including a neutral echelon where there is no bonus nor penalization associated with. Formally, if $2ns + 1$ echelons are considered in a given parametrization, their levels should be $(e_1, e_2, ..., e_{ns-1}, e_{ns}, e_{ns+1}, ..., e_{2ns+1})$, where $e_i = -e_{2(ns+1)-i}$ and, the middle echelon should be neutral, that is, $e_{ns+1} = 0$. To be clear, the scale of the echelons will indicate the percentage of increase/decrease in the regular tax of SS Contribution. Although other options could have been made, we decided, in order to keep the model simple and transparent, that each echelon should impact the same proportion of contributions to the SS system. In other words, each echelon should impact $\frac{100}{2ns+1}\%$ of the total contributions to the SS system. Let nc be the number of Portuguese companies that contribute for Portuguese Social Security System and that may be included on this new system. Consider I as a given productivity Index of interest and, for each company C_i, $i = 1, ..., nc$, $SC_I(i)$ as its score under the index I. Now, among all the nc companies labelled, consider their inverse rank $r_1, ..., r_{nc}$ against I, that is, $1 \leq r_i \leq n_c$, where $r_i = 1$ means that the company has the worst performance accordingly to the Index I and $r_i = nc$ represents the rank of the best company w.r.t. I. Let $o(i)$, $i = 1, ..., nc$ be a given ordered sequence of companies, such that $r_{o(1)} \leq r_{o(2)} \leq \dots \leq r_{o(nc)}$, that is, the companies sorted in an ascending order accordingly with the Index I. Define also for each company, its wage volume as $w(i)$ and its contribution to the SS system, $SSC(i)$. Define $AcSSC(o(i))$ as the cumulative sum of contributions for all the companies j which rank is $r_j \leq o(i)$ and $TotalSSC = \sum_{i=1}^{nc} SSC(i)$. Finally, consider $F(x)$ the empirical cumulative distribution function of the Social Security Contributions with respect to the Index I. As the model is supposed to be applied to a large subset of the Portuguese companies, is reasonable to assume that given a quantile α,

$$\exists i \in 1, ..., nc : F^{-1}(\alpha) \approx SC_I(i). \tag{1}$$

Given these definitions, the problem to assign a given echelon j to each company i, can be addressed by:

$$ECH(i) = e_j, \tag{2}$$

Where j is the smallest element in the set $\{1, \dots, 2ns + 1\}$ that verifies the condition: $SC_I(i) \leq F^{-1}\left(\frac{k}{2ns+1}\right)$.

In conclusion, given a productivity Index I and one parametrization of the echelons $(e_1, e_2, ..., e_{ns-1}, e_{ns}, e_{ns+1}, ..., e_{2ns+1})$ with each of their bonus/malus rates, the described model assigns one and only one of those echelons to each company C_i, which will depend on its relative performance against the whole set of Portuguese companies regarding the Index I. Figure 1 intents to present a geometric representation of the model. It is important to notice that if there

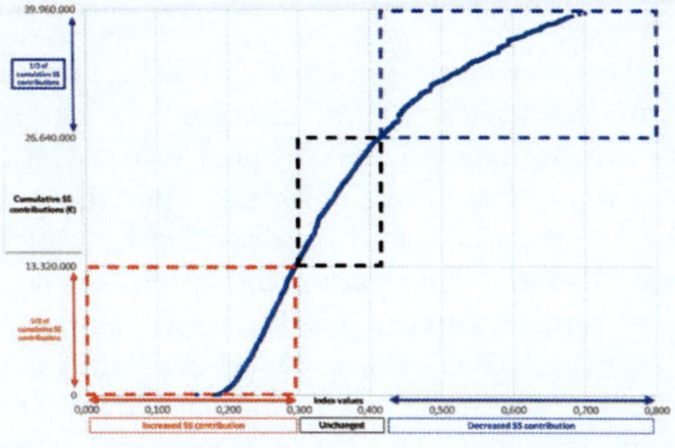

Fig. 1. Geometric interpretation of the proposed mathematical model (example with 3 echelons).

is a considerable gap between the two companies that have consecutive ranks in the neighborhood of the echelon limits, that is, if Eq. 1 does not hold when $\alpha = \frac{j}{2ns+1}$, the total sum of the contributions to the SS system may not remain constant. In fact, the inner product

$$(|e_1|, |e_{2ns+1}|, |e_2|, |e_{2ns}|, ..., |e_{ns-1}|, |e_{ns+1}|) \cdot (SSC(M_1), SSC(M_2), SSC(M_3), SSC(M_4), ..., SSC(M_{2ns-1}), SSC(M_{2ns})) \tag{3}$$

is an upper bound for this deviation, where $\cdot$ stands for the inner product and $SSC(M_k)$, $1 < k < 2ns$ represents the wages amount of the $2ns$ largest companies contributions to the SS system. It should be noticed that, however, as the number of companies where the model is intended to be applied correspond to almost all the Portuguese companies, this should not present a significant deviation. The presented mathematical model may be used to run any simulations, given that both echelons rates and the Index I is properly defined. With an accurate database as support, it will return the changes to each Portuguese company according to the model/index selected. Obviously, the performance of each company will depend on the Index that is considered to rank the companies in terms of added value to the Portuguese economy. Besides the theoretical and economic reasons, that are slightly explained on Sect. 3 for the indicators considered, we propose a new Key Performer Indicator, FKPI, that may be used to infer numerically the fairness of any index. To do so, consider a partition of the whole set of Portuguese companies (e.g., regarding company size or activity sector). Formally, let $P = \{P_1, P_2, ..., P_m\}$ be a partition of all the Portuguese companies (e.g., Psize = {micro, small, medium, big} or Psector = {Primary, Secondary, Tertiary}). A fair system should distribute the benefit/penalties echelons across all the companies without benefit or penalize any particular subset (e.g. SME vs Large companies, or primary sector vs secondary and/or tertiary). If so, a given sector or company size is not "a-priori" penalized against the others. As so, a measure of fairness of an index I may be accessed by evaluating the ratio

$$\text{FKPI}(I, P_j) = \frac{\text{Nr. of companies with decreased contributions in } P_j \text{ subset}}{\text{Nr. of companies with increased contributions in } P_j \text{ subset}} \tag{4}$$

for each partition/subset considered (where F stands for fairness). The closer this ratio is to 1, the more the benefits will be distributed across all the sub-sectors. In the Sect. 4, this will be denoted as "Fairness KPI". With this update, the proposed model is now applicable to any indexes that the decision makers may want to test, beside the ones proposed on this report, being ready to simulate and compare different scenarios and parameterizations. As so, it may be seen as a very helpful tool to support further political decisions regarding this subject. A final remark to point out that we also propose a limit for which the salaries in each company should be affected by this measure. In order to benefit companies where the salaries gap is not to wide, it is proposed to limit the maximum value of wage that may be subject to an increasing or decreasing contribution. In our different scenarios comparison this value was settled to the maximum of 2 national minimum wages. This means that, if a company has a high/low index rate that will imply a discounted/augmented contribution to SS and if a employee earns 3 national minimum wages, only 2/3 of his wage will be taxed at the discounted/augmented rate, while the remaining 1/3 will be taxed at the usual rate. Regarding the comparison of indexes, a dataset of 255,585 portuguese companies was downloaded on 03-07-2019, from the SABI database provided by Bureau van Dijk Electronic Publishing. Only active Portuguese companies were selected, and with known values for year 2017 in Turnover, HR costs, Gross Value Added and Total Assets. We excluded companies operating in the CAE 84 "Public Administration and Defense; Social Security" and 99, "Activities of international organizations and other overseas institutions. The following variables were collected from the database: Name; VAT number; Sector of Activity (Main CAE (Rev3)); Number of Workers; Turnover (m EUR); Total Assets (m EUR); Gross Value Added (GVA) (m EUR); Human Resources (HR) Costs (m EUR); EBITDA (m EUR); EBIT (m EUR); Cost of Goods Sold and Materials Consumed (m EUR); and Supplies and External Services (m EUR). As the available database didn't had the value of the wage per worker, for comparison purposes the average wage (company total wages amount/number of workers) was used as an indicator.

3 Indicators

In this section, we present four different indexes, with a brief description of their main drivers and contextualization. Additionally, in Sect. 4, a summary of the results concerning the application of each index to the considered database of Portuguese companies will be presented.

GVApW Index Definition

$$GVApW(i) = \frac{GVA(i)}{\#Workers(i)} \quad (5)$$

where, $GVA(i)$ - Gross Value Added from company i and $\#Workers(i)$ - Number of full-time workers from company i.

Contextualization: Gross Value Added (GVA), at company level, is a metric used to represent the value added by a particular product or service or corporate unit that the company currently produces/provides. In other words, the GVA reveals the bottom-line profit. Once the consumption of fixed assets is considered in the GVA calculation, the company knows how much net value a particular operation adds to its bottom line. In the present work, and due to database limitations, we defined the company i GVA as,

$$GVA(i) = OperatingRevenues(i) - OperatingCosts(i) - FixedAssetsConsumptions(i)$$

Using GVA over the number of workers (GVApW) we aim to represent a proxy to measure the company's productivity. Hence, higher values of this ratio are associated with higher productivity rather than with higher absolute contribution to the nominal Gross Domestic Product (GDP). Additionally, using the value added by worker, we take into account national accounting practices in terms of GDP calculation based on the supply-side approach. For that reason, at least theoretically, this Index favors (penalize) companies with higher (lower) productivity. Nonetheless, using the GVApW Index, we ignore some important dimensions of competitiveness, such as: Structure of Operational Costs, Structure of Production (i.e., Labor Intensive versus Capital Intensive) or Efficiency.

LACIM Index Definition

$$LACIM(i) = w_1\, I_{GVA}(i) + w_2\, p(i) \tag{6}$$

where, $I_{GVA}(i) = \frac{GVA(i)-GVA_{min}}{GVA_{Max}-GVA_{min}}$, $p(i) = \frac{Cp(i)}{Cp(i)+Cm(i)+Cf(i)}$ and $w_1 + w_2 = 1$, $w_1, w_2 \geq 0$ and, also, GVA_{min} and GVA_{Max} are, respectively the minimum and maximum value of GVA among all the database companies, $Cp(i)$ - Human Resources Costs from company i, $Cm(i)$ - Costs of sold goods and consumed material from company i and $Cf(i)$ - Supplies and external services costs from company i.

Contextualization: This index is a weighted combination of two parcels: a normalized version of GVA, $I_{GVA}(i)$, that represents the relative rank of company i GVA, over the national panorama, and the $p(i)$ term that represents the weight of wage's in the company i total expenses. As, for each company i, both parcels return a value in $[0, 1]$, w_1 and w_2 are adjustable weights for each parcel. In the comparisons presented in Sect. 4, it was taken $w_1 = w_2 = 0.5$. The $p(i)$ term takes into account the Structure of Operational Costs in order to differently weights companies based on different human resources (HR) costs relative to the operational costs. Using this metric, we take into account national accounting practices in terms of GDP calculation based on supply-side approach but in a relative way. Such difference to GVA does not mean that companies with higher absolute contribution to GDP are not beneficiated in terms of a tax rate reduction. Nonetheless, by the normalization of the first term of the model, this Index takes into consideration the relation between GVA of a particular company over the remain ones. Additionally, the second term that we add allows us to control the weight of total HR costs over the remain operational costs. Hence, theoretically, LACIM index tends to favour (penalize) companies with higher (lower) GVA and/or companies with higher ratio of HR costs over operational costs. If so, using this Index, we overcome some limitations of GVApW. Nonetheless, under this model we are still ignoring efficiency metrics. Additionally, since we do not include the consumption of fixed capital on the second metric of this model we also ignore differences between labour-intensive and capital-intensive companies. Finally, comparing with GVApW, this metric only indirectly takes into account the differences between productivity levels across companies.

FLACIM Index Definition

$$FLACIM(i) = w_1\, I_2(i) + w_2\, p(i) \tag{7}$$

where, $I_2(i) = \frac{GVA(i)}{TSSR(i)}$ and $TSSR(i)$ - Total Sales and Services Revenues from company i.

Contextualization: With the first term of this Index, we aim to build a proxy to measure the efficiency of a company's operational activity. Higher values on the first term are associated with higher operational margins and, indirectly, with higher operational return of invested capital. Since we find companies with negative GVA and with low sales and services revenues but with higher operational costs, the ratio of term 1 ranges between 0 and 1. This left and right truncations allow us to overcome some biased results, that may be associated with some odd accountability practices. Additionally, since we are using non-consolidated accounts this truncation tends to avoid biased results related to holding companies. Overall, this Index takes into account the contribution of each company to the GDP, its efficiency and, in line with LACIM Index, the structure of Operational Costs. Theoretically, FLACIM Index tends to favour more efficient companies and/or companies with higher ratio of HR costs over operational costs. This Index is still ignoring two important competitiveness metrics: Productivity and Structure of Production.

FLACIMT Index Definition

$$FLACIMT(i) = w_1 I_3(i) + w_2 p(i) \tag{8}$$

where, $I_3(i) = \frac{GVApNw(i) - GVApNw_{min}}{GVApNw_{Max} - GVApNw_{min}}$ and $GVApNw(i) \equiv \frac{GVA(i)}{\#Workers(i)}$ - Gross Value Added per Number of Workers from company i, $GVApNw_{min}$ and $GVApNw_{Max}$ are, respectively the minimum and maximum value of GVApNw among all the database companies.

Contextualization: This model aims to take into account the most important features provided by each of the previously reported models. Higher values on the first term are associated with higher value added by worker, based on the arguments developed in GVApW Index. In line with the arguments developed to explain the first term of the second Index, in this index we also normalized the GVA per worker. Overall, this Index takes into account the contribution of each worker to the GDP, controlling the productivity of each company and how it relates to the global entrepreneurial ecosystem. In line with the LACIM and FLACIM indexes, this Index also controls for the structure of Operational Costs. Theoretically, Index 4 tends to favor more productive companies and/or companies with higher ratio of HR costs over operational costs. This Index does not take into account two important competitiveness metrics: Efficiency and Structure of Production.

4 Results and Discussion

In order to apply the model presented on this report within the Portuguese scenario, the database described on the previous section was used to run the model with respect to the 4 indexes established on Sect. 3. Results of the scenario with 3 echelons, $(-1\%, 0\%, 1\%)$, will be presented on the following lines (the analysis for 5 echalons is not presented due to space restrictions but available under request). To apply the developed model, it was also necessary to establish some parameters such as w_1 and w_2, the weighting that appears in the index **LACIM**, **FLACIM** and **FLACIMT**. We decided to choose it uniformly, $w_1 = w_2 = 0.5$, although the model may have other values for w_1 and w_2, since the condition $w_1 + w_2 = 1$ is verified. As mentioned, we also decided to propose a limit for which the salaries in each company should be affected by this measure. In the following comparisons that value was settled as the double of the national minimum wage (in 2017, 557 EUR) and was denoted in the following tables as 2. With all of these parameters settled, it is now important to observe if the two main initial constraints were fulfilled. According to the observation of the third and fourth columns of Tables 5–12, we may conclude that the proposed model obeys to the first constrain imposed.

The variation between the present scenario and the proposed model, presented on the last two lines of the 3rd and 4th columns, is almost 0, considering or not the restriction of this application for 2 national minimum salaries (4th and 3rd column, respectively). As so, it is reasonable to admit that the proposed mathematical model guarantees the first constraint. Relatively to the second constraint it was settled that a fair system should distribute the benefit/penalties echelons across all the companies without benefit or penalize any particular subset (e.g. SME vs Large companies, or primary sector vs secondary and/or tertiary). This because, in that case, a given sector or company size is not "a-priori" penalized against the others. Regarding the KPI proposed, It is straightforward that the closer this ratio is to 1, more the benefits will be distributed across all the sub-sectors. This KPI is presented on the last line of each comparison table, with respect to either sector of activity (columns 5–8) as well as to the company size (columns 9–12) (Table 1).

Applying the mathematical model considering three echelons defined by the index GVApW, it is possible to observe that 67% of the total number of firms would be penalized and only 11% would be benefit, resulting on the remaining 21% to stand without change on their SS contributions. As so, using this index in the model, we would have a large number of companies with their contributions raised while just a small number of companies would see their contributions to SS reduced. Moreover, the medium and the large companies are the ones which have a bigger number of companies that would benefit from this scheme, 27% and 41% respectively, and only a smallest number of companies belonging to

Table 1. % of companies affected by changes in rates of SS constribution: GVA per worker.

GVApW	Original SSC	Change in SSC		Nb. firms affected				Nb. firms affected			
		All	2*MW	Total	Sector1	Sector2	Sector3	Micro	Small	Medium	Large
1% change	13 394 074€	133 941€	130 421€	67%	53%	60%	67%	70%	55%	39%	28%
0% change	13 396 229€	–€	0€	21%	27%	22%	21%	19%	30%	34%	31%
−1% change	13 398 776€	−133 988€	−71 526€	11%	20%	10%	13%	10%	15%	27%	41%
Sums	**40 189 079€**	−47€	58 896€	Fairness KPI (:1)				Fairness KPI			
% (Total SSC)		0%	0,15%	5,84	2,68	7,06	5,28	6,8	3,71	1,42	0,7

this category would be penalized (39% and 28%, respectively). So, it is possible to conclude that using this index, the large companies would be benefited by this model (FKPI = 0.7). When we analyze the fairness of this index presented in the last row of the table, with respect to all the companies in the database, we get FKPI = 5.84. This means that for each company that would have their SS contributions reduced, 5.84 others would have their contribution increased. As stated before, as closer this ratio is to 1, the more the benefits will be distributed uniformly across the Portuguese companies. Notice that, with this index, this ratio only is near to 1 for the medium and the large companies. After all the forthcoming comparison tables are presented with common purposes. In Table 2 one can check the results regarding the application of the mathematical model using three echelons defined by the index LACIM. As it may been observed, 36% of the total number of firms would be penalized (also 36% on the echelon without change) and 27% would be benefited. So, using this index, although more companies are penalized than benefited the difference is not high (9%). However, when we do the sector analysis this is only true for Sector 1 and Sector 2, because in Sector 3 we have the opposite: 22% of the companies would be penalized and 38% would be benefited. So, we can conclude that using the index LACIM, the Sector 3 would be benefited by this model. When we analyse the Fairness of this index, we get FKPI = 1.33, considering once again the total number of firms. This ratio is near to 1 which is a good indicator, but within a by Sector comparison that is not verified. For Sector 1 this ratio is equal to 1.95, for the Sector 2 is equal to 2.87 and for Sector 3 is only 0.59. Table 3 concerns to the application ofthe mathematical model using three echelons defined by the index FLACIM. As we can observe, 36% of the total number of firms would be penalized and 29% would be benefited. Observing the sector analysis we have in the Sector 2, 50% of the companies that would be penalized and only 17% would be benefited whereas in Sector 3 and also Sector 1 we have the opposite: only 22% (respectively 29%) of the companies that would be penalized and 40% (respectively 37%) would be benefited. So, we can conclude that using the index FLACIM, the Sector 2 would be penalized by this model. When we analyse the Fairness of this index, we get FKPI = 1.23 (considering the total number of companies). As expected for the Sector 2 this ratio is equal to 2.89 and for the Sectors 3 and 1 only 0.54 (resp. 0.79). Lastly, by the application of the mathematical model using three echelons defined by the index FLACIMT, 31% of the total number of firms would be penalized and also 31% would be benefited. So, using this index although the

Table 2. % of companies affected by changes in rates of SS constribution: LACIM.

LACIM	Original SSC	Change in SSC		Nb. firms affected				Nb. firms affected			
		All	2*MW	Total	Sector1	Sector2	Sector3	Micro	Small	Medium	Large
1% change	13 386 930€	133 869€	107 096€	36%	43%	50%	22%	37%	34%	40%	46%
0% change	13 405 422€	–€	0€	36%	35%	33%	39%	36%	39%	34%	26%
−1% change	13 396 727€	−133 967€	−98 658€	27%	22%	17%	38%	28%	26%	26%	28%
Sums	**40 189 079€**	−98€	8 438€	Fairness KPI (:1)				Fairness KPI			
% (Total SSC)		0%	0,02%	1,33	1,95	2,87	0,59	1,33	1,3	1,58	1,64

Table 3. % of companies affected by changes in rates of SS constribution: FLACIM.

FLACIM	Original SSC	Change in SSC		Nb. firms affected				Nb. firms affected			
		All	2*MW	Total	Sector1	Sector2	Sector3	Micro	Small	Medium	Large
1% change	13 396 354€	133 964€	105 101€	36%	29%	50%	22%	36%	31%	36%	41%
0% change	13 395 544€	−€	0€	35%	33%	33%	38%	35%	40%	35%	29%
−1% change	13 397 180€	−133 972€	−102 477€	29%	37%	17%	40%	29%	29%	29%	30%
Sums	**40 189 079€**	−8€	2 625€	Fairness KPI (:1)				Fairness KPI			
% (Total SSC)		0%	0,01%	1,23	0,79	2,89	0,54	1,26	1,09	1,22	1,37

Table 4. % of companies affected by changes in rates of SS constribution: FLACIMT.

FLACIMT	Original SSC	Change in SSC		Nb. firms affected				Nb. firms affected			
		All	2*MW	Total	Sector1	Sector2	Sector3	Micro	Small	Medium	Large
1% change	13 395 508€	133 955€	102 805€	31%	38%	44%	18%	31%	29%	35%	44%
0% change	13 397 151€	−€	0€	38%	37%	36%	40%	38%	41%	37%	27%
−1% change	13 396 419€	−133 964€	−103 220€	31%	25%	20%	42%	31%	30%	28%	29%
Sums	**40 189 079€**	−9€	−415€	Fairness KPI (:1)				Fairness KPI			
% (Total SSC)		0%	0,00%	1,03	1,51	2,2	0,43	1,02	0,98	1,26	1,52

total number of penalized companies be the same than the total number of benefited companies when we do the sector analysis the Sector 3 stands out because only 18% of the companies that would be penalized and 42% would be benefited. So, we can conclude that using the index FLACIMT, the Sector 3 would be benefited by this model. Nevertheless, when we analyse the Fairness of this index, we get FKPI(FLACIMT) = 1.03, considering the total number of companies. This ratio is approximately 1, which is a good mark, although taking into account each activity Sector these values are not verified. The ratio for Sector 1 is equal to 1.51, for the Sector 2 is equal to 2.2 and for the Sector 3 is only 0.43. It is also important to observe that, in general, this index has a very similar performance to the LACIM index (Table 4).

5 Conclusions and Recommendations

The challenge proposed at the 155th European Study Group with Industry consisted in the development of a mathematical model which allowed to calculate, in a transparent and clear way, the variable amount of fees that each Company has to pay to the Portuguese Social Security in order to take into account its contribution to the added value of the different products. The model developed is able to be parametrized against any different (odd) number of echelons with respect to decreased/increased Social Security contribution of each Portuguese firm. For its implementation, we considered the empirical cumulative distribution function of the Social Security contributions with respect to four different indexes. Regarding the model created and results of the different scenario comparisons based on a real dataset, the challenge main questions were addressed. Additionally, new indexes were proposed and tested with a population of Portuguese firms ($N = 3 \times 10^5$) and several comparisons with respect to different

Key Performance Indexes (KPI's) were considered. We can conclude that this is a good model in the sense that the overall contribution made by the companies for the SS System does not suffer any large change. This remains true when we consider the restriction of this application to 2 national minimum wages (fourth column). Regarding results, we can conclude that the index GVApW is the one that shows worst results in this matter. For future work and recommendations, we believe that a more refined analysis should be done. For example: a) preform a simulation study in order to apply different weights in the proposed indexes in order to find the better weights that do not penalize sectors or type of companies; b) taking into account profits versus non-profits firms; c) testing the differences between labour intensive versus capital intensive companies. Furthermore, the model output may be used in the subsequent year (credit tax) avoiding unnecessary forecast and/or prediction. The proposed model is resilient and developed in a way that is applicable to any indexes that the users may want to test, beside the ones proposed on this report, being ready to simulate and compare different scenarios and parameterizations. As so, we believe that several political options can be taken supported on parameterizations of this MSO model.

Acknowledgement. The authors A. Borges and F. Duarte are supported by national funds through FCT - Fundação para a Ciência e Tecnologia through project UIDB/04728/2020. The research of the C. Nogueira author was partially financed by Portuguese Funds through FCT (Fundação para a Ciência e a Tecnologia) within the Projects UIDB/00013/2020 and UIDP/00013/2020.

References

1. Chu, H., Cheng, C.: Optimal social security tax with myopic agents. Econ. Model. **80**, 320–327 (2019)
2. Neumann, M.: Earnings responses to social security contributions. Labour Econ. **40**, 55–73 (2017)
3. Adam, S., Phillips, D., Roantree, B.: 35 years of reforms: a panel analysis of the incidence of, and employee and employer responses to, social security contributions in the UK. J. Public Econ. **171**, 29–50 (2019)
4. Nielsen, I., Smyth, R.: Who bears the burden of employer compliance with social security contributions? Evidence from Chinese firm level data. China Econ. Rev. **19**(2), 230–244 (2008)
5. Empresas em Portugal em 2017. Instituto Nacional de Estatística, Lisboa (2019)

Check for updates

A View on Harmonization of Interaction of Business Entities in Conditions of Change

Viktor Petrenko, Alla Polyanska, Lesya Verbovska(✉), and Galyna Bodnar

Ivano-Frankivsk National Technical University of Oil and Gas, Karpatska Street 15, Ivano-Frankivsk 76000, Ukraine

Abstract. The actuality of the research of the issue of interaction of economic entities in conditions of change has been grounded and expediency of determining the level of harmonization of this interaction on the basis of the study of personal collective interests' balance of economic entities has been substantiated. The methods of estimation of interaction level based on training technology has been offered, which allows studying the opinion of the training participants on the basis of their answers to the questions presented in the formed map of analysis and assessment of the status of problem elements of the process of interaction with usage of the criteria of the interaction level and the level of their processing. For this purpose the methods for researching the balance of interests of economic entities has been developed in the article, the technology of harmonization based on the calculation of the level of harmonization of economic entities interests and proposes have been developed and a map of measures to harmonize their interaction has been proposed. It has been summed up that the described view on harmonization of business entities in conditions of changes can be used to identify ways for determination of ways for improving the interaction on the basis of harmonization of personal and collective interests in different fields of activity.

Keywords: Changes · Interaction · Business entities · Harmonization · Team · Staff · Training · Decision map

1 The Actuality of the Problem

The actuality of the study of the issue on harmonization of interaction between business entities is determined by the fact that in the conditions of dynamic changes in which the world economy is developing it is important not only to identify and manage of changes in the process of their implementation, but also to insure the harmonization of interaction between the entities of this process. For this purpose it is important to take into account the conditions of harmonization of actions of business entities as the main subdivisions which carry out the basic functions of changes and also the subdivisions and organizations that serve or maintain realization of these changes. Achieving the state of harmonization as the basic principle of economic entities interaction in the processes of change cannot be isolated from other processes and events both in organizations participating in the change and in society in general, so it is important to understand

J. Machado et al. (Eds.): icieng 2021, LNME, pp. 342–359, 2022.
https://doi.org/10.1007/978-3-030-78170-5_30

which obstacles may arise and act on the ways of their interaction harmonization in the change management process, as well as to determine what advantages and disadvantages of this process may arise for business entities and what activities it is advisable to plan their successful implementation.

2 Analysis of the Recent Research and Publications

One of the main tasks of the domestic practice of change management in the context of high dynamics of changes in the external environment of business entities is to find the ways to adapt business entities to these changes. The experience of the world schools of management published as a result of research many scientists demonstrates that the ability generating, developing and implementing of the necessary changes is a major factor in improving the efficiency and success of any organizational unit.

Various problems and aspects of proper change management have been the subject of interest of such well-known foreign scholars as I. Adizes, I. Ansoff, F. Guillard, L. Greiner, R. M. Kanter, J. Kotter, E. Cameron, M. Mescon, K. Levin [1–4] and many others, by [1–4], who indicated the expediency of cultivation in the processes of change of "mutual trust", "respect" and "communications", "leadership".

Coulter, N [5], paying attention to deepening of specialization and emergence in our time, and in this connection, to communication problems and difficulties of interaction between its various branches, drew attention to demonstrate the importance of manifestations of synergetic, the main leitmotif of which (Kanter R.) was considered by proving the need of the fact that people and organizations, the state and business in this process to necessarily cooperate with each other.

Adapting foreign experience and developing their own methods, models, technologies and tools for harmonization and change management, Ukrainian researchers and scholars draw attention to a wide range of issues aimed at harmonizing the interaction of business entities in the conditions of change. In particular, the issue of the importance of culture in the process of change management (A. Polyanska, I. Zapuhlyak [6]), human resources development (L. Verbovska [8]), (V Khytr, [10]), (V. Petrenko, [11]), (V. Vartsaba [15]), as well as features of harmonisation in management at enterprises (N. Popova [12], S. Bai [13], I. Turchina, I. Khovrak [21], M. Kalinichenko, [23], S. Sysojeva [24], by [6, 8–13, 15, 23, 24]).

In the context of the research topic of this article it is necessary to pay attention to the fact that in their research on change management in their works V. Vartsaba and O. Khytra emphasize on the need for "... development, creation and application of a qualitatively new system-synergetic model for managing transformational processes with perfect a mechanism for defining and coordination of subjects and objects interests of management,..." [9], by [9].

However, the coordination of these interests occurs at the point of balance of economic interests of the subject with taking into account of interests of its market counterparties. Thus, V. Vartsaba considers coordination of economic interests through the continuum of differentiation of "economic interest" of the subject according to the criterion of orientation for meeting of one's own or other's needs.

An indisputable fact is that all organizations, business and state are also represented by people. Therefore, it is necessary to realize that people themselves are the first pillars of social and economic systems of any complexity and at any level of hierarchical system, who are capable of high creativity and innovation and who are interacting with each other and influencing each other, form a complex system as a single unit capable of displaying synergism. Therefore, it requires additional research to study and substantiate ways to achieve a balance of interests of economic relations subjects taking into account the interests of each by harmonizing relationships, which, according to the authors, is achieved by identifying and harmonizing the interests of economic entities, determining their compliance with certain requirements, values ethics, culture, management, and understandings of how to use these tools to harmonize interactions.

The views of scholars on the questions of change management and the expediency of harmonizing the interaction of business entities in the process of change have been widely researched and undoubtedly they have weighty importance for the development of management science. But in practice there are often situations of isolation of scientific developments from the vital problems arise. The attempt to overcome this shortcoming and to bring the results of scientific exploration closer to solving of applied problems have been worked out in the article.

2.1 The Purpose of the Article

To determine the sense of harmonisation for better understanding of the "harmonization of interaction" term and as the result of this work to consider the technology of "harmonization of interaction" with using of the coefficient of harmonization for evaluation the economic entities in conditions of change management on the basis of identifying their economic interests by using the analysis map and making recommendations for development and implementation of appropriate actions aimed at harmonization of individual and collective interests of economic entities at different levels of management: individual, group/team, organizational.

3 Research Methodology

As a rule by harmonisation we mean "the act of making different people, plans, situations, suitable for each other, or the result of this". Within people activity we consider the harmonisation as dynamic process leading to the relationship. In the turn, the relationship depends on interaction based on the people trust, and the last is connected with the balancing of people interests' like individual and collective.

The problem of harmonization of interests, goals and actions in the processes of business entities activity management under conditions of necessary changes has received some development in plenty management scientific works in which it was demonstrated that any changes are successful only under conditions of unity of staff in assessment of their needs, goals and tasks, as the latter determine the level of interaction of people in groups.

At the same time, the research of issues and assessment of the level of harmonization of interaction of different subjects, in particular, members of staff, become apparent in the results obtained during the conducting of questioning of a number of the staff of the subjects which becomes possible and reaches its most effectiveness when using of such a form of training of the staff as training. However, the possibility to identify the factors that hinder harmonization of interaction between subjects and members of their staff becomes apparent in the results of the survey of a plurality of subjects' staff which becomes possible and it achieves its greatest effectiveness when using such a form of teaching of staff as training. The main advantage of training technology is the fact that in the process of training of participants, except the focused on a particular problem area, at the same time there also exists the opportunity to consolidate and develop appropriate skills for their practical application with further survey of participants for obtaining information about the level of usefulness of both training in general and its individual elements.

The object of the study became the employees of different levels of management of a certain territorial organizational entity, at which the role of implementation of already planned changes was partly assigned; partly it was decision-making on the expediency of any changes, as well as the personnel who will directly make decisions in the process of managing their implementation. The participants of the training were partly familiar with each other, in particular the heads of regional structural units were well familiar with each other, the heads of departments of these units were only partially familiar, and the rest of participants had only experience of extramural communication via e-mail and telephone.

In the course of the training, the groups analysed the current state and prerequisites for conducting active business; the survey of the participants was focused on identifying their vision and decisions on how to achieve harmonization of the subjects' cooperation in the process of reaching a common solution for solving a problem situation which was formed in the organization; simulation of the situation foresaw participation in teamwork during which the participants tried to agree on their roles, to establish communication in the team and to formulate a joint, most harmonized decision.
At the same time, teamwork gave the participants of the training an opportunity to understand the needs for change, to justify their expediency in specific cases and to try to harmonize their own vision of the situation with the vision of the team.

To investigate the interaction of the participants of the training and to work out the obtained results of the survey in 12 groups (5 participants in each group), the organizers of the training used the $sgn[x_i]$, function which determines the most popular indicators in the groups of respondents:

$$sgn[x_i] = \begin{cases} 1, & if\ x_i \gg n/2 \\ 0, & if\ x_i < n/2 \end{cases} \quad (1)$$

According to the methods of applying the $sgn[x_i]$ function, each of the questions is assigned a double value whether 1 or 0, with observance of the following conditions:

- if the majority of the participants of the formed group (12 groups of 5 participants) agree with the dominant importance of the suggested for assessing criterion during performing of production tasks which were connected with interaction, then this criterion gets a value of 1;
- if the majority of the group members do not agree with the dominance of this criterion, it was assigned a score of 0.

For determining the evaluation criterion for each group of evaluation criteria which characterizes the level of interaction, we determine the value of the evaluation of each criterion by summarizing the results obtained by the result matrix:

$$X_{ij}^{g} = \sum\nolimits_{ij}^{n} x_i, \tag{2}$$

where X_{ij}^{g} - the index of the evaluation; *x* is the evaluation criterion; *g* is the index of the criterion by which the assessment was conducted; *n* is the number of criteria by which the assessment was performed; *j* is the index of the group participant, and; *m* is the number of participants in the n group.

After determining the sum of the estimates, we determine the most influential (in the respondents' opinion) criterion as the maximum value of the evaluation of the evaluation criteria, which is determined by:

$$X_i^n = \max\left\{\sum\nolimits_{i=1}^{n} x_{ij}\right\}. \tag{3}$$

The results of the survey for obtaining the estimates of the indicators by which the level of interaction is characterized, is represented by a binary matrix.

The obtained results of the evaluation are the output data for assessing the level of harmonization as a defined continuum of interests of the interaction participants which is proposed to be considered using the model shown in Fig. 1.

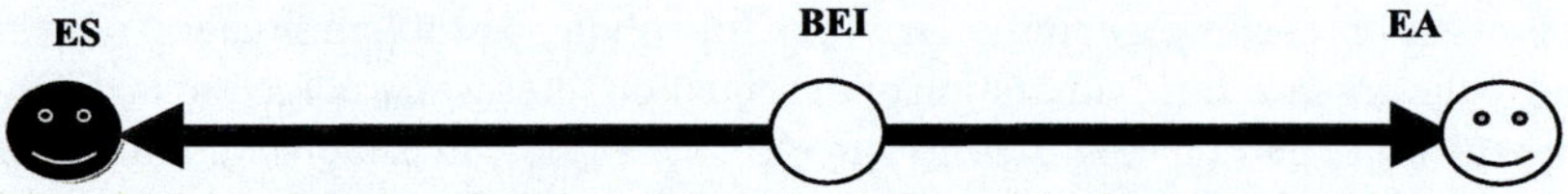

Fig. 1. Continuum of differentiation of options of the "economic interest" concept of the subject according to the orientation criterion for meeting their own or someone else's needs. Source: Varzaba V.

At the same time, the characteristic points of the continuum of interests are the following:

- (ES) "economic egoism", which is defined in literature as "... the hypertrophied economic interests of different economic entities, as well as the areas and regions of the country that put them above the interests of the respective entities" is one the extreme manifestation of desire of economic interest of the entity and appropriate behaviour in the market;
- (EA) "economic altruism" as "... the ability... to selflessly sacrifice one's interests..." by the same subject with appropriate behaviour [17], by [17];
- (BEI) demonstrates the point of balance of economic interest of an entity with taking into account of interests of its market counterparties.

On the basis of this model, we propose to determine the level of harmonization based on the study of the interaction of economic entities by determining the point of the actual level of balance of interest (BI) of the subject with taking into account the point of view of other participants in the interaction using the following calculations:

$$BI_g = \begin{cases} balanced, if \ N_{X^g_{ij}} \uparrow > N_{X^g_{ij}} \downarrow]; \\ non \ balanced, if \ N_{X^g_{ij}} \uparrow \leq N_{X^g_{ij}} \downarrow, \end{cases} \tag{4}$$

where BI_g is the level of balance of interests according to the g group of criteria; $N_{X^g_{ij}\uparrow}$ is the number of values of estimates greater than the average value; $N_{X^g_{ij}}$ is the value of the estimates less than and equal to the average value of the estimates.

From here, we can define the coefficient of conformity as the ratio of the value of the estimates that are greater than the average value to the number of the value of the estimates, smaller and equal to the average value of the estimates:

$$BI_g = \frac{N_{X^g_{ij}\uparrow}}{N_{X^g_{ij}\downarrow}}, \tag{5}$$

where $N_{X^g_{ij}\uparrow}$ is the number of the values of the estimates that are greater than the average value, smaller and equal to the average value of the estimates.

On the basis of the obtained results, we will determine the level of harmonization of interaction of economic entities on the basis of taking into account the level of balance of their interests:

$$\Delta BI_g = \max BI - \min BI, \tag{6}$$

$$BI_{harm} = \frac{\Delta BI_g}{2}, \tag{7}$$

$$BI_{harm\ EE} = \max BI_{EE} - BI_{harm}, \tag{8}$$

$$BI_{harmEA} = \min BI_{EA} + BI_{harm}. \tag{9}$$

where ΔBI_g is the difference in the balance of interests; $BI(harm)$ is the corrective coefficient of harmonization; $BI_{harm\ EE}$ is the estimated level of harmonization.

4 Research Results

Foreign and domestic practice of managing the processes of change in teams of enterprises shows that for their successful implementation are not only specific methods are important but also observance of the following set of guideline principles:

- the purpose of change – increase of efficiency (it is necessary to establish strong ties between the goals for making changes in an enterprise and rigorous performance criteria);
- the choice of a strategy and reorganization of an organizational structure are important (a properly oriented strategy and a viable organizational and economic structure should be the basis for change);

- teams of employees are the basic structural elements (it is necessary to identify opportunities for creating teams and to ensure their effective activity);
- the process of change should be based on the values of an enterprise;
- changes must be based on the development of new competences;
- it is necessary to focus efforts on a limited number of goals.

To investigate how the stated issues are understandable for participants of changes, to assess how participants of changes are fulfilling their potential in the process of management of their implementation, and whether they are willing to spend efforts to achieve interaction it is suggested by the results of conducting training for government leaders and employees of government organizations who are already involved in the process of changes and on the activity of which the success of changes is depended on, as reform processes force to react on these changes not only management but also employees of other categories.

The choice of training technology for investigation of its understanding of interaction and harmonization by the participants, the need for such activities, assessing the ability of participants to harmonize interaction in solving production issues was stipulated by the fact that this educational training technology has several advantages, among which are the following:

- coverage of participants who are representatives of different levels of management and performers which allows each participant to get acquainted and understand how the process of joint decision-making and implementation of harmonized solutions happens;
- the possibility of obtaining additional knowledge in the sphere of activity which interests all the participants, in particular in this case concerning the process of management of changes.
- obtaining additional knowledge by the participants and putting it into practice by means of attraction of them for solving practical situations and obtaining additional skills;
- involvement in the process of implementation of changes through decision-making in situations as close to real as possible;
- raising of qualification level of participants regarding the content and current tendencies that are occurring in the society at different levels of conducting of reforms;
- the use of participants of training by trainers as primary sources for collecting scientific generalization, processing, interpretation and further use of information on evaluations of their training results, the acquired knowledge, the formed positions, attitudes and practical usefulness of the training topics.

That is, according to the results of any training, in addition to the expansion of the necessary knowledge and experience, information about the assessment of the training participants concerning their readiness and ability to cooperate in the processes of conducting changes can also be obtained.

To clarify to which extent the knowledge acquired in the training process and the assessment of their necessity and expediency are "harmonized" with the environment and the subjects of change management by us (by the analogy of studies of processes of harmonization of interests using "cards" or "matrixes" of stakeholders (Beth A. Simmons [18], R. Florea, R. Florea [19], by [18, 19]) it has been proposed to use an "analysis map", in four quadrants, which presents the problematic issues of the change process (Fig. 2).

Economic altruism ← **LEVEL OF INTERACTION** → Economic egoism	
1.1. Not well known people in the team; 1.2. Personal dislike; 1.3. Distrust of others; 1.4. Negative past experience; 1.5. Different views on situations; 1.6. Preconceived attitude towards the members of the team; 1.7. Network and methods of communication; 1.8. Presentation of false information; 1.9. Unwillingness to take into account the opinion of others; 1.10. Individual features of the individual; 1.11. Not interested in solving common issues; 1.12. Functioning of cooperation channels; 1.13 Participation in the goal-setting process 1.14. High level of competition between team members; 1.12. Possibility of understanding inside the team; 1.13. Increasing of the level of communication through joint exercises, trainings; 1.14 Addiction to informal communication	3.1. Level of motivation and stimulation; 3.2. Adherents of the team; 3.3. Each pursues its own goal and sets it above the collective; 3.4. Team members have no common goal, have different vision of development strategy; 3.5. The level of empathy in the team; 3.6. Understanding your own role in achieving goals; 3.7. Observance of traditions 3.8. Lack of motivation and stimulation; 3.9. Distrust due to negative previous experience; 3.10. Commitment to the team by the leader; 3.11. Level of support of each other by team members; 3.12. Conformity with collective and individual goals; 3.13. Lack of a leader with a clear position; 3.14. Understanding culture and its inheritance
2.1. Uncertain actions which are perceived as incompetence; 2.2. The opportunity to express one's own opinion. 2.3. Decision-making support; 2.4. No ability / desire to hear team members; 2.5. Different interests and goals; 2.6. Misunderstandings between team members; 2.7. Opportunities for public discussion of needs and decisions; 2.8. Differences in ways or tools for achieving goals; 2.9. Emotional instability; 2.10. Unprofessionalism	4.1. Perception of control as a way of punishment; 4.2. Reluctance to be controlled by other team members; 4.3. Understanding of reducing the efficiency of work of the team through poor cooperation; 4.4. The need for a head as a leader, not as an administrator; 4.5. Mutual support status; 4.6. Understanding of individual responsibility; 4.7. Status of distribution of responsibilities and tasks among team members; 4.8. Condition of organization of work; 4.9. The desire to take responsibility for the actions of others; 4.10. Documents circulation status, desire of team members to report; 4.11. Personal responsibility for the final result
Economic altruism ← LEVEL OF TRUST → Economic selfishness	

Fig. 2. Map analysis and evaluation of the problematic elements of interaction process based on the study of the criteria of the level of trust and level of interaction (Source: formed and generated by authors on the base of training questionary).

This map was created by distributing of the estimation questions into the quadrants for two criteria – the level of interaction and the level of trust, and the estimation of which was made according to the "economic interest": economic altruism or economic egoism.

Table 1. The results of evaluation of the training participants are the subjects of change

Evaluation criterion	Group number											
	1	2	3	4	5	6	7	8	9	10	11	12
1.1	1	1	0	1	0	0	1	0	0	1	1	0
1.2	0	1	1	1	0	0	0	1	1	0	1	1
1.3	1	1	0	1	1	0	1	0	1	0	1	1
1.4	1	0	1	1	0	0	0	0	1	1	1	1
1.5	1	1	1	0	0	0	1	1	1	1	0	0
1.6	0	1	1	1	1	1	0	1	0	1	0	1
1.7	0	1	1	1	0	1	1	0	1	1	1	1
1.8	0	1	1	0	1	1	0	0	0	1	1	1
1.9	1	1	1	1	0	0	1	0	1	1	1	0
1.10	0	1	0	0	1	1	0	1	0	0	1	1
1.11	1	1	1	1	0	1	1	0	1	0	0	0
1.12	0	1	0	0	1	0	0	1	0	0	1	1
1.13	1	0	1	1	1	1	1	0	1	0	1	0
1.14	0	0	0	0	1	0	0	1	0	1	0	1
2.1	1	0	1	1	1	1	1	0	1	0	0	1
2.2	0	1	1	1	1	1	0	1	1	0	1	0
2.3	1	0	1	0	1	1	1	1	1	1	1	0
2.4	1	1	1	1	1	1	1	0	1	1	1	1
2.5	1	0	1	1	0	0	1	0	0	1	1	1
2.6	0	1	1	0	1	1	1	1	1	1	1	1
2.7	0	0	0	1	0	0	1	1	0	1	1	0
2.8	0	0	1	0	1	1	0	0	1	1	1	0
2.9	0	0	0	0	0	0	0	0	0	1	0	1
2.10	1	1	0	0	0	1	1	0	0	0	1	1
3.1	1	1	1	1	1	1	0	1	0	1	1	1
3.2	1	1	1	1	1	1	1	1	1	0	1	1
3.3	0	0	0	1	1	1	1	1	0	1	1	0
3.4	1	0	1	0	1	1	1	1	1	0	1	1
3.5	0	1	0	0	1	0	1	1	0	1	1	1
3.6	1	0	0	0	0	1	1	1	1	1	1	0
3.7	0	1	0	0	0	0	1	1	0	1	1	0

(*continued*)

Table 1. (*continued*)

Evaluation criterion	Group number											
	1	2	3	4	5	6	7	8	9	10	11	12
3.8	1	0	1	1	1	1	0	0	1	0	0	0
3.9	0	1	1	1	1	0	1	1	0	1	1	1
3.10	1	0	1	0	1	1	0	0	1	1	0	1
3.11	0	1	1	1	0	1	1	1	1	1	0	1
3.12	1	0	0	0	1	1	0	0	1	1	1	1
3.13	0	1	1	1	1	0	1	1	1	0	1	1
3.14	1	0	1	1	1	1	1	0	1	1	1	1
4.1	0	1	0	1	1	0	1	1	1	1	0	1
4.2	1	1	1	0	1	1	0	1	1	1	1	1
4.3	0	1	1	1	0	1	1	1	1	1	0	1
4.4	1	1	0	0	1	1	1	1	1	1	1	1
4.5	0	1	1	1	0	0	0	1	1	1	0	1
4.6	1	1	0	1	1	1	1	1	0	1	1	0
4.7	0	1	1	1	0	1	0	1	0	1	0	1
4.8	1	1	0	0	1	0	0	1	1	0	0	0
4.9	0	1	1	1	0	0	0	0	1	0	0	1
4.10	1	1	1	0	0	1	0	1	0	1	1	0
4.11	1	1	0	1	0	1	0	0	1	1	0	1

Such disposition allows better understanding of what interests dominate in the groups and due to the further deeper analysis will help to determine the conditions of their balancing towards achieving desired level of interaction and trust. The suggested map is formed in the coordinates "the level of interaction – the level of trust", and in the formed quadrants the lists of questions have been presented, the answer to which the participants of training are called to identify the existing obstacles to the harmonized interaction of the training participants. At the same time, the received answers can and should be the starting point for the development of a list of measures aimed at improving the professional level of training participants directly in their workplaces and, finally, become valuable information on the directions and possibilities for their development in them, the so called "soft skills", which are necessary to maintain an effective level of harmonized interaction.

For this purpose, in the quadrants of the analysis maps that characterize the training of participants' ability for interaction, have been divided into four, and grouped into two subgroups: the first subgroup determines the level of interaction as the ability to work together to resolve common issues based on self-interest (EE) and collective interests

(EA), and the second – determines the level of trust between the participants also on the basis of consideration of personal interests (EE) and collective interests (EA).

The results of application of the techniques of evaluation by function $sgn[x_i]$ are presented in Table 1.

- if the majority of participants in the formed group (12 groups of 5 participants in each) agree with the dominant importance of the criterion proposed for evaluation of the performance tasks related to the interaction, then this criterion is given a value of 1;
- if the majority of the group members do not agree with the dominance of this criterion, then it is assigned a score of 0.

In this case, the obtained survey evaluation criteria of interaction are in the form of evaluation determinants each of which is obtained on the basis of summation of the participants' assessments with the subsequent determination of the most important criteria of the participants' interaction in training. According to the results of the surveys in this study, we have formed four evaluation matrices according to the stated four groups of criteria for assessing the level of interaction for the participants of training $\mathrm{X}_{\mathrm{ij}}^{g}\mathrm{g}$.

$$\mathrm{X}_{ij}^{1}\begin{bmatrix} x_{ij}^{1} \\ x_{ij}^{1} \\ x_{ij}^{1} \\ x_{ij}^{1} \\ x_{5}^{1} \\ x_{6}^{1} \\ x_{7}^{1} \\ x_{8}^{1} \\ x_{9}^{1} \\ x_{10}^{1} \\ x_{11}^{1} \\ x_{12}^{1} \\ x_{13}^{1} \\ x_{14}^{1} \end{bmatrix} = \begin{bmatrix} 1\,1\,0\,1\,0\,0\,1\,0\,0\,1\,1\,0 \\ 0\,1\,1\,1\,0\,0\,0\,1\,1\,0\,1\,1 \\ 1\,1\,0\,1\,1\,0\,1\,0\,1\,0\,1\,1 \\ 1\,0\,1\,0\,0\,0\,0\,1\,1\,1\,1\,1 \\ 1\,1\,1\,0\,0\,0\,1\,1\,1\,1\,0\,0 \\ 0\,1\,1\,1\,1\,1\,0\,1\,0\,1\,0\,1 \\ 0\,1\,1\,1\,0\,1\,1\,0\,1\,1\,1\,1 \\ 0\,1\,1\,0\,1\,1\,0\,0\,0\,1\,1\,1 \\ 1\,1\,1\,1\,0\,0\,1\,0\,1\,1\,1\,0 \\ 0\,1\,0\,0\,1\,1\,0\,1\,0\,0\,1\,1 \\ 1\,1\,1\,1\,0\,1\,1\,0\,1\,0\,0\,0 \\ 0\,1\,0\,0\,1\,0\,0\,1\,0\,0\,1\,1 \\ 1\,0\,1\,1\,1\,1\,1\,0\,1\,0\,1\,0 \\ 0\,0\,0\,0\,1\,0\,0\,1\,0\,1\,0\,1 \end{bmatrix} = \begin{bmatrix} 6 \\ 7 \\ 8 \\ 7 \\ 7 \\ 8 \\ 9 \\ 7 \\ 8 \\ 6 \\ 7 \\ 5 \\ 8 \\ 4 \end{bmatrix}$$

Hence, we determine the most influential (according to respondents' opinion) evaluation criterion:

$$X_i^n = \max_{1-14}\{9\} = x_7^{14}.$$

Thus, in the first group for assessing the level of interaction of the respondents of the survey the criterion X_7^{14} has become the weightiest which draws attention to the need for effective communication.

$$X_{ij}^2 \begin{bmatrix} x_1^2 \\ x_2^2 \\ x_3^2 \\ x_4^2 \\ x_5^2 \\ x_6^2 \\ x_7^2 \\ x_8^2 \\ x_9^2 \\ x_{10}^2 \end{bmatrix} = \begin{bmatrix} 1&0&1&1&1&1&1&1&0&1&0&0 \\ 0&1&1&1&1&1&0&1&1&0&1&0 \\ 1&0&1&0&1&1&1&1&1&1&1&0 \\ 1&1&1&1&1&1&1&0&1&1&1&1 \\ 1&0&1&1&0&0&1&0&0&1&1&1 \\ 0&1&1&0&1&1&1&1&1&1&1&1 \\ 0&0&0&1&0&0&1&1&0&1&1&0 \\ 0&0&0&1&0&0&1&1&0&1&1&0 \\ 0&0&0&0&0&0&0&0&0&1&0&1 \\ 1&1&0&0&0&1&1&0&0&0&1&1 \end{bmatrix} = \begin{bmatrix} 8 \\ 8 \\ 9 \\ 11 \\ 7 \\ 10 \\ 5 \\ 6 \\ 2 \\ 6 \end{bmatrix}$$

Hence, we determine the most influential (according to respondents' opinion) evaluation criterion:

$$X_i^n = \max_{1-10}\{11\} = x_4^{10}.$$

Therefore, in this group of criteria for assessing the level of interaction, the X_4^{10} criterion which characterizes the ability to listen to another team member, became the most important.

$$X_{ij}^3 \begin{bmatrix} x_1^3 \\ x_2^3 \\ x_3^3 \\ x_4^3 \\ x_5^3 \\ x_6^3 \\ x_7^3 \\ x_8^3 \\ x_9^3 \\ x_{10}^3 \\ x_{11}^3 \\ x_{12}^3 \\ x_{13}^3 \\ x_{14}^3 \end{bmatrix} = \begin{bmatrix} 1&1&1&1&1&1&0&1&0&1&1&1 \\ 0&1&1&1&1&1&1&1&1&0&1&1 \\ 0&0&0&1&1&1&1&1&0&1&1&0 \\ 1&0&1&0&1&1&1&1&1&0&1&1 \\ 0&1&0&0&1&0&1&1&0&1&1&1 \\ 1&0&0&0&0&1&1&1&1&1&1&0 \\ 0&1&0&0&0&0&1&1&0&1&1&0 \\ 1&0&1&1&1&1&0&0&1&0&0&0 \\ 0&1&1&1&1&0&1&1&0&1&1&1 \\ 1&0&1&0&1&1&0&0&1&1&0&1 \\ 0&1&1&1&0&1&1&1&1&1&0&1 \\ 1&0&0&0&1&1&0&0&1&1&1&1 \\ 0&1&1&1&1&0&1&1&1&0&1&1 \\ 1&0&1&1&1&1&0&1&1&1&1&1 \end{bmatrix} = \begin{bmatrix} 10 \\ 11 \\ 7 \\ 9 \\ 7 \\ 7 \\ 5 \\ 6 \\ 9 \\ 7 \\ 9 \\ 7 \\ 9 \\ 10 \end{bmatrix}$$

Hence, we determine the most influential (according to respondents' opinion) evaluation criterion:

$$X_i^n = \max_{1-14}\{11\} = x_2^{14}.$$

Therefore, for the stated group of criterion of estimating the level of interaction the most important for this group of criteria became the X_2^{14} which attracts attention to the necessity of upholders in the team;

$$X_{ij}^4 \begin{bmatrix} x_1^4 \\ x_2^4 \\ x_3^4 \\ x_4^4 \\ x_5^4 \\ x_6^4 \\ x_7^4 \\ x_8^4 \\ x_9^4 \\ x_{10}^4 \\ x_{11}^4 \end{bmatrix} = \begin{bmatrix} 0\,1\,0\,1\,1\,0\,1\,1\,1\,0\,1\,1 \\ 1\,1\,1\,0\,1\,1\,0\,1\,1\,1\,1\,1 \\ 0\,1\,1\,1\,0\,1\,1\,1\,1\,1\,0\,1 \\ 1\,1\,0\,0\,1\,1\,1\,1\,1\,1\,1\,1 \\ 0\,1\,1\,1\,0\,0\,0\,1\,1\,1\,0\,1 \\ 1\,1\,0\,1\,1\,1\,1\,1\,0\,1\,1\,0 \\ 0\,1\,1\,1\,0\,1\,0\,1\,0\,1\,0\,1 \\ 1\,1\,0\,0\,1\,0\,0\,1\,1\,0\,0\,0 \\ 0\,1\,1\,1\,0\,0\,0\,0\,1\,0\,0\,1 \\ 1\,1\,1\,0\,0\,1\,0\,1\,0\,1\,1\,0 \\ 1\,1\,0\,1\,0\,1\,0\,0\,1\,1\,0\,1 \end{bmatrix} = \begin{bmatrix} 8 \\ 10 \\ 9 \\ 10 \\ 7 \\ 9 \\ 7 \\ 5 \\ 5 \\ 7 \\ 7 \end{bmatrix}$$

Hence, we determine the most influential (according to respondents) evaluation criterion:

$$X_i^n = \max_{1-11}\{10\} = x_{2,4}^{11}.$$

Therefore, for this group of criteria for evaluating the level of interaction, the most important ones became two criteria X_2^{11}, X_4^{11}: the unwillingness to be controlled by other team members and the desire to see the leader as a leader rather than an administrator. Thus, the conducted evaluation of the results of the survey of the participants of the training allowed identifying the criteria that were selected as the most important to ensure interaction. The obtained results of evaluation will be used for determining how balanced and harmonized the responses received are, i.e. how well the interests of business entities are coordinated. In accordance with the chosen methods of determining the continuum of interests of the participants of the interaction, which is proposed to be considered using the model presented in Fig. 1 and Formula 4, we will calculate the level of harmonization of interaction between business entities, determined on the basis of the results of the survey. We will determine the level of balance of interests of economic entities for each group of criteria (EE) and (EA). Accordingly, for the criterion of interaction, the calculations will be as follows:

$$BI_{interaction} = \frac{10}{14} = 0{,}71,$$

$$BI_{trust} = \frac{8}{10} = 0{,}80,$$

$$\Delta BI = |0{,}71 - 0{,}80| = 0{,}09,$$

$$BI_{harm} = \frac{\Delta BI}{2} = \frac{0{,}09}{2} = 0{,}045,$$

$$BI_{harm\ EE} = BI_{\max} - BI_{harm} = 0{,}80 - 0{,}045 = 0{,}755,$$

$$BI_{harmEA} = BI_{\min} + BI_{harm} = 0{,}71 + 0{,}045 = 0{,}755,$$

For the trust criterion, the calculations will be as follows:

$$BI_{interaction} = \frac{12}{14} = 0{,}86,$$

$$BI_{trust} = \frac{9}{11} = 0{,}82,$$

$$\Delta BI = |0{,}86 - 0{,}82| = 0{,}04,$$

$$BI_{harm} = \frac{\Delta BI}{2} = \frac{0{,}04}{2} = 0{,}02,$$

$$BI_{harm\ EE} = BI_{max} - BI_{harm} = 0{,}86 - 0{,}02 = 0.84,$$

$$BI_{harmEA} = BI_{min} + BI_{harm} = 0{,}82 + 0{,}02 = 0{,}84.$$

The results of the calculations are presented graphically in Fig. 3 (for the interaction criterion) and Fig. 4 (for trust criterion).

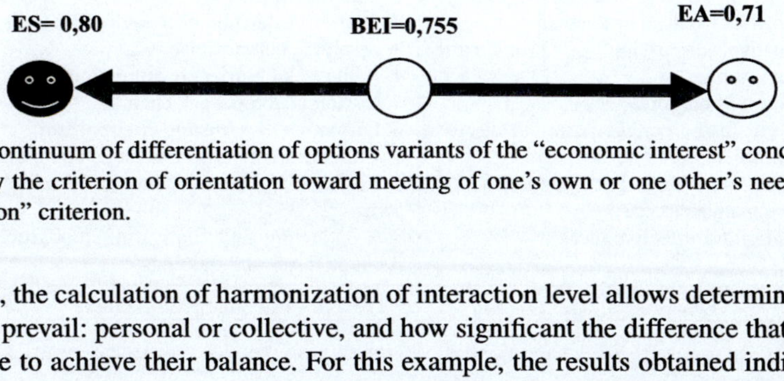

Fig. 3. Continuum of differentiation of options variants of the "economic interest" concept of the subject by the criterion of orientation toward meeting of one's own or one other's needs for the "interaction" criterion.

Thus, the calculation of harmonization of interaction level allows determining what interests prevail: personal or collective, and how significant the difference that must be overcome to achieve their balance. For this example, the results obtained indicate that collective interests are more consistent than individual ones in achieving interaction. In the case of the criterion of trust, collective and individual interests are more balanced, but collective interests also have an advantage.

Thus, the calculation of harmonization of interaction level allows determining what interests prevail: personal or collective, and how significant the difference that must be overcome to achieve their balance. For this example, the results obtained indicate that collective interests are more consistent than individual ones in achieving balanced state. In the case of the criterion of trust, collective and individual interests are more balanced, but collective interests also have an advantage.

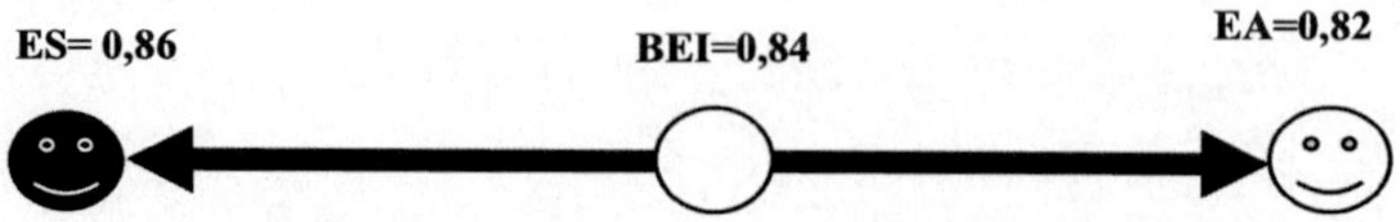

Fig. 4. Continuum of differentiation of options variants of the "economic interest" concept of the subject by the criterion of orientation toward meeting of one's own or one other's needs for the "trust" criterion. Source: Based on our own research

Economic altruism ← LEVEL OF INTERACTION → Economic selfishness	
1. Encouragement and motivation in various forms; 2. Compromise policy; 3. The leader who unites the team is motivating; 4. Interest, stimulate team members; 5. There must be a leader; 6. Ability to convey the purpose assigned to the team members; 7. It is necessary to interest team members (maybe even financially); 8 Ability to listen to the views of other members and reach a common goal; ; 9. Self-organization and self-development; 10. Improving the quality of communications; 11. Formation of team spirit;	1. Effective distribution of responsibilities among team members; 2. Awareness that you work in a team to achieve results; 3. Discussion of problem situations; 4. Trust in a manager; 5. The authority of a leader; 6. Developing clear rules (strategies); 7.. Increase in bonuses and rewards 8. Distribution of responsibilities by profession; 9. Professionalism; 10. Creation of a reporting system (oral or documented); 11. Personal control of a manager at each stage of the task, coordination of team members;
1. To develop the skills of clear communication 2. Ability to express one's own opinion without fear of reaction of the team; 3. Collective discussion of problematic situations; 4. Formation of corporate spirit; 5. T get out of the comfort zone, to discuss existing problems; 6. Development and encouragement for alternative team solutions; 7. Motivation for effective ideas;	1. Communication, the ability to make concessions, setting for positive communications; 2. Learn to listen and hear each other in the team; 3. Applying compromise solutions 4. Removal of barriers to communication; 5. Customized feedback channels 6. Formation of a trusting environment; 7. Collective meetings, meetings, communication 8. Establishing understanding in the middle of the team; 9. Conducting and motivating to participate in joint exercises and trainings;
Economic altruism ← LEVEL OF TRUST → Economic selfishness	

Fig. 5. Map of actions for harmonization of interaction between business entities (Source: generated by the authors)

Except of that, processing of the results of the conducted inquiry of its participants during the training process allowed us identifying the most influential criteria that prevented them from achieving a harmonious cooperation of business entities, and which have been highlighted in Fig. 4. Then it becomes apparent that mastering of skills necessary for managing by the participants, individual and group changes which are presented in the appropriate quadrants of the "action map" (Fig. 5) will increase the effectiveness of the communication process, reduce resistance to change, increase understanding of the essence of implementation of changes, and the leadership will effectively come to management of groups during implementation of changes.

That is, inserted lists of the measures for improvement of participants' propensity for harmonized interaction into the quadrants of matrix in Fig. 2 will be used consciously by trainers (and in real life – by leaders of processes of change at real business entities) for improvement of management of changes by means of eliminating the causes of resistance in the staff environment levels.

5 Discussion Questions

Thus, the article presents the results of the study of the possibilities of harmonization of relationships between managers and executors of real organizational entities of one of the subsystems of the regional system of public administration in the processes of implementation of the proposed to them higher levels of change management.

Undoubtedly, the answers provided by the participants of the training took into account the peculiarities of their activities, namely the administrative nature, focusing on the professional sphere of activity, which is limited by the current legislation, instructions, procedure of activities, clearly defined hierarchical subordination and determination of the decision-making environment. Accordingly, the lists of proposals for harmonizing the interaction of business entities in the process of change management proposed by the Map of actions (Fig. 4) also take into account this feature of the activity and take into account the expectations and wishes of the training participants themselves. It is obvious that these provisions and recommendations can be successfully used to harmonize the interests, goals and objectives of changes and changes needed in society and in other spheres of human activity, in particular – in public, regional and corporate management, in management of production enterprises, in education, etc.

We agree that the amount of people participated in the conducted research wasn't enough to strongly ensure the adequate getting results. This investigation was provided by the one training. But on it basis we got the ability to develop the previous worked out theory of "harmonisation of interaction" and applied it for accomplishing the practical task to harmonise the interaction in the conditions of change.

6 Conclusions and Prospects for Further Research

Thus, changes are an objective reality of the present, requiring adequate action on the part of business entities involved in these changes. The diversity of change participants and the uncertainty and variability of the environment in which they occur requires certain tools to balance the interests of change participants and to explore ways and methods to harmonize their interaction.

In this article, harmonization is determined as activity of getting balance between collective and individual people's interests and as result of such consideration of this term the harmonization technology is proposed as a tool for improving the interaction of business entities, which are presented by different people, as a way of balancing, first and foremost, the interests of those involved in change. Coordination of interests and increasing confidence are defined as determinants of achieving active engagement. It has been determined that such criteria are actual for the entities implementing changes at the regional level, which have been determined on the basis of the results of evaluation of the opinion of the respective economic entities during the training. In particular, the training has offered a methodological toolkit for assessing the level of harmonization of business entities potentially involved in the change management process through a survey during the training.

In the basis of evaluation of thoughts of the training participants lies the use of the $sgn[x_i]$ function, which defines the most popular evaluation criteria of harmonization of interests in the groups of respondents, it allowed to process the results of the survey and to determine the measures and methods of harmonization of achievement of harmonization of interaction (approximation of interests) of the training participants – subjects of changes.

Using of the coefficient of harmonisation has allowed determining the continuum of differentiation of options variants of the "economic interest" concept of the subject by the criterion of orientation toward meeting of one's own or one other's needs according to the "interaction" and "trust" criteria.

The identification of problematic elements of knowledge, qualification and psychological readiness of human resources of organizational entities is necessary for properly managing, planning and implementation of changes which are relevant for these entities.

In our opinion, the suggested and approved in the process of training approach to such identification was offered and tested in the course of training, and nowadays has some prospects for large-scale of it usage, not only for evaluating the current state of entities, but also for promptly formulating and implementing of appropriate actions to improve this state, aimed at harmonization of interests, goals, objectives and aspirations that affect the dynamics and quality of the results of changes.

A further focus of our research is to expand the elements of the "map of actions" to harmonize the interaction of business entities through conducting of further researches and studies of the criteria for harmonisation of interests of business entities in the process of managing changes in business.

References

1. Rosabeth, M.: Change Masters Kanter. Free Press (1985)
2. Adizes, I.: Mastering Change. The Power of Mutual Trust and Respect in Personal Life, Family Life, Business and Society (1992)
3. Kotter, J.: The Heart of Change: Real-Life Stories of How People Change Their Organizations. Harvard Business Review Press, Boston (2012)
4. Lewin, K.: Field, Dorwin: Cartwright Theory in Social Science: Selected Theoretical Papers. Harper & Brothers, New York (1951)
5. Arthur Coulter, N.: Human: Synergetics, p. 201. Prentice-Hall Inc., Englewood Cliffs (1976)
6. Zapukhliak, I., Zaiachuk, Ya., Alla Polyanska, A., Kinash, I.: Applying fuzzy logic to evaluate enterprise readiness for change. Manag. Sci. Lett. **9** (2019)
7. Polyanska, A., Zapukhliak, I., Oksana, D.: Culture of organization in conditions of change as an ability of efficient transformations: the case of gas transportation companies in Ukraine. Oeconomia Copernicana **10**(3), 561–580 (2019)
8. Verbovska, L.: Management of Personnel Development in Change. Springer, Cham (2019)

9. Verbovska, L., Dzvinchuk, D., Petrenko, V., Yasinska, Y.: Nine essays on change management, Ivano-Frankivsk (2018)
10. Khytra, O.: Synergetic aspect of the study of the essence of intellectual capital, Kharkiv (2012)
11. Petrenko, V., Bodnar, G.: Graphoanalytic modeling of processes and conditions of harmonization of interests of parties – participants of public-private partnerships. Scientific Works of the National University of Food Technologies, Kujv (2010)
12. Popova, N., Zenina, Ya., Vinnik, M.: The concept of "harmonization" and its use in management activities at enterprises, Lugansk (2010)
13. Bai, S., Tymtsunyk, V.: Teleological foundations of enterprise management harmonization. State administration. Improvement and development (2011). http://www.dy.nayka.com.ua. Accessed 11 Sept 2019
14. Vartsaba, V., Petrenko, V.: Harmonized management and management – current direction of development of management theory and practice, Kyjv (2014)
15. Vartsaba, V.: Synergetic paradigm of harmonized management of human resources of social and economic systems. Problems of Economy, Lviv (2015)
16. Mochernyi, S., et al. (ed.): Economic Encyclopedia: In Three Volumes, vol. 1. Academia Publishing Center, Kujv (2000)
17. Simchenko, V., Ablyalimov, E., Chachua, Yu.: Modern approaches to coordination of interests of an enterprise with economic entities (2012). http://www.economy.nayka.com.ua. Accessed 12 Jan 2020
18. Simmons, B.A.: The international politics of harmonization: the case of capital market regulation. https://dash.harvard.edu/bitstream/handle/1/3382976/international_politics.pdf?sequence=2. Accessed 15 Sept 2019
19. Florea, R., Florea, R.: Stakeholders interests analyze and harmonization - [starting point of strategic approach]. http://www.ugb.ro/etc/. Accessed 12 Dec 2019
20. Polyanskiy, V.: Constitutional foundations for the harmonization of social interests in the public authorities system. Kutafin Univ. Law Rev. **4**(1), 28–39 (2017)
21. Trunina, I., Khovrak, I.: Harmonization of interests of employers and institutions of higher education as a basis for sustainable regional development (2019). https://ieeexplore.ieee.org/document/8896688/metrics#metrics. Accessed 10 Jan 2020
22. Baiur, D.: Development of socially responsible corporate governance in the context of harmonization of interests of stakeholders. https://echas.eenu.edu.ua/index.php. Accessed 11 Nov 2019
23. Kalinichenko, M.: Harmonization of economic interests in the system of marketing management of an industrial enterprise **3**(63) (2013). http://www.pmdp.org.ua/images/. Accessed 15 Jan 2020
24. Sysoeva, S.: Society and personality: harmonization of development, no. 12. Ridna Shkola, Kujv (2012)

A VRP Model to Support Last Mile Maritime Containers

M. Teresa Pereira[1,2](✉), Luís Ferreirinha[1], F. A. Ferreira[3], and Marisa Oliveira[1]

[1] CIDEM, School of Engineering of Porto (ISEP), Polytechnic of Porto, 4200-072 Porto, Portugal
mtp@isep.ipp.pt

[2] INEGI - Instituto de Ciência e Inovação em Engenharia Mecânica e Engenharia Industrial, Rua Dr. Roberto Frias, 400, 4200-465 Porto, Portugal

[3] Applied Management Research Unit (UNIAG), School of Hospitality and Tourism of Polytechnic Institute of Porto (ESHT), Rua D. Sancho I, 981, 4480-876 Vila do Conde, Portugal

Abstract. Due to the constant and hasty development of the process of economic globalization, the transport network has become increasingly complex. This causes significative changes in the manner of cargo transport, which leads to satisfy the costumer's needs first and only then try to achieve an effective and efficient low-cost distribution. The decision support system developed allows to allocate the closest truck, with capacity, and give the respective quotes automatically, based on predefined criteria and optimization. The result will have a high impact on the parties' performance: better management of infrastructures used, on the footprint, time of service, customer' level service, reduction of empty kilometres, less cost and reduction of the receipt period.

This investigation has the main goal of developing an online Inland Container Transport (ICT) Decision Support System (DSS). Maritime container distribution and collecting processes is a routing problem, in regions which are oriented to container seaports or inland terminals. Therefore, the problem to be investigated is closely related to the vehicle routing problem with backhauls (VRPB) that finds an optimal set of orders (or routes) with deliveries (linehauls) and pickups (backhauls).

Keywords: Last mile distribution · Maritime container · Multi-Criteria · Linear programming

1 Introduction

Due to the constant and hasty development of the process of economic globalization, the transport network has become increasingly complex. This causes significative changes in the manner of cargo transport, which leads to satisfy the costumer's needs first, and only then try to achieve an effective and efficient low-cost distribution.

Short Sea Shipping (SSS) refers to maritime transport of goods, much of them containerized, from one or more origin to one or more destinations in a uniform flow [1]. The SSS has been experiencing a significant increase since the end of nineties [2],

J. Machado et al. (Eds.): icieng 2021, LNME, pp. 360–370, 2022.
https://doi.org/10.1007/978-3-030-78170-5_31

in Portugal increased from 4% to nowadays 40%, continuous growth of global volumes of containers puts increasing pressure on the in-land road [3] assuming more and more importance due to externalities, problems of road congestion, environmental concerns and traffic safety [4].

Last mile transportation of containers is a process that can be optimized and brings mutual benefits to stakeholders: shipping agents, freight forwarders, carriers, companies and ports. This problem currently is based on historical analysis [3] and haulers' price tables of transportation per covered area, followed by negotiation and agreement, which can take hours or days.

The decision support system developed allows to allocate the closest truck, with capacity, and give the respective quotes automatically, based on predefined criteria and optimization. The result will have a high impact on the parties' performance: better management of infrastructures used, on the footprint, time of service, customer' level service, reduction of empty kilometres, less cost and reduction of the receipt period.

This investigation as the main goal of develop an online Inland Container Transport (ICT) Decision Support System (DSS). ISO container distribution and collecting processes is a routing problem, in regions which are oriented to container seaports or inland terminals. Therefore, the problem to be investigated is closely related to the vehicle routing problem with backhauls (VRPB) that finds an optimal set of orders (or routes) with deliveries (linehauls) and pickups (back-hauls). The specificity of the routing problem analysed lies in the fact that a truck may simultaneously carry one 40 ft, or two 20 ft containers, using an appropriate trailer type. This means that in one route two, three or four nodes, can be visited, which is equivalent to the problem of matching pickup and delivery nodes in single routes which provide a total travel distance shorter than, in the case, when nodes are visited separately. This work aims to contribute to the economic competitiveness of operators and ICT and contribute, in general, for the social and economic development of the region and for a sustainable logistics.

The remaining sections of this paper are organized as follows: in Sect. 2 some literature review about short sea shipping, inland transportation and dynamic vehicle routing problems is analysed. The description of the problem can be viewed in Sect. 3 as well as the results for a given data. In Sect. 4 is where the paper finally presents some conclusions and provides some ideas for future works.

2 Literature Review

2.1 Short Sea Shipping

SSS refers to maritime transport of goods, much of them containerized, from one or more origin to one or more destinations in a uniform flow [1]. The SSS has been experiencing a significant increase since the end of nineties [2], in Portugal increased from 4% to nowadays 40%, continuous growth of global volumes of containers puts increasing pressure on the in-land road [3] assuming more and more importance due to externalities, problems of road congestion, environmental concerns and traffic safety [4].

Containers of different sizes, mostly ISO of 20 ft, and 40 ft empty and/or loaded should be delivered to, or collected from the customers, defining the existing inland networks feeding the shippers lines, pre- or post-carriage, usually defined as haulage in ports and drayage in intermodal terminals. The general problem of haulage operations is its cost effectiveness, require reliable inland connections due shippers just in time supply chain, and environmental impact of the ITC is bound by restrictions from governments and from shippers themselves [3]. Despite the relative short distance of the truck movements compared to other mode, as rail, drayage accounts for a large percentage (between 25% and 40%) of origin to destination expenses [2].

2.2 Inland Container Transportation

Inland container transportation (ICT) refers to the container movements among customer locations, container terminals, and depots by different means of transportation in one transportation chain. This last one can be divided into three sections: (i) the pre-haulage, (ii) the main-haulage and (iii) the end-haulage. The pre-haulage is based on transport fully loaded containers from the senders to terminals, which is usually done by trucks. The main-haulage is basically covered by the rails or sea shipping's deliveries between terminals, since it covers the longest distances. Finally, the end-haulages refers to the transportations of the containers from the terminals to the receivers, which, in a similar way to the first one, this operation is carried out by the trucks [5, 6]. A schematic of what has been said can be seen in Fig. 1.

Fig. 1. Sections of transportation chain

As mentioned before, in an intermodal transportation chain, between 25% and 40% of the total intermodal container transportation costs are from the trucking sections, i.e., from the pre and end-haulage [7]. As for the sea transportation, the inland costs range between 40% and 80% of the total intermodal transportation [8].

A port or intermodal terminal may be used by a dozen or more of haulage companies, where each of them faces a trip scheduling problem with trips between shippers, receivers and one or more terminals meeting several requirements, such as costumer's pre-specified pickup and delivery times (time-windows), on road travel times, and realistic limit on the length of a realistic of the working day [2]. Most shipments are known about in advanced, only a small fraction of loads in a given set are short notice. However, loads have to be efficiently reassigned due to the traffic, docs or terminal delays. This subject is related to several identified problems, such as haulage's cost, empty kilometres, poor coordination of containers pickup and deliver, high footprint due congestion of trucks in port areas, where this last one has created serious environmental and traffic problems in many countries [2, 9]. One of the main reasons for congestion is the concentrated arrivals of road trucks at peak hours [9]. At the same time, container terminals pay more and more attention to the service quality of inland transport modes such as trucks, trains and barges [10]. Hinterland networks for maritime ISO (International Organization for Standardization) container transportation require planning methods in order to increase efficiency and reliability of the inland road, rail and waterway connections [3]. Fostered by geographic information system (GIS) that yields new possibilities for the modelling of large multi-modal freight networks [2]. ICT is a problem that can be optimized. Multi-criteria routing and scheduling in a multimodal fixed scheduled network with time-dependent travel times involves the determination of the non-dominated itineraries. This type of problem emerges from many transportation planning decisions including ITC [11]. Several combinatory methods and heuristics can be applied, i.e. window partition [12], genetic algorithms [5], integer programming [13]. Truck appointment systems are a common approach to reduce truck turnaround times, [10] provides a tool to use the truck appointment system to increase not only the service quality of trucks, but also of trains, barges and vessels. [10] propose a mixed integer linear programming model to solve this type of problems. [9] developed a negotiation process for smoothing truck arrivals in peak hours among multiple trucking companies and a terminal. The negotiation process is to be used for the appointment system for road trucks in container terminals. [12] developed a heuristic based on the window partition method to obtain near optimal schedules. [3] developed a real-time decision rules for suitable allocations of containers to inland services by analysing the solution structure of a centralized optimization method used offline on past data. On other side, container liner shipping companies only partially alter their shipping networks to cope with the changing demand, rather than entirely redesign and change the network [14]. In view of the practice, [14] proposes an optimal container liner shipping network alteration problem based on an interesting idea of segment, which is a sequence of legs from a head port to a tail port that are visited by the same type of ship more than once in a given time.

2.3 Vehicle Routing Problems

The Vehicle Routing Problem (VRP) concerns the transportation of goods from one place to another, whether they are from the factory to the warehouse or from the warehouse to the customer, and it is a classical NP-hard problem that was first introduced by Dantzig and Ramser as a generalization of the Traveling Salesman Problem (TSP) [8, 15, 16]. This problem is usually defined on a graph $G = (V, A, C)$, where $V = \{v_0, \ldots v_n\}$ represent the vertices, i.e., the customers that need to be served except v_0, since this one stands for the *depot/warehouse*. The $A = \{(v_i, v_j)_{i,j} \in V, i \neq j\}$, represent the arcs that connect the vertices, i.e., the possibilities that exist to go from v_i to v_j. Finally, $C = \{C_{ij}\}_{(vi, vj) \in A}$ represents the cost matrix that is defined by the arcs, where it can be traveling distances, time or even traveling costs. This type of problem is certain a well-know and widely studied in the domain of Complex Combinatorial Problems (COPs), that basically tries to find a set of routes that connects all the *n* customers whereas each customer is visited exactly once, while minimizing the overall *routing cost* [17–19].

Despite its simple and classical definition, the real-world environment forced the analysis of other variants inherent to the problem, such as the Capacitated VRP (CVRP), where each customer has a demand of products and the available trucks have limit capacity, whether physical or legal. In this kind of problem, it is assumed that all the trucks have the same capacity, which originated the Heterogeneous fleet VRP (HFVRP), which is basically the same as the previous one, but in this case, all the trucks have different capacities. The specification of the time delivery window by the customers is another variant of the problem, which concerns VRP with Time Windows (VRPTW). Another common variant of this classical problem is when the trucks have to deliver and also pick-up freight along the vertices which is known as VRP with Pick-up and Delivery (VRPPD). There are much more variants than the ones described that have been deeply studied in the literature [20–22]. In a real-world environment, most of them are together, which convert this problem seemingly simple problem into a complex one [19].

Despite the exponential growth of complexity in solving this problem with the various variants, the same can become even more complicated when we tried to replicate it into the real-world environment. Most of the effort to solve this kind of problems has been applied to traditional ones with deterministic operational environment where all the information needed to solve it is known at the beginning of the problem. This is, before assigning the vehicles to the routes that best fit the performance measure, which is usually the minimum cost, we have all the necessary information to solve the problem optimally, like the number of clients to serve and the respectively demand, the traveling time, and many other variables related to the specific problem, and where it is also assumed that such information will not change as time passes. However, the practical experience shows us the other side of the problem, which is the uncertain environment, where breakdowns occur, unpredicted customers' requests keep arriving, travel times are affected from random factors such as bad weather conditions, car accidents, road works, or other special events that might cause traffic congestion, occur frequently, which forces the revision of pre-established routes. This kind of environment is best known as the dynamic environment, which is an ongoing fashion that implies that some information besides the one we initially have is revealed as time passes, like the traveling times, the vehicle locations, and so on [23, 24]. Despite its complexity, the Dynamic VRP (DVRP) have emerged as an area of intense investigations mostly due to recent advances in communication and information technologies, that nowadays allow to be obtained and processed in real-time in an easier and friendly way [25].

3 Problem Description and Modeling Approach

This paper as the main goal of develop an online ICT DSS. ISO container distribution and collecting processes is a routing problem, in regions which are oriented to container seaports or inland terminals. Therefore, the problem to be investigate is closely related to the vehicle routing problem with back-hauls (VRPB) that finds an optimal set of orders (or routes) visiting deliveries (linehauls) and pickups (backhauls). The specificity of the routing problem analysed lies in the fact that a truck may simultaneously carry one 40 ft, or two 20 ft containers, using an appropriate trailer type. This means that in one route two, three or four nodes, can be visited, which is equivalent to the problem of matching pickup and delivery nodes in single routes which provide a total travel distance shorter than in the case when nodes are visited separately. The problem of matching customer nodes in the container distribution and collecting processes can be solved as a multiple matching IP. For problems of larger sizes Heuristic approach is recommended [13]. Being a decision support system (DSS) for optimization of the haulage of containers the major CAEs are 52291 and 49410, with 50% and 30%, respectively, which makes the shipping agents or forwarders and transporters the main beneficiaries of the DSS.

This problem is currently based on historical analysis [3] and haulers' price tables of transportation per covered area, followed by negotiation and agreement. This process can take hours, even days and offers great potential for optimization. Consequently, the streamlined this process is central to the competitiveness of stakeholders: shipping agent, forwarding agent, carrier and enterprises.

This process can be optimized using techniques of combinatorics where inserts the dynamic routes, multi-criteria decision support models, modelling, management and geographic positioning systems and simulation models, which bring mutual benefits to the parties.

3.1 VRP Model

A VRP model was developed combining a MCDA model, that previously we calculated the performance using a multicriteria method taking into consideration: (i) distance in kilometres to destination; (ii) travel time to their destination; (iii) cost per kilometre; (iv) ecological footprint (CO_2 emissions g/km); (v) available capacity of the truck in tonnes; (vi) available truck capacity in TEU, and the model formulation presented as follows:

Parameters

n Number of containers;
m Number of trucks;
T_j Truck j;
CT_i Container i;
a_i TONs of $\mathrm{CT_i}$;
b_i TEUs of $\mathrm{CT_i}$;
c_j TONs available on $\mathrm{T_j}$;
d_j TEUs available on $\mathrm{T_j}$;
P_j Performance of $\mathrm{T_j}$;
S Number of containers
K Coefficient to give more importance to the container assignment

DV

$$x_{ij} = \begin{cases} 1 \;\; \textit{if the } CT_i \textit{ is transported on the } T_j \\ \quad 0, \text{ otherwise} \end{cases}$$

$$y_i = \begin{cases} 1, & T_j \text{ is used} \\ 0, & \text{otherwise} \end{cases}$$

OF

$$Max\, z : f(x) = K.\left(\sum\nolimits_{i=1}^{m} \sum\nolimits_{j=1}^{n} x_{ij}\right) - \sum\nolimits_{j=1}^{n} y_j P_j \tag{1}$$

ST

$$\sum_{i=1}^{m} x_{ij}.a_i \leq y_j.c_j \tag{2}$$

$$\sum_{i=1}^{m} x_{ij}.b_i \leq y_j.d_j \tag{3}$$

$$\sum_{j=1}^{n} x_{ij} \leq 1 \tag{4}$$

$$\sum_{i=1}^{m} \sum_{j=1}^{n} x_{ij} \leq S \tag{5}$$

$$\sum_{j=1}^{n} x_{ij} \leq y_j.M \tag{6}$$

$$M.\left(\sum_{j=1}^{n} x_{ij}\right) \geq y_j \tag{7}$$

$$i \in M,\ j \in N \tag{8}$$

$$N = \{1, 2, \ldots, n\},\ M = \{1, 2, \ldots, m\} \tag{9}$$

The objective function (1) is to maximize the number of containers to be assigned using the smallest number of trucks possible with the best appraisal. That is, the function forces the model to aggregate the largest number of containers to the set of trucks available using the trucks with the best MCDA performance in the first place considering the distance of the available truck. Thus, if it is possible to assign all containers without using all available trucks, it is ensured that the best-rated lorries were assigned firstly, since the closer to 1 is the evaluation, the better the truck concerned, so the fewer trucks are used and the more trucks with better performance if they assign, the greater the value of the objective function. Also, how many fewer containers exist in the container park, fewer costs will be involved. The constraints (2) relate to the capacities in tonnes or of the set of containers assigned by the truck. In other words, this restriction prevents the set of loads assigned by the truck from exceeding the available tonnes of it. The same applies to the following restrictions (3), and in this case is referring to TEU containers, i.e., this restriction prevents the set of containers affected by the truck from exceeding TEU available ones. The restrictions (4) Prevent the container *i* from being transported/assigned in several trucks. The restrictions (5) Prevent the number of containers assigned from exceeding the number of containers available. Subsequently, both restrictions in (6) and (7) serve to prevent a truck from being used without cargo and that the cargo is assigned by a truck that is not in use. Lastly, the latest constraints enclose the set of containers and trucks in the model.

3.2 VRP Model Results

To test the developed model, real data from a shipping company was used, using ten containers and select ten trucks from available ones, that previously performance appraisal

Table 1. VRP models results.

Container ID	T0001	T0002	T0003	T0004	T0005	T0006	T0007	T0008	T0009	T00010	Tons Required
CT0001	0	0	0	0	1	0	0	0	0	0	20
CT0002	0	0	0	0	0	0	1	0	0	0	21
CT0003	1	0	0	0	0	0	0	0	0	0	16
CT0004	0	0	1	0	0	0	0	0	0	0	4
CT0005	1	0	0	0	0	0	0	0	0	0	8
CT0006	0	1	0	0	0	0	0	0	0	0	20
CT0007	0	0	0	0	0	0	0	0	0	1	23
CT0008	0	0	0	0	0	0	0	1	0	0	9
CT0009	0	0	0	1	0	0	0	0	0	0	20
CT0010	0	0	1	0	0	0	0	0	0	0	4
(a)	24	20	8	20	20	0	21	9	0	23	
(b)	40	20	40	24	20	3	21	18	3	24	
(c)	40	20	40	40	40	0	20	20	0	40	
(d)	40	20	40	40	40	20	20	20	20	40	
(e)	1.72	1.78	1.94	2.06	2.56	**0.00**	5.67	5.78	**0.00**	6.39	**27.89**

Container ID	Width	Quantity	Demand	Nr Containers	FO
CT0001	40	1	10	10	9999962.111
CT0002	20	1			
CT0003	20	1			
CT0004	20	1			
CT0005	20	1			
CT0006	20	1			
CT0007	40	1			
CT0008	20	1			
CT0009	40	1			
CT0010	20	1			

where:

(a) CurrentCapacity(TON)	The tonnage inherent in the affected load
(b) MaxCapacity(TON)	The maximum capacity of truck j in TONES
(c) CurrentCapacity(UNI)	TEU associated with the affected load
(d) MaxCapacity(UNI)	The maximum capacity of truck j in TEU
(e) Performance	The performance associated with the truck j

was calculated, using as previously referred a multicriteria method taking into consideration: (i) distance in kilometres to destination; (ii) travel time to their destination; (iii) cost per kilometre; (iv) ecological footprint (CO_2 emissions g/km); (v) available capacity of the truck in tonnes; (vi) available truck capacity in TEU.

From Table 1, can be foreseen that the ten select maritime containers (CT0001 till CT0010) were affected to eight trucks. It should be noted that although the T0006 truck has better performance than the T0007 and T0008 trucks, it was not selected. This is due to the available capacity in it, i.e., the T0006 truck has only three tonnes free for a container of 1 TEU, however none of the selected containers fall into these characteristics, hence the non-selection of the respective truck. The same goes for the T0009 truck. The interpreter that containers were affected to each truck is as follows: the CT0003 and CT005 containers were affected to the T0001 truck so the T0001 truck takes two containers of one TEU each (total of two TEUs) with a total weight of 24 TONS, and the maximum the truck could take was 40 TONS. The same reasoning is applied to the other trucks. As can still be seen in Table 1, the model does not take into account the distances that trucks will make with the containers that have been affected by them, so it is not known whether the T0001 truck goes to customer C0001 and C0002, and in turn the truck T0003 also goes to the client C0001 and C0002. If this happened, the loads for their customers could be aggregated in another way that avoided two destinations for each truck. Hence a Second model is needed to overpass this situation. As future work we propose to use a meta-heuristic, using as initial solution, the presented solution to explore the solutions space.

4 Conclusions

The main aim of this paper is to design an innovative (read-to-implement) online platform for negotiating and operationalize pickup and deliveries of containers to/from clients from/to a given port. Therefore, a VRP model was developed associated to a multicriteria model to compute the trucks performance for a more sustainable solution.

The model consists of a mathematical linear model in which its objective function is the allocation of the largest number of containers to the smallest number of trucks available, giving the trucks with the best performance rating. However, this model does not take into account either distance or transport time, and in turn the associated costs, hence the need to develop a second model, using a meta-heuristic, to overpass that as future work. As a future work, it is also intended to validate the tool with real data of several carrier companies working for the same propose. Moreover, it is in the interests of the authors to encourage the possibility of transporting containers of customers to the depot. Another important aspect to consider in the future concerns the need to return to the depot for containers cleaning, before carrying out a second cargo.

Acknowledgments. Fernanda A. Ferreira acknowledges the financial support of UNIAG, R&D unit funded by the FCT - Portuguese Foundation for the Development of Science and Technology, Ministry of Science, Technology and Higher Education, under the Project UIDB/04752/2020.

References

1. Crainic, T.G., Kim, K.H.: Intermodal transportation. In: Handbooks in Operations Research and Management Science, vol. 14, pp. 467–537 (2007)
2. Bontekoning, Y., Macharis, C., Trip, J.: Is a new applied transportation research field emerging? A review of intermodal rail–truck freight transport literature. Transp. Res. Part A Policy Pract. **38**(1), 1–34 (2004)
3. Riessen, B., Negenborn, R.R., Dekker, R.: Real-time container transport planning with decision trees based on offline obtained optimal solutions. Decis. Support Syst. **89**, 1–16 (2016)
4. Pereira, T., Rocha, J., Telhada, J., Carvalho, M.S.: Characterization of the Portuguese SSS into the Europe: a contribution. In: Corman, F., Voß, S., Negenborn, R.R. (eds.) ICCL 2015. LNCS, vol. 9335, pp. 252–266. Springer, Cham (2015). https://doi.org/10.1007/978-3-319-24264-4_18
5. Wang, W.F., Yun, W.Y.: Scheduling for inland container truck and train transportation. Int. J. Prod. Econ. **143**(2), 349–356 (2013)
6. Funke, J., Kopfer, H.: A model for a multi-size inland container transportation problem. Transp. Res. Part E Logist. Transp. Rev. **89**, 70–85 (2016)
7. Macharis, C., Bontekoning, Y.: Opportunities for OR in intermodal freight transport research: a review. Eur. J. Oper. Res. **153**(2), 400–416 (2004)
8. Notteboom, T.E., Rodrigue, J.P.: Port regionalization: towards a new phase in port development. Marit. Policy Manag. **32**(3), 297–313 (2005)
9. Phan, M.H., Kim, K.H.: Negotiating truck arrival times among trucking companies and a container terminal. Transp. Res. Part E Logist. Transp. Rev. **75**, 132–144 (2015)
10. Zehendner, E., Feillet, D.: Benefits of a truck appointment system on the service quality of inland transport modes at a multimodal container terminal. Eur. J. Oper. Res. **235**(2), 461–469 (2014)
11. Androutsopoulos, K.N., Zografos, K.G.: Solving the multi-criteria time-dependent routing and scheduling problem in a multimodal fixed scheduled network. Eur. J. Oper. Res. **192**(1), 18–28 (2009)
12. Zhang, R., Yun, W.Y., Kopfer, H.: Heuristic-based truck scheduling for inland container transportation. OR Spectr. **32**(3), 787–808 (2010). https://doi.org/10.1007/s00291-010-0193-4
13. Vidović, M., Radivojević, G., Raković, B.: Vehicle routing in containers pickup up and delivery processes. Procedia Soc. Behav. Sci. **20**, 335–343 (2011)
14. Wang, S., Liu, Z., Meng, Q.: Segment-based alteration for container liner shipping network design. Transp. Res. Part B Methodol. **72**, 128–145 (2015)
15. Dantzig, G.B., Ramser, J.H.: The truck dispatching problem. Manag. Sci. **6**(1), 80–91 (1959)
16. Monekosso, N., Remagnino, P.: The analysis and performance evaluation of the pheromone-Q-learning algorithm. Expert Syst. **21**(2), 80–91 (2004)
17. Santos, A.S., Madureira, A.M., Varela, M.L.R.: Study on the impact of the NS in the performance of meta-heuristics in the TSP. In: IEEE International Conference on Systems, Man, and Cybernetics (SMC), Budapest, Hungary, pp. 1110–1115 (2016)
18. Soysal, M., Çimen, M.: A simulation based restricted dynamic programming approach for the green time dependent vehicle routing problem. Comput. Oper. Res. **88**, 297–305 (2017)
19. Pillac, V., Gendreau, M., Guéret, C., Medaglia, A.L.: A review of dynamic vehicle routing problems. Eur. J. Oper. Res. **225**(1), 1–11 (2013)
20. Lin, C., Choy, K.L., Ho, G.T.S., Chung, S.H., Lam, H.Y.: survey of green vehicle routing problem: past and future trends. Expert Syst. Appl. **41**(4), 1118–1138 (2014)

21. Eksioglu, B., Vural, A.V., Reisman, A.: The vehicle routing problem: a taxonomic review. Comput. Ind. Eng. **57**(4), 1472–1483 (2009)
22. Laporte, G.: The vehicle routing problem: an overview of exact and approximate algorithms. Eur. J. Oper. Res. **59**(3), 345–358 (1992)
23. Lin, C., Choy, K.L., Ho, G.T.S., Lam, H.Y., Pang, G.K.H., Chin, K.S.: A decision support system for optimizing dynamic courier routing operations. Expert Syst. Appl. **41**(15), 6917–6933 (2014)
24. Çimen, M., Soysal, M.: Time-dependent green vehicle routing problem with stochastic vehicle speeds: an approximate dynamic programming algorithm. Transp. Res. Part D Transp. Environ. **54**, 82–98 (2017)
25. Gendreau, M., Guertin, M., Potvin, J.Y., Séguin, R.: Neighborhood search heuristics for a dynamic vehicle dispatching problem with pick-ups and deliveries. Transp. Res. Part C Emerg. Technol. **14**(3), 157–174 (2006)

Industry 4.0 in the Wine Sector – Development of a Decision Support System Based on Simulation Models

Jéssica Sá[1], Luís Pinto Ferreira[1,2(✉)], Teresa Dieguez[3,4], José Carlos Sá[1,2], and Francisco José Gomes da Silva[1,2]

[1] ISEP - School of Engineering, Polytechnic of Porto, Porto, Portugal
lpf@isep.ipp.pt
[2] INEGI – Instituto de Ciência e Inovação em Engenharia Mecânica e Engenharia Industrial, Porto, Portugal
[3] CiTUR – Centro de Investigação, Desenvolvimento e Inovação em Turismo, Leiria, Portugal
[4] IPCA - Polytechnic Institute of Cávado and Ave, Barcelos, Portugal

Abstract. As a pillar of Industry 4.0, simulation embodies an extremely useful tool in the resolution of issues related to decision-making and the choice of strategies. Since these are problems which exist in the wine sector, the purpose of this work was to build a decision support tool based on system dynamics, which would encompass the main factors that influence the price of wine. Its main objective is to assist managers in the definition of the strategies which are best suited to the company's mission. In order to build this tool, several interviews were conducted with various experts in the industry, thus gaining an insight into the factors which most influence the price of wine, both according to pertinent literature or according to these specialists.

Keywords: Industry 4.0 · System dynamics · Wine sector · Decision support system

1 Introduction

The use of simulation as a form of understanding the issues involved in decision-making in the context of organizations has gained increased vitality [1]. Indeed, by means of computer simulation, decision-makers are able to learn more about the structure of the system they intend to study, particularly with regard to how the various factors are related, and what the effects of this interaction might be [2]. Simulation also allows for a comparison between alternative hypotheses, as well as providing explanations and support for the recommendations of the study at hand [3–5]. The focus of this study resides in the issue of decision-making, more specifically in the context of companies in the wine sector. This is a crucial aspect for any company: when managers are required to make choices, they must be able to do so through access to information and careful analysis. Therefore, the objective of this study is to create a management support tool by means of system dynamics; this will facilitate decision-making and be flexible enough

J. Machado et al. (Eds.): icieng 2021, LNME, pp. 371–384, 2022.
https://doi.org/10.1007/978-3-030-78170-5_32

to adapt to the reality of each company operating in the wine sector. In fact, given that one of the main uses of system dynamics is focused on providing support to decision-making when selecting strategies and policies [6], the principal objective of this process of system modelling is to compare the results of various scenarios, thus supplying more information to those who are responsible for making decisions regarding the company's strategic policies [7, 8]. It was to this end that simulation methodology has been used to develop the decision support tool. System dynamics and its principles of feedback and side effects enable assisting many managers to think about how a strategy might or might not be successful, and what types of consequences – whether intentional or accidental – may arise [9]. Models based on system dynamics are thus able to handle and monitor the assumptions related to the systems' structures, and the effects of the changes implemented to these subsystems, subsequently generating scenarios which depend on the variations of the variables used [10]. This methodology has been used as a simulation tool in different areas of activity. This article is divided into four main sections: the first one is an introduction to the theme of this study; the second section presents a review of literature; the third deals with the topic of industry 4.0 in the wine sector and the decision support system developed in the context of this study. The conclusions of this paper are discussed in the last section.

2 Review of Literature

Agriculture 4.0 represents a great opportunity to address the uncertainty associated to the agricultural sector and its production chain [11]. In fact, some European and American countries are already developing strategies to support digital agriculture [12]. Since simulation is one of the pillars of Industry 4.0 [13], and due to the applicability of the field of system dynamics, this is often used as a decision support tool when solving the various problems related to the agricultural sector. An example of this applicability is the use of models based on system dynamics to measure the potential impact of water use and its association to climate and population change [14]. Also, in the context of climate change, and by means of models based on system dynamics, an analysis was undertaken about the economic impact of severe storms on Germany's forest resources, even before they occurred. The purpose was to assist public and private forest landowners in deciding on the best strategies for forest rescue and management [15]. The use of system dynamics is also useful in the food industry. It was in this context that the methodology was implemented in a study conducted in a province of Iran, which aimed to assess the impact on revenue, as well as on the efficiency of rework, of the main factors inherent to the development of new green products in this sector. The study concluded that factors such as scheduled completion time, the team itself, and the facilities required for each project, all play a key role in the development of new products [16]. The use of models based on system dynamics has also proved to be extremely suitable in the olive oil industry. They are used to investigate the effect of speculation, as well as dynamics and stocks, along the supply chain of olive oil production, and to analyze possible policies of implementation [17]. With regard to the specific case of wine production, the methodology based on system dynamics is, for example, particularly useful in building models which explain the relationship between the factors that influence the wine industry, namely: sales, demand

and price. Its ultimate aim is that of providing a clear insight into the dynamics of wine production and consumption [18]. Other important issues for wine products, which can be analyzed from the perspective of system dynamics, pertain to the relationships between the causes and effects of relevant variables: wine supply and demand; soil release rate; land cultivation; vine lifecycle; and wine production yield [19]. In the long term, the effects of climate change and policies, international price fluctuations and exchange rates can all be anticipated by system dynamics, thus facilitating strategic decision-making for producers and winemakers [20].

3 Industry 4.0 in the Wine Sector

The objective of this study was to build a decision support tool which would assist company managers operating in the wine sector to make strategic choices, based on the comparison of several scenarios. The input variables of the tool were defined, not only by analyzing pertinent literature, but also by means of interviews held with nine specialists in the area, with the purpose of understanding which factors influence the price of wine. The decision support tool was developed through Vensim simulation software provided by Ventana Systems. Simulation is a very useful tool to support the decision-making process within Industry 4.0 [21].

3.1 Identification of the Main Problems in the Wine Sector

As in other agricultural activities, wine production has been confronted with obstacles which hinder management and decision-making in the various companies operating in the area [21]. A key issue for the agricultural sector resides in climate change and, accordingly, investments made to address this problem have generated great impact in the sector [22]. In the specific context of the wine industry, there are other issues associated to climate and the environment surrounding the grape growing area, which greatly influence the quality of wine, such as: soil characteristics, the crop culture history and the topography of the terrain [23]. Since climate is a determining factor in the quality of wine [24] and, given its unpredictable nature, this constitutes a serious problem for growing wine grapes. Despite the importance of climate, soil conditions are also an important factor to consider and, as such, strategies must be studied to surpass them [25]. In addition to these aspects, there are also problems related to consumers' drinking preferences and their reduced interest in the beverage [26].

3.2 Identification of the Parameters Which Influence the Production, Quality and Price of Wine in Portugal

Climate is widely known as a factor which impacts on the success of all agricultural systems. It is able to determine whether a crop is suitable for a given region, and exercises great control over the production and quality of a crop, ultimately bolstering economic sustainability [27]. The wine sector is no exception, as wine production and quality are greatly influenced by climate [24, 27]. Vines respond to countless climatic factors; yet it is temperature which greatly influences the growth of vines and the productivity of wine grape varieties [28]. Besides climate, one must also consider the characteristics of the soil itself, as well as the topography and culture history of the land, which all generate a considerable impact on the perception of a wine's quality [23, 29]. This leads one to the concept of terroir, a French term which describes the entire network of environmental and natural factors that cannot be modified by the producer in a simple manner [30]. This combination of natural features is reflected in the final product and, in conjunction with the decisions made by management, gives rise to distinct wines whose source can be identified. One may thus conclude that the suitable selection of a cultivation site plays a major role in ensuring that an end product of higher value is achieved [31]. The quality of a wine is therefore determined by a combination of numerous factors and, in turn, parameters such as quality, appeal, availability and wine price all influence the sales in this sector [18]. Indeed, when deciding on a purchase, consumers have expectations regarding the quality of the product; once the product has been bought, it generates an experience of the quality in question. In literature, some authors claim that the relation between quality expectation and quality experience is what determines consumer satisfaction regarding the product and, ultimately, the likelihood of buying it once again [32]. However, as wine cannot be tasted at the moment of purchase, consumers tend to make a choice based on the information displayed on the label of the bottle itself; however, it is the price, region of origin and brand, which constitute its most important attributes [33]. With regard to the issue of pricing, several empirical studies have suggested that wine prices depend on quality, reputation and objective characteristics (such as the wine harvest year, the region of origin of grapes and their variety) [34]. Consumers tend to process various perceived signs of quality in order to deal with a sense of uncertainty, which are mainly of an extrinsic nature, such as: price, producer, brand, region, awards, ratings and recommendations [33]. Accordingly, another feature which determines a consumer's decision to buy a bottle of wine is that of the region where the wine was produced [35]. Indeed, when the intrinsic quality of a wine cannot be established objectively, consumers use the region of origin as an indicator to replace quality [36].

3.3 Decision Support System Developed

Figure 1 presents the stock-flow diagram of the decision support system developed by means of *Vensim* software, which represents the relationships between the variables supporting the system. This model was developed based on the literature and the nine interviews conducted with experts in the wine sector. Given the subjectivity of some factors inherent to the model, such as the impact of wine reputation on its price, percentage variables were included to allow each user to adjust the model to his company's strategy and reality. In this model, there are percentage variables which are simultaneously associated to the variables of: soil, harvest year, climatic conditions, innovation and technology, region of origin, awards and marketing. These are organized on a *Likert scale*, which consists of a response scale used primarily in questionnaires to obtain preferences or the degree of a participant's agreement with a given subject [37].

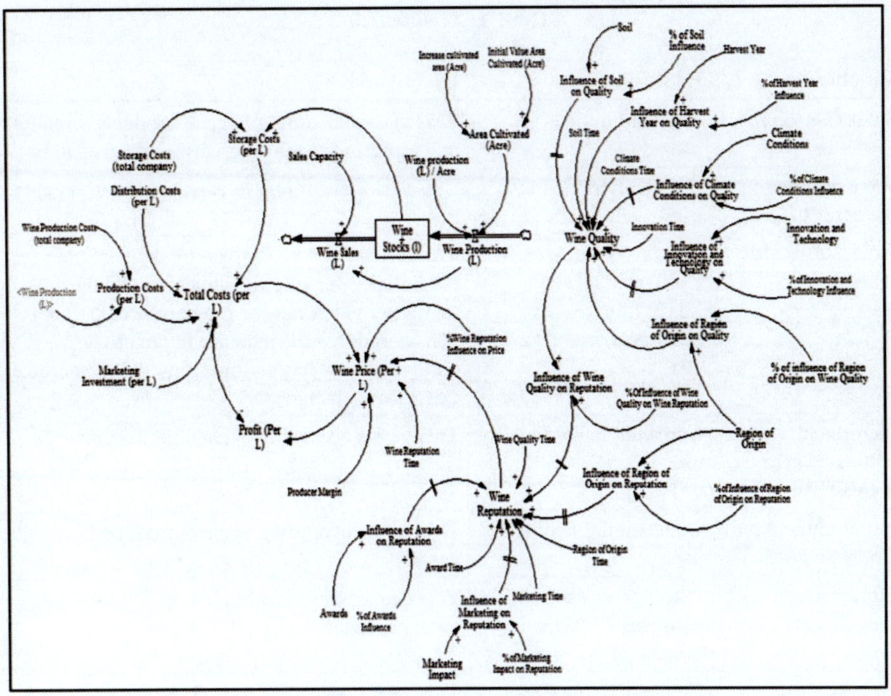

Fig. 1. Stock-flow diagram.

Stocks and flows form the basis for the construction of models grounded on system dynamics. They depict the state of the system under study, and provide the information on which decisions and actions are based. Stocks can change by means of their flows, which can be described as amounts that are added (entry) or subtracted (exit) from stock over a period of time [38]. Tables 1 and 2 present the variables considered, as well as their respective mathematical equations.

Table 1. Detail of model variables.

Variable	Equation	Description
Wine Sales (l)	IF THEN ELSE (Sales Capacity > "Wine Production (l)", "Wine Production (l)", Sales Capacity)	Since each company's sales capacity is not unlimited, the Wine Sales variable is equal to the amount of wine production if the sales capacity is higher than production; it is equal to sales capacity if this is lower than production
Wine Production (l)	Wine production per Ha (L/Ha) * "Area Cultivated (Ha)"	It is considered that all cultivated grapes are used for the production of wine, and that no grapes are purchased from third parties

(*continued*)

Table 1. (*continued*)

Variable	Equation	Description
Sales Capacity	Constant defined by user	This varies according to each producer's reality and represents each company's sales capacity
Wine production per acre (L/acre)	Constant defined by user	This varies according to each producer's reality
Area Cultivated (Ha)	"Initial Value Area Cultivated (Ha)" + STEP ("Increase in cultivated area (Ha)", 12)	The cultivated area is given by the sum between the initial value of cultivated acre and the increase (which may or may not occur). It was considered that the increase in area to be cultivated only has an impact on the next crop, i.e., 12 months later
Increase in cultivated area (acre)	Constant defined by user	This varies according to each producer's capacity
Initial Value Area Cultivated (acre)	Constant defined by user	This varies according to each producer's capacity
Storage Costs (per liter)	"Storage Costs (total company)"/"Wine Stocks (l)"	This determines the storage cost for each liter of wine produced
Storage Costs (total Company)	Constant defined by user	This varies according to each producer's characteristics
Distribution Costs (per liter)	Constant defined by user	This varies according to each producer's characteristics
Wine Production Costs (total company)	Constant defined by user	This varies according to each producer's characteristics
Production Costs (per liter)	"Wine Production Costs (total company)"/"Wine Production (liters)"	This determines the cost of production for each liter of wine produced
Marketing Investment (per liter)	Constant defined by user	This varies according to each producer's capacity
Total Costs (per liter)	"Distribution Costs (per liter)" + "Marketing Investment (per liter)" + "Production Costs (per liter)" + "Storage Costs (per liter)"	All non-itemized costs need to be included in the production costs
Profit (Per liter)	"Wine Price (Per liter)" − "Total Costs (per liter)"	This determines the profit for each liter of wine produced

(*continued*)

Table 1. (*continued*)

Variable	Equation	Description
Wine Price (per liter)	"Total Costs (per liter)" + (1 + Produtor Margin) + DELAY1("%Wine Reputation Influence on Price" * Wine Reputation, Wine Reputation Time)	It is considered that the influence of wine reputation on price is not immediate; there is thus a delay, which is defined by the user
Producer Yield	Constant defined by user	This varies according to each producer's characteristics
% Wine Reputation Influence on Price	Constant defined by user	This varies according to each producer's characteristics
Wine Reputation	(DELAY1 (Influence of Awards on Reputation, Awards Time) + DELAY1 (Influence of Marketing on Reputation, Marketing Time) + DELAY1 (Influence of Region of Origin on Reputation, Region of Origin Time) + DELAY1 (Influence of Wine Quality on Reputation, Wine Quality Time))/4	All the variables which influence Wine Reputation are defined by means of a Likert scale, in this context. For the sake of coherence, the variable concerned must also be defined in the same way. As such, it must be divided by the number (4) of variables concerned (the influence of awards, marketing, region of origin and of the quality of wine on reputation). It was considered that the effect of these four variables is not immediate, hence the introduction of a period (delay), so that the variables take effect
Wine Reputation Time	Constant defined by user	Delay (number of months required for the wine's reputation to produce an effect on its price)
Influence of Awards on Reputation	"% of Awards Influence" * Awards	This variable depends on the percentage of importance defined by the user regarding the influence of awards on the reputation of the wine, and of the value that is attributed to the variable Awards
Awards	Constant defined by user	Variable represented on a Likert scale (ranging from 1 to 5, where 1 represents a wine which has received no awards and 5 a wine which has received many awards)
% of Awards Influence	Constant defined by user	This varies according to the weight defined by each producer/specialist in the area regarding the importance of awards in wine reputation

(*continued*)

Table 1. (*continued*)

Variable	Equation	Description
Awards Time	Constant defined by user	Delay (number of months required for the awards won to generate an effect on the wine's reputation)
Influence of Marketing on Reputation	"% of Marketing Impact on Reputation" * Marketing Impact	This variable depends on the percentage of importance defined by the user for Marketing and its influence on wine reputation, as well as the value attributed to the variable Marketing Impact
Marketing Impact	Constant defined by user	Variable represented by means of a Likert scale (ranging from 1 to 5, where 1 represents very little marketing and 5 represents a great deal of marketing regarding the wine considered)
% of Marketing Impact on Reputation	Constant defined by user	This varies according to the weight defined by each producer/specialist in the area regarding the importance of marketing in wine reputation
Marketing Time	Constant defined by user	Delay (number of months required for marketing to impact on wine reputation)
Region of Origin Time	Constant defined by user	Delay (number of months required for the Region of Origin to impact on wine reputation)
Wine Quality Time	Constant defined by user	Delay (number of months required for wine quality to impact on its reputation)
Influence of Region of Origin on Reputation	"% of Influence of Region of Origin on Reputation" * Region of Origin	This variable depends on the percentage of importance defined by the user for Region of Origin to impact on wine reputation, and on the value assigned to the Region of Origin variable
Influence of Wine Quality on Reputation	Wine Quality * "% Of Influence of Wine Quality on Wine Reputation"	This variable depends on the percentage of importance defined by the user for wine quality to impact on wine reputation, and on the value assigned to the Wine Quality variable
% of Influence of Wine Quality on Wine Reputation	Constant defined by user	This varies according to the weight defined by each producer/specialist in the area, regarding the importance of wine quality in its reputation
% of Influence of Region of Origin on Reputation	Constant defined by user	This varies according to the weight defined by each producer/specialist in the area regarding the importance of Region of Origin in wine reputation
Region of Origin	Constant defined by user	Variable represented by means of a Likert scale (ranging from 1 to 5, where 1 means that the region of origin is rather undervalued and 5 that the region of origin is highly valued)

(*continued*)

Table 1. (*continued*)

Variable	Equation	Description
Wine Quality	(SMOOTH3 (Influence of Climate Conditions on Quality, Climate Conditions Time) + SMOOTH3 (Influence of Soil on Quality, Soil Time) + Influence of Harvest Year on Quality + SMOOTH3 (Influence of Innovation and Technology on Quality, Innovation Time) + Influence of Region of Origin on Quality)/5	All the variables which influence Wine Quality are defined by means of a Likert scale in this context. For the sake of coherence, the variable in question should also be defined in the same way; hence, it must be divided by the number (5) of variables concerned (climatic conditions, soil characteristics, harvest year, innovation and technology, and region of origin). One considered that the effect of the variables does not impact immediately on climatic conditions and soil, nor on innovation and technology; hence the introduction of a time period (delay), so that the variables can actually generate effect
% of influence of Region of Origin on Wine Quality	Constant defined by user	This varies according to the weight defined by each producer/specialist in the area regarding the importance of Region of Origin in wine quality
Influence of Region of Origin on Quality	"% of influence of Region of Origin on Wine Quality" * Region of Origin	This depends on the percentage of importance defined by the user regarding Region of Origin and its impact on wine quality; it also depends on the value assigned to the Region of Origin variable
% of Innovation and Technology Influence	Constant defined by user	This varies according to the weight defined by each producer/specialist in the area regarding the importance of Innovation and Technology in wine quality
Innovation and Technology	Constant defined by user	Variable represented by means of a Likert scale (ranging from 1 to 5, where 1 represents very little investment in innovation and technology and 5 represents little investment in this area)
Innovation Time	Constant defined by user	Delay (the number of months required for investment in innovation and technology to impact on wine quality
Influence of Innovation and Technology on Quality	"% of Innovation and Technology Influence" * Innovation and Technology	This depends on the percentage of importance defined by the user regarding Innovation and Technology and its impact on wine quality; it also depends on the value assigned to the Innovation and Technology variable

(*continued*)

Table 1. (*continued*)

Variable	Equation	Description
% of Climate Condition Influence	Constant defined by user	This varies according to the weight defined by each producer/specialist in the area regarding the importance of climatic conditions in wine quality
Climate Conditions	Constant defined by user	Variable represented by means of a Likert scale (ranging from 1 to 5, where 1 means that climatic conditions for the year under analysis were very unfavorable and 5 means they were very favorable)
Influence of Climate Conditions on Quality	Climate Conditions * "% of Climate Conditions Influence"	This depends on the percentage of importance defined by the user regarding climatic conditions and their impact on wine quality; it also depends on the value assigned to the Climate Conditions Influence variable
Climate Condition Time	Constant defined by user	Delay (the number of months required for climatic conditions to impact on wine quality)
% of Harvest Year Influence	Constant defined by user	This varies according to the weight defined by each producer/specialist in the area regarding the importance of the harvest year in wine quality
Harvest Year	Constant defined by user	Variable represented by means of a Likert scale (ranging from 1 to 5, where 1 means that the harvest year was very bad and 5 means it was very good)
Influence of Harvest Year on Quality	% of Harvest Year Influence * Harvest Year	This depends on the percentage of importance defined by the user regarding the harvest year and its impact on wine quality; it also depends on the value assigned to the Harvest Year variable
% of Soil Influence	Constant defined by user	This varies according to the weight defined by each producer/specialist in the area regarding the importance of soil characteristics in wine quality
Soil	Constant defined by user	Variable represented by means of a *Likert scale* (ranging from 1 to 5, where 1 means that the soil characteristics are very bad and 5 means they are very good)

(*continued*)

Table 1. *(continued)*

Variable	Equation	Description
Influence of Soil on Quality	% of Soil Influence * Soil	This depends on the percentage of importance defined by the user regarding soil characteristics and their impact on wine quality; it also depends on the value assigned to the Soil variable
Soil Time	Constant defined by user	Delay (the number of months required for soil characteristics to impact on wine quality)

Table 2. Detail of model stocks.

Stock	Equation	Comments
Wine stocks (l)	"Wine Production (l)" - "Wine Sales (l)"	Each company's stock consists of unsold production

In the decision support system developed, one can parametrize all the variables pointed out as adjustable by the user in Table 1. In addition, and in order to make the interpretation of the model's results clearer, the tool also provides the graphs of the variables which the user intends to analyze (Fig. 2).

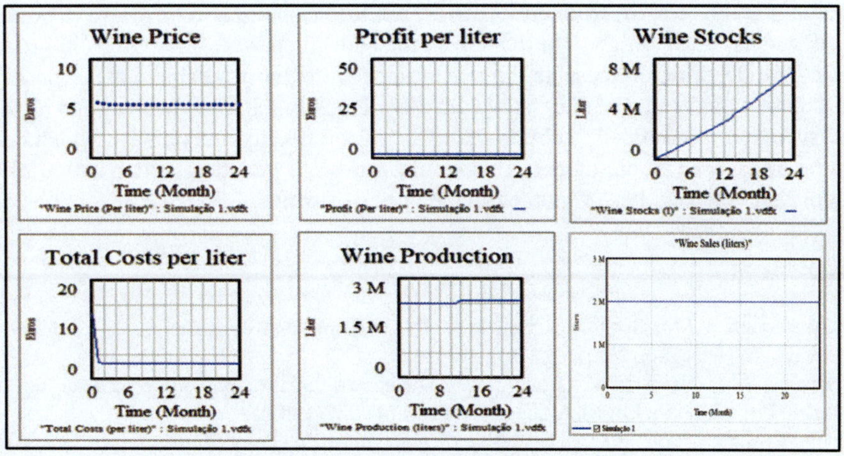

Fig. 2. Outputs provided by the model.

4 Conclusions and Future Work

The primary objective of this study was to analyze and develop a decision support system, based on system dynamics, to help producers and managers operating in the wine sector.

Intuitive, quick and generic, this assistance will enable professionals in the sector to define strategies which act upon and respond to the variation of the many factors which influence the price, production and quality of wine. Users can thus test the impact of their decisions and strategies on the price and profits earned. The belief that the model would be useful to producers and managers in the wine sector led to interviews with people operating in the field, so that they, too, could make a practical contribution, and help to build the support tool. Still in the context of system dynamics and Vensim software, and upon the conclusion of this study, one saw the subsequent development of a model which combines the main factors that influence the price of wines. Accordingly, after each simulation procedure, the user is able to analyze the progress and behavior of the model by studying the graphs and tables furnished by the tool. The perspective presented by the tool provides assistance in the selection of an appropriate strategy to achieve the objectives established by the respective company. The construction of this tool, as well as the entire study prior to its development, intends to provide users with graphic and visual utilities, which will allow them to choose a strategy to achieve the profits they aim to make, either by reducing costs or by increasing their profit margin. Additionally, these actions can be combined with the reputation of the wine in question amongst its consumers. Some limitations can be pointed out in this study, namely the fact that the model's results are influenced by the amount of knowledge the user possesses, both of the market and of the product's consumers. Since the definition of some parameters was undertaken on the basis of percentage, this can confer some subjectivity on the model. Furthermore, the model does not consider the evolution in performance of beverages which compete with wine, such as beer, sparkling wines and liqueurs. Finally, the model does not include cases in which wine-producing companies resort to the purchase of raw material (namely grapes) from third parties. Attention should also be drawn to possible future studies. These might aim to address the abovementioned limitations in order to obtain a model which represents the real system in the most reliable manner possible. This solution could also be integrated with other industry 4.0 tools: sensors and big data analysis may be combined to obtain real data on soil quality and climate, which could then be transposed to the model. This would provide a quantitative translation of the relationship between these variables and the price of wine.

References

1. Swaminathan, J.M., Smith, S.F., Sadeh, N.M.: Modeling supply chain dynamics: a multiagent approach. Decis. Sci. **29**(3), 607–632 (1998)
2. Disney, S.M., Naim, M.M., Towill, D.R.: Dynamic simulation modelling for lean logistics. Int. J. Phys. Distrib. Logist. Manag. **27**(3/4), 174–196 (1997)
3. White, K.P., Ingalls, R.G.: The basics of simulation. In: Proceedings of the 2017 Winter Simulation Conference, Dezembro 2017, pp. 1–15 (2017)
4. Ferreira, L.P., Gómez, E.A., Lourido, G.C.P., Quintas, J.D., Tjahjono, B.: Analysis and optimisation of a network of closed-loop automobile assembly line using simulation. Int. J. Adv. Manuf. Technol. **59**(1–4), 351–366 (2012)
5. Ferreira, L.P., Gómez, E.A., Lourido, G.P., Tjahjono, B.: Modeling a complex production line using virtual cells. In: Ao, S.I., Gelman, L. (eds.) Electrical Engineering and Intelligent Systems, LNEE, vol. 130, pp. 361–373. Springer, New York (2013). https://doi.org/10.1007/978-1-4614-2317-1_30

6. Santos, P., Ferreira, L.P., Dieguez, T., Tjahjono, B.: Supporting strategic management decisions at a casino using system dynamics. In: Juan, M.G. (ed.) Modeling and Simulating Business Dynamics. Selected Papers on System Dynamics Collection, pp.156–170. ISBN: 978-1686-997-556 KDP Publishers (2019)
7. Currie, D.J., Smith, C., Jagals, P.: The application of system dynamics modelling to environmental health decision-making and policy - a scoping review. BMC Public Health **18**(1), 1–11 (2018)
8. Eren Şenaras, A.: Structure and behavior in system dynamics: a case study in logistic. J. Bus. Res-Turk **9**(4), 321–340 (2017)
9. Hjorth, P., Bagheri, A.: Navigating towards sustainable development: a system dynamics approach. Futures **38**(1), 74–92 (2005)
10. Al-Khatib, I.A., Eleyan, D., Garfield, J.: A system dynamics approach for hospital waste management in a city in a developing country: the case of Nablus. Palestine Environ. Monit. Assess. **188**(9), 503 (2016)
11. Zambon, I., Cecchini, M., Egidi, G., Saporito, M.G., Colantoni, A.: Revolution 4.0: Industry vs. agriculture in a future development for SMEs. Processes **7**(1), 1–36 (2019)
12. Ozdogan, B., Gacar, A., Aktas, H.: Digital agriculture practices in the context of agriculture 4.0. J. Econ. Fin. Account. **4**(2), 184–191 (2017)
13. Vaidya, S., Ambad, P., Bhosle, S.: Industry 4.0 - a glimpse. Procedia Manuf. **20**, 233–238 (2018)
14. Chiu, Y.: Water consumption impact under climate and population scenarios. In: 35th International Conference of the System Dynamics Society and 60th Anniversary of System Dynamics Celebration. Cambridge, MA, USA Conference, pp. 1–11 (2017)
15. Murshed, S.M., Werner, U.: A system dynamics approach to assess economic impacts of extreme winter storms in forestry. In: Proceedings of the 34th International Conference of the System Dynamics Society Delft, Netherlands, pp. 17–21 (July 2016)
16. Ardakani, D.A., Eslamieh, V.: Designing the green new product development model: system dynamics approach (case study: small and medium food industries in Iran). In: Proceedings of the 34th International Conference of the System Dynamics Society*Delft, Netherlands, pp. 17–21, 3(2), 54–67 (2016)
17. Balkan, B.A., Meral, S.: Olive oil industry dynamics: the case of Turkey. In: Proceedings of the 35th International Conference of the System Dynamics Society, pp. 1–26 (2017)
18. Zamora, M.C., Acosta, C.M.M., Goldner, M.C.: The wine chain in Argentina: the influence of production-consumption dynamics. In: Proceedings of the 21st International Conference of the System Dynamics Society, pp. 20–24. Nova York, EUA (July, 2003)
19. Bivona, E.: Applying System dynamics to the management of a small firm a case study in the wine industry. In: 16th Conference on International System Dynamics Conference, Quebec (1998)
20. Chester, J., Cooper, J.F.S., Cavana, R.Y.: A dynamic model for the New Zealand wine industry. In: Proceedings of the 29th Annual Conference of the Operational Research Society of New Zealand, 23–24 August 1993, University of Auckland, New Zealand (1993)
21. Sá, J., Ferreira, L.P., Dieguez, T., Sá, J.C., Silva, F.J.G.: Role of the Industry 4.0 in the Wine Production and Enotourism Sectors. In: de Carvalho, J.V., Rocha, Á., Liberato, P., Peña, A. (eds.) ICOTTS 2020. SIST, vol. 208, pp. 171–180. Springer, Singapore (2021). https://doi.org/10.1007/978-981-33-4256-9_16
22. Vaughan, C., Buja, L., Kruczkiewicz, A., Goddard, L.: Identifying research priorities to advance climate services. Clim. Serv **4**, 65–74 (2016)
23. Charters, S.: Wine and Society: The Cultural and Social Context of a Drink. ISBN-13: 978-0750666350 (2006)

24. Graça, A.: The MED-GOLD project: advanced user-centric climate services for higher resilience and profitability in the grape and wine sector. In: BIO Web of Conferences, vol. 12, p. 1005. EDP Sciences (2019)
25. Bernetti, I., Menghini, S., Marinelli, N., Sacchelli, S., Sottini, V.A.: Assessment of climate change impact on viticulture: economic evaluations and adaptation strategies analysis for the Tuscan wine sector. Wine Econ. Policy **1**(1), 73–86 (2012)
26. Goncharuk, A.G.: The challenges of wine business in research. J. Appl. Manag. Invest. **6**(4), 253–259 (2017)
27. Jones, G.V., Alves, F.: Impact of climate change on wine production: a global overview and regional assessment in the Douro valley of Portugal. Int. J. Glob. Warm. **4**(3–4), 383–406 (2012)
28. Mullins, M.D., Bouquet, A., Williams, L.E.: Biology of the Grapevine. Cambridge University Press, Cambridge (1992)
29. Bodin, F., Morlat, R.: Characterization of viticultural terroirs using a simple field model based on soil depth I. Validation of the water supply regime, phenology and vine vigour, in the Anjou vineyard (France). Plant Soil **281**(1–2), 37–54 (2006)
30. Carey, V.A.: Spatial characterisation of natural terroir units for viticulture in the Bottelaryberg-Simonsberg-Helderberg winegrowing area. Doctoral dissertation, Stellenbosch University, Stellenbosch (2001)
31. Lopo, M., Teixeira dos Santos, C.A., Páscoa, R.N.M.J., Graça, A.R., Lopes, J.A.: Near infrared spectroscopy as a tool for intensive mapping of vineyards soil. Precision Agric. **19**(3), 445–462 (2017). https://doi.org/10.1007/s11119-017-9529-2
32. Grunert, K.G.: Current issues in the understanding of consumer food choice. Trends Food Sci. Technol. **13**(8), 275–285 (2002)
33. Lockshin, L., Jarvis, W., d'Hauteville, F., Perrouty, J.P.: Using simulations from discrete choice experiments to measure consumer sensitivity to brand, region, price, and awards in wine choice. Food Qual. Prefer. **17**(3–4), 166–178 (2006)
34. Oczkowski, E.: Hedonic wine price functions and measurement error. Econ. Record **77**(239), 374–382 (2001)
35. Santos, J.F., Ribeiro, J.C.: Product attribute saliency and region of origin: some empirical evidence from Portugal product. Paper presented at the 99th seminar of the EAAE 'The Future of Rural Europe in the Global Agri-Food system', Copenhagen, Denmark (2005)
36. Ribeiro, J.C., Santos, J.D.F.: Portuguese quality wine and the region-of-origin effect: consumers and retailers' perceptions. Núcleo de investigação em políticas Económicas Universidade do Minho (2008)
37. Bertram, D.: Likert scales are the meaning of life (2017). pointcare.matf.bg.ac.rs/~kristina/topic-dane-likert.pdf. Acedido em jan. 01 2020
38. Duggan, J.: An introduction to system dynamics. In: System Dynamics Modeling with R. LNSN, pp. 1–24. Springer, Cham (2016). https://doi.org/10.1007/978-3-319-34043-2_1

Probability Laws for Nearly Gaussian Random Variables and Application

Rui Gonçalves(✉)

Faculdade de Engenharia do Porto, Porto, Portugal
rjasg@fe.up.pt

Abstract. In an earlier work we described and applied a methodology to find an adequate distribution for Nearly Gaussian (NG) random variables. In this work, we compare two different methods, $m1$ and $m2$ to estimate a power transform parameter for NG random variables. The $m1$ method is heuristic and based on sample kurtosis. Herein, we describe and apply it using a new reduced data set. The second method $m2$ is based on the maximization of a pseudo-log-likelihood function. As an application, we compare the performance of each method using high power statistical tests for the null hypothesis of normality. The data we use are the daily errors in the forecasts of maximum and minimum temperatures in the city of Porto. We show that the high kurtosis of the original data is due to high correlation among data. We also found that although consistent with normality the data is better fitted by distributions of the power normal (PN) family than by the normal distribution. Regarding the comparison of the two parameter estimation methods we found that the $m1$ provides higher p-values for the observed statistics tests except for the Shapiro-Wilk test.

Keywords: Nearly Gaussian · Power normal · power transform · Box-Cox · Temperature distribution

1 Introduction

In [1] NG random variables are characterized as showing approximate symmetry, a bell shaped density curve and a kurtosis close to the gaussian (gaussian = 3). This kind of variables may be the sum of independent and identically distributed (i.i.d.) random variables but in a number not enough for the full consequences of the central limit theorem. In [1] is proposed a simple heuristic method that we call $m1$ to find a suitable distribution for NG random variables. It consists in comparing the kurtosis of the data with that of a random variable X^c where X is normally distributed. If the sample kurtosis is close to the one of X^c then the distribution of this variable is a good model for the data and the transformation for normality consists in raising data to the exponent $1/c$. Note that data should be centered before this transformation. In Sect. 2, we present some results about the moments of a random variable with a distribution obtained by a power transformation of the normal on which the method $m1$ is based. The Box-Cox (BC)

J. Machado et al. (Eds.): icieng 2021, LNME, pp. 385–395, 2022.
https://doi.org/10.1007/978-3-030-78170-5_33

transformation [2] of a positive random variable is defined by $Y = (X^\lambda - 1)/\lambda$ when $\lambda \neq 0$ and $Y = \log(X)$ if $\lambda = 0$ where λ is the transformation parameter. The use of BC transformation is intended to obtain data with a distribution closer to the gaussian. The positiveness requirement of the BC transformation, $X > 0$ implies that $Y > -1/\lambda$ which means that the transformed data is left-truncated normal and not normal. But, when $P(Y < -1/\lambda)$ is very small then data can be considered gaussian for all purposes. If that is not the case, then we can add a constant $(X + d)$ such that data becomes positive, see [3] and [4]. In Sect. 3 we present in detail a method that we call $m2$, based on maximum pseudo-log-likelihood estimation to find a parameter for the BC transformation of data into normality. In Sect. 4, we study the daily error in maximum and minimum temperature forecasts data. This kind of data has been considered to be a candidate for NG random variables [1,5]. We statistically analyze the data and we eliminate observations to reduce the effect of correlation. The obtained reduced data is found to be consistent with the normal distribution. Nevertheless, we apply methods $m1$ and $m2$ to see if a distribution of the power law family could provide a better fit than the normal. In the case of the $m2$ method we use several values of d to evaluate the influence of this constant on the results. We use the Shapiro-Wilk (SW), Anderson-Darling (AD) and the Lilliefors (Lf) goodness-of-fit tests on the power transformed data sets and we comment the results. In Sect. 5 we present the conclusions of the study and the final remarks.

2 Heuristic Method Based on Kurtosis

In this section we follow [1]. Consider the following proposition:

Proposition 1. *Let c be a real number. If $Y = X^c$ is gaussian, $Y \sim N(\mu, \sigma^2)$, then the probability density function (pdf) of the power transformation of a Gaussian variable, $X = Y^{1/c}$ is*

$$f_X(x) = \frac{1}{\sqrt{2\pi}\sigma} c|x^{c-1}| \exp\left[-\frac{1}{2}\left(\frac{x^c - \mu}{\sigma}\right)^2\right]. \tag{1}$$

This transformation is related to the BC transformation. In Fig. 1 is presented, as an example, the graphic of f for $c = 0.7$.

If X is a Gaussian random variable with parameters $\mu = 0$ and variance σ^2, then for $c > 0$ we have

$$E(X^c) = \int_{-\infty}^{\infty} \frac{x^c}{\sqrt{2\pi}\sigma} e^{-\frac{x^2}{2\sigma^2}} dx = \frac{2^{c/2}}{2\sqrt{\pi}} [1 + (-1)^c] \Gamma\left(\frac{c}{2} + \frac{1}{2}\right), \tag{2}$$

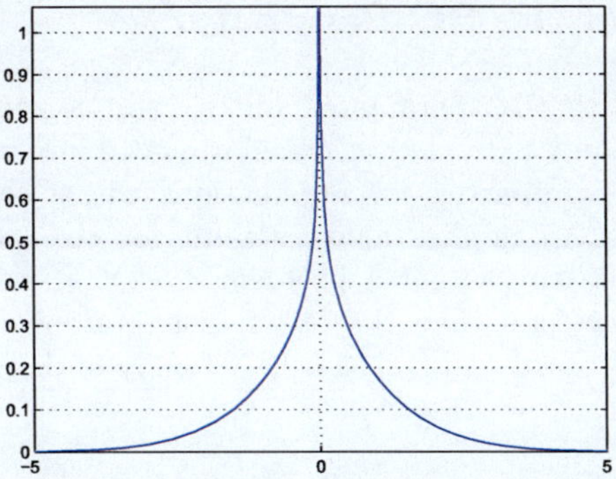

Fig. 1. Pdf of a Gaussian variable raised to the power 0.7.

Table 1. Kurtosis of the random variable $Y = X^c$ where $X \sim N(0,1)$ for a few values of c.

c	$\beta_2(c)$	c	$\beta_2(c)$
7/9	2.237	17/15	3.58
9/11	2.358	15/13	3.68
11/13	2.447	11/15	2.11
13/15	2.514	19/21	2.643
15/17	2.566	23/25	2.697
17/19	2.6	29/31	2.753
13/11	3.828	31/33	2.767
23/19	3.979	21/23	2.672
11/9	4.042	35/37	2.79
9/7	4.404	43/41	3.2056

where Γ is the gamma function. The kurtosis coefficient (β_2) is given by

$$\beta_2(c) = \frac{\int_{-\infty}^{\infty} \frac{1}{\sqrt{2\pi}} x^{4c} e^{-x^2/2} dx}{\left(\int_{-\infty}^{\infty} \frac{1}{\sqrt{2\pi}} x^{2c} e^{-x^2/2} dx\right)^2} = \sqrt{\pi} \frac{\Gamma(2c + \frac{1}{2})}{\Gamma^2(c + \frac{1}{2})}. \tag{3}$$

In Table 1 we present the value of $\beta_2(c)$ for a few values of $c = (2k+1)/(2j+1)$ where $k, j \in \{0, 1, \ldots\}$.

The adequate c parameter is found by comparing data kurtosis to the $\beta_2(c)$. Don't forget that the identification of the c exponent requires that the variable has mean zero and skewness close to zero. If such a c exists then values of the sample elevated to $1/c$ are approximately Gaussian distributed.

3 Estimation of the BC Parameter

In the following λ is the BC parameter and we will consider that $0 < \lambda < 1$. The BC transformation is intended to transform data into normality and consequently with suitable properties for applications where normality is required. Positiveness of data (X) implies a lower limit for the transformed data, Y. But as we said earlier, this is not a problem if $P(Y < -1/\lambda)$ is small. If the transformation is successful then the exact distribution of X is the so called (PN) distribution, see [6–9] and [10]. The PN family of distributions has its pdf given by

$$f_X(x) = \frac{x^{\lambda-1}}{\sigma A(\lambda,\mu,\sigma)}\phi\left(\frac{x^{(\lambda)}-\mu}{\sigma}\right) \tag{4}$$

where $x > 0$ and $x^{(\lambda)} = \frac{x^{\lambda}-1}{\lambda}$. We define

$$A(\lambda,\mu,\sigma) = \begin{cases} \Phi(k) & ,\lambda > 0 \\ 1 & ,\lambda = 0 \\ \Phi(-k) & ,\lambda < 0 \end{cases} \tag{5}$$

where k is the standardized truncation point of the truncated normal distribution, $k = (\lambda\mu + 1)/(\lambda\sigma)$, ϕ is the standard normal pdf and Φ is the standard normal cumulative distribution function (cdf). Note that $A(k)$ works as a normalizing constant for f. Note that the for $\lambda = 0$ the distribution associated is the log-normal that we don't address in this work. Consider $X_1, X_2, ..., X_n$, as a collection of n independent and identically distributed (i.i.d.) random variables with PN distribution. The log-likelihood function is given by

$$l = \ln(L(x;\lambda,\mu,\sigma)) = -\frac{n}{2}\ln(2\pi) - \frac{n}{2}\ln(\sigma^2) - \frac{1}{2}\sum_{i=1}^{n}\left(\frac{x_i^{(\lambda)}-\mu}{\sigma}\right)^2 + (\lambda-1)\sum_{i=1}^{n}\ln(x_i) - n\ln(A(k)). \tag{6}$$

As we stated earlier, if μ is large then $A(k) \approx 1$ so we will remove $A(k)$ from the log likelihood[1]. So we have

$$l = \ln(L(x;\lambda,\mu,\sigma)) = -\frac{n}{2}\ln(2\pi) - \frac{n}{2}\ln(\sigma^2) - \frac{1}{2}\sum_{i=1}^{n}\left(\frac{x_i^{(\lambda)}-\mu}{\sigma}\right)^2 + (\lambda-1)\sum_{i=1}^{n}\ln(x_i). \tag{7}$$

---

[1] See [7] for an analysis with $A(k) \neq 1$.

The values of λ, μ and σ that maximize the log-likelihood are found by solving the system composed by the derivative of that function with respect to each parameter equated to zero. The system is non-linear and solutions can only be obtained numerically. Usually this systems have more than one solution and to obtain the one we want we need to choose adequate initial values for the parameters. To overcame this difficulty we can consider a simplification of the log-likelihood function. For example, assuming a fixed λ, we can solve the system in order to the parameters μ and σ. That gives us the estimators of that parameters. Then we go back to the log-likelihood and replace the parameters μ and σ by their estimators. That was the approach in [11] that we describe next. So, fixing λ we consider only the derivatives with respect to μ and σ. Equating them to zero we have

$$l = \begin{cases} -\frac{n}{2}\ln(2\pi\sigma^2) - \frac{1}{2\sigma^2}\sum_{i=1}^{n}\left(\frac{x_i^\lambda - 1}{\lambda} - \mu\right)^2 + (\lambda-1)\sum_{i=1}^{n}\ln(x_i) = 0 & ,\lambda \neq 0 \\ -\frac{n}{2}\ln(2\pi\sigma^2) - \frac{1}{2\sigma^2}\sum_{i=1}^{n}(\ln(x_i) - \mu)^2 - \sum_{i=1}^{n}\ln(x_i) = 0 & ,\lambda = 0. \end{cases} \tag{8}$$

Considering the case $\lambda \neq 0$ and differentiating the system 8 the solutions for the resulting system are

$$\begin{cases} \frac{\partial l}{\partial \mu} = 0 \\ \frac{\partial l}{\partial \sigma^2} = 0 \end{cases} \Longleftrightarrow \begin{cases} \hat{\mu} = \frac{1}{n}\sum_{i=1}^{n}\frac{x_i^\lambda - 1}{\lambda} \\ \hat{\sigma^2} = \frac{1}{n}\sum_{i=1}^{n}\left(\frac{x_i^\lambda - 1}{\lambda} - \frac{1}{n}\left(\sum_{i=1}^{n}\frac{x_i^\lambda - 1}{\lambda}\right)\right)^2 \end{cases}.$$

Note that we fixed the λ parameter so the estimators of μ and σ are only the best for a specific λ. The estimators of the parameters are used to build a pseudo-log-likelihood function (l^*) that is easier to handle. The term pseudo is justified by the fact that the function is not the log-likelihood but an approximation of it instead. Using an initial value for λ, 1 for example, then we can set the Newton-Raphson recurrence scheme to find the value of λ that maximizes the pseudo-log-likelihood function. So, replacing the parameters by the estimators, we have

$$l^* = -\frac{n}{2}\left[\ln(2\pi) + \ln\left\{\frac{1}{n}\sum_{i=1}^{n}\left(\frac{x_i^\lambda - 1}{\lambda} - \frac{1}{n}\sum_{j=1}^{n}\left(\frac{x_j^\lambda - 1}{\lambda}\right)\right)^2\right\} + 1\right] + (\lambda - 1)\sum_{i=1}^{n}\ln(x_i) \tag{9}$$

Let's call $\hat{\lambda}_L$ to an initial estimate of λ parameter and consider the variables P_i and the mean of all P_i, $i = 1, \cdots, n$, $\bar{P}$ defined by

$$P_i = \left(\frac{x_i^{\hat{\lambda}_L} - 1}{\hat{\lambda}_L}\right)\ \bar{P} = \frac{1}{n}\sum_{j=1}^{n}\left(\frac{x_j^{\hat{\lambda}_L} - 1}{\hat{\lambda}_L}\right). \tag{10}$$

Consider also the variables Q_i and the mean of all Q_i, $i = 1, \cdots, n$,$\bar{Q}$ defined by

$$Q_i = \left(\frac{x_i^{\hat{\lambda}_L}(\ln(x_i))x_i^{\hat{\lambda}_L} - x_i^{\hat{\lambda}_L} + 1}{\hat{\lambda}_L^2}\right),\ \bar{Q} = \frac{1}{n}\sum_{j=1}^{n}\left(\frac{x_j^{\hat{\lambda}_L}(\ln(x_j))x_j^{\hat{\lambda}_L} - x_j^{\hat{\lambda}_L} + 1}{\hat{\lambda}_L^2}\right). \tag{11}$$

Note that Q_i is the partial derivative of P_i with respect to λ. The derivative of l^* with respect to λ, $\frac{\partial l^*}{\partial \lambda}|_{\lambda=\hat{\lambda}_L}$ to which we call $h(\hat{\lambda}_L)$ equated to zero can be written as

$$h(\hat{\lambda}_L) = 0 \equiv -n\frac{\sum_{i=1}^{n}(P_i - \bar{P})(Q_i - \bar{Q})}{\sum_{i=1}^{n}(P_i - \bar{P})^2} + \sum_{i=1}^{n}\ln(x_i) = 0 \tag{12}$$

The Newton-Raphson recurrence scheme is the following,

$$\hat{\lambda}^{(k+1)} = \hat{\lambda}^{(k)} - \frac{h(\hat{\lambda}_L^{(k)})}{h'(\hat{\lambda}_L^{(k)})} \tag{13}$$

To obtain a condensed form to represent $h'(\lambda)$, consider first the derivative of Q_i with respect to λ that is given by

$$\frac{\partial Q_i}{\partial \lambda} = \frac{\lambda^3(\ln(x_i))^2 x_i^{\lambda} - 2\lambda^2\ln(x_i)x_i^{\lambda} + 2\lambda x_i^{\lambda} - 2\lambda}{\lambda^4}. \tag{14}$$

Let's call R_i to this derivative. The mean of R_i for $i = 1, \cdots, n$ is given by

$$\bar{R} = \frac{1}{n}\sum_{j=1}^{n}\left(\frac{\lambda^3(\ln(x_j))^2 x_j^{\lambda} - 2\lambda^2\ln(x_j)x_j^{\lambda} + 2\lambda x_j^{\lambda} - 2\lambda}{\lambda^4}\right). \tag{15}$$

Using R_i, the derivative of h with respect to λ is as follows,

$$\begin{aligned} h'(\hat{\lambda}) = -n & \left[\frac{\sum_{i=1}^{n}(P_i - \bar{P})(R_i - \bar{R}) + \sum_{i=1}^{n}(Q_i - \bar{Q})^2}{\sum_{i=1}^{n}\left(P_i - \bar{P}\right)^2}\right. \\ & \left. -2\frac{\left\{\sum_{i=1}^{n}(P_i - \bar{P})(Q_i - \bar{Q})\right\}^2}{\left\{\sum_{i=1}^{n}(P_i - \bar{P})^2\right\}^2}\right] \end{aligned} \tag{16}$$

Note that variance of an unbiased estimator $\hat{\theta}$ of θ is, in certain regularity conditions, bounded below by the reciprocal of the Fisher information

$I(\theta) = -nE\left[\frac{\partial^2 l(x;\theta)}{\partial\theta^2}\right]$. The Fisher information is the expected value of the second partial derivative of the likelihood function with respect to the parameter multiplied by -n, which we called h'. So we can say that the variance of $\hat{\lambda}$ is approximately $-1/h'(\hat{\lambda}_L)$.

4 Application: Error in Temperature Forecasts

In this section we present an analysis of the error in maximum and minimum temperature forecasts data for the city of Porto, Portugal during the year 2011. We start with the errors in maximum temperatures. We define $X = T_F - T_O$ where T_F is the forecast and T_O the observed maximum temperatures on the same day. We call Y to the error in the minimum temperatures forecast, $Y = t_F - t_O$ where t_F is the forecasted minimum temperature and t_O is the minimum temperature observed on the same day. The size of our data set is 347 because some days are missing. We need to calculate some statistics of the data. So taking a sample as $x_1, ...x_n$ drawn from a population distributed with mean μ and standard deviation σ, the usual statistics and corresponding notation we use are

$$\bar{x} = \sum_{i=1}^{n} \frac{x_i}{n}, \tag{17}$$

which is the sample mean. The statistic,

$$\hat{\mu}_k = \sum_{i=1}^{n} \frac{(x_i - \bar{x})^k}{n-1}, \tag{18}$$

for $k = 1, 2, ...$, is the estimated k-th order central moment. The sample standard deviation $\hat{\sigma} = s_x = \sqrt{\hat{\mu}_2}$. The sample skewness coefficient is

$$b_1 = \frac{\hat{\mu}_3}{s_x^3}, \tag{19}$$

and the sample kurtosis is

$$b_2 = \frac{\hat{\mu}_4}{s_x^4} \tag{20}$$

The x data sample yields $b_1 = 0.035$ meaning that x data is close to symmetry. We also get $b_2 = 3.54$ which is a kurtosis higher than that of the normal distribution. As for y data we have $b_1 = 0.038$ and $b_2 = 2.84$. We can say that the y data set is rather symmetrical and its kurtosis is closer to the normal curve than the x data set. We can see the corresponding histograms in Fig. 2 (Fig. 3).

We must note that the observations, being daily, are correlated and that can distort data so we have to eliminate observations for a more realistic analysis of the temperature prediction process, see [1,5]. The autocorrelation function (ACF), ρ_k is a measure of temporal correlation often use in time series analysis. It can be used to evaluate the intensity of linear correlation in a time series

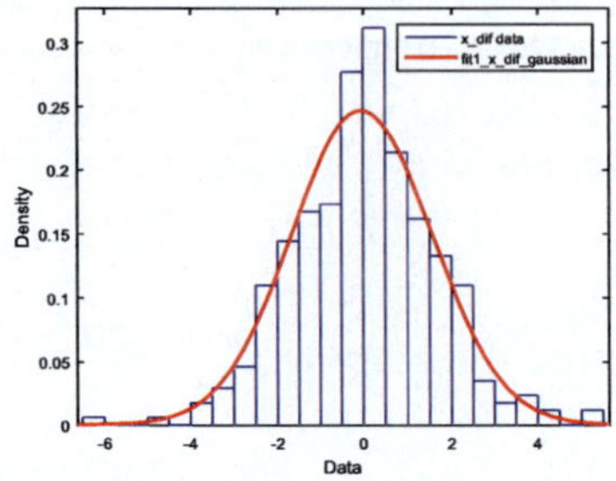

Fig. 2. Histogram and gaussian fit for the X data.

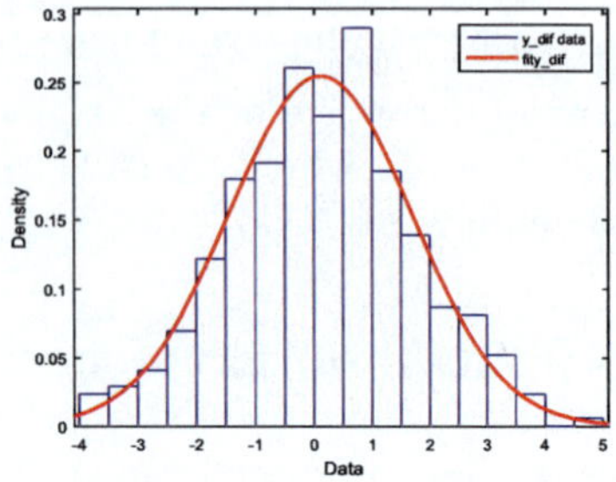

Fig. 3. Histogram and gaussian fit for the Y data.

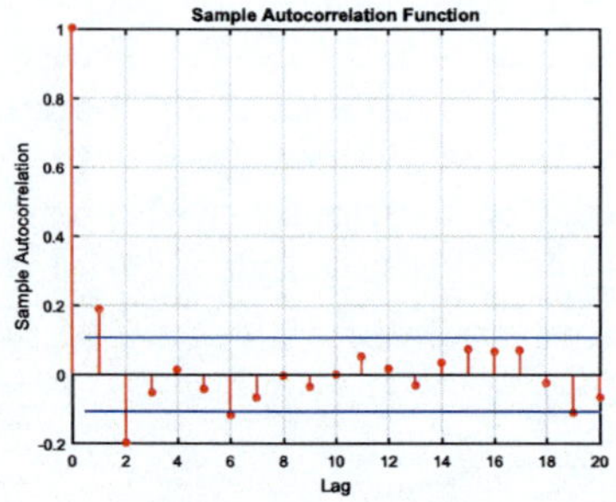

Fig. 4. Sample ACF for the X data set.

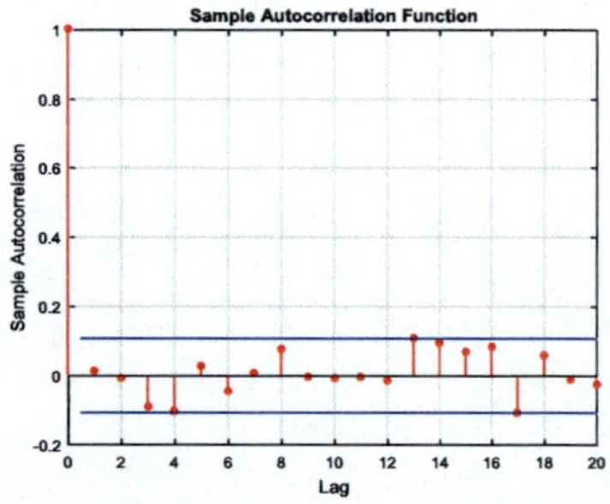

Fig. 5. Sample ACF for the Y data set.

because it measures the linear correlation between observations separated by k time instants. The ACF for white noise which is a sequence of independent and identically distributed (i.i.d.) normal random variables with mean 0 and variance σ^2 is zero. That is the ACF we seek for our data sets. In practice, we look to the last k value for which the sample ACF $\hat{\rho}_k$ is inside the 95% confidence interval for a random variable drawn from a gaussian distribution of mean zero and unit variance. If $k = m$ we will only consider one observation in each m.

Looking at the sample ACF of the X and Y data sets in Figs. 4 and 5, we see that for the first, ρ_2 is close but not below the lower 95% limit and because of that in [5] we used only one observation in each two. But herein we are going to eliminate the second observation so we will use only one observation in each 3 and consequently the data will be even less correlated. We call this data set as the X reduced data set. We can see that the Y data is less correlated so we would not need to eliminate more observations but anyway we are also going to keep one in each 3 observations. This data set will be referred as the Y reduced data set. In Figs. 6 and 7 we show the histograms of both X and Y reduced data sets.

The skewness of the X reduced data set ($\hat{b}_1 = 0.22$) is higher that of the X data so the data reduction caused a small deviation from symmetry. The kurtosis

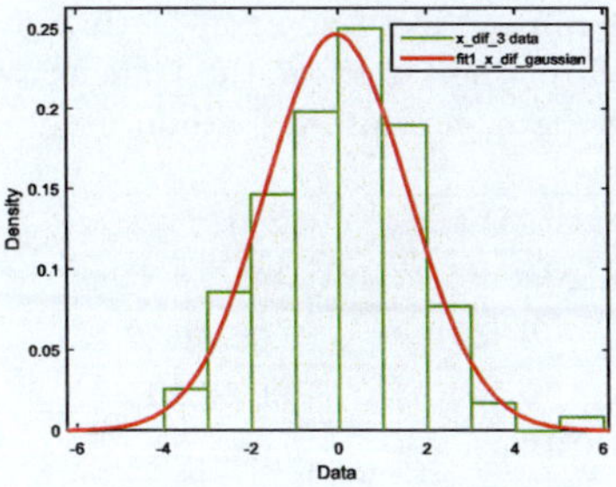

Fig. 6. Histogram and gaussian fit for the X reduced data.

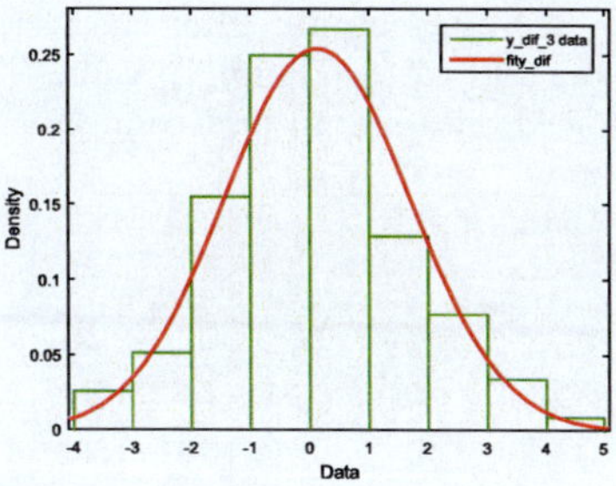

Fig. 7. Histogram and gaussian fit for the Y reduced data.

Table 2. Values obtained for the tests statistics Shapiro-Wilk, Anderson-Darling and Lilliefors performed on the X reduced data set and corresponding p-values.

	SW	AD	Lf
Statistic	0.99	0.44	0.063
p-value	0.24	0.29	0.30

is $\hat{b}_2 = 3.21$ which is closer to 3 and consequently to the gaussian curve. For the Y reduced data the skewness is $\hat{b}_1 = 0.26$ which is higher than that of X and the kurtosis is $\hat{b}_2 = 3.13$ which is closer to the normal curve than the Y data. We can say that the excessive observations at the center of the X data when compared to the normal were due to the excessive correlation.

In Table 2 we present the observed statistics values and the corresponding p-values of tests SW, AD and Lf performed on the X reduced data set to test the null hypothesis of normality. We can see that there is no statistical evidence to say that the X reduced data sets is not normal. The results for the Y reduced data show that the p-values of the observed statistic values are already very high so there is no reason to consider transformations. The question that we may ask is whether a power transformation of a gaussian random variable can provide a better fit to the X reduced set than the normal. Taking that in mind, we apply the $m1$ method to find an exponent to transform the X reduced data set. Using Table 1 we see that a random variable elevated to the exponent $43/41$ has a kurtosis $(\hat{b}_2)$ close to the sample kurtosis calculated from the X reduced data. Let us call X_{m1} to the X reduced data set transformed with this exponent. The method m_2 requires positiveness of data and to achieve that it is proposed in [4] to add a constant d to data. This process is the so called shift power transformation, [2,12]. To the constant to be added to data is only required that the summed data becomes positive. In [12] is pointed out that the likelihood function of the shifted data maybe poorly behaved often having no local maximum as shown in [13]. Taking that in mind we will use different constants. The minimum value for d must be $> -min(X)$ and there is not an indication for the upper limit. To see if the results may have a dependence on d we decided to use several values

Table 3. Values obtained for the tests statistics Shapiro-Wilk, Anderson-Darling and Lilliefors and corresponding p-values. The data used for the tests was the transformed X reduced data set, in first column is mentioned the transformation method used.

Data set	d	$\hat{\lambda}$	$\hat{\mu}$	$\hat{\sigma}^2$	$\hat{b}_2$	SW test		AD test		Lf test	
						stat.	p-value	stat.	p-value	stat.	p-value
X_{m1}	–	$\frac{43}{41}$	0	2.35	2.99	0.99	0.41	0.34	0.41	0.061	0.37
X_{m2}	d_{min}	0.82	2.32	1.64	3.08	0.99	0.40	0.41	0.34	0.072	0.15
X_{m2}	5	0.77	3.16	1.22	3.03	0.99	0.40	0.51	0.2	0.083	0.17
X_{m2}	10	0.54	4.56	0.31	2.99	0.99	0.55	0.50	0.20	0.069	0.19
X_{m2}	15	0.31	4.27	0.062	2.97	0.99	0.55	0.50	0.20	0.069	0.19
X_{m2}	20	0.089	3.43	0.011	2.97	0.99	0.55	0.50	0.21	0.069	0.19

for this constant. We call X_{m2} data set for the reduced data transform by this method. In Table 3 we present the observed statistics values and the corresponding p-values for the SW, AD and Lf tests as well as the estimates $\hat{\lambda}$, $\hat{\mu}$, $\hat{\sigma}^2$ and $\hat{b}_2$ for the X_{m1} and X_{m2} data sets for several values of d.

We can see that the fit provided by the $m1$ method has higher p-values for all performed tests and consequently gives a better fit to the X_{m1} data set than the normal distribution. The X_{m1} data fit is also better than that of X_{m2} for all values of d except for the SW test that performs better on the X_{m2} data set for higher values of d. We must also note that the gaussian fit gave better results than the ones on the X_{m2} data set except for the SW test for higher values of d. Regarding the dependence on d we can say that there is an increase on the p-value of the SW statistic when d goes from 5 to 10 but it stops around this number. There is also an increase of the p-value of the Lf statistic but even for $d = 20$ the p-value is still much less that the one observed using the $m1$ method and even to that of the normal fit.

5 Conclusions and Final Remarks

This study has two goals. The first goal is to compare methods, $m1$ and $m2$, to find distributions for NG random variables. Both methods consist in the estimation of exponents for power data transformations. The $m1$ method provides an exponent for a simple power transformation and the other provides an estimate of the λ parameter of the classic BC transformation. The method $m1$ is heuristic and based on kurtosis comparison. The method $m2$ is based on the maximization of the so called pseudo-log-likelihood function that results from the log-likelihood function by replacing the μ and σ parameters by their estimators. The second goal is the application of this methods to study the distribution of errors in temperatures prediction which are considered to be good candidates to be qualified as NG random variables. In our statistical analysis we found that the high kurtosis revealed by the original data when compared to the normal distribution, and reported in [1] and [5] was due to temporal correlation and it disappears after data reduction. The reduced data sets X and Y of errors in maximum and minimum temperatures, respectively, revealed to be consistent with normality. Nonetheless it was put on test if a distribution of the PN family could provide a better fit than the normal on the X reduced data set. We concluded that using $m1$ method we obtain distributions of the power normal family that fit the data better than the normal distribution. The $m2$ method, that is studied for several values of d, only performs better than the fits of the $m1$ method and normal distribution for the SW test and for higher values of d. That was not the case for the AD and Lf tests. It was observed that the added constant for positiveness, d has an effect in the performance of the $m2$ method. It was observed an increase on the p-value of the SW test when d increases but that effect stabilizes for values of d around 10. It was also observed that the higher the d the higher the p-values of the Lf test get. However the observed p-values for the $m2$ method were significantly worst that the ones observed by the fits provided by the $m1$ method and also by the normal distribution. This study is based on a year of daily errors and a lot of data had to be eliminated because of correlation. It would be interesting to see if using larger data sets the conclusions are still the same.

References

1. Lefebvre, M.: Nearly gaussian distributions and application. Commun. Stat. Theory Methods **39**, 823–836 (2010)
2. Box, G.E.P., Cox, D.R.: An analysis of transformations. J. Roy. Stat. Soc. Ser. B **26**, 211–243 (1964)
3. Freeman, J., Modarres, R.: Inverse box-cox: the power-normal distribution. Stat. Probab. Lett. **1**, 764–772 (2006)
4. Gnanadesikan, R.: Methods for Statistical Analysis of Multivariate Observations, vol. 143. Wiley, New York (1977)
5. Gonçalves, R., Modeling errors in temperature forecasts. In: Skiadas, C.H. (ed.) Proceedings of the SMTDA Conference, vol. 27, pp. 307–314 (2014)
6. Goto, M., Inoue, T.I.: On estimation of parameters in power-normal distribution. Bull. Inform. Cybern. **21**, 41–53 (1984)
7. Goto, M., Inoue, T.: Some properties of power normal distribution. Jpn. J. Biometrics **1**, 28–54 (1980)
8. Gonçalves, R.: Inverse box-cox and the power normal distribution. In:: Machado, J., Soares, F., Veiga, G. (eds.) Innovation, Engineering and Entrepreneurship, vol. 505, pp. 805–810. Springer, Cham (2019). https://doi.org/10.1007/978-3-319-91334-6_110
9. Gonçalves, R.: The power normal distribution. J. Phys. Conf. Ser. **1334**, 12–14 (2019)
10. Gonçalves, R.: The power normal distribution. In: AIP Conference Proceeding, vol. 2116 (2019)
11. Rahman, M., Pearson, L.: Anderson-darling statistic in estimating the Box-Cox transformation parameter. J. Appl. Probab. Stat. **3**, 23–35 (2008)
12. Atkinson, A., Pericchi, L.: Grouped likelihood for the shifted power transformation. J. Roy. Stat. Soc. Ser. B **53**, 473–482 (1991)
13. Atkinson, A.: Plots, Transformations, and Regression, pp. 196–216. Oxford University Press, Oxford (1985)

The Use of Principal Component Analysis and Logistic Regression for Cutter State Identification

Edward Kozłowski[1], Dariusz Mazurkiewicz[2(✉)], Jarosław Sęp[3], and Tomasz Żabiński[3]

[1] Faculty of Management, Lublin University of Technology, Nadbystrzycka 38, 20-618 Lublin, Poland
{e.kozlovski,d.mazurkiewicz}@pollub.pl
[2] Faculty of Mechanical Engineering, Lublin University of Technology, Nadbystrzycka 36, 20-618 Lublin, Poland
[3] Faculty of Mechanical Engineering and Aeronautics, Rzeszów University of Technology, Al. Powstańców Warszawy 12, 35-959 Rzeszów, Poland
{jsztmiop,tomz}@prz.edu.pl

Abstract. It is well known that due to Industry 4.0 requirements and challenges, future research directions in production engineering will focus on the creation of intelligent sensors and their integration by means of intelligent platforms. Therefore, the key skill will be appropriate analysis and processing of signals recorded by these sensors, which may relate to manufacturing process parameters. The application of principal component analysis and logistic regression enables effective data processing. This has been shown using a real-world numerical example - the data related to cutter state identification based on signals generated during machining. This way, it has been proven, that the above methods may find practical application in condition monitoring systems. In particular, they may be highly helpful in real-time cutter state identification.

Keywords: Logistic regression · Principal component analysis · Cutter state identification

Mathematics Subject Classification (2010) 62N05 · 90B25

1 Introduction

Data analysis methods that are widely used by researchers for large detasets include statistical methods (linear regression, multiple regression, variance analysis, contingency tables) and methods based on artificial intelligence, machine learning and deep learning (classification trees, regression trees, reinforced trees, fuzzy trees, random forests, artificial neural networks, genetic algorithms, evolutionary algorithms, fuzzy sets, rough sets, association rules, support vector machines or Bayes classifiers). The most effective methods for large data sets

J. Machado et al. (Eds.): icieng 2021, LNME, pp. 396–405, 2022.
https://doi.org/10.1007/978-3-030-78170-5_34

are usually those from the latter group. As a result, they are called modern or advanced data analysis methods, because they support researchers in finding solutions to the most difficult tasks that cannot be solved by statistical tools.

The literature review [4–8, 14, 16–18] shows that future research directions in production engineering will focus on the creation of intelligent sensors and their integration by means of intelligent platforms. Therefore, the key skill will be appropriate analysis and processing of signals recorded by these sensors, which may relate to manufacturing process parameters. Moreover, several studies [9, 15, 19] suggest that in order to cope with the dynamic data growth, there is a need to create methods and tools that not only automatically collect data, but also select the most relevant data and use appropriate analyses to extract knowledge from them. According to this identified research challenge, the possibility of applying major component analysis and logistic regression for machine element state identification was analyzed and verified based on industrial time series data. The structure of this article consists of description of logistic regression that was used to design the classifier, followed by Principal Component Analysis and numerical example based on industrial data.

2 Logistic Regression

Let $D = \left\{\left(x_{(i)}, y_i\right) : x_{(i)} \in \mathbb{R}^m,\ y_i \in \{0,1\},\ 1 \leq i \leq n\right\}$ be the learning data set. For i–th observation, $1 \leq i \leq n$ the vector $x_{(i)} \in \mathbb{R}^m$ denotes the realizations of independent variables (usually processed signals from the sensors) but $y_i \in \{0,1\}$ denotes the appropriate class of an object. In the analyzed case, we assume $y_i = 1$ when the cutter is identified as blunt, while if the cutter is sharp we assume that $y_i = 0$. Observing the signal $x \in \mathbb{R}^m$ received from the sensors installed on the cutting machine, our task is to classify the cutter state. In the analyzed case, logistic regression was used to design the classifier $f : \mathbb{R}^m \to \{0,1\}$ (see e.g. [1–3]).

Let $(\Omega, \mathcal{F}, P)$ be a probabilistic space and Y a discrete random variable, where $Y : \Omega \to \{0,1\}$. Logistic regression describes the probability distribution of the dependent variable Y on the basis of the realization of independent variables $X \in \mathbb{R}^m$ (in other words $P\,(Y = y\,|X\,)$ should be determined, where $y \in \{0,1\}$).

Odds is the probability of success to the probability of failure ratio:

$$\theta\,(X) = \frac{P\,(Y = 1\,|X\,)}{1 - P\,(Y = 1\,|X\,)} \tag{1}$$

The task of logistic regression is to estimate the probability of success $P\,(Y = 1\,|X\,)$ based on the realization x and we denote that:

$$P\,(Y = 1\,|X\,) = p\,(X)\,.$$

Since the probability of success is equal to $p\,(X) \in (0,1)$, therefore from the Eq. (1) it follows that odds $\theta\,(X) \in (0,\infty)$, while $\ln\theta\,(X) \in (-\infty,\infty)$. The natural algorithm of the odds is known in the literature as *log-odds* or *logit*.

In logistic regression, we analyze the linear dependence of the logarithm of odds based on the realization of independent variables X. For this purpose, we consider a dependence described by the equation:

$$\ln\theta(X) = \ln\left(\frac{p(X)}{1-p(X)}\right) = X\beta + \varepsilon. \tag{2}$$

where ε is a random variable with normal distribution $N(0,\sigma^2)$ and $\beta = (\beta_1, \beta_2, ..., \beta_m) \in \mathbb{R}^m$. If there is an intercept in the model (2) then the matrix contains the column of ones corresponding to the intercept.

From the Eq. (2) we obtain:

$$p(\beta, X) \stackrel{def}{=} p(X) = \frac{e^{X\beta}}{1+e^{X\beta}}. \tag{3}$$

To estimate the parameters β, we apply the Maximum Likelihood Method. Let $Y \in \{0,1\}^n$ and $X \in \mathbb{R}^{n\times m}$, where:

$$Y = \begin{bmatrix} y_1 \\ y_2 \\ \vdots \\ y_n \end{bmatrix}, \quad X = \begin{bmatrix} x_{11} & x_{12} & \dots & x_{1m} \\ x_{11} & x_{12} & \dots & x_{1m} \\ \vdots & \vdots & \vdots & \vdots \\ x_{n1} & x_{n2} & \dots & x_{nm} \end{bmatrix} = \begin{bmatrix} x_{(1)} \\ x_{(2)} \\ \vdots \\ x_{(n)} \end{bmatrix}$$

and y_i means i–th realization of a dependent variable describing the probability of success or lack thereof, however $x_{(i)}$-th describes the vector of realization of independent variables during i-th observations, $1 \le i \le n$.

Estimation of the logistic regression parameters (3) consists in solving the problem:

$$\max_{\beta} L(\beta, Y, X), \tag{4}$$

where the likelihood function $L(\beta, Y, X)$ is given by the formula:

$$L(\beta, Y, X) = \prod_{i=1}^{N} \left(p(\beta, x_{(i)})^{y_i}(1-p(\beta, x_{(i)}))^{1-y_i}\right). \tag{5}$$

The problem (4) is replaced by an auxiliary task:

$$\max_{\beta} \ln L(\beta, Y, X). \tag{6}$$

From the formulas (2) and (3), we obtain that a natural logarithm of the likelihood function (5) is equal to:

$$l(\beta) = \ln L(\beta, Y, X) = \sum_{i=1}^{N} \left(y_i \ln p(\beta, x_{(i)}) + (1-y_i)\ln(1-p(\beta, x_{(i)}))\right)$$

$$= \sum_{i=1}^{N} \left(y_i \ln \frac{p(\beta, x_{(i)})}{1-p(\beta, x_{(i)})} + \ln(1-p(\beta, x_{(i)}))\right)$$

$$= \sum_{i=1}^{N} \left(y_i x_{(i)} \beta - \ln \left(1 + e^{x_{(i)} \beta} \right) \right) \tag{7}$$

The necessary condition of optimality for the task (6) is given by the formula:

$$\frac{\partial l}{\partial \beta}(\beta) = \mathbf{0}$$

where $\mathbf{0} = col(0, 0, ..., 0) \in \mathbb{R}^m$ and:

$$\frac{\partial l}{\partial \beta}(\beta) = X^T z(\beta) \text{ oraz } z(\beta) = \begin{bmatrix} y_1 - p(\beta, x_{(1)}) \\ y_2 - p(\beta, x_{(2)}) \\ \vdots \\ y_n - p(\beta, x_{(n)}) \end{bmatrix}. \tag{8}$$

The second order derivative matrix is equal to:

$$\frac{\partial^2 l}{\partial \beta \partial \beta^T}(\beta) = -\sum_{i=1}^{n} x_i^T x_i h_i(\beta) = -X^T Z(\beta) X, \tag{9}$$

where $Z(\beta) = diag(h_1(\beta), h_2(\beta), ..., h_n(\beta))$ and $h_i(\beta) = p(\beta, x_{(i)}) \left(1 - p(\beta, x_{(i)})\right)$ for $1 \leq i \leq n$. The second order derivative matrix given by the formula (9) is nonnegative. For $1 \leq j, k \leq m$

$$\frac{\partial l(\beta)}{\partial \beta_j} = \sum_{i=1}^{n} \left(y_i - p(\beta, x_{(i)}) \right) x_{ij} = 0$$

and

$$\frac{\partial^2 l(\beta)}{\partial \beta_j \partial \beta_k} = \sum_{i=1}^{n} x_{ij} x_{ik} p(\beta, x_{(i)}) \left(1 - p(\beta, x_{(i)})\right)$$

The values of parameters β are estimated iteratively. To determine the estimators of unknown parameters in the model (2)(solution of the Eq. (8)), we use the Newton-Raphson algorithm where in step $k+1$ the values of parameter estimators β are calculated using the formula:

$$\beta_{k+1} = \beta_k - \left(\frac{\partial^2 l}{\partial \beta \partial \beta^T}(\beta_k) \right)^{-1} \frac{\partial l}{\partial \beta}(\beta_k), \tag{10}$$

The first and second derivatives of the objective function (7) are given by the formulas (8) and (9), respectively.

3 Principal Component Analysis

Principal Component Analysis (PCA) is based on the identification of factors (components) appearing in the dataset X. The dataset consists of n observations for m variables (in R^m space it is a cloud of n points). The goal of PCA is to ensure the rotation of the coordinate system in order to first maximize the variance of the first coordinate, then the variance of the second coordinate, and so forth. The coordinates of the new system are called loads and are generated by principal components. The initial components explain most variances in a new space. PCA is often used for reducing the size of statistical data by rejecting factors, which do not explain much of the variance present in the data.

The decomposition of the matrix X is based on Singular Value Decomposition. Every real matrix X can be described as:

$$X = UDV^T + \varepsilon_X, \tag{11}$$

where $U \in R^{n\times m}$ is the matrix of left singular vectors, $D \in R^{m\times m}$ - is the diagonal matrix containing singular values, and $V \in R^{m\times m}$ - is the matrix of right singular vectors. The matrix V is orthogonal and satisfies the equation: $V^T = V^{-1}$. Decomposition (11) of the matrix X can be presented in the form:

$$X = TP^T + \varepsilon_X, \tag{12}$$

where $T = UD$ denotes *scores* in the orthogonal system and $P = V$ presents loadings. The matrix T in the new space we determine by multiplying the equation $X = TP^T$ by P and we get:

$$XP = TP^TP = T. \tag{13}$$

According to formula (13), the matrix columns V contain weights used in a linear combination to create new dimensions. Coordinate variances in the new space are calculated by the formula:

$$\lambda_i = \frac{d_i^2}{n-1},$$

where d_i, $0 \le i \le k$ are singular values of the diagonal matrix D. The proportion of variance explained by i–th principal component is equal to $\frac{\lambda_i}{\sum_{j=1}^{m}\lambda_j}$, while the proportion of variance explained by k principal components is equal to $\frac{\sum_{j=1}^{k}\lambda_i}{\sum_{j=1}^{m}\lambda_j}$ for $1 \le k \le m$.

4 Numerical Example

For every observation $1 \le i \le n$ we create elements of a learning dataset: for the blunt cutter we assume $y_i = 1$ and for the sharp cutter (no warnings) $y_i = 0$. This was done for 2173 samples that were attached to the data set. To this end, we analyzed signals AccSignal1 and P2y ($x_{(i)} = (x_{i1}, x_{i2}, \dots, x_{im})$ where $m = 213$)

coming from the experiments performed in testbed consisting of the Haas VM-3 CNC machine equipped with an inline direct-drive spindle and a set of sensors: 7 accelerometers (sensitivity 100 mV/g) integrated with temperature sensors, 1 acoustic emission sensor (sensitivity 53 mV/Pa) and 1 force and torque sensor (3-axis, sensitivity: 1 mV/N, 10 mV/Nm). Two accelerometers were mounted on the lower bearing of the spindle, two ones were mounted on the higher bearing of the spindle, another two accelerometers were mounted on the Z-axis and one accelerometer was mounted on the workpiece. The acoustic emission sensor was mounted in the machine cabin. The force and torque sensor was mounted in the chuck. In this study, AccSignal1 from the accelerometer mounted on the lower bearing of the CNC machine spindle and the P2y force signal parallel to the Y-axis of the force and torque sensor were used.

The stationary property of AccSignal1 was examined using the Augmented Dickey-Fuller (ADF) test (see for example [10,11]). At the significance level $\alpha = 0.01$ the null hypothesis of unit root existence was rejected. In addition, the significance of the correlation was examined $\{r_\tau\}_{1\leq\tau\leq200}$ by the Ljung-Box test (see e.g. [12,13]). Using the Herglotz's theorem for functional time series, AccSignal1 should be identified by the autocorrelation function. Thus, in the training set for AccSignal1 signals the correlation values for lags $1 \leq \tau \leq 200$ were implemented as $x_{i1}, x_{i2}, \ldots, x_{i200}$. Based on the ADF test, it was shown for some P2y signals that the unit root exists. These signals were identified by means of ARIMA class models, where the autoregression order and the moving average order do not exceed 5. In the training set, the following were implemented: integration degree x_{i201}, autoregression coefficients values $x_{i202}, x_{i203}, \ldots, x_{i206}$, coefficients values $x_{i207}, x_{i208}, \ldots, x_{i211}$, intercept x_{i212} and the variance of external disorders x_{i213}.

Prior to SVD implementation, the values of explanatory variables were standardized (centered and scaled). For such defined training dataset, the correlation properties were examined and the impact of each component was analyzed to explain the variability of independent variables. Figure 1 shows the chart of correlation coefficients for both explanatory variables $X_1, X_2, \ldots, X_{213}$ and principal components $T_1, T_2, \ldots, T_{213}$. There exists a correlation between the explanatory variables (Fig. 1a), but no such correlation was observed between the principal components (Fig. 1b).

Figure 2 shows the proportion of the first 100 principal components in explaining the variance of explanatory variables and the cumulative proportion. The figure reveals that practically the first 40 principal components explain almost 100% of variability.

Logistic regression was used to identify the cutter state, with scores of principal components being selected as explanatory variables in Eq. (3). The maximum likelihood method was employed to estimate structural parameters in model (3), and the cutter state probability was calculated for each case. For comparison, 15 and 30 components were used in model (3). Figure 3 shows the cutter probability distributions estimated with the use of 15 and 30 principal components, respectively. For 15 principal components the proportion of the explained variance is equal to 0.9793, while for 30 of them this value is 0.9977.

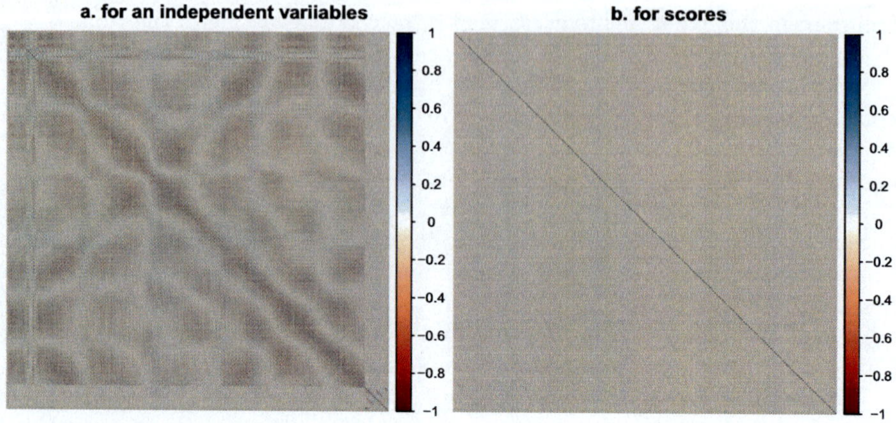

Fig. 1. Correlation chart of explanatory variables and principal components

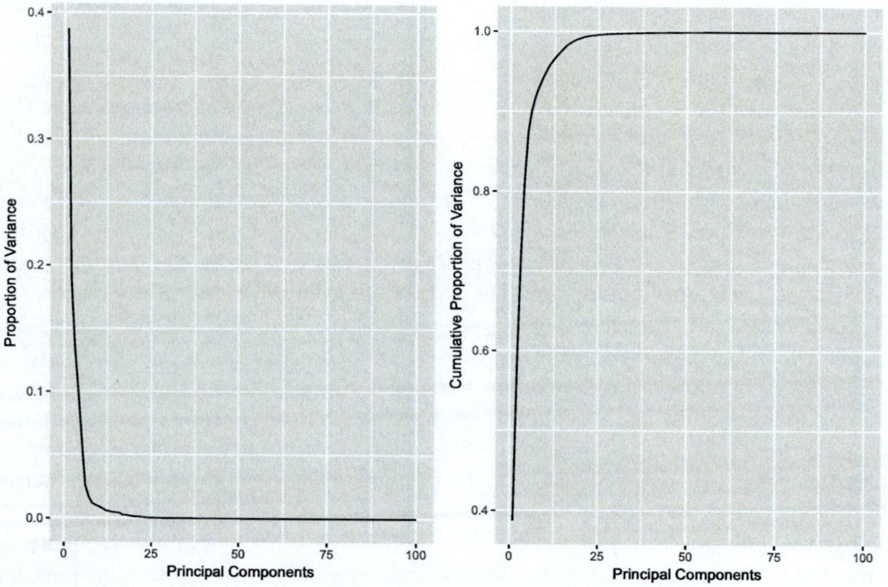

Fig. 2. Share of principal components in explaining the variance of explained variables

Figure 3 reveals that by applying logistic regression for cutter state identification and taking into account only 30 principal components (instead of 213 variables) we can achieve the expected result. The cutter state can be predicted with a very small error. Table 1 presents the confusion matrix of recognition. Only 82 samples were not correctly recognized for 2173 cases (i.e. the recognition error is 3.77%). Table 2 gives that classifier operating characteristics. The

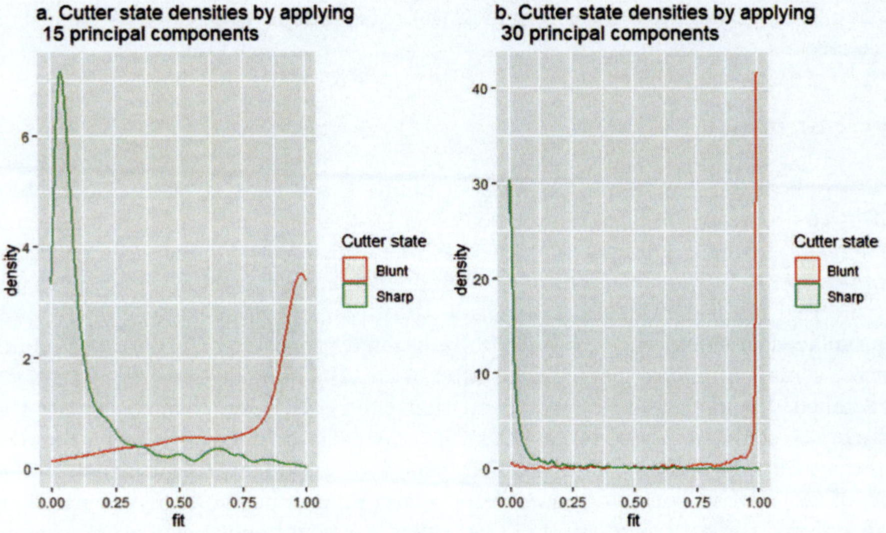

Fig. 3. Probability distributions

Table 1. Confusion matrix

	Sharp	Blunt
Sharp	1215	61
Blunt	21	876

Table 2. Classifier Operating characteristics of cutter state

Characteristics	Values
Sensitivity	0.9348986
Specificity	0.9830097
Pos Pred Value	0.9765886
Neg Pred Value	0.9521944
Precision	0.9765886
Prevalence	0.4312011
Detection Rate	0.4031293
Detection Prevalence	0.4127934
Balanced Accuracy	0.9589542

true positive rate (sensitivity, hit rate) is equal to 0.9349 and describes the proportion of real positive cases that have correctly identified the blunt cutter as blunt. The true negative rate (specificity) is equal to 0.983 and denotes the proportion of real negative cases in which the sharp cutter was correctly identified as sharp.

5 Summary

The application of principal component analysis and logistic regression enables effective processing of data related to cutter state identification based on signals generated during machining. As a results, the above methods may find practical application in condition monitoring systems. In particular, they may be helpful in the real-time identification of cutter state [20,21]. In effect, the cutter will be replaced in the moment it loses its functional properties. Currently, in many cases, a machining tool is replaced after a certain number of parts have been machined. In practice, this may mean that the replacement is made either too late or too soon. In industrial realities this may cause problems with ensuring the product quality and/or the cost effectiveness of manufacturing processes. The use of an effective condition monitoring system based on the analyzed methods can prevent these adverse effects. The future work will be concentrated on non-technical and sustainable aspects of technical infrastructure and their influence on availability and reliability taken into consideration while developing analytical approach of time series modelling.

References

1. Hastie, T., Tibshirani, R., Friedman, J.: The Elements of Statistical Learning. Springer-Verlag, New York (2009)
2. Wehrens, R.: Chemometrics with R, Springer-Verlag GmbH, New York (2011)
3. Tibshirani, R.: Regression shrinkage and selection via the Lasso. J. R. Stat. Soc. Ser. B **58**, 267–288 (1994)
4. Kozłowski, E., Mazurkiewicz, D., Żabiński, T., Prucnal, S., Sęp, J.: Machining sensor data management for operation-level predictive model. Expert Syst. Appl. **159**, 949–953 (2017)
5. Daniewski, K., Kosicka, E., Mazurkiewicz, D.: Analysis of the correctness of determination of the effectiveness of maintenance service actions. Manag. Prod. Eng. Rev. **9**(2), 20–25 (2018)
6. Jasiulewicz-Kaczmarek, M., Antosz, K., Żywica, P., Mazurkiewicz, D., Sun, B., Ren, Y.: Framework of machine criticality assessment with criteria interactions. Eksploatacja i Niezawodnosc - Maintenance Reliab. **23**(2), 207–220 (2021)
7. Teti, R., Jemielniak, K., O'Donnell, G., Dornfeld, D.: Advanced monitoring of machining operations. CIRP Ann. **59**(2), 717–739 (2010)
8. Kozłowski, E., Mazurkiewicz, D., Żabiński, T., Prucnal, S., Sęp, J.: Assessment model of cutting tool condition for real-time supervision system. Eksploatacja i Niezawodnosc - Maintenance Reliab. **21**(4), 679–685 (2019)
9. Hedberg, T.D., Hartman, N.W., Rosche, P., Fischer, K.: Identified research directions for using manufacturing knowledge earlier in the product life cycle. Int. J. Prod. Res. **55**(3), 819–827 (2017)
10. Charemza, W.W., Syczewska, E.M.: Joint application of the Dickey-Fuller and KPSS tests. Econ. Lett. **61**(1), 17–21 (1998)

11. Dickey, D.A., Fuller, W.A.: Distribution of the estimators for autoregressive time series with a unit root. J. Am. Stat. Assoc. **74**, 427–431 (1979)
12. Box, G.E.P., Pierce, D.A.: Distribution of residual autocorrelations in autoregressive-integrated moving average time series models. J. Am. Stat. Assoc. **65**(332), 1509–1526 (1970)
13. Ljung, G.M., Box, G.E.P.: On a measure of lack of fit in time series models. Biometrika **65**(2), 297–303 (1978)
14. Borucka, A., Grzelak, M.: Application of logistic regression for production machinery efficiency evaluation. Appl. Sci. **9**, 4770 (2019)
15. Gevorkyan, E., Prokopiv, M., Rucki, M., Morozow, D.: Durability and exploitation performance of cutting tools made out of chromium oxide nanocomposite materials. Eksploatacja i Niezawodnosc - Maintenance Reliab. **21**(4), 686–691 (2019)
16. Loska, A.: Patterning as a way to identify and assessment of the exploitation policy of selected technical systems. MAPE **2**(1), 213–223 (2019)
17. Kozłowski, E., Rymarczyk, T., Kłosowski, G.: Logistic regression application to image reconstruction in UST. In: 2019 Applications of Electromagnetics in Modern Engineering and Medicine (PTZE) (2019)
18. Rymarczyk, T., Kozłowski, E., Kłosowski, G., Niderla, K.: Logistic regression for machine learning in process tomography. Sensors **19**(15), 3400 (2019)
19. Fang, Y., Tao, W., Tee, K.F.: A new computational method for structural reliability with big data. Eksploatacja i Niezawodnosc - Maintenance Reliab. **21**(1), 159–163 (2019)
20. Żabiński, T., et al.: CNC milling tool head imbalance prediction using computational intelligence methods. In: Rutkowski, L., Korytkowski, M., Scherer, R., Tadeusiewicz, R., Zadeh, L., Zurada, J. (eds.) Artificial Intelligence and Soft Computing. Proceedings of the 14th International Conference on Artificial Intelligence and Soft Computing ICAISC, Zakopane, Poland, pp. 503–514 (2015)
21. Żabiński, T., Mączka, T., Kluska, J., Madera, M., Sęp, J.: Condition monitoring in Industry 4.0 production systems - the idea of computational intelligence methods application. In: Teti, R., Doriana, M., D'Addona, D. (eds.) Procedia CIRP, Volume 72. Proceedings of the 12th CIRP Conference on Intelligent Computation in Manufacturing Engineering, Naples, Italy, pp. 63–67 (2019)

The Impact of Industry 4.0 Paradigm on the Pharmaceutical Industry in Portugal

Ana Correia Simões[1(✉)], Joana Teixeira Mendes[2], and José Coelho Rodrigues[1,2]

[1] INESC TEC, Campus FEUP, Porto, Portugal
ana.c.simoes@inesctec.pt
[2] Faculty of Economy, University of Porto, Porto, Portugal

Abstract. Technological evolution has continuously driven the development of industries and consequently of society. The fourth industrial revolution consists in the combination of a set of physical and digital technologies that has been changing systems' operations within industries.

The pharmaceutical industry has a considerable impact on well-being and has been strongly challenged with this new reality, not only by those that are transversal to all industries but also due to the fact that it is a highly regulated sector, which creates additional barriers for industry 4.0 (I4.0) initiative's implementation. However, it is due to the fact that this revolution provides high growth opportunities to the industry, and consequently for the improvement of population's quality of life, that this topic has been subject to so much research at a global level. This study's main purpose is to understand the impact of I4.0 paradigm implementation in the pharmaceutical industry (mainly in the production area), to analyze the technological readiness of Portuguese pharmaceutical companies to implement I4.0 technologies and to understand the role of the I4.0 paradigm to fight the pandemic situation caused by the COVID-19. To achieve this purpose, an exploratory multiple-case study based on semi-structured interviews was conducted in two Portuguese pharmaceutical companies. It is expected that the results of this work lead to recommendations that help the Portuguese pharmaceutical industry to be better prepared to face the challenges that are coming with this revolution.

Keywords: Industry 4.0 (I4.0) · Digitalization · Technology · Innovation · Pharmaceutical industry · Impacts · Challenges

1 Introduction

The development of technologies has continuously driven the advancement of industries. In recent years, the fourth industrial revolution (Industry 4.0–I4.0) has been a topic that has attracted attention of companies and researchers around the world. [1]. This revolution occurs with the development and integration of digital and physical technologies, and aims to support human work, streamlining processes and making them more effective [1]. There is a great responsibility, to the pharmaceutical industry, for the health of human beings, ensuring the well-being of the population and contributing to improve society's quality of life [2]. For the companies in this industry the challenges

J. Machado et al. (Eds.): icieng 2021, LNME, pp. 406–419, 2022.
https://doi.org/10.1007/978-3-030-78170-5_35

of implementing I4.0 technologies are more complex by the nature of their business when compared to other industries [2]. The impacts of the implementation of these technologies, such as considerable increase in operational efficiency, cost minimization and greater flexibility in the production processes, have aroused the interest of pharmaceutical companies, as well as the willingness to overcome the barriers associated with its implementation [3, 4].

Therefore, the pharmaceutical sector is responsible for social well-being and the notion that the I4.0 paradigm can bring great growth opportunities to the sector [15, 16] reinforces the importance of studying the major challenges that this paradigm poses to the industry. During the conference entitled "Digitalization in Pharma" held in 2017 [5] some challenges were presented to the pharma industry regarding the necessity of improvements in efficiency, flexibility and speed of the production processes, as well as, the necessity of compliance with the specific requirements of the sector.

Although it has been possible to verify the existence in Portugal of programs to support companies in the implementation of I4.0 technologies, there is no detailed information to conclude with empirical evidence i) what is the current context/scenario of the Portuguese pharmaceutical companies regarding the implementation of I4.0 paradigm in the production process; and ii) to measure their readiness to the adopt I4.0 technologies. Thus, this study has two main objectives. First, to understand the impact of I4.0 paradigm implementation on the production area of the Portuguese pharmaceutical companies, as well as the main implementation opportunities and challenges. Second, to analyze the companies' readiness to implement I4.0 technologies. This research was conducted during the pandemic crisis caused by COVID-19 in 2020. Therefore, it was considered important to adjust the research objectives taking this context into consideration. Consequently, in order to understand the relevance of I4.0 technologies implemented in the pharmaceutical industry to fight the virus, a third objective was added, which wasto understand the role of the I4.0 paradigm in combating COVID-19 in the pharmaceutical sector.

The objective of this research is to scope out the current level of I4.0 adoption by Portuguese pharma companies, and to provide information to both the management of those companies and the academia. The results presented in this study can be considered relevant as a first approach to this topic and used as a first approach to deduce the future strategy and direction that this industry may take in the Portuguese context.

2 Pharma 4.0 Impacts and Challenges

The pharmaceutical industry occupies a prominent place in the global economy due to the considerable impact it has on social well-being by improving citizens' health and quality of life. The industry is responsible for providing medicines that contribute to the improvement of health and quality of life of society, as well as, innovating and developing new therapies to satisfy the treatment needs and to prevent new pathologies [6].

The pharmaceutical industry distinguishes itself from the other industries by its high standards in terms of innovation, technological solutions and qualification of human resources. For these reasons, this industry generates the most added value per worker.

Furthermore, as it is a very sensitive sector, closely linked to the well-being and health of the population, it requires a high level of documentation and is subject to very strict regulations, ranging from the research and development process to the production and distribution of medicines [2].

In 2019, the Portuguese pharma industrial companies (producers of pharmaceutical products) represented 1.2% of the total sales of products and services provided by all Portuguese industrial companies, with a turnover of 1,130 M€. Regarding the product sales (942.9 M€), 68.7% was for foreign countries [21].

The implementation of the emerging technologies of Pharma Industry 4.0 supports sustainable value creation, leads to more agile, smart and personalized pharma industry, thus allowing to obtain competitive advantages [15, 18]. Some of the Pharma I4.0 concepts are cyberphysical systems and dark factories, which require data science tools as technological core components [17]. The International Society for Pharmaceutical Engineering (ISPE) is a non-profit organization in the USA, which aims to support companies to implement technological solutions [3]. With the main purpose of supporting the pharmaceutical companies in the implementation of I4.0 technologies and to overcome the obstacles created during this process, ISPE created the "Pharma 4.0" initiative [7]. The operating mode underlying this initiative is illustrated in Fig. 1 which describes the key points to be considered in the process.

This model consists of four elements: i) Resources: includes human capital, machines, equipment and tools; ii) Organization and Processes: refers to the internal organization of the company regarding the structure of its processes; iii) Information System: responsible for preparing, processing, storing and transferring data and information; iv) Culture: refers to the organic structure of the company, with the different departments and the culture being completely aligned [7].

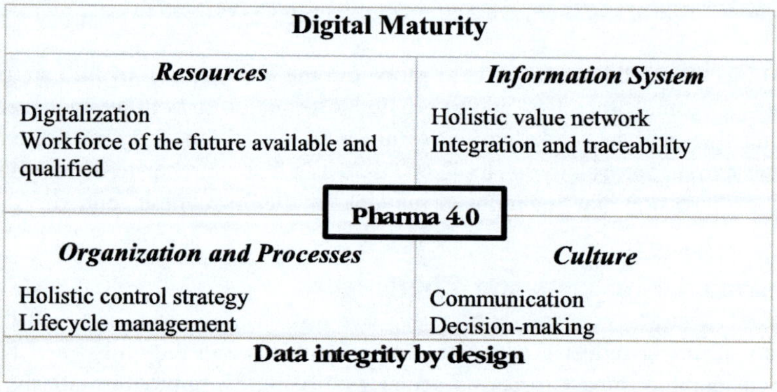

Fig. 1. From I4.0 to Pharmacy 4.0: Operating Model. (Source: adopted from [7])

The two driving factors in this model are: i) Digital Maturity - the ability to produce according to I4.0 parameters; ii) Data Integrity by design - data flows in several directions throughout the production chain, so data must follow a roadmap that is transparent in all its steps, in order to ensure integrity and accuracy in production [7].

Despite the great advantages embraced by Pharma 4.0, the readiness of the industry for this digitalization move is still unsatisfactory [16]. The challenges that the implementation of I4.0 technologies presents to the pharmaceutical industry are more complex, when compared to other industries, by the nature of its business [20]. This greater complexity stems from the importance of the pharmaceutical industry's impact on society, which makes it a highly regulated sector with complex processes that entail strict quality requirements to be met [2, 20].

According to the literature there are several challenges faced by pharmaceutical companies to deal with the implementation of I4.0 technologies:

- The need to empower human resources: the connectivity and volume of data generated by I4.0 paradigm require workers capable of identifying and analyzing information in order to maximize the technologies' potential [4];
- The revenue gains are not visible at the beginning [4];
- Data security and privacy concerns (Cyber security): due to the increased data connectivity, there is a need to protect industrial systems from cyber-attacks that could jeopardize data and the information security [8];
- Lack of a clear vision for digital operations and leadership support [4];
- Large financial investment required: many companies refer to the cost of implementation as one of the major barriers to adopt I4.0 technologies. The investment needed to adapt the factory to the principles of Pharma 4.0 is very high. Nevertheless, the investment is quickly recovered by productivity gains and reduced downtime and waste [9].

Moreover, while the implementation of I4.0 is playing an increasingly significant role in the modernization of the pharmaceutical industry, the big challenge is the holistic fusion of I4.0 into the culture of the organizations [4, 16]. The lack of people with the skills and competences to deal with this transformation is also a considerable challenge.

I4.0 technologies are able to generate, analyze and communicate data between machines and humans, providing more transparent and effective information exchange resulting in faster responses [4]. Many pharmaceutical companies around the world have already started to implement I4.0 initiatives: i) Pfizer has already started the digitalization processes in all of its plants, making all data visible and transparent along the entire production line; ii) Merck is making its plants "smarter", with a focus on more responsive and adaptive production processes; iii) Sanofi has started the implementation of collaborative robots and software that have enabled to reduce the paper use [10].

3 Materials and Methods

The research method used in this research was a qualitative multiple case study, using interviews as the main source for collecting data.

3.1 Case Design

The targets of this case study were the Portuguese pharmaceutical companies. After meticulous research, using the companies' website, news published in the media in 2020 and a social network (LinkedIn), a list of 19 companies in the pharmaceutical sector, and the respective contacts, was prepared. The contacts were made during March and June 2020. During this period, Portugal faced the most critical phase of the COVID-19 pandemic. That fact can explain the low adherence to the research by the companies. In fact, only two companies were able to collaborate. The two cases (companies) are characterized in Table 1.

Table 1. Companies (cases) description (Source: Authors own elaboration)

Company	Activity	Turnover (million €) (2019)	Nr. of employees	Typology
Company A	Drugs production for human use	42	280	Medium enterprise
Company B	Drugs production for human use	50	700	Medium enterprise

3.2 Data Collection and Data Analysis

For data collection, multiple sources of evidence were used. One of the data sources was documents provided by the companies' websites and social media, as well as, the public and private companies' reports. Other data collection method used were the interviews, one of the most important sources of evidence in case studies [11]. For the semi-structured interview, an interview guide was prepared based on literature review findings.

In these companies, the criteria to select participants for the study took into consideration their functions in the industrial management departments, as well as, the key role in the I4.0 strategy implementation. Two managers with this profile were selected: one in company A (I1 CA) and another one in company B (I2 CB). A third interview was done to the director of the Digitization and Innovation Department for the Pharmaceutical Industry in Germany (I3). This manager works in a company that designs and provides technological solutions for industrial companies, in particular to the pharmaceutical industry. Due to the expertise, knowledge and experience of this manager his inclusion in this study was considered a crucial value added. The interviews were conducted using a video call platform due to the pandemic situation previously referred. A brief description of the interviewees is presented in Table 2.

After the data collection, the data analysis started with a coding scheme suggested by Strauss and Corbin [12]. Relevant words and expressions were marked and memos were written. Then, related expressions were summarized into categories and subcategories. Secondly, a within-case analysis was performed accompanied by a pattern matching with the theoretical background. Finally, a cross-comparison of the cases was made to search for similarities and differences between them, with the intention of reaching more generalizable conclusions [11].

Since reliability and validity in case studies are crucial for the quality of the research, during the research design and implementation these aspects were considered [11]. For external validity reasons, a multiple case study (and a replication logic) was the research method chosen, enhancing the robustness of the study. Likewise, in order to establish correct operational measures during data collection, i.e., to ensure construct validity, multiple sources of evidence (documents and interviews) were used. Finally, and towards ensuring reliability, for the semi-structured interview, an interview guide was developed and followed during the interviews.

Table 2. Interviewees description (Source: Authors own elaboration)

Interviewee	Function	Interview duration (min.)	Interview date
I1 CA	Chief Executive	23	July 2020
I2 CB	Quality Manager	38	July 2020
I3	Director of the Digitization and Innovation Department	33	December 2020

4 Results

4.1 Case Study A: Description

The case A (CA) is a Portuguese pharmaceutical company that belongs to a group of leading companies in the manufacture of drugs in Portugal. This company has been in the market for more than 20 years. It has a portfolio of more than 650 products and exports 75% of its production. Its core business is the production of small and medium series of drugs for the European market. Therefore, this company must have highly flexible production facilities, producing different types of pharmaceutical products: granulates, tablets, coated tablets, pills and capsules. The product portfolio also includes the production of pharmaceutical packaging, development of packaging, development of formulas and processes, development and validation of analytical methods.

This company has already started the implementation of some technologies such as collaborative robots. CA is currently developing a project of *Big Data Analytics* (BDA), more specifically *Electronic Batch Records* (EBR), which is a program that makes detailed and continuous records throughout the production process. The company intends to continue the implementation of I4.0 technologies until it has a fully automated plant.

4.2 Results by Category: Case Study A

In Table 3 a synthesis of the main results of CA per category are presented.

Table 3. Results per category – Case study A

Knowledge of the Industry 4.0 paradigm
The interviewee masters the concept of the I4.0 paradigm and also proved to be enthusiastic about the benefits it generates, stating that "*the first to embrace the 4.0 Industry, will be the first to take advantage of it sooner*" (I1 CA). According to the interviewee, employees are aware that technology is changing some processes
Industry 4.0 paradigm implementation
The company is currently working on a BDA project to handle a high volume of data collected to make better use of it. More specifically to be able to make more efficient the operations, packaging and quality processes. The company has already implemented robots in order to assist the workers, with greater collaboration between robots and humans, and also to ensure humans' safety. The company has already improved the operational results by reducing operating costs since the implementation of these technologies. The interviewee also believes that the investment in EBR technology has been recovered in approximately one year The implementation of the I4.0 initiatives in the shop floor has been both conservative and gradual. After an awareness of the impacts associated with I4.0 technologies, contacts have been made with several technology suppliers: i) to understand the impact of the technologies in the production processes performance; ii) to design solutions tailored to the needs of the company. The technologies already implemented were preceded by pilot tests
Implementation opportunities and challenges
Opportunities: When technologies are well implemented and adopted, there are visible improvements in efficiency levels (production) and the company becomes more competitive, which in turn will be reflected in cost efficiencies and consequently in company results. Since the implementation of EBR, there has been a high reduction in paper use. The interviewee reinforces that "*apart from the reputation gains, there are economic gains that allow more to be done with less.*" (I1 CA) **Challenges**: One of the most challenging phases in the implementation is convincing people within the company of the project merits. Resistance to change has to be overcome through effective communication, explaining to all employees that these technologies will have a positive impact in the company, highlighting the implementation results, and convincing people that the technologies will not reduce jobs. I1 CA explains that "*there is a well-established idea that by implementing (the technologies) jobs will be taken away from operators*". Another challenge is the shortage of skilled labor. According to I1 CA there is a need for more specialized skills to work with these technologies and investments are also needed in vocational training schools and universities
Company readiness to implement the technologies
Although the company has already taken the first steps to the implementation of BDA and robots in the production area, the company will only be fully prepared when the resistance to change is reduced and employees feel more motivated and willing to work with the technologies
Industry 4.0 technologies and the combat the COVID-19
The BDA and collaborative robots were an added value during the pandemic period as they allowed the company to continue to operate without risking employee safety. The BDA allowed for the reduction of paper and the robots allowed for the reduction of employee concentration on the shop floor, which considerably reduced the likelihood of contagion while avoiding the company to stop the production

4.3 Case Study B: Description

The Case B (CB) is one of the most entrepreneurial and innovative Portuguese companies dedicated to the production of generic drugs in the national pharmaceutical sector. This company exports 85% of its production to 45 different countries. The company has activities in the entire value chain (from R&D to the market).

CB has already started a project to implement I4.0 technologies, through pilot tests, to manage all quality processes. This project consists of a system that is interconnected with the Enterprise Resource Planning (ERP) and involves a wide range of processes, such as preventive and corrective actions, supplier management, audit management. In addition to ERP, the implementation of other quality control technologies is also in place, such as Empower Chromatography Data System and Warehouse Management Systems (WMS). The company intends to continue the implementation of I4.0 technologies focusing on production processes and also intends to have the production area fully automated within 10 years.

4.4 Results by Category: Case Study B

In Table 4 a synthesis of the main results of CB per category are presented.

Table 4. Results per category – Case study B

Knowledge of the Industry 4.0 paradigm
The interviewee (I2 CB) is familiar with the I4.0 concept and applies a technical language that demonstrates a high level of knowledge. The company is currently training employees to work on projects that involve the I4.0 paradigm
Industry 4.0 paradigm implementation
The company (CB) started gradually the implementation of the technologies by departments. It started with the implementation of ECDS, which is an example of a digital quality control technology applied to control the processes' quality. The implementation of this technology was carried out by installing sensors on the machines that transmit data in real time to the database, which, together with the reference numbers of each package, enables the optimization and increases the efficiency of production. The warehouse is also almost fully automated via WMS. As far as the implemented technologies are concerned, it is possible to conclude that all processes are already digitalized through the ERP system. Currently, all processes are carried out with the support of information systems and RFID, including material order forms and the delivery of materials in the shop floor The company is currently studying how to implement other I4.0 technologies in the production area because it will have to comply with the legal requirements of the authorities and the production rules of each country (more than 40 countries) to which the company exports. However, the ERP is implemented in all areas of the company including production. The implementation of pilot tests in the production area started in 2019 before the pandemic. The company intends to implement the MES system, which will enable to link machines with ERP. The company had to make difficult choices between i) acquiring software to manage in real time what is happening in the production area and ii) to better exploit the database they already have through tools such as powerBI (one of the BDA tools for data processing)

(*continued*)

Table 4. (*continued*)

Implementation opportunities and challenges
Opportunities: Reduced use of paper - According to I2 CB "*until recently I had 4 or 5 files full of dossiers and now I don't have a single one*". Prior to the implementation of ERP, the documental process was made in *conectograms* that had to be printed, for visualization, analysis and signature. With the automation, the printing process is no longer necessary, thus allowing a significant reduction in the use of paper. Additionally, the processes' automation also increased the efficiency of the machines in the packaging process, which led to positive differences in the operational results ***Challenges***: Pharma is a highly regulated industry, so every process must be documented and signed in order to follow the *Good Manufacturing Practices* (GMPs) and to be approved by the regulatory authorities of each country to which the company exports Resistance to change is one of the big challenges, I2 CB said "*people always resist*" and adds that "*digitization will not reduce employment, it will redirect it. This is one of the points of our (technological) roadmap: not to reduce, but to redirect people to other tasks*". Being a producer of generic drugs increases the investment and implementation challenge, since generic producers have a lower profit margin compared to other companies that produce branded drugs and buying a MES system for a plant of this size is a high investment (including annual licenses)
Company readiness to implement the technologies
The company is aware of the positive impact of process automation, and in order to minimize the consequences of the natural resistance to change, the company has already started to train employees to better communicate the importance of automation for the company. Although the processes automation is not yet implemented in all company areas, a roadmap has already been designed with the objective of full automation within 10 years. The company's current plans include investments in the maintenance area. The idea is to use AI to perform certain interventions in the equipment, such as preventive maintenance or calibrations. The maintenance operators receive notifications and information about the maintenance intervention on electronic devices (smartphones and tablets). All the documentation associated with that equipment and respective intervention is registered in the system
Industry 4.0 technologies and the combat the COVID-19
I2 CB refereed that the company did not stop the production during the pandemic period (2020) and the employees that have functions compatible with remote work, performed their tasks at their homes

4.5 Results by Category: Cross Case Analysis

In Table 5 a synthesis of the main results of cross case analysis per category are presented:

Table 5. Results per category – cross case analysis

Knowledge of the Industry 4.0 paradigm
Both interviewees showed to be well informed about the I4.0 paradigm and about its impacts in the production area of their companies. There is, however, a difference in relation to the perceptions of the employees in both companies: in company A although many employees are not fully aware of the digital transformation, they are aware of the rapid technological evolution that is occurring in the company and its impacts, not only in their lives but also in the way they work; company B is already at a more advanced stage, and is training the employees for better handling with these technologies
Industry 4.0 paradigm implementation
The late start I4.0 technologies implementation (when compared to other companies in the same sector) is common in both companies and is justified by the limitations imposed by legal regulations. Both companies are already implementing pharma I4.0 technologies, not only to some production processes but also in quality control. However, both companies recognize that they still have a long journey to full production automation Both interviewees agreed that, in their companies, before implementing any technology it is necessary to develop and apply roadmaps, do pilot tests, as well as to implement the technologies in phases. The two companies are at a different level of digital maturity: company A has started to implement an EBR in production while company B is already at a higher maturity level, having implemented ERP and is planning the implementation of the MES
Implementation opportunities and challenges
Opportunities: Both companies reported: i) having efficiency gains in production, making the company more competitive (more competitive prices in the market); ii) reduce the use of paper, making them more sustainable. According to I1 CA, the technological solutions of I4.0 enable companies to increase their revenues through cost reduction, increase flexibility, improve speed and quality **Challenges**: i) Both interviewees reported situations of resistance to change, with the prevailing idea among employees that technologies will eliminate jobs. In both companies the intention is to redirect people, whose tasks were eliminated by technologies, to other areas of the company. ii) The lack of qualified labor to work with I4.0 technologies, which implies task changes, is another challenge identified by the both interviewees. They also acknowledge that it is necessary to train people so that they are better prepared for this new reality; iii) Both interviewees refer that heavy investments are needed for the acquisition and implementation of I4.0 technologies, recognizing that economic incentives from the Government would allow companies to feel more confident (lower risk) in doing the investments. Another challenge that arises, according to I1 CA, is the need for pharmaceutical companies to start producing strategically and locally. The interviewee states that COVID-19 reinforced this idea because in a pandemic situation the raw materials to make medicines may not be accessible and it is necessary that companies are prepared to produce them locally. The same interviewee further states that if companies were prepared to produce the raw material locally, at this time the production and distribution of the vaccine discovered by Pfizer for COVID-19 would not be so challenging A challenge is also posed to companies that produce generic drugs (ex. company B) where I1 CB mentions that, although these companies are receptive to the adoption of I4.0 technologies, in general its implementation is very difficult due to the nature of the business (recovering the investment)

(*continued*)

Table 5. (*continued*)

Company readiness to implement the technologies
Both companies i) are still at an early stage of implementation and doing pilot tests for the new technologies; ii) although they are already adopting some technologies such as EBR, MES and robots, they are not yet fully automated. iii) many operators are not yet fully aware of the transformation that the sector is undergoing and there is still a great resistance to change; iv) lack of skills of some workers to work with the new technologies. However, both managers are informed about the I4.0 paradigm and are committed to define strategies to continue the implementation of the I4.0 paradigm in their companies
Industry 4.0 technologies and the combat the COVID-19
The companies did not stop production during the pandemic. In the case of CA the use of collaborative robots on the shop floor and the use of EBR avoided the concentration of people on the shop floor and also allowed a drastic reduction of paper which substantially reduces the probability of contagion within the company. In addition, in both companies, the people who could perform their tasks remotely (thanks to the digital technologies already implemented) started to work remotely In an external context, I3 refers that the pandemic situation caused by COVID-19 accelerated the advancement of these technologies, since this situation caused a lot of pressure to quickly find a vaccine. Therefore, through technologies, such as mRNA it was possible to accelerate the process and produce a vaccine with a high degree of reliability. The pharmaceutical company Pfizer is an example of a large company with which the technology provider (where I3 works) cooperates, which allows them to use the technological solutions in an early stage

4.6 Recommendations

Both interviewees I1 CA and I2 CB mentioned that implementation processes of I4.0 technologies require time. PwC [4] defined six steps to implement I4.0 technologies: i) Define the strategy for I4.0; ii) Develop pilot projects; iii) Define the core competences needed; iv) Develop data analysis capabilities; v) Proceed with the transformation and vi) Plan the approach to the digital ecosystem. These recommendations are reinforced by interviewees I1 CA and I2 CB who state that this transition should happen according to a well-defined strategy. To this end, it is advisable to use pilot tests as suggested also in the literature [13]. It is necessary to have an effective internal communication in order to inform and raise awareness among employees about this new reality. This process may start by presenting the reasons for the need for the digital transformation and what are the positive and negative impacts of this transformation for the company and for the employees [4].

Interviewee I1 CA indicates that the knowledge of technological solutions presented by technology providers is made through the company's participation in conferences. In addition to the aspects previously mentioned, these conferences also present the contexts/scenarios where technologies can be applied, which helps companies become more aware of the applicability and impact of technologies. Thus, it is recommended that pharmaceutical companies seek to identify technology suppliers (consultants or non-profit organizations such as ISPE) and, with the support of these companies, to set the goals for the technological roadmap.

5 Discussion and Conclusion

The conclusions indicate that the national pharmaceutical companies covered by this study are managed by teams with experience in innovation projects and know-how of the I4.0 paradigm. However, there is still a lack of knowledge of this reality among employees with lower levels of education. Therefore, these employees are not well informed about the positive impacts that these technologies present to them and to their companies, which leads them to consider that adopting these technologies will put their jobs at risk. However, as previously presented, the employees with the tasks that will be extinct as a consequence of automation will be redirected to other tasks in the company and for which they have to be trained.

Through the implementation of I4.0 technologies it is possible to achieve a reduction in costs, an increase in efficiency and speed in some processes, greater transparency and quality of data, a significant reduction in the amount of paper used, and better communication and visibility of information within the company. The pharmaceutical companies included in this study, by using automation and digital tools, were able to maintain production during the pandemic period, minimizing face-to-face contacts and allowing a significant reduction of the staff present in the company.

This study intends to contribute to increase the awareness, among companies, regarding the importance of the I4.0 paradigm, especially the pharmaceutical companies that produce drugs. In Portugal there are few companies in this segment. Those that exist are small and have low production levels when compared to other companies in other countries. Therefore, it is critical to move forward and increase the investment in I4.0 technologies and to train people, aligning their competences/skills with what I4.0 needs. As Darwin [14] stated it is not the strongest who survive, but those who best adapt to change. For pharmaceutical companies in Portugal the question regarding the adoption of I4.0 technologies is not "what if?" but "when?". The impacts of I4.0 in the pharmaceutical industry in Portugal, as this study concludes, are real. However, it is necessary that companies invest in the transition from a more traditional production process to an automated one and do it with proper planning. Companies that do not follow this transformation will not be able to remain competitive in the market.

This study presents some limitations, mainly in relation to the methodology. It only includes a sample of two pharmaceutical companies and therefore the results are very limited to these two realities, not allowing for the generalization of the results. In addition, only three people were interviewed, which increases the risk of single informant bias. Furthermore, data collection using interviews is also subject to different interpretations of concepts on both sides (interviewer and interviewee). The pandemic context made it difficult for companies to participate in this study. However, with the data collected it was possible to reach relevant conclusions, which can be considered significant as a first approach to this topic. It was a great challenge to contact and involve pharmaceutical companies during the pandemic phase experienced in 2020 because companies had their efforts focused on combating the virus. Thus, as future work, it is suggested to increase the number of companies, as well as the number of interviews per company in order to have a broader perspective on this theme.

Acknowledgments. This work is financed by National Funds through the Portuguese funding agency, FCT - Fundação para a Ciência e a Tecnologia within project UIDB/50014/2020.

References

1. Liao, Y., Deschamps, F., Loures, Rocha E., Ramos, L.: Past, present and future of Industry 4.0-a systematic literature review and research agenda proposal. Int. J. Prod. Res. **55**(12), 3609–3629 (2017). https://doi.org/10.1080/00207543.2017.1308576
2. Escária, V.: Estudo de Diagnóstico Estratégico da Indústria Farmacêutica em Portugal-Relatório Final (2016)
3. Binggeli, L., Heesakkers, H., Wölbeling, C., Zimmer, T.: Pharma 4.0™: Hype or Reality? Pharmaceutical Engineering (2018). https://ispe.org/pharmaceutical-engineering/july-august-2018/pharma-40tm-hype-or-reality (2018).
4. Correia, A., Deus, P., Baptista, J.: Indústria 4.0: Construir a Empresa Digital. Price Price Waterhouse Coopers Portugal (2016). https://www.pwc.pt/pt/temas-actuais/2016/pwc-industria-40.pdf
5. Siemens: Digitalization in Pharma. Presentation Made by Siemens at the Digitalization in Pharma Forum (2017)
6. Apifarma. https://www.apifarma.pt/apifarma/Paginas/Missão.aspx (2020)
7. ISPE. Pharma 4.0. https://ispe.org/initiatives/pharma-4.0. (2017)
8. Reis, J.: Indústria 4.0: Estado da arte, análise de casos de estudo e proposta de um serviço de consultoria, Dissertação *de* Mestrado. Mestrado Integrado em Engenharia Mecânica. Faculdade de Engenharia da Universidade do Porto (2019)
9. Ehrhardt, M.: Is pharma ready for the future? (2015). https://www.strategy-business.com/article/00363?gko=06d07
10. Markarian, J.: Modernizing pharma manufacturing. Pharm Technol **42**(4), 20–25 (2018)
11. Yin, R.K.: Case Study Research and Applications Design and methods, 6th edn. Sage Publications, Thousand Oaks (2018)
12. Strauss, A., Corbin, J.: Basics of Qualitative Research: Techniques and Procedures for Developing Grounded Theory, 2nd edn. Sage Publications, London (1998)
13. Barata, J., Cunha, P.R.D., Coyle, S.: Guidelines for using pilot projects in the fourth industrial revolution. In: ACIS 20 Proceedings (2018). https://aisel.aisnet.org/acis2018/80
14. Darwin, C.: On the Origin of Species by Means of Natural Selection, or the Preservation of Favoured Races in the Struggle for Life. John Murray, London (1859)
15. Ding, B.: Pharma Industry 4.0: literature review and research opportunities in sustainable pharmaceutical supply chains. Process Saf. Environ. Prot. **119**, 115–130 (2018). https://doi.org/10.1016/j.psep.2018.06.031
16. Reinhardt, C., Oliveira, J., Ring, D.: Current perspectives on the development of Industry 4.0 in the pharmaceutical sector. J. Ind. Inf. Integr. **18** (2020). https://doi.org/10.1016/j.jii.2020.100131
17. Steinwandter, V., Borchert, D., Herwig, C.: Data science tools and applications on the way to Pharma 4.0. Drug Discov. Today **24**(9), 1795–1805 (2019). https://doi.org/10.1016/j.drudis.2019.06.005
18. Barenji, R., Akdag, Y., Yet, B., Oner L.: Cyber-physical-based PAT (CPbPAT) framework for Pharma 4.0. Int. J. Pharma. **567** (2019). https://doi.org/10.1016/j.ijpharm.2019.06.036
19. Kumar, S., Talasila, D., Gowrav, M., Gangadharappa, H.: Adaptations of pharma 4.0 from industry 4.0. Drug invent. Today **14**, 405–415 (2020)

20. Chen, Y., Yang, O., Sampat, C., Bhalode, P., Ramachandran, R., Ierapetritou, M.: Digital twins in pharmaceutical and biopharmaceutical manufacturing: a literature review. Processes **8**(9), 1088 (2020)
21. Instituto Nacional de Estatística - Estatísticas da Produção Industrial: 2019. Lisboa: INE (2020). ISSN: 0872-9298, ISBN: 978-989-25-0548-0. https://www.ine.pt/xurl/pub/467890702

Digital Customer Network Strategy Influences on Hotel Business

Teresa Dieguez[1,2], Nguyen Thi Ly[3], Luis Pinto Ferreira[4,5(✉)], and Francisco José Gomes da Silva[4,5]

[1] IPCA - Polytechnic Institute of Cávado and Ave, Vila Frescainha, Portugal
[2] CiTUR – Centre for Tourism Research, Development and Innovation, Leiria, Portugal
[3] ESHT, School of Hospitality and Tourism, Polytechnic of Porto, Vila do Conde, Portugal
[4] ISEP - School of Engineering, Polytechnic of Porto, Porto, Portugal
lpf@isep.ipp.pt
[5] INEGI – Instituto de Ciência e Inovação em Engenharia Mecânica e Engenharia Industrial, Porto, Portugal

Abstract. The new digital era is highly transforming and influencing the hotel industry. Customer' satisfaction impacts future business and customer networks emanate from digital platforms. Digital customer networks influence consumer perceptions, attitudes, and decisions to purchase. However, these digital customer networks are applicable as strategies to improve sales, reduce costs, and build loyal relationships through customers' access, connection, engage and collaboration. The present research slicked to assess the influence of digital customer network strategy on hotel business. For this purpose, it has been conducted an exploratory study, through interviews, using remote videoconference, in July 2020 to near five General Manager of five different hotels based in Portugal. The study shows that nowadays the digital is effectively transforming the way stakeholders do business, innovate, work and cooperate. It also adds that in future this influence will be higher, which demands an accurate attention from the management and an incorporation of digital on hotel strategies. In fact, the study shows that digital marketing enhances customer service, it is cheaper and reaches the market and consumers more effectively. Social media promotes the hotel's image and brand, but it can also blacken it quickly. The SEO (Search Engine Optimization) strategy is impactful and if wisely managed, it can have a long-lasting impact with free-of-cost and an attractive website. The website helps the business to get to know about its customers and represents the major operations of the hotel.

Keywords: Digital marketing · Business strategy · Digital customer network · Marketing strategy · Hotel business

1 Introduction

The digital customer network strategy is increasingly becoming a focus for business seeking to expand their operations and profitability [1]. Almost all industries in the global business environment are exploring ways to harness the online business environment and

J. Machado et al. (Eds.): icieng 2021, LNME, pp. 420–431, 2022.
https://doi.org/10.1007/978-3-030-78170-5_36

reach new market and clients. The hotel industry is one of those which have joined the online business community, an Internet marketplace not very different from the physical business environment in what concerns challenges, competitiveness and rules. Current research on hotel business shows the increasing significance of digital customer network strategy, representing a target group that hotel focus on for their many products and offers in order to increase their market share, target and serve new business markets [2]. Every hotel business today must set up a working website that creates and builds on a community that will remain loyal to the brand [3]. Current research approves the use of digital media in marketing the different offers to their customers [4]. Every company has a defined strategy for its growth and an increase in profitability, being crucial to increase specific marketing strategies to particular customer networks, especially when facing low seasons [5]. Most of the traffic generated in online websites is driven by the marketing strategies employed by the company [6]. Digital media can help hotel business to influence the behaviors of its customers through a community chat on their website [7] and also through the creation of a large community that maintains a steady stream of clients throughout different seasons, like international hotel chains already do through Facebook Group, Twitter's chats, and Instagram posts [8]. In this way, hotels can conveniently inform their customers of new and upcoming products [9], while customers can plan their time around the purchasing and financing of their products. It allows reaching a broader population scale, reaching potential customers scattered around the world [10]. Secondly, digital customer networking strategies are inexpensive compared to traditional mass media channels of marketing and advertising [11]. It usually requires the company's marketing team to create active and interactive media profiles to reach out to customers and attract new numbers. From a digital marketing perspective, the company only requires one staff member to make a relevant post that announces the product to the market [12]. Sharing this information to millions of potential customers does not require financial investment, except for the organization's foundational infrastructure. Compared to traditional marketing strategies, digital marketing is less expensive and accommodate different scales of business. Current Covid-19 pandemic provides unique opportunities for hotel businesses across the world. Being one of the most affected industries by the pandemic [13], the hotel industry has to find solutions to attract customers and get visibility and competitive advantages [14]. The customers are now more than a mass of isolated individuals, and hotels can develop their customer network strategy by knowing their behaviors and generating real value. They also must provide online and app-based services to customers, producing attractive content on websites and social media, as well as engaging customers in various campaigns. Digital customer network strategy or digital marketing strategy is, then, an inevitable part of the marketing and branding [2]. The present study is divided in 4 sections. After the introduction is presented a literature review about the state-of-the art, related to the concept, role and elements of digital marketing, strategy of digital marketing in business development and customer network. Section 3 explains the research methodology and finally Sect. 4 highlights the main ideas, limitations and suggests future research.

2 Review of Literature

Digital marketing can be defined as management of digital technologies, which are used to create channels to reach potential recipients, aiming to achieve the enterprise's goals, through more effective fulfillment of the consumer needs [15]. The Digital marketing is oftentimes considered as a synonym of Internet marketing or e-marketing, but in fact is a digital identity of a company. It consists in promotion of products or brands through one of diverse electronic forms, with which companies can send personalized contents to particular recipients [15]. In reference to the earlier development of fast and efficient transaction technologies, digital marketing as a concept majorly outlines a group of processes that integrates all the digital channels around to promote a brand, product, or service. In the past decades, digital marketing revolved around websites, but later diversified into the advertising industry and various bidding sites like Alibaba among others [16]. Digital marketing is viable through the internet both as working and displaying platform [17]. To unveil the real usefulness of Digital marketing, it is essential to consider the work scope of e-commerce specialists, which majorly involves the processes that are established and developed in an entity to identify, attract and build customer loyalty. The customization and peculiarity of digital marketing are created by some specific and relational characteristics in various operational functions such as customer service, promotion of sales, confidentiality, security, personality, and the community, among others [18]. Digital marketing creates and leads to various dynamics not only among businesses but also among customer behavior [19]. For such an environment to be operational, digital marketing gives a unique platform for companies to identify and understand customer requirements and create various prospects for them based on the place and time. It is also useful in reducing the cost of marketing by doing away with unnecessary transactions. In the last decades, Digital marketing has rapidly undergone positive changes and developments; this is evidenced by the widespread use of the Internet by various companies in the world, majorly for product marketing and brand promotion. Other entities have fully integrated Internet facilities into their departments, which are considered as the modern platform that is necessary for the development and growth of business in the current era [20].

2.1 Elements of Digital Marketing

Over the past few decades, global industries faced technological changes that led to opportunities such as greater flexibility, reactivity and product individualization. However, they also presented various challenges such as rapid technological change, increased complexity, changing customer preferences and legal requirements [21]. In such a situation, shaping digital markets involves the success factors of the traditional marketing channels by moving from the publishing of ads in magazines and newspapers to social media platforms, among other virtual platforms [22]. In this case, the tools that are connected to the internet marketing also facilitating the design and development of different transactions within different supply chains, namely: Websites [23], Search Engine Marketing [24], Social Media Marketing [25], Affiliate Marketing [26], Email Marketing [27], Mobile Marketing [28], Video Marketing [29], Corporate Marketing [30] and Blogging [31].

2.2 The Strategy of Digital Marketing in Business Development

From earlier discussions, digital marketing is among the essential types of marketing in the current era and has become popular due to the low costs associated with the processes. Using this platform, enterprises can accurately reach a targeted group of clients at the lowest customers' attracting cost and reduced loss in marketing campaigns [32]. Digital marketing has gained its popularity because it allows marketers to target and track various aspects of the businesses, such as the customer satisfaction index and the return on investment (ROI). Satisfaction is the consumer's assessment of a product or service in terms of the extent to which that product or service has met his/her needs or expectations [33]. The customer satisfaction index is self-weighting to maximize the explanation of customer satisfaction on customer loyalty. Looking at the indexes and impacts, users can determine which drivers of satisfaction, if improved, would have the most effect on customer loyalty [34]. ROI is the ratio of net profit from any investment. In the field of marketing, ROI could be derived by getting the revenue received by the business minus the strategic investment cost divided by the investment cost [35]. By applying the method of financial formula, ROI could lead to errors in the data because it cannot easily substitute the values only because of the marketing campaign earned revenue. Most researchers have resorted to separating the expenditure for marketing. Therefore, in this regard, the subsequent sections will examine how the strategies of digital marketing lead to business development, namely access, engagement, connection and collaboration.

2.3 Customer Network

Any marketing campaign aims to help the organization to identify, satisfy and retain their current customers. Such activities are necessary for the formation of strategic customer recruitment methodologies and relationship management. For marketing practitioners, digital marketing presents an automatic outgrowth of the marketing concept, which majorly orients the entire organization around understanding and addressing the necessary customer needs [36]. Technology has enabled companies to collect and analyze client information in a meaningful and great extent. The Internet has provided the necessary platforms for customers and businesses to discover each other and communicate together. The customer network helps the firm to achieve the following outcomes: maximizing the client´s lifetime value [37], creating competitive advantage [38], turning customers into Marketers [36], network science [39], social network analysis [40, 41], and capital network analysis [42].

In the current market, most consumers tend to learn about brands through the Internet [43, 44]. Consumers have the potential of generating more than 500 billion impressions on a product or service in a year through social networks, which is about a quarter of the total number of ideologies that companies create via online advertising [45]. According to [46], the development of digital technology has rendered the traditional factors of advertisement obsolete. Besides, the media and market have become more fragmented. The evolution of digital platforms has rendered consumers more empowered and paved a modern era for prompt producer-consumer interaction. Besides, the Internet has turned consumers into knowledgeable people. The difference between the new consumer and the traditional consumer is that the current consumer is more knowledgeable about various goods and services and could play a crucial part in enhancing these services, products, or brands. Moreover, due to the existence of a close community of consumers and widespread knowledge would be exploited by a business only in cases where they know that the exploiter would give value-added services or products [46]. This new breed of consumers could use the Internet and tell the brand owners what they want, their expectations, and their thoughts [47]. In short, the client makes clearly their demands and needs, and it is up to the business to deliver. Instead of pushing a product's market onto consumers, marketers should consider dialoguing with their clients to gain their trust. Generally, this is the future of marketing since it hugely recognizes that consumers are in control of the whole process as opposed to the traditional forms of marketing. Rao et al. [48] stated that information technology has resulted in established consumer-producer collaborative relationships. The authors outlined that, as customers are continuously becoming less trusting of companies, the remaining trust should direct in those companies that work in tandem with them when developing products and services. However, consumers trust each other a lot as compared to the manufacturer; they tend to find the information given by their fellow consumers as credible. Producers and marketers should never ignore the influence that clients have on their fellow consumers in terms of passing various messages. Marketers should note that an individual's social network does not finish with their direct network. However, users can pass communications from one system to another if there is commonality of members in one or two more networks. Consumers trust in each other, and social network platforms have the potential as a word-of-mouth communications medium regarding the services, brands, and products [49]. The main concern for marketers and firms should be how to generate this faith and dialogue with various consumers and inspire them to influence other consumers by passing on the marketing/promotional messages and recommendations to each other.

Consumers that use social platforms view these spheres as their personal space, where they can freely air out their opinions, beliefs, insights, and thoughts with those that have similar interests. The capacity to become broadcaster and share vital information has allowed consumers to become more influential and command their needs and anticipate from the brand, business, or the marketers [50]. Empowered consumers expect that companies should meet their demands. With the appropriate use of social networks, platforms can be a means for entities to engage with consumers and encourage them to pass on the firm's messages to their networks that the company established on-site. Since consumers spend most of their time on social networks, engaging them through these platforms is an essential act.

3 Research Methodology

The Research methodology refers to "a system of beliefs and assumptions about the development of knowledge" [51]. It plays a "vital role in academic research" [52] and its aim is to provide a structure for planning and do operational activities that follow up the carrying out of the procedures [53]. For the purpose of the present study, the research Question is expressed as follows: "What is the impact of digital customer network strategy on the hotel industry?". For the present research the specific objectives are additionally intended:

- To investigate the overall influence of digital technology on the hotel industry;
- To examine the impact of social media marketing on the hotel business;
- To explore whether SEO strategy is impactful in the hotel industry;
- To explore the role of the website in boosting the hotel industry;
- To investigate the challenges and risks in the use of social network.

3.1 Research Steps

The research methodology consists of several steps, which have been illustrated in six different layers by Saunders et al. [51], namely: philosophy, approach, choice, strategy, time horizon and techniques & procedures. In this study, it was used the interpretivism as philosophy research. Researchers try to analyze human emotions and the social role through interviews, observation, and analysis of the existing texts [54]. Researchers interacted with other people (interviewees) and constructed a meaningful reality by collaborating with the people through long five interview sessions. As research approach, the study is inductive, in which the researchers intend to use a thematic analysis. One of the popular methods of qualitative approach, i.e., coding, was used for the analysis. It was hard to maintain objectivity in the research considering its objectives, for data analysis, the narrative type. Every theme derived from the coding was analyzed narratively. As research strategy, it is an ethnographic study as it examines the behavior and base on the observation of the group. It is a cross-sectional or short-term research.

In the present study, researchers conducted interviews with five general managers (GMs) of five hotels. It is a convenience sampling, also known as haphazard sampling, because the interviewed were chosen due their easy accessibility, availability at a given time and readiness to participate. Although frequently used in research, it is neither determined nor strategic [55]. The main belief associated with convenience sampling is that the members of the target population are consistent. That is, that there would be no variance in the investigation results attained from a random sample, a nearby sample, a co-operative sample, or a sample gathered in some unreachable fragment of the population [56]. One of obvious disadvantage of convenience sampling is that it is likely to be biased [57], but the authors of the present study are aware of it and presents it as an exploratory study. For the interviews, an interview guided was prepared. The interview questions are divided into four different parts. In the first part, the researchers asked interviewees some demographic questions such as their years of experience in the hotel industry, their educational background and their duties and responsibilities. The second section of the interview is the web and social network, which is one of the key concepts of our research aim. Eleven questions are formulated under this section. In the third part, the questions are based on the key goals of social media in the context of the hotel industry. A total of six questions are included in this section. The fourth and last part of the interview consists of the questions related to the impact of using social media.

3.2 Data Analysis

As earlier mentioned, the researchers relied on a thematic analysis of the collected data. The researchers conducted interviews with five general managers (GM) of five different hotels, located in Portugal: two five stars and three four stars hotels. Due confidential reason they were referred in the study as GM1, GM2, GM3, GM4, GM5. GM1 is in Carvoeiro, offers custom-designed premium services for the families and claims to be the most romantic resort in Europe. The website is updated, dynamic and appealing. GM2 is in Lisbon, has a subtle combination of tradition and trend and strives to attain perfection. Comfort, confidence and centrality have been the three commitments of the hotels to its guests. GM3, GM4 and GM5 are in the North of Portugal. GM3 offers rural tourism experience to the guests and claims that it provides personalized treatment to them. GM4 is for sports lovers and offers the service of swimming and football. GM5 provides the same facilities as the other hotels are providing such as gym, spa, restaurant. The hotel also has prepared himself for delivery clean, protected and efficient service to its customers during COVID pandemic. All the general managers have been working in their respective hotels for several years, and they have extensive experience of more than 15 years in the hotel and hospitality industry. It was a semi-structured interview in which the respondents were asked questions related to the research topic. Interviews were recorded, extracted, and the paraphrased content was highlighted in different colors. Out of the extract, researchers obtained a total of six themes, summarized on Table 1.

In what concerns Digital Customer Strategy, customer-oriented strategy is the effective method to catch customers for present and future by the way of using creativity, so that they can attract market? Hotels choose to conduct customer profile review to know the customer in a better way. The company obtains required information about the customers based on two main clusters: data set and customer's segmentation. Data set: five types of data helps make segmentation decides whether products or service suit to customers, namely demographic (age, gender, education, income), psychographic (attitude, personal traits), behavioral (engagement in business, behavior of buying products or service), financial and transactional (payment history, buying preference).

In what concerns customer´s segmentation: four categories of analysis help decide whether products or service are suitable to the financial results of the business: negative (more cost than profit), low (less profit, pose high risk), medium (moderately profit), high (high profitable). Employee-friendliness is of utmost importance in the hotel industry as it is a people-oriented service sector. Attitudes and behavior of the hotel staff are the cornerstones of the hotel industry. It is important to keep the employees satisfied and happy. Social networking sites are the Internet or digital-based platforms where the people can make friends and stay connected with family, friends, colleagues, customers and other stakeholders. There are benefits of social media and how benefits it brings. Besides benefits, social media also causes risk for users that sum from the interviews. The risks include two main types: customers' complaint and negative feedback, and blackmailing. Based on information of participants, there have two types of communication models: Verbal or Explicit Communication and Non-verbal or implicit Communication. Nowadays, Non-verbal or implicit Communication is concerned. While talking about digital networking, content is one of the crucial factors. The types of public accountability are shown via content that marketer posted in social media: fulfillment of promise (be

Table 1. Themes and coding.

Themes <=> coding
i) Digital customer strategy for present and future <=> Video marketing/Interaction on social media / Market demand and trend/Market situation/Social networking tools/Customer's Profile Review/Content analysis/Appointment of an experienced marketing manager/Customer interaction/Brand Building/Making customer change/Right interaction/Tools and software/Artificial Intelligence
ii) Employee-friendly reputation on social media <=> Familial relationship with employees/Involvement of employee families/Social media usage/Employee recognition
iii) Risks of social media <=> Threatening from customers/Fake pictures/Spoiling reputation/Blackmailing by using social media/Complaining/Writing negative reviews/Wrong message
iv) Language and Communication (Verbal and non-verbal) safety Interaction <=> Language/Communication/Messages/Portuguese/English/French/Italian/Customer experience/Customer feedback/Reaching tothe audience/Public opinion/Image building/Directinteraction/Customersafety
v) Importance and essentials of social media <=> Social media/Free-of-the-charge/To inform/spreading message/Adherence to safety measures/Assurance of safety and cleanliness/Use of masks/company's existence/Promotion points/Low-cost option/Knowing the customer/Competitors' actions update/Helping customers more effectively/Analysing customers' feedback on social media
vi) Public Accountability <=> Fulfilment of promises/Responsibility regarding content/Awareness about the customers

surely safe and securing during stay at hotel especially during the recent pandemic) and responsibility regarding content.

4 Conclusions and Future Work

The main goal of the present study was to examine how the digital network strategy influences the hotel business. For that purpose, an exploratory research was carried out, in July 2020, through qualitative techniques. Five interviews were conducted, between five general managers of five different hotels. To investigate the overall influence of digital technology on the hotel industry, the study shows that it enhances customer service, market their services at a surprisingly low cost comparing to traditional tools, provides the online booking system and it is a convenient communication between hoteliers and guest. To examine the impact of social media marketing on the hotel business, the research points out the growing of the hotel business in a very short time span, the building hotel image quickly than before and the tarnishing hotel image by posting negative reviews. To explore whether SEO strategy is impactful in the hotel industry, the study shows that digital marketing helps to gain higher rank in business via using the appropriate and precise use of keywords, to have a long-lasting impact with Free-Of-Cost and to make website more attractive. To explore the role of the website in boosting the hotel industry, digital marketing helps learning about the hotel as the customers' first reaction, is the key platform to communicate with customers and represents the major operations of the hotel. To investigate the challenges and risks in the use of social network, digital marketing aids sustaining the image of the business, may help losing Internet Protocol (IP) and sensitive data. Apart from the threats and risks of social media, hotels cannot boost and sustain their business without using digital technology (including social media). The strategic and careful use of social media ensures quality service to customers. It also has made the operation of this industry easy and hassle-free. As limitations, researchers identify the primary data, only derived from five participants. If researchers had included more participants, more information could have been obtained. Moreover, it is a qualitative method in which researchers relied only on the five interviews. Future studies can also use quantitative methods with many participants. A case study method can also be used for future studies. Validity is the central issue of many of the scholarly studies. It means that the conclusion and findings derived from the research may not be applicable after some years. It is because of the market trend, technology, and other factors that change over time. That is why future researchers should conduct ongoing research in this area.

References

1. Nylén, D., Holmström, J.: Digital innovation strategy: A framework for diagnosing and improving digital product and service innovation. Bus. Horiz. **58**(1), 57–67 (2015)
2. Verhoef, P.C., et al.: Digital transformation: a multidisciplinary reflection and research agenda. J. Bus. Res. **122**, 889–901 (2019)
3. Gielens, K., Steenkamp, J.B.E.: Branding in the era of digital (dis) intermediation. Int. J. Res. Mark. **36**(3), 367–384 (2019)
4. Kafley, S., Mojumder, K.: Customer experience and digital marketing in the hotel business. Degree programm in restaurant entrepreneurship bachelor's thesis. Laurea University of Applied Sciences (2018)
5. Rogers, D.L.: The Network Is Your Customer: Five Strategies to Thrive in a Digital Age. Yale University Press, New Haven (2014)
6. Ritter, T., Pedersen, C.L.: Digitization capability and the digitalization of business models in business-to-business firms: past, present, and future. Ind. Mark. Manage. **86**, 180–190 (2020)
7. Hofacker, C.F., Belanche, D.: Eight social media challenges for marketing managers. Span. J. Market ESIC **20**(2), 73–80 (2016)
8. Hashim, K.F., Fadhil, N.A.: Engaging with customer using social media platform: a case study of Malaysia hotels. Procedia Comput. Sci. **124**, 4–11 (2017)
9. De. Pelsmacker, P., Van Tilburg, S., Holthof, C.: Digital marketing strategies, online reviews, and hotel performance. Int. J. Hosp. Manag. **72**, 47–55 (2018)
10. Bolton, R.N., et al.: Customer experience challenges: bringing together digital, physical and social realms. J. Serv. Manag. **29**(5), 776–808 (2018)
11. Galati, A., Crescimanno, M., Tinervia, S., Fagnani, F.: Social media as a strategic marketing tool in the Sicilian wine industry: evidence from facebook. Wine Econ. Policy **6**(1), 40–47 (2017)
12. Kaakinen, H., Purkayastha, E.: Online marketing of hospitality services: tourist satisfaction with online accommodation booking. Bachelor's thesis Centria University of Applied Sciences Degree Program in Tourism. Finland (2016)
13. Tidey, A.: Coronavirus in Europe: Tourism Sector 'hardest Hit' by Covid-19. Euronews. https://www.euronews.com/2020/04/16/coronavirus-in-europe-tourism-sector-hardest-hit-by-covid-19. Accessed 12 May 2020
14. Dwivedi, Y.K., et al.: Setting the future of digital and social media marketing research: perspectives and research propositions. Int. J. Inf. Manag. (2020). https://doi.org/10.1016/j.ijinfomgt.2020.102168
15. Sawicki, A.: Digital marketing. World Sci. News **48**, 82–88 (2016)
16. Aslam, B., Karjaluoto, H.: Digital advertising around paid spaces, E-advertising industry's revenue engine: a review and research agenda. Telemat. Inform. **34**(8), 1650–1662 (2017)

17. Ghotbifar, F., Marjani, M., Ramazani, A.: Identifying and assessing the factors affecting skill gap in digital marketing in communication industry companies. Independ. J. Manage. Prod. **8**(1), 1–14 (2017)
18. Ahmed, R.R., Streimikiene, D., Berchtold, G., Vveinhardt, J., Channar, Z.A., Soomro, R.H.: Effectiveness of online digital media advertising as a strategic tool for building brand sustainability: evidence from FMCGs and services sectors of Pakistan. Sustainability. **11**(12), 3436 (2019)
19. Baumöl, U., Hollebeek, L., Jung, R.: Dynamics of customer interaction on social media platforms. Electron. Mark. **26**(3), 199–202 (2016). https://doi.org/10.1007/s12525-016-0227-0
20. United Nations Conference on Trade and development (UNCTAD) Digital Economy Report 2019: Value Creation and Capture: Implications for developing countries. United Nations Publications, New York (2019)
21. Rachinger, M., Rauter, R., Müller, C., Vorraber, W., Schirgi, E.: Digitalization and its influence on business model innovation. J. Manuf. Technol. Manag. **30**(8), 1143–1160 (2019)
22. Espinoza, M., Cevallos, N., Tusev, A.: Changes in media consumption and its impact in modern advertising: a case study of advertising strategies in Ecuador. INNOVA Res. J. **2**(6), 120–135 (2017)
23. Bala, M-, Verma, D.: A critical review of digital marketing. Int. J. Manage. IT Eng. **8**(10), 321–339 (2018)
24. Gramatikova, E.: Trends in courier services in terms of digitalization of commerce. Izvestia J. Union Sci. Varna Econ. Sci. Ser. **8**(3), 17–27 (2019)
25. Saura, J.R., Palos-Sanchez, P.R., Correia, M.B.: Digital marketing strategies based on the e-business model: literature review and future directions. In: Guerra, A.G. (eds.) Organizational Transformation and Management Innovation in the Fourth Industrial Revolution, pp. 86–103, IGI Global, Hershey (2019)
26. Kumar, V., Pradhan, P.: Reputation management through online feedbacks in e-business environment. Int. J. Enterp. Inf. Syst. **12**(1), 21–37 (2016)
27. Datta, S.K., Sanyal, S.N.: Integrated marketing channels. In: Andaleeb, S.S. Hasan, K. (eds.) Strategic Marketing Management in Asia, Emerald Group Publishing Limited, Bingley, pp. 403–424 (2016). https://doi.org/10.1108/978-1-78635-746-520161014
28. Costa, T., Borges-Tiago, T., Tiago, F.: Pharmaceutical communication over social media channels: 24/7 management challenges. Digit. Commun. Manag. **6** 77–96 (2018)
29. Hänninen, N., Karjaluoto, H.: The effect of marketing communication on business relationship loyalty. Mark. Intell. Plan. **35**(4), 458–472 (2017)
30. Järvinen, J.: The Use of Digital Analytics for Measuring and Optimizing Digital Marketing Performance. Jyväskylä University School of Business and Economics (2016). ISBN: 978-951-39-6776-5 (nid.), ISBN: 978-951-39-6777–2 (PDF)
31. Järvinen, J., Taiminen, H.: Harnessing marketing automation for B2B content marketing. Ind. Mark. Manage. **54**, 164–175 (2016)
32. Atshaya, S., Rungta, S.: Digital marketing VS internet marketing: a detailed study. Int. J. Novel Res. Market. Manag. Econ. **3**(1), 29–33 (2016)
33. Gitomer, J.: Customer Satisfaction is Worthless, Customer Loyalty Is Priceless: How to Make Customers Love You, Keep Them Coming Back, and Tell Everyone They Know. Bard Press, Austin (1998)
34. Ilieska, K.: Customer satisfaction index – as a base for strategic marketing management. TEM J. **2**(4), 327–331 (2013)
35. Negoiţă, O.D., Purcărea, A.A., Popescu, M., Niculescu, A., Coman, C.N.: Digital marketing tools used by companies. FAIMA Bus. Manag. J. **6**(4), 66–76 (2018)

36. Samran, Z., Wahyuni, S., Misril, M., Nabila, R., Putri, A.: Determination of digital marketing strategies as effective communication techniques for go on travel brand awareness. J. Res. Market. **9**(3), 752–757 (2019)
37. Berghofer, F., Hofbauer, G., Sangl, A.: Indicators to choose a suitable marketing automation platform. Int. J. Manag. Sci. Bus. Admin. **4**(3), 52–60 (2018)
38. Mejía-Trejo, J.: Designing a digital marketing model innovation to increase the competitiveness. First Insight. Mexico Nova Sci. **10**(20), 569–591 (2018)
39. Centobelli, P., Cerchione, R., Esposito, E., Raffa, M.: Digital marketing in small and medium enterprises: the impact of web-based technologies. Adv. Sci. Lett. **22**(5–6), 1473–1476 (2016)
40. Herttua, T., Jakob, E., Nave, S., Gupta, R., Zylka, M.P.: Growth hacking: exploring the meaning of an internet-born digital marketing buzzword. Designing networks for innovation and improvisation. In: Proceedings of the 6th International COINs Conference, pp. 151–161, Springer Proceedings in Complexity (2016)
41. Stankovska, I., Josimovski, S., Edwards, C.: Digital channels diminish SME barriers: the case of the UK. Econ. Res-Ekonomska istraživanja **29**(1), 217–232 (2016)
42. Taneja, G., Vij, S.: Dynamics of a digital advertising campaign (January 5, 2019). In: Proceedings of 10th International Conference on Digital Strategies for Organizational Success (2019)
43. Hill, S., Provost, F., Volinsky, C.: Network-based marketing: identifying likely adopters via consumer networks. Stat. Sci. **21**(2), 256–276 (2006)
44. Barreto, S., Barbosa, R.J.V., Barbosa, B.: Optimization models in google ads campaigns. In: Semerádová, T., Weinlich, P. (eds.) Impact of Online Advertising on Business Performance, pp. 138–176. IGI Global, Hershey (2020)
45. Li, S., Vlassis, N., Kawale, J., Fu, Y.: Matching via dimensionality reduction for estimation of treatment effects in digital marketing campaigns. In: International Joint Conferences on Artificial Intelligence, pp. 3768–3774 (2016)
46. Swist, T., Hodge, B., Collin, P.: 'Cultural making': how complexity and power relations are modulated in transdisciplinary research. Continuum **30**(4), 489–501 (2016)
47. Kim, C.M.: Social Media Campaigns: Strategies for Public Relations and Marketing. Routledge, London (2016)
48. Rao, S., Srivatsala, V., Suneetha, V.: Optimizing technical ecosystem of digital marketing. In: Dash, S.S., Arun Bhaskar, M., Panigrahi, B.K., Das, S. (eds.) Artificial Intelligence and Evolutionary Computations in Engineering Systems. AISC, vol. 394, pp. 691–703. Springer, New Delhi (2016). https://doi.org/10.1007/978-81-322-2656-7_63
49. Ridge-Newman, A., Mitchell, M.: Digital political marketing. In: Political Marketing and the 2015 UK General Election, pp. 99–116, Palgrave Macmillan, London (2016)
50. Dodson, I.: The Art of Digital Marketing - The Definitive Guide to Creating Strategic, Targeted and Measurable Online Campaigns. Digital Marketing Institution. https://www.nima.today/wp-content/uploads/2018/11/The-Art-Of-Digital-Marketing-Ian-Dodson.pdf. Accessed 21 Jan 2020
51. Saunders, M., Lewis, P., Thornhill, A.: Research Methods for Business Students, 7th edn. Pearson Education, Harlow (2016)
52. Crisan, M., Borza, A.: Research methodology - a quality assurance instrument analysis of internationalized masters of Babes-Bolya University. Rev. Econ. Stud. Res. Virgil Madgearu **8**(1), 37–49 (2015)
53. Mwangi, I.K., Njoka, J.M.: Research methodology. In: Kumssa, A., Williams, J.H., Jones, J.F. (eds.) Conflict and Human Security in Africa, pp. 15–27. Palgrave Macmillan US, New York (2011). https://doi.org/10.1057/9780230119260_2
54. Rowlands, B.H.: Grounded in practice: using interpretive research to build theory. Electr. J. Bus. Res. Methodol. **3**(1), 81–92 (2005)

55. Palinkas, L.A., Horwitz, S.M., Green, C.A., Wisdom, J.P., Duan, N., Hoagwood, K.: Purposeful sampling for qualitative data collection and analysis in mixed method implementation research. Admin. Policy Ment. Health Ment. Health Serv. Res. **42**(5), 533–544 (2015)
56. Etikan, I., Musa, S.A., Alkassim, R.S.: Comparison of convenience sampling and purposive sampling. Am. J. Theor. Appl. Stat. **5**(1), 1–4 (2016)
57. Mackey, A., Gass, S.M.: Second Language Research: Methodology and Design. Routledge, London (2015)

Work Environment as a Factor in the Conduct of Manufacturing Processes

Adam Górny(✉)

Institute of Safety and Quality Engineering, Poznań University of Technology, Prof. J. Rychlewski Street No. 2 (room 357), 60-965 Poznań, Poland
adam.gorny@put.poznan.pl

Abstract. To make their manufacturing processes more efficient, manufacturers need to create a proper environment complete with adequate working conditions. Such an environment forms a crucial part of the set of environmental factors that affect worker performance. Its immense importance becomes evident as soon as the environment is recognized as part of manufacturing systems in which processes are designed to help transform inputs into outputs, regardless of the manner in which such inputs and outputs are identified.

Manufacturing processes are viewed as the way a manufacturer undertakes to achieve its objectives. Factors in the process environment may be seen as either disrupting or facilitating process flows. The classification of such factors depends on their nature and the severity of their impacts.

The article focuses on selected essential impacts of such factors. It identifies the place and significance of disruptors and facilitators within process structure. It also describes the way in which the work environment affects process efficiency and defines the conditions necessary for environmental factors to augment processes.

Keywords: Manufacturing processes · Work environment · Safety · Dominants and disruptions of processes

1 Introduction

The ability to conduct manufacturing processes is linked inextricably with the provision of adequate working conditions for workers [1]. A key component of such conditions is the work environment, which affects workers' ability to perform work as well as the safety of workers as well as any other persons who enter production floor [2]. Incompatibilities in the work environment usually negatively affect of the employees performance. Such an environment forms a crucial part of the overall set of those environmental factors that affect manufacturing performance. It is viewed in particular through the lens of the criteria that need to be met to ensure the proper functioning of the workforce.

The work environment is a vital component of any manufacturing system and one of critical importance for the proper conduct of activities [3], especially where new challenges arise [4]. The successful completion of processes depends crucially on the

J. Machado et al. (Eds.): icieng 2021, LNME, pp. 432–440, 2022.
https://doi.org/10.1007/978-3-030-78170-5_37

conditions in which they are conducted. For best results, manufacturers are advised to actively seek the continuous improvement of such conditions. In many sources, for a long time has been pointed that working conditions affect the satisfaction of doing work and the increase in efficiency [5–7]. This area of improvement support efforts to satisfy the expectations of internal process customers.

Studies of factors affecting manufacturing processes suggest that an adequate work environment is a vital part of the manufacturing environment. In particular, it is critical to empower workers and internal process customers [8], who are the two primary target groups for related improvement measures.

2 Manufacturing Process

2.1 Conditions in Which Manufacturing Processes Are Conducted

Every process can be defined as a set of interrelated measures that affect one another and that are taken to transform inputs into outputs [8, 9]. As a precondition for process effectiveness, companies should minimize all existing and potential disruptions. In the realm of manufacturing, this pertains to the transformation of materials and semi-finished products into finished goods. Measures are taken deliberately to effect gradual desired changes in the processed items. As such changes accumulate, the item is brought ever closer to its intended final shape and characteristics. The manufacturing process ends when all the required characteristics have been achieved and all the required actions have been completed.

Such processes can also be viewed as a way in which enterprises perform their tasks. In such a view, all actions can be represented as a set of intertwined processes conducted with an eye to producing specified final outcomes. A sequence of such interrelated activities is provided in Fig. 1.

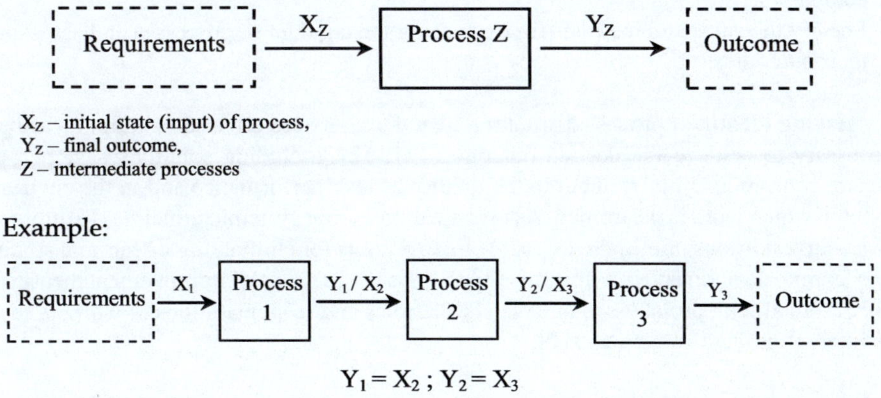

Fig. 1. Sequence of actions taken to transform requirements into outcomes in overall process structure.

Regardless of the adopted view and the measures that a manufacturer takes, success in ensuring a proper course of the manufacturing process depends on creating adequate conditions for its efficient conduct [8]. Any disruptions in the process may either prevent its conduct or hamper it to the point where it is no longer financially viable. In particular, the magnitude of the problem can be made ascertained by viewing all measures taken by an enterprise as a whole by way of recognizing mutual links among processes.

The interconnections affecting the course of multiple processes are outlined in Fig. 2.

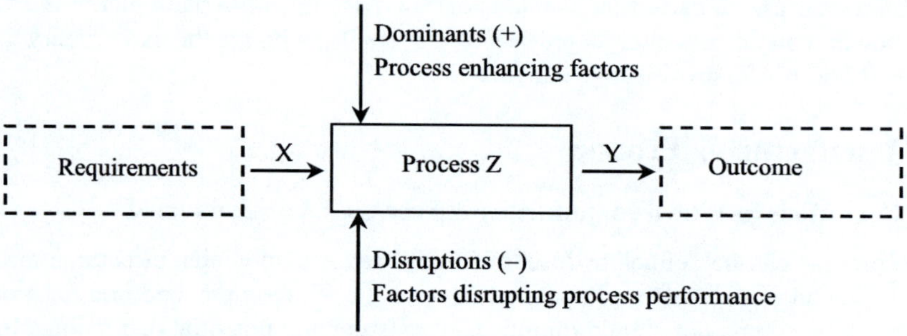

Fig. 2. Impacts of process affecting factors.

For tasks to be carried out effectively under such conditions, it is required that the relevant manufacturer [10]:

- Identify all of its task-critical processes along with the order in which they are to be conducted,
- Identify mutual interactions among processes by specifying the initial state, the final outcome and mutual relationships,
- Assign priorities to processes with respect to their significance for effective task completion,
- Specify the rules and methods for overseeing the conduct of processes and assessing their outcomes.

Having identified process disruptors, manufacturers are advised to adopt measures necessary to achieve the desired outcomes [11, 12], including solutions designed to ensure continuous improvement in the quality of task performance and in the environment in which work is performed. As organizations adopt systemic principles to improve working conditions, they gain access to effective tools for eliminating threats and strains and learn to develop and improve their operations [2, 10, 13]. All improvement measures are designed with proper account taken of the roles and responsibilities of workers seen as internal process customers [10].

Meanwhile, regarding the capacity to complete processes, due note needs to be taken of the dominants that enhance manufacturers' ability to ensure their correct conduct. Such dominants are instrumental in:

- Increasing the efficiency of process-related actions,
- Conserving the resources expanded to achieve intended process outcomes,
- Updating links among processes,
- Reducing process disruptions.

2.2 Work Environment as a Precondition for Process Completion

The work environment is a key factor for manufacturing efficiency. It is crucial for the ability to perform work and achieve the desired outcomes of any organization [5, 14]. Every work-environment deficiency will undermine an organization's ability to complete its tasks. Irregularities not only cause companies to default on regulatory obligations but also to fail to meet stakeholder expectations, in particular with respect to occupational well-being [15]. Incorrect working conditions usually lead to physical and mental exhaustion. In a situation where the employee is exposed to adverse environmental conditions for a long time, this may lead to adverse changes in his state of health. This is the reason of reduced productivity at work. Usually, in an organization where working conditions worsen, there is an increase of the number of accidents, and thus absence caused by diseases. This is accompanied by a lack of commitment to work and not identifying with the company, which negatively affects the efficiency and quality of production [14, 16].

Work environment parameters describe the conditions in which work is performed. Such conditions include social, psychological, physical and environmental factors, as summarized in Table 1.

Table 1. Work environment factors.

Factor category	Factor category description
Social	All formal and informal interactions among workers
Psychological	Factors related to human behaviors and workers' perceived ability to function in the workforce
External-environment related	Factors affecting development and operation in external environment and the ability to complete assigned work tasks
Physical-work-environment related	Factors affecting workers' ability to function in work environment with account taken of deleterious, hazardous and strenuous impacts on worker health and well-being

Some of the key factors of this kind relate to the physical work environment, including such physical factors as thermal comfort (thermal environment parameters), workstation/workplace lighting, noise and vibrations affecting workers, the impact of electromagnetic fields and e.g. laser radiation and chemical factors, including toxicity resulting from the use of chemical compounds and their mixtures at work, as well as the effects of exposure to non-toxic dust. A whole separate category of factors are psycho-physical in nature, pertaining to static and dynamic loads on worker bodies and nervous systems.

3 Guidelines Concerning Work Environment Impacts on Processes

Assessing the work environment (conditions for work performing) and its impacts on the performance of manufacturing tasks is a daunting endeavor. Working conditions have a fundamental impact on shaping the efficiency of production processes. Their impact on the employee is comprehensive. The extent of this impact depends on the nature and level of its severity.

The work environment can be described in terms of both its negative and positive impacts on workers. Thus, work environment factors can be viewed as work impediments as well as facilitators of work that help manufacturers achieve better results. Examples of both types of impacts are provided in Table 2.

Table 2. Work environment factors affecting work performance.

Work environment factor	Impact
Adverse impacts	
Noise at workstation	Noise levels disrupting worker efficiency
Vibration level	Workstation vibrations compromising work performance
Non-toxic dust	Workstation exposure to dust undermining worker health and work performance
Positive impacts	
Workstation lighting	Enhanced lighting helping to reduce error rates
Thermal comfort	Thermal environment in line with comfort guidelines helping to cut power consumption while boosting worker performance

Workstation environment characteristics can be viewed as indications of the conditions in which processes are conducted. They can also serve as criteria for ascertaining whether humans are able to operate and perform work in a given environment [17]. Note should also be taken of the fact that individual factors may produce impacts ranging from positive to negative. Such impacts are known to vary from one case to the next. Hence, in assessing the overall impact of such factors, due account needs to be taken of the type of work environment in question, the nature of relevant impacts, impact levels and conformity with regulatory requirements.

An analysis focused on conditions in the work environment in which processes are conducted, described in terms of the pertinent factors, may also point to the significance of work environment parameters for assessing the ability of humans to function in specific conditions [18]. Such conditions are vital for worker safety and productivity [19]. In such an analysis, untoward factors are seen as potentially endangering worker health.

Adverse factors should be viewed as directly compromising performance capacities and work efficiency.

It is every employer's duty to both keep its work environment specifications in conformity with regulatory requirements and meet worker expectations. This is particularly critical where workers become internal process customers, which makes them essential for the measures taken. Additionally, manufacturers should account for their organization's environment comprised of internal and external conditions affecting their company's ability to achieve its goals and develop appropriate ways of dealing with their stakeholders. The key objective behind the measures they take is to ensure safe and healthy working conditions and comply with ergonomic requirements, thereby optimizing loads [14, 20] and improving workers' ability to function in the workplace.

The manufacturers that embark on improving the work environment should create conditions that ensure the effective conduct of processes. This is crucial for achieving outcomes that will satisfy expectations. To that end, companies need to identify the needs and expectations of process customers [5, 9]. The key factors in the work environment are the threats and strains that require corrective and preventive actions and that increase companies' capacities to perform their tasks. Improvement measures should therefore be seen as integral parts of the process to help companies enhance their abilities to perform tasks at all stages of process implementation. Any improvement measures should reflect the specific nature of the tasks at hand and remain in line with the overall operating strategy of the concerned organization.

4 Effect of Work Environment Improvements on Process Efficiency

To conduct their processes efficiently, companies need to improve their manufacturing conditions.

By selecting and effectively implementing adequate measures aimed at developing their work environments, companies target potential irregularities, including those that affect company performance and operation [5, 14, 20]. To that end, they need to identify specific issues and improvements affecting multiple functions of their enterprise, and especially ongoing processes [21]. The improvements of process environments, especially in the process approach, should extend to [9, 10]:

- The ability to effectively utilize resources,
- Measures adopted on the basis of objective evidence,
- Recognizing the satisfaction of all concerned parties as a valid criterion for the assessment of measures designed to enable the company to meet stakeholder needs and expectations.

The measures taken should help enterprises accomplish their goals of developing their work environments. Their adoption requisite for achieving desired outcomes and ensuring solution effectiveness. Such measures and outcomes include [2, 9, 10, 22, 23]:

- The company's ability to continuously improve its working conditions,
- Conformity with regulatory requirements and the satisfaction of worker expectations regarding the work environment,

- The allocation of resources that the organization needs to ensure the required process quality, including the elimination or mitigation of factors that undermine its ability to ensure the safety of all parties involved in the performance of work,
- Improvements in workers' knowledge and skills that are relevant for the work they perform and that enable them to contribute to work environment management and ultimately perform their work properly,
- The achievement of improved working conditions, including lower accident rates (expressed as the number of accident and occupational diseases relative to the overall headcount and working time),
- A reduction of adverse factors, including accidents and occupational diseases that adversely affect workers' ability to perform work and that result from failures to ensure working conditions that satisfy worker needs,
- The ability to make continuous improvements in all aspects of work performance, including the performance of tasks associated with safety, aimed at securing steady and consistent reductions in the work-related risks that result from failures to establish the required work environment and working conditions,
- Greater involvement of all workers, including the management, in efforts to improve working conditions,
- Having worker roles accounted for in efforts to improve the work environment.

A reduction in the number of non-conformities and the severity of their impacts is essential for the ability to complete manufacturing tasks [22, 24, 25]. Improvements aimed at either eliminating or mitigating disruptions should be seen as a way to boosting a company's ability to perform work.

Some of the most notable benefits of such improvements include:

- Better work efficiency, including faster performance of work,
- Improved work effectiveness,
- Lesser adverse biological impacts of work performance,
- Improved work quality,
- More effective and efficient performance of work.

Note that the above outcomes translate into a wide range of improvements in the conditions in which processes are conducted [14, 15]. The benefits result directly from accounting for and managing the work environment seen as a way to prevent process disruptions and achieve greater progress in process enhancement. The improvements should reflect both types of benefits of specific solutions. This is essential for companies' ability to review issues objectively and for ensuring that all possible factors has been recognized.

Neither can an organization ignore the variability of the impacts of the solutions it adopts. Excessive focus on any specific improvement factor may have adverse consequences for the workforce. An improvement factor selected to remedy a misdiagnosed problem may actually undermine a company's ability to conduct processes.

One example of the above is workstation lighting. Insufficient workstation lighting or any other failure to ensure proper visual conditions will negatively affect manufacturing performance. However, overcompensation through excessive illuminance leading to glare that goes beyond the adaptive capabilities of human sight will negatively impact work performance.

5 Summary

In order to conduct work processes and create conditions in which work can be carried out properly, companies need to account for factors in the work environment. Such factors are critical for the effective completion of tasks by workers, especially where workers are viewed as internal process customers. To recognize the significance of the work environment, enterprises should identify its impact on the way in which processes are conducted. To that end, they need to describe the positive and negative influences of the environment on both the ability to effectively complete processes and the final outcomes. This will allow undertakings to take actions informed by objective knowledge regarding the available resources and ultimately satisfy all process stakeholders.

To ensure that processes are conducted effectively, companies should identify the conditions in which such effectiveness is assured. Of particular importance for such identification is the ability to continuously improve the manner in which activities are conducted. This view may be labeled a process approach to corporate responsibilities.

To successfully ensure effectiveness, enterprises should approach each factor in their environment individually by recognizing its specific nature and identifying the way it disrupts processes, improves effectiveness, reduces lead times, improves quality, increases efficiency, reduces the impacts of work on worker biology and enhances the work environment.

References

1. Park, S., Xu, S., Rogers, W., Pasman, H., El-Halwagi, M.M.: Incorporating inherent safety during the conceptual process design stage: a literature review. J. Loss Prevent. Process Ind. **63**, 104040 (2020)
2. Górny, A.: The elements of work environment in the improvement process of quality management system structure. In: Karwowski, W., Salvendy, G. (eds.) Advances in Human Factors, Ergonomics and Safety in Manufacturing and Service Industries, pp. 599–606. CRC Press, Taylor & Francis Group, Boca Raton (2011)
3. Aqlan, F., Al-Fandi, L.: Prioritizing process improvement initiatives in manufacturing environments. Int. J. Prod. Econ. **196**, 261–268 (2018)
4. Telukdarie, A., Buhulaiga, E., Bag, S., Gupta, S., Luo, Z.: Industry 4.0 implementation for multinationals. Process Saf. Environ. Prot. **118**, 316–329 (2018)
5. Pires, M.L.: Working conditions and organizational support influence on satisfaction and performance. Eur. J. Appl. Bus. Manag. Special Issue of ICABM 2018, 162–182 (2018)
6. Kahya, E.: The effects of job characteristics and working conditions on job performance. Int. J. Ind. Ergon. **37**(6), 515–523 (2007)
7. Massoudi, A.H., Hamdi, S.S.A.: The consequence of work environment on employees productivity. IOSR J. Bus. Manag. **19**(1), 35–42 (2017)
8. Górny, A.: Man as internal customer for working environment improvements. Procedia Manuf. **3**, 4700–4707 (2015)
9. Jeston, J., Nelis, J.: Business Process Management, 3rd edn., Routledge, Taylor and Francis Group, London, New York (2014)

10. EN ISO 9001: Quality Managements Systems. Requirements (ISO 9001:2015). European Committee for Standardization, Brussels (2015)
11. EN ISO 9004: Quality Management. Quality of an Organization. Guidance to Achieve Sustained Success (ISO 9004:2018). European Committee for Standardization, Brussels (2018)
12. Amyotte, P.R,. Goraya, A.U., Hendershot, D.C., Hendershot, D.C., Khan, F: Incorporation of Inherent Safety Principles in Process Safety Management. Process Safety Progress 26(4), 333–346 (2007).
13. Leppänen, A., Hopsu, L., Klemola, S.: Can improvement of work and work process knowledge support well-being at work? Int. Congr. Ser. **1280**, 377–381 (2005)
14. Górny, A.: Use of quality management principles in the shaping of work environment. Commun. Comput. Inf. Sci. **529**, 136–142 (2015)
15. Górny, A.: Improvements in the production environment made using quality management tools. In: Hamrol, A., Grabowska, M., Maletic, D., Woll, R. (eds.) Manufacturing 2019. LNME, pp. 267–276. Springer, Cham (2019). https://doi.org/10.1007/978-3-030-17269-5_20
16. Chen, P.Y., Cooper, C. (eds.): Wellbeing: A Complete Reference Guide, Work and WellBeing. Wiley Blackwell, Hoboken (2014)
17. Aggarwal, S., Devi, A., Kaur, D.: Impact of working conditions on employees performance. Int. Res. J. Commerce Arts Sci. **5**(4), 93–98 (2014)
18. Chaib, R., Verzea, I., Cozminca, I., Benidir, M.: Preserving and improving the safety and health at work: case of Hamma Bouziane cement plant (Algeria). Saf. Sci. **76**, 145–150 (2015)
19. Veldman, J., Gaalman, G.: A model of strategic product quality and process improvement incentives. Int. J. Prod. Econ. **149**, 202–210 (2014)
20. Bernevig-Sava, M.-A., et al.: Improving safety in the workplace using checklists legal requirements Calitatea. Suppl. Qual. Access **18**(S1), 31–34 (2017)
21. Butlewskj, M., Misztal, A., Tytyk, E., Walkowiak, D.: Ergonomic service quality of the elderly on the finical market. Occup. Saf. Hyg. II (Selected Extended and Revised Contributions from the International Symposium Occupational Safety and Hygiene, SHO 2014), 579–583 (2014) (Guimaraes).
22. Cepeda-Carrion, I., Martelo-Landroguez, S., Leal-Rodríguez, A.L., Leal-Millán, A.: Critical processes of knowledge management: an approach toward the creation of customer value. Eur. Res. Manag. Bus. Econ. **23**(1), 1–7 (2017)
23. Wachter, J.K., Yorio, P.L.: A system of safety management practices and worker engagement for reducing and preventing accidents: an empirical and theoretical investigation. Accid. Anal. Prev. **68**, 117–130 (2014)
24. Górny, A.: Total quality management in the improvement of work environment – conditions of ergonomics. In: Goossens, R.H.M. (ed.) AHFE 2017. AISC, vol. 605, pp. 91–100. Springer, Cham (2018). https://doi.org/10.1007/978-3-319-60828-0_10
25. Ridley, J., Channing, J.: Safety at Work, 7th edn., Elsevier, Butterworth - Heinemann, Burlington (2013)

University-Industry Partnerships in the Development of the Academic Patents: Factors for Building Trust

Liliana Alves(✉) and Ana Dias Daniel

Department of Economics, Management, Industrial Engineering and Tourism, University of Aveiro, Aveiro, Portugal

lalves@ua.pt

Abstract. Our research aims to analyze the role of trust in partnerships of academic inventors that developed patented inventions. We studied the influence of similarity in the area of research, reciprocal communication, decision process similarity, team size, contact frequency, and relationship maturity in trust building. We used a mixed method approach to data collection: questionnaire and interview. We observed that patents developed in partnership are still a minority: from the 104 analyzed patents, only 37 were developed in partnership: 18 University-Industry partnerships and 19 Academic partnerships. Partnerships are mostly national. We conclude that decision process similarity and contact frequency are significant for building trust within a partnership. Academic inventors with extensive experience in patenting and in patent commercialization stressed that University-Industry partnerships are important for a better co-development of patents with a focus on the market.

Keywords: Academic patents · Trust · University-Industry partnership

1 Introduction

The University-Industry (UI) collaboration is critical to academic and companies performance [1–4]. From the perspective of academia, there has been reported a positive effect of these collaboration types on the research performance. More specifically, it was found that academics that benefit from industrial funding performed more applied research, collaborated more with other researchers both from academia and industry, and reported a higher number of scientific publications and entrepreneurial outputs [5]. Also, those were more enthusiastic about engaging in revenue-generating opportunities than others who do not collaborate [6]. Additionally, academic researchers who already have informal interactions with industry are more prone to engage in further collaborative research, as well as to spend larger proportion of their research time working with industrial researchers [7]. The UI partnerships take different forms and outputs, among which patenting an invention is a common output in the engineering field. Considering the importance of patenting as a way of academic knowledge transfer and revenue source

J. Machado et al. (Eds.): icieng 2021, LNME, pp. 441–452, 2022.
https://doi.org/10.1007/978-3-030-78170-5_38

for further research, we study the effects of UI partnerships throughout the development of the academic patented inventions within the Portuguese context.

On the other hand, trust is a core concept in partnerships, since trust is the link that connects partners and makes partnerships work, namely in knowledge-sharing networks [8]. Trust helps in the exchange of information, encourages informal contacts and makes the partnership more flexible and responsive to challenges [9]. In our research, we use the recent Bstieler et al.'s [10] approach to analyze the trust in partnerships of university's patented inventions. In their model, the authors study factors such as demographic similarity, reciprocal communication, decision process similarity and relationship maturity. By adapting this model, we aim to find the most influential factors in building trust in partnerships that worked on the development of patented academic inventions and contribute to better university-industry approach policies.

2 Literature Review

UI collaboration and the promotion of knowledge transfer may assume different modes [3]. Gulbrandsen and Smeby [5] studied the influence of industry funding in the research performance of professors in Norwegian universities and has identified four different engagement modes: patents, commercial products, establishment of firms, and consulting contracts. In turn, Boardman [11] assessed the different ways that U.S. academic researchers engage with industry and found out six different types: consultancy, students internships, working for a company, development of patent/copyright with industrial partners, commercialization of research, development of co-authored papers with industrial researchers. Similar conclusions were obtained both by Grimpe and Fier [12] and Haeussler and Colyvas [13] that had studied respectively the practices in German and in the United Kingdom universities. In a review of the literature on university-industry relations, Perkmann et al. [3] highlight the main differences between academic engagement and commercialization, stating that those are distinct types of relations with different antecedents and outputs.

The UI relationships are exposed to a high amount of uncertainty and fluctuations in the way the relationship evolve over time [14–16]. In order to build a long-term relationship and overcome uncertainties, trust between the parties is crucial. Colquitt et al. [17] stated that the research of trust is composed by two fields. The first field follows the viewpoint of Mayer et al. [18] and focuses on the issue of vulnerability, while the second field follows the viewpoint of Rousseau et al. [19] and focuses on the question of expectation. Mayer et al. [18] defined trust as a unitary construct that shows willingness to be vulnerable based on ability, integrity and benevolence of others. The assessment of ability and integrity is performed by reason based on the success of trust and consistency between words, actions and values. The evaluation of benevolence is performed by emotion, advising past situations of affection and concern. On the other hand, Rousseau et al. [19] defined trust as a psychological state of willingness to be vulnerable based upon positive expectations of the intentions or behavior of another party in uncertain situations.

De Jong and Elfring [20] mentioned that trust is fundamental for reducing uncertainty: starting from the idea of vulnerability to another and based on their knowledge of each other, each person creates positive expectations regarding their actions and reduces the degree of uncertainty about something unpredictable - future actions. Coote et al. [21] stated that trust exists when one partner believes in the honesty, reliability and integrity of other partners. Dirks and Ferrin [22] defined interpersonal trust as the psychological state of individuals related to confidence and positive expectations in the actions of others.

Thus, trust is a key asset in collaboration between organizations and in networking creation, particularly in knowledge-sharing networks [8]. Trust is a critical asset as it increases strategic flexibility, predictability and adaptability; reduces management and acquisition costs, as well as social complexity; and encourages informal collaborative networks and collaborative innovation [9]. Schaubroeck et al. [23] observed that the reduction of individual uncertainty, triggered by trust, reinforces the quality of social exchanges. Das and Teng [24] argued that trust supports partner integration, decreases concerns about opportunistic behavior, and reduces formal contracts. Rodríguez-Pose and von Berlepsch [25] stated that the greater the interaction between partners, the greater the likelihood of building trust. The greater the trust, the easier the information sharing, open communication and conflict management [8, 26, 27].

Bstieler et al. [10] argued the mutual trust as key factor in university-industry research collaborations and examined how trust in inter-organizational relationship develops over time. The authors analyzed the influence of trust bases – demographic similarity, reciprocal communication and decision process similarity – in the mutual trust of UI research collaborations, seeing the moderating effect of relationship maturity. Bstieler et al. [10] found that relationship maturity moderates the associations of reciprocal communication and decision process similarity with trust. Considering the relevance and depth of Bstieler et al. [10]'s approach, we used in our study a similar strategy to assess the trust in university research partnerships in development of patented invention in Portuguese public academia. So, our starting question is: what are the most relevant factors to trust building in the partnerships that developed patented invention?

In the next section we present our mixed methods approach and data collection; sample and descriptive statistics; results from ordered probit regressions and interview analysis; as well as the main conclusions.

3 Methodology

3.1 Mixed Methods Approach and Data Collection

We used a mixed method approach – questionnaire and interviews – for data collection [28, 29]. The main objective of the questionnaire was to collect quantitative data related to university partnerships and its influence in the development of the academic patented inventions. In turn, the interviews allowed to collect qualitative data regarding the type and characteristics of UI collaborations.

First, we considered the questionnaire validated by Bstieler et al. [10], namely the items related to trust, reciprocal communication and decision process similarity. We also revised the items related to demographic similarity and relationship maturity. The demographic similarity was adapted to our questionnaire as research similarity and the relationship maturity was measured by previous collaboration with partners. Additionally, we included questions about the team size and the contact frequency, given the reference in the literature to these two variables for building trust [3, 7].

Second, we requested an interview with several academic inventors with great experience in patenting. The aim was to collect a personal perspective about the relevance of UI collaborations, mainly in the development of patented inventions. Thus, the interviews with academic inventors enriched the research approach, aided the interpretation of the data and pointed out relevant suggestions.

For data collection about patents, we used the databases of Portuguese National Institute of Industrial Property (INPI) and the European Patent Office (EPO) (overlapping patents in databases were verified and eliminated). We identified the patents currently granted and in force submitted by Portuguese public universities – 570 - and we sent by e-mail the questionnaire to all first (as leader) academic inventors of each patent. It was mentioned in the email that if inventors considered more suitable, they could resend the questionnaire to another inventor of that specific patent. From these databases, we also identified the top academic inventors, i.e. who stand out as inventor of academic patented inventions. Seeing the national list of university patents and their academic inventors, we contacted inventors who had three or more university patents granted and in force. In total, we contacted a total of 43 top academic inventors to schedule an interview.

3.2 Sample and Descriptive Statistics

From the 570 questionnaires sent, we had 104 valid answers, corresponding to 104 university patents. As can be seen in Table 1, only 37 university patents were developed within partnerships: 18 UI partnerships and 19 academic partnerships.

Table 1. Types of partnerships

Partners	Frequency
National company	5
National company + Other National university	10
National company + European university	1
European company + Other National university	1
European company + European university	1
UI partnerships	18
Other National university	16
Other National university + European university	2
European university	1
Academic partnerships	19
Total partnerships	37

Partnerships are mostly national.16 resulted from collaborations with only other national university, 10 resulted from collaborations with other universities and national companies, and 5 resulted from collaborations with only national companies. Partnerships involving other national and European partners were few and partnerships outside Europe were non-existent.

In Table 2, we present the descriptive statistics of the factors' variables analyzed in a likert scale of 7 points. As can be seen in the table, most variables present a mean of 5 points. However, we highlight the difference between the averages in research similarity: the inventors answered that have more research similarity with academic partners than with industry partners. This may be related to complement the knowledge about the development of a patented invention. On the other hand, we highlight the tolerance of risk mean in the decision process similarity factor: 4.95. Risk-taking appears to be a critical topic in decision process and eventually makes it difficult to choose more innovative paths.

Table 2. Descriptive statistics

	N	Minimum	Maximum	Mean	Std. deviation
Trust					
Frank in dealing with us	37	3	7	5,92	1,187
Promises reliable	37	3	7	5,81	1,309
Honest	37	3	7	5,84	1,236
Partner on our side	37	2	7	5,81	1,330
Research similarity					
Academic partners	32	1	7	5,13	1,862
Industry partners	18	1	7	3,94	2,164
All partners	37	1	7	4,68	1,857
Reciprocal communication					
Timely	37	1	7	5,38	1,622
Accurate	37	1	7	5,49	1,465
Adequate	37	1	7	5,78	1,456
Complete	37	1	7	5,54	1,538
Decision process similarity					
Time to decision	37	2	7	5,30	1,412
Decision-making style	37	1	7	5,03	1,554
Tolerance of risk	37	1	7	4,95	1,615
Understanding of how things should be done	37	2	7	5,22	1,397

In 27 cases, the collaboration with partners was prior to the development of patented invention. Regarding the team size and contact frequency, most of the teams were small (1–4 people) and with weekly/monthly contact frequency.

Additionally, from the 43 top academic inventors contacted to schedule an interview, only 19 were available. We interviewed 4 inventors from NOVA Lisboa University, 4 inventors from University of Aveiro, 2 inventors from University of Lisbon, 4 inventors from University of Minho and 5 inventors from University of Porto. The number of patents developed in partnerships was low: among the 63 patents in force of all interviewed inventors, only 11 were developed in partnership.

The interviews took a broad approach to the academic inventors' experience in patenting, not restricting only to the experience with patents in force. The analysis of partnerships in patent development, specifically the importance of collaborations with industry, was the main topic of the interviews. All the interviews were recorded and transcribed. After, we reduced raw data into structured and quantifiable data by the cataloging of the answers, identification of units of thought and frequency of response [4, 30–33].

4 Results

First, we performed an exploratory factorial analysis on the variables proposed by Bstieler et al. [10] as being part of the trust, reciprocal communication and decision process similarity factors. The analysis confirmed in our sample the existence of the 3 factors proposed by the authors and with the same associated variables. Thus, the variables "promises reliable", "partner on our side", "frank in dealing with us" and "honest" constitute the factor trust. In its turn, the factor reciprocal communication is composed by the variables "communication adequate", "communication accurate", "communication complete" and "communication timely". Finally, the variables "tolerance of risk", "decision-making style", "time to decision" and "understanding of how things should be done" constitute the factor decision process similarity. As can be seen in Table 3, the exploratory factorial analysis showed the KMO value is .842 and a total variance explained is 89.284%. The Bartlett's Test of Sphericity disclosed a 583.804 chi square and a .000 significance level. Considering the satisfactory values obtained in the exploratory factorial analysis, the factors were used in the following analysis.

Second, the analysis of the data collected through the questionnaires was analyzed through ordered probit regressions in order to understand how research similarity, reciprocal communication, decision process similarity, team size, contact frequency and relationship maturity influence the probability of feeling trust in the partnerships. We tested in the partnerships as a whole, and in each type of partnerships. The ordered probit regressions disclosed that there are significant associations in the partnerships as a whole (model 1) and in the UI partnerships (model 2). As can be seen in model 1 of Table 4, the decision process similarity is positively related with the probability of feel trust in the partners (p value: 3.48; 0.001 significance level). In its turn, the results showed that low contact frequency (bi-annual) is negatively related with the probability of feeling trust (p value: −3.37; 0.001 significance level). Finally, previous collaboration with partners before patent development seems to be positively related with the likelihood of feeling trust in partners (p value: 1.98; 0.05 significance level).

Table 3. Results of exploratory factor analysis

Variables	Commonality	Trust	Reciprocal communication	Decision process similarity
Promises reliable	.963	**.919**	.172	.297
Partner on our side	.946	**.915**	.243	.223
Frank in dealing with us	.968	**.891**	.343	.236
Honest	.943	**.835**	.202	.452
Communication adequate	.851	.244	**.881**	.122
Communication accurate	.934	.242	**.876**	.328
Communication complete	.920	.366	**.847**	.262
Communication timely	.867	.084	**.801**	.467
Tolerance of risk	.892	.260	.200	**.886**
Decision-making style	.863	.241	.297	**.847**
Time to decision	.894	.410	.404	**.750**
Understanding of how things should be done	.674	.335	.258	**.704**
Variance explained (%)		31.990	29.227	28.067
Total variance explained (%)		89.284		
Kaiser-Meyer-Olkin measure of sampling adequacy		.842		
Bartlett's test of sphericity	Approx. Chi Square	583.804		
	df	66		
	Sig.	.000		

The ordered probit regression in model of UI partnerships (model 2) showed similar results in terms of decision process similarity and contact frequency. It seems that the decision process similarity is positively related to the probability of feeling trust in UI partners (p value: 2.05; 0.05 significance level). Like model 1, the bi-annual contact frequency is negatively related with the probability of feeling trust in UI partners (p value: −2.32; 0.05 significance level), but it goes further. The model 2 reveal that a quarterly contact frequency is positively related with trust (p value: 2.14; 0.05 significance level).

Table 4. Results of ordered probit regressions

	Model 1 trust in partnership	Model 2 trust in UI partnership
Research similarity	−0.402	0.622
	(−1.79)	(1.16)
Reciprocal communication factor	0.141	−3.876*
	(0.64)	(−2.19)
Decision process similarity factor	1.207***	1.353*
	(3.48)	(2.05)
Team size		
1–4 members	−0.255	−8.125*
	(-0.42)	(−2.04)
5–9 members	0.998	3.724*
	(1.41)	(2.48)
10–14 members	0	0
	(.)	(.)
Contact frequency		
Bi-annual	−3.131***	−6.733*
	(−3.37)	(−2.32)
Quarterly	0.676	9.696*
	(0.76)	(2.14)
Monthly	0.101	0
	(0.17)	(.)
Weekly	0	2.198
	(.)	(1.45)
Daily	−0.868	−0.227
	(−1.02)	(−0.09)
Relationship maturity		
No previous collaboration	0	0
	(.)	(.)
With previous collaboration	1.354*	−0.705
	(1.98)	(−0.52)
Pseudo R^2	0.3121	0.6254
Number of observations	37	18

* 0.05 significance level/** 0.01 significance level/*** 0.001significance level

Additionally, the model presented significant associations between the team size and the probability of feeling trust in UI partners. Given the results, the small teams trust less (p value: −2.04; 0.05 significance level) than bigger teams (5–9 members) (p value: 2.48; 0.05 significance level). Finally, the reciprocal communication factor is negatively related with the probability of feeling trust in UI partners (p value: −2.19; 0.05 significance level). It seems that in some cases the inventors have a good communication with partners but do not trust them, and in other cases the difficulties in communication do not undermine trust in partners. So, the ordered probit regressions disclosed that the decision process similarity seems to be related to higher trust felt in the partners, while the low contact frequency and being in small team is associated with feeling less trust in partners.

Third, we completed our study with the analysis of interviews with top academic inventors. This sample was divided into two groups: inventors who often collaborate with industry; and inventors who do not always collaborate with industry. This analysis allows the identification of common patterns in the two groups, namely it was found that those who collaborate frequently with the industry tend to consider a patent as a process of co-development. On the other hand, inventors who do not collaborate or collaborate sporadically with the industry tend to refer to a patent as something developed academically which is then sold to the industry.

Inventor 10: Companies are fundamental to research. If we want to reach the market, we must have some company interested and associated with the project. (…) It is fundamental. Do you know why? Because we don't do business at the university.

Inventor 19: From my experience, no company wants to buy patents. They want to co-develop solutions that can be patented, and they are available to recognize in monetary terms the university's ownership and to pay for it. They do not want you to contact them and say: "Look, we have a solution here that is very good. You're going to make a lot of money from it." The answer is: "No, thanks." When solutions are co-developed from a certain stage, they internalize those solutions, it becomes theirs and they are available to pay for it. They can pay more or less has negotiation is done, but they are available to pay for it.

The inventors who collaborate more with the industry highlighted the importance of trust in partners, the time needed to build trust and the bridges to connect academia to industry.

Inventor 2: One of the most important things is the process of mutual trust that must be established between academics and business people. You must trust each other. (…) It is not always easy, and it is a gradual process. You don't create it overnight. But, to start this collaboration, a high degree of trust between people is necessary. It's between people. It is not between institutions. (…) academics cannot be eternally suspicious in relation to entrepreneurs. We need to attract people who work in companies, give them understanding, do some workshops, things like that. And we should also go to companies, visit the factory, listen to people - because sometimes academics do not listen either.

Although few patents are being commercialized, we observe that those that are being commercialized are the result of partnerships with industry. The inventors stated advantages associated with collaboration with the industry, such as assertive identification of market gaps, perception of technology viability and access to facilities and networks.

Inventor 3: The things we do with the industry are obviously much more valued because the industry has an exact notion of what they want. The great advances that we made (…) the company knew what they needed. An academic does not know. In fact, it is this disconnection from the world… we aim to do things for the industry, but we never work with the industry. (…) It is very important to dialogue with those who make and produce. Because I can often get a brilliant idea, but I don't know the costs, the process complexity and reliability of what I want to do. And, therefore, this effort is lost. (…).

Inventor 11: Companies know what they want. They have this very clearly. (…) They do not radically change what they are doing. (…) If they say it is not interesting, then it is not of interest. It is so clear. It is not for technological reasons or for economic reasons…

Despite the inventors mentioned difficulties in approaching the industry and in knowledge and experience engagement, those who have frequent collaboration with industry are unanimous in recognizing the crucial role of partnerships with the industry.

5 Discussion and Conclusion

We analyzed the university partnerships, namely the factors that influence the development of trust, such as similarity in the area of research, reciprocal communication, decision process similarity, team size, contact frequency, and relationship maturity. We used a mix method to data collection: questionnaire and interview.

We identified the patents currently granted and in force submitted by Portuguese public universities – 570 – and sent a questionnaire to academic inventors to identify inventions developed in partnerships. Among 104 patents analyzed, we identified 37 patents developed in collaboration in partnerships: 18 UI partnerships and 19 academic partnerships. The ordered probit regressions performed in our sample showed significant associations between the factors analyzed and the trust felt in partners, as a whole, and in partners of UI partnerships. The decision process similarity and the low contact frequency showed influence in trust on both models. The results reveal that decision process similarity is positively related to trust and the bi-annual contact frequency is negatively related to trust. In addition, we interviewed the 19 top academic inventors with extensive experience in patenting at five Portuguese public universities. We analyzed 63 patents of which only 11 were developed in partnership. We highlight that the 6 academics who commercialized patents also collaborated with the industry. These academics emphasized the long-term relationship with industrial partners and the time and willingness to build trust. We observed a common pattern among these inventors: they refer to the development of the patent as a process of co-development, and not as something to be developed by academics and sold to industry. These inventors highlighted that close relationship with the industry is fundamental to assertive identification of market gaps, the perception of technological viability and access to facilities and networks.

Although our sample is not representative, the results obtained suggest that there are still few partnerships in the development of patented inventions, that is, UI partnerships. Even though in the last two decades in Portugal there has been an approximation between academia and industry, it is necessary that these organizations collaborate even more. Collaboration with the industry has benefits, namely in the development of patented invention, however, most collaborations are still strictly academic. Therefore, public policies should continue to promote the UI partnerships through funded projects that necessarily include university and industry partners, but they need to go even further. Given the benefits of long-term relationships, it seems important to take steps to strengthen existing UI partnerships. It is necessary to bring the university and industry even closer together and to encourage the creation of long-term relationships with successful results over the years. Longitudinal research on this topic is necessary to better understand the Portuguese case and to be able to take more appropriate policies.

References

1. D'Este, P., Patel, P.: University–industry linkages in the UK: what are the factors underlying the variety of interactions with industry? Res. Policy **36**(9), 1295–1313 (2007)
2. Etzkowitz, H., Leydesdorff, L.: The dynamics of innovation: from National systems and Mode 2 to a Triple Helix of university–industry– government relations. Res. Policy **29**(2), 109–123 (2000)
3. Perkmann, M., et al.: Academic engagement and commercialisation: a review of the literature on university–industry relations. Res. Policy **42**(2), 423–442 (2013)
4. Hermans, J., Castiaux, A.: Contingent knowledge transfers in university–industry R&D projects. Knowl. Manag. Res. Pract. **15**(1), 68–77 (2017)
5. Gulbrandsen, M., Smeby, J.C.: Industry funding and university professors' research performance. Res. Policy **34**, 932–950 (2005)
6. Campbell, T.I.D., Slaughter, S.: Faculty and administrators' attitudes toward potential conflicts of interest, commitment, and equity in university-industry relationships. J. Higher Educ. **70**, 309–352 (1999)
7. Ponomariov, B., Boardman, P.C.: The effect of informal industry contacts on the time university scientists allocate to collaborative research with industry. J. Technol. Transfer **33**, 301–313 (2008)
8. Abrams, L.C., Cross, R., Lesser, E., Levin, D.Z.: Nurturing interpersonal trust in knowledge-sharing networks. Acad. Manag. Exec. **17**(4), 64–77 (2003)
9. Seppanen, R., Blomqvist, K., Sundqvist, S.: Measuring inter-organizational trust—a critical review of the empirical research in 1990–2003. Ind. Mark. Manage. **36**, 249–265 (2007)
10. Bstieler, L., Hemmert, M., Barczak, G.: The changing bases of mutual trust formation in inter-organizational relationships: A dyadic study of university-industry research collaborations. J. Bus. Res. **74**, 47–54 (2017)
11. Boardman, P.C.: Beyond the stars: The impact of affiliation with university biotechnology centers on the industrial involvement of university scientists. Technovation **28**, 291–297 (2008)
12. Grimpe, C., Fier, H.: Informal university technology transfer: a comparison between the United States and Germany. J. Technol. Transfer **35**, 637–650 (2010)
13. Haeussler, C., Colyvas, J.A.: Breaking the ivory tower: academic entrepreneurship in the life sciences in UK and Germany. Res. Policy **40**(1), 41–54 (2011)
14. Schilke, O., Cook, K.S.: A cross-level process theory of trust development in interorganizational relationships. Strateg. Organ. **11**(3), 281–303 (2013)
15. Vanneste, B.S., Puranam, P., Kretschmer, T.: Trust over time in exchange relationships: meta-analysis and theory. Strateg. Manag. J. **35**(12), 1891–1902 (2014)
16. Daniel, A.D., Alves, L.: University-industry technology transfer: the commercialization of university's patents. Knowl. Manag. Res. Pract. **18**(3), 276–296 (2020)

17. Colquitt, J.A., Lepine, J.A., Zapata, C.P., Wild, R.E.: Trust in typical and high-reliability contexts: building and reacting to trust among firefighters. Acad. Manag. J. **54**(5), 999–1015 (2011)
18. Mayer, R.C., Davis, J.H., Schoorman, F.D.: An integrative model of organizational trust. Acad. Manag. Rev. **20**, 709–734 (1995)
19. Rousseau, D.M., Sitkin, S.B., Burt, R.S., Camerer, C.: Not so different after all: a cross-discipline view of trust. Acad. Manag. Rev. **23**, 393–404 (1998)
20. De. Jong, B.A., Elfring, T.: How does trust affect the performance of ongoing teams? The mediationg role of reflexivity, monitoring, and effort. Acad. Manag. J. **53**(3), 535–549 (2010)
21. Coote, L., Forrest, E.J., Tam, T.W.: An investigation into commitment in non-Western industrial marketing relationships. Ind. Mark. Manage. **32**(7), 595–604 (2003)
22. Dirks, K.T., Ferrin, D.L.: Trust in leadership: meta-analytic findings and implications for research and practice. J. Appl. Psychol. **87**, 611–628 (2002)
23. Schaubroeck, J.M., Peng, A.C., Hannah, S.T.: Developing trust with peers and leaders: impacts on organizational identification and performance during entry. Acad. Manag. J. **56**(4), 1148–1168 (2013)
24. Das, T.K., Teng, B.S.: Trust, control and risk in strategic alliances: an integrated framework. Organ. Stud. **22**, 251–283 (2001)
25. Rodríguez-Pose, A., von Berlepsch, V.: Social capital and individual happiness in Europe. J. Happiness Stud. **15**(2), 357–386 (2013). https://doi.org/10.1007/s10902-013-9426-y
26. Blomqvist, K., Hurmelinna, P., Seppanen, R.: Playing the collaboration game right-balancing trust and contracting. Technovation **25**, 497–504 (2005)
27. Chua, R.Y.J., Ingram, P., Morris, M.W.: From the head and the heart: locating cognition- and affect- based trust in managers' professional networks. Acad. Manag. J. **51**(3), 436–452 (2008)
28. Creswell, J.W., Plano, C.V.: Designing and Conducting Mixed Methods Research. Sage, Thousand Oaks (2007)
29. Johnson, R.B., Onwuegbuzie, A.J., Turner, L.A.: Toward a definition of mixed methods research. J. Mixed Methods Res. **1**(2), 112–133 (2007)
30. Miles, M., Huberman, A.: An Expanded Sourcebook Qualitative Data Analysis California. Sage Publications, Estados Unidos da América (1994)
31. Siegel, D.S., Waldman, D.A., Atwater, L.E., Link, A.N.: Commercial knowledge transfers from universities to firms: improving the effectiveness of university-industry collaboration. J. High Technol. Manag. Res. **14**(1), 111–133 (2003)
32. Noblet, J.P., Simon, E., Parent, R.: Absorptive capacity: a proposed operationalization. Knowl. Manag. Res. Pract. **9**(4), 367–377 (2011)
33. Valkokari, K., Passi, J., Rantala, T.: Managing knowledge within networked innovation. Knowl. Manag. Res. Pract. **10**(1), 27–40 (2012)

A Generalization of the Krein Parameters of a Symmetric Association Scheme

Vasco Moço Mano[1(✉)] and Luís Almeida Vieira[2(✉)]

[1] University of Porto, Porto, Portugal
[2] University of Porto, Faculty of Engineering, Porto, Portugal
lvieira@fe.up.pt

Abstract. In this paper we formulate a generalization of the Krein parameters of a symmetric association scheme which, in turn, allows us to deduce some necessary conditions for the existence of these schemes. This is achieved by applying some matrix analysis results over the elements of the unique basis of minimal orthogonal idempotents of the underlying algebra of a symmetric association scheme.

Keywords: Algebraic combinatorics · Association scheme · Matrix analysis

1 Introduction

This paper is devoted to devising a generalization of the Krein parameters of a symmetric association scheme. Such a generalization enables us to deduce some necessary conditions for the existence of these combinatorial structures. The work presented is an extension of the ones published in [1,2].

The first appearance of the concept of association scheme is credited to a 1952 paper of Bose and Shimamoto [3]. Association schemes are complex combinatorial structures that can be regarded as generalizations of undirected graphs. Indeed, the particular case of an association scheme with two classes matches the set constituted by a strongly regular graph and its complement. Association schemes rapidly arose as powerful combinatorics tools with a wide range of applications. We can cite applications in the field of statistics and combinatorial designs, [3–5], coding theory, [6], group theory, [7,8], or character theory, [9].

In this work we consider symmetric association schemes with k classes, denoted by $(A, \{S_i\}_{i=0}^k)$, and the corresponding Bose-Mesner algebra, denoted by $\mathcal{B}$, that is, the algebra spanned by the matrices of $(A, \{S_i\}_{i=0}^k)$. This algebra $\mathcal{B}$ has an associated unique basis of minimal orthogonal idempotents, $\{F_0, F_1 \ldots, F_k\}$. In Sect. 2 we present the fundamental concepts involved in the development of our work: a short introduction on association schemes is presented in Subsect. 2.1 and some properties on symmetric matrices, which are relevant to our work, are presented in Subsect. 2.2. There are different classes of parameters associated to an association scheme. One of these classes of parameters, the Krein parameters, are generalized in Sect. 3. Some properties of the generalized Krein parameters are established and some conclusions drawn.

J. Machado et al. (Eds.): icieng 2021, LNME, pp. 453–460, 2022.
https://doi.org/10.1007/978-3-030-78170-5_39

2 Theoretical Fundamentals

The fundamental theoretical concepts and results that are used in the development of our work are presented next. We start with a brief survey of symmetric association schemes.

2.1 Symmetric Association Schemes

Along the following paragraphs, we present a short introduction to symmetric association schemes. A detailed survey can be found, for instance, in [10].

A *symmetric association scheme* with k classes is a pair $(A, \{S_i\}_{i=0}^k)$, where A is a finite set of order n and each S_i, $i = 0, 1, \ldots, k$ is a nonempty subset of $A \times A$ verifying the axioms:

(1) $S_0 = \{(a, a) : a \in A\}$;
(2) if $(a, b) \in S_i$, then $(b, a) \in S_i$, for all a, b in A and i in $\{0, 1, \ldots, k\}$;
(3) for all i, j, l in $\{0, 1, \ldots, k\}$ there is an integer α_{ij}^l such that, for all (a, b) in S_l,

$$|\{c \in A : (a, c) \in S_i \text{ and } (c, b) \in S_j\}| = \alpha_{ij}^l.$$

The parameters α_{ij}^l are called the *intersection numbers* of $(A, \{S_i\}_{i=0}^k)$ and can be seen as the entries of the so called *intersection matrices* $L_0, L_1, \ldots, L_k$, with $(L_i)_{lj} = \alpha_{ij}^l$, where $L_0 = I_n$.

The reason why $(A, \{S_i\}_{i=0}^k)$ is called symmetric is the fact that, by axiom (2), each S_i, $i = 0, 1, \ldots, k$ is a symmetric relation of A. This definition is credited to Bose and Shimamoto, [3], but a more general definition of association schemes can be found in [6]. In this paper, however, we only deal with association schemes which are symmetric.

The sets $S_0, S_1, \ldots, S_k$ of a symmetric association scheme can be identified by their adjacency matrices $M_0, M_1, \ldots, M_k$, where each M_i is a matrix of order n such that $(M_i)_{ab} = 1$, if $(a, b) \in S_i$, and $(M_i)_{ab} = 0$, otherwise. The matrices of $(A, \{S_i\}_{i=0}^k)$ satisfy the following properties:

(1) $M_0 = I_n$;
(2) $\sum_{i=0}^k M_i = J_n$;
(3) $M_i = M_i^\top$, $\forall i \in \{0, 1, \ldots, k\}$;
(4) $M_i M_j = \sum_{l=0}^k \alpha_{ij}^l M_l$, $\forall i, j \in \{0, 1, \ldots, k\}$;

where I_n and J_n denote the identity and the all ones matrices of order n, respectively, and $M^\top$ denotes the transpose of M. Observe that, from (2), we can deduce that $\{M_i\}_{i=0}^k$ is a free set. It is also known (see, for instance, [10]) that the symmetry of the scheme implies that $\alpha_{ij}^l = \alpha_{ji}^l$ and, therefore, the product of the matrices of $(A, \{S_i\}_{i=0}^k)$ is commutative.

We can acknowledge $M_1, M_2, \ldots, M_k$ as the matrices of adjacency of the respective simple and undirected graphs $G_1, G_2, \ldots, G_d$, with a common vertex set V.

The more trivial association schemes are the ones with only one class. They corresponds to the matrices $M_0 = I_n$ and $M_1 = J_n - I_n$. Since G_1 is the complete graph this case is irrelevant.

The next case contemplates symmetric association schemes with two classes. This kind of scheme corresponds to strongly regular graphs. In short, a *strongly regular graph* is just an undirected simple graph for which the number of common neighbors to any pair of vertices is fixed and depends only on the adjacency or non-adjacency of that pair of vertices. As we were saying, for a symmetric association scheme with two classes, we have $M_0 = I_n$, M_1, $M_2 = J_n - M_1 - I_n$, where M_1 and M_2 are the adjacency matrices of a strongly regular graph and it's complement, respectively, the *complement* of a graph being the graph whose edges and vertices correspond to the vertices and edges of the starting graph, respectively. We also have that, if M is the adjacency matrix of a strongly regular graph, then $I_n, M, J_n - M - I_n$ make up an two class association scheme.

The symmetric matrices $M_0, M_1, \ldots, M_k$ of a symmetric association scheme have the property of having constant diagonal and they span a commutative algebra, $\mathcal{B}$, with dimension $k+1$. This algebra is called the *Bose-Mesner algebra* of $(A, \{S_i\}_{i=0}^k)$, introduced in [5]. Note that $\mathcal{B}$ is an algebra with respect not only to the usual matrix product but also to the *element-wise product* of matrices, introduced next:

$$(X \circ Y)_{ij} = X_{ij} Y_{ij}, \tag{1}$$

for any two matrices X and y of the same order. With the product defined in (1), $\mathcal{B}$ is an associative and Abelian algebra with unit J_n.

If F in $\mathcal{B}$ has the property $F^2 = F$, it is called an *idempotent* of $\mathcal{B}$. If F_1 and F_2 in $\mathcal{B}$ are such that $F_1 F_2 = 0$, then they are called *orthogonal.* It turns out that the Bose-Mesner algebra $\mathcal{B}$ has a unique basis of minimal orthogonal idempotents $\{F_0, F_1, \ldots, F_k\}$. Also, the elements of $\{F_0, F_1, \ldots, F_k\}$ satisfy the equality

$$\sum_{i=0}^{k} F_i = I_n.$$

Another class of parameters associated to $(A, \{S_i\}_{i=0}^k)$ are the *Krein parameters* discovered by Scott [13]. These are the numbers β_{ij}^l, with $0 \leq i, j, l \leq k$, such that

$$F_i \circ F_j = \sum_{l=0}^{k} \beta_{ij}^l F_l. \tag{2}$$

The Krein parameters of $(A, \{S_i\}_{i=0}^k)$, β_{ij}^l, can be interpreted as the dual parameters of the intersection numbers of $(A, \{S_i\}_{i=0}^k)$, α_{ij}^l. The well known fact that their value is bounded from above, specifically $\beta_{ij}^l \geq 0$, see [10], produces a set of necessary conditions for the existence of a symmetric association scheme $(A, \{S_i\}_{i=0}^k)$. In Sect. 3 the Krein parameters will be object of a thorough generalization and some properties will be deduced.

2.2 Matrix Analysis Results

In order to prove our main results, presented in Sect. 3, we make use of some symmetric matrices properties. In this section we present the adopted mathematical symbology and the most important results needed.

The set of all square real matrices of order n is represented by $\mathcal{M}_n(\mathbb{R})$ and the set of rectangular real matrices of type $m \times n$ is denoted by $\mathcal{M}_{m,n}(\mathbb{R})$. The space of real symmetric matrices of order n is denoted by $Sym_n(\mathbb{R})$.

In the last section we introduced in (1) the element-wise multiplication of matrices, also referred to as the Hadamard product or Schur product of matrices denoted by $\circ$. Now we introduce another important type of multiplication of matrices.

For any natural numbers m, n, p and q, the *Kronecker product*, that we denote by $\otimes$, is defined for matrices $X = [x_{ij}] \in \mathcal{M}_{m,n}(\mathbb{R})$ and $Y = [y_{ij}] \in \mathcal{M}_{p,q}(\mathbb{R})$, as

$$X \otimes Y = \begin{pmatrix} x_{11}Y & \cdots & x_{1n}Y \\ \vdots & \ddots & \vdots \\ x_{m1}Y & \cdots & x_{mn}Y \end{pmatrix}.$$

For $X \in Sym_n(\mathbb{R})$, let $\lambda_1(X), \lambda_2(X), \ldots, \lambda_n(X)$ be the eigenvalues of X in increasing order. Also, a *principal submatrix* of X, represented by X_r, is a matrix obtained by eliminating $n-r$ rows and the corresponding columns from the matrix X.

In our results we will make use of the *eigenvalues interlacing theorem* [11]. This result states that if $X \in Sym_n(\mathbb{R})$ and if X_r is a principal submatrix of X, then the eigenvalues of X_r are interlaced with the eigenvalues of X in the following manner:

$$\lambda_i(X) \leq \lambda_i(X_r) \leq \lambda_{n-r+i}(X),$$

for any i such that $1 \leq i \leq r$.

Finally, another important result that we will use is that, given two matrices $X, Y \in \mathcal{M}_{m,n}(\mathbb{R})$, then $X \circ Y$ is a principal submatrix of $X \otimes Y$, see [12, Lemma 5.1.1].

3 Generalized Krein Parameters

In this section we generalize the Krein parameters of an association scheme. In the following we consider a symmetric association scheme with k classes, $(A, \{S_i\}_{i=0}^k)$, the corresponding Bose-Mesner algebra, $\mathcal{B}$, and the associated base of minimal orthogonal idempotents, $\{F_0, F_1, \ldots, F_k\}$.

We start by introducing the following notation for the element-wise and Kronecker powers of square matrices. Let $X \in \mathcal{M}_n(\mathbb{R})$ be a square matrix of order

n and m be a natural number. Then, $X^{\circ m}$ and $X^{\otimes m}$ denote the *element-wise power* of order m of X and the *Kronecker power* of order m of X, respectively, with the following needed assumptions: $X^{\circ 0} = J_n$, $X^{\circ 1} = X$ and $X^{\otimes 1} = X$. The notation introduced above will be important to simplify the calculations that we will make next.

We are now in conditions to generalize the Krein parameters of an association scheme.

Definition 1. *Let $(A, \{S_i\}_{i=0}^k)$ be a symmetric association scheme with k classes, $\mathcal{B}$ the Bose-Mesner algebra spanned by $(A, \{S_i\}_{i=0}^k)$ and $\{F_0, F_1, \ldots, F_k\}$ the associated base of minimal orthogonal idempotents. The generalized Krein parameters of $(A, \{S_i\}_{i=0}^k)$ are the real numbers $\beta^l_{n_0,n_1,\ldots,n_k}$, with $l \in \{0, 1, \ldots, k\}$ and $n_0, n_1, \ldots, n_k \in \mathbb{N}$, such that*

$$F_0^{\circ n_0} \circ F_1^{\circ n_1} \circ \cdots \circ F_k^{\circ n_k} = \sum_{l=0}^{k} \beta^l_{n_0,n_1,\ldots,n_k} F_l.$$

Note that the generalized Krein parameters of $(A, \{S_i\}_{i=0}^k)$ are well defined due to the fact that $\mathcal{B}$ is closed under the element-wise product and $\{F_0, F_1, \ldots, F_d\}$ is a basis of $\mathcal{B}$.

Observe that the generalized Krein parameters presented in Definition 1 correspond to the usual Krein parameters defined in (2) when: all but two of the n_i's, with $i \in \{0, 1, \ldots, k\}$, are 0 and the two which are non-null are equal to 1; or if all but one of the n_i's, with $i \in \{0, 1, \ldots, k\}$, are equal to 0 and the one that is non-null is equal to 2.

The generalized Krein parameters of an association scheme satisfy the following properties. Firstly, the generalized Krein parameters are bounded from above and from below.

Theorem 1. *Let $(A, \{S_i\}_{i=0}^k)$ be a symmetric association scheme with k classes. Then, for $l \in \{0, 1, \ldots, k\}$ and $n_0, n_1, \ldots, n_k \in \mathbb{N}$, the generalized Krein parameters of $(A, \{S_i\}_{i=0}^k)$ satisfy the double inequality:*

$$0 \leq \beta^l_{n_0,n_1,\ldots,n_k} \leq 1. \tag{3}$$

Proof. Let $\mathcal{B}$ be the Bose-Mesner algebra spanned by the matrices of $(A, \{S_i\}_{i=0}^k)$ and $\{F_0, F_1, \ldots, F_k\}$ be the underlying base of minimal orthogonal idempotents of $\mathcal{B}$.

The properties presented in Subsect. 2.2 allow us to conclude that the matrix $F_0^{\circ n_0} \circ F_1^{\circ n_1} \circ \cdots \circ F_k^{\circ n_k}$ is a principal submatrix of the matrix given by $F_0^{\otimes n_0} \otimes F_1^{\otimes n_1} \otimes \cdots \otimes F_k^{\otimes n_k}$.

Here we must establish that, whenever $n_i = 0$, for $i \in \{0, 1, \ldots, k\}$, then the element $F_i^{\otimes n_i}$ is removed from the Kronecker product $F_0^{\otimes n_0} \otimes F_1^{\otimes n_1} \otimes \cdots \otimes F_k^{\otimes n_k}$.

Note that the matrix $F_0^{\otimes n_0} \otimes F_1^{\otimes n_1} \otimes \cdots \otimes F_k^{\otimes n_k}$ is an idempotent matrix and, as it is well known, the eigenvalues of an idempotent matrix are, necessarily, 0 or 1.

Applying the eigenvalues interlacing theorem, we conclude that the eigenvalues of $F_0^{\circ n_0} \circ F_1^{\circ n_1} \circ \cdots \circ F_k^{\circ n_k}$ are interlaced with the eigenvalues of the matrix $F_0^{\otimes n_0} \otimes F_1^{\otimes n_1} \otimes \cdots \otimes F_k^{\otimes n_k}$ and, since the eigenvalues of the former are precisely the generalized Krein parameters of $(A, \{S_i\}_{i=0}^k)$ and the eigenvalues of the later are 0 and 1, inequality (3) follows. □

Another property attained by the generalized Krein parameters of an association scheme is presented next.

Theorem 2. *Let $(A, \{S_i\}_{i=0}^k)$ be a symmetric association scheme with k classes. Then, for any $l \in \{0, 1, \ldots, k\}$ and $m \in \mathbb{N}$, we have that*

$$\sum_{n_0+n_1+\cdots+n_k=m} \binom{m}{n_0, n_1, \ldots, n_k} \beta^l_{n_0,n_1,\ldots,n_k} = 1, \tag{4}$$

where $0 \leq n_i \leq m$, for each $i \in \{0, 1, \ldots, k\}$, and

$$\binom{m}{n_0, n_1, \ldots, n_k} = \frac{m!}{n_0! n_1! \ldots n_k!}.$$

Proof. Let $\mathcal{B}$ be the Bose-Mesner algebra spanned by the matrices of $(A, \{S_i\}_{i=0}^k)$ and $\{F_0, F_1, \ldots, F_k\}$ be the underlying base of minimal orthogonal idempotents of $\mathcal{B}$. Let m be a natural number.

We have that

$$\begin{aligned}\left(\sum_{j=0}^k F_j\right)^{\circ m} &= \sum_{n_0+\cdots+n_k=m} \binom{m}{n_0, \ldots, n_k} F_0^{\circ n_0} \circ \cdots \circ F_k^{\circ n_k} \\ &= I_n,\end{aligned}$$

where $0 \leq n_i \leq m$, for each $i \in \{0, 1, \ldots, k\}$. Then, for any $l \in \{0, 1, \ldots, k\}$, we have

$$\begin{aligned}\left(\sum_{j=0}^k F_j\right)^{\circ m} F_l &= \sum_{n_0+\cdots+n_k=m} \left[\binom{m}{n_0, \ldots, n_k} F_0^{\circ n_0} \circ \cdots \circ F_k^{\circ n_k}\right] F_l \\ &= F_l\end{aligned}$$

and, therefore, by Definition 1, we have

$$\sum_{n_0+\cdots+n_k=m} \binom{m}{n_0, \ldots, n_k} \beta^l_{n_0,\ldots,n_k} = 1.$$

□

Theorem 2 establishes in (4) that the sum of all the generalized Krein parameters, in a given direction and with a given total sum of exponents, of a symmetric association scheme is always equal to 1. Besides the lower and upper bounds presented in Theorem 1, this constitutes an extra condition that the generalized Krein parameters must comply for the symmetric association scheme to be able to exist.

In this paper we presented a thorough generalization of the Krein parameters of a symmetric association scheme and we proved two properties that these parameters must satisfy. The properties presented on these generalized Krein parameters open up new possibilities regarding the establishing of admissibility conditions for the existence of certain symmetric association schemes. In particular, we are currently trying to apply these conclusions to prove the non-existence of certain strongly regular graphs which, as we have seen, correspond to symmetric association schemes with two classes. To analyze the existence of some of these graphs constitutes a very challenging problem in the field of combinatorics on which, hopefully, these results will shed a new light upon.

Acknowledgments. In this work Luís Vieira was partially supported by the Center of Research of Mathematics of University of Porto (UID/MAT/00144/2013), which is funded by FCT (Portugal) with national (MEC) and European structural funds through the program FEDER, under the partnership agreement PT2020.

References

1. Mano, V., Vieira, L.: Symmetric association schemes and generalized Krein parameters. Int. J. Math. Models Methods Appl. Sci. **9**, 310–314 (2015)
2. Mano, V.M., Martins, E.A., Vieira, L.A.: Some results on the Krein parameters of an association scheme. In: Bourguignon, J.P., Jeltsch, R., Pinto, A., Viana, M. (eds.) Dynamics, Games and Science. CIM Series in Mathematical Sciencess, vol. 1, Springer, Cham (2015). https://doi.org/10.1007/978-3-319-16118-1_23
3. Bose, R.C., Shimamoto, T.: Classification and analysis of partially balanced incomplete block designs with two associate classes. J. Am. Statist. Assoc. **47**, 151–184 (1952)
4. Bose, R.C.: Strongly regular graphs, partial geometries and partially balanced designs. Pacific J. Math **13**, 389–419 (1963)
5. Bose, R.C., Mesner, D.M.: On linear associative algebras corresponding to association schemes of partially balanced designs. Ann. Math. Statist. **30**, 21–38 (1959)
6. Delsarte, P.: An algebraic approach to the association schemes of coding theory. Philips Research Reports Supplements 10 (1973)
7. Higman, D.G.: Coherent configurations part I: ordinary representation theory. Geometriae **4**, 1–32 (1975)
8. Higman, D.G.: Coherent configurations part II: weights. Geometriae **5**, 413–424 (1976)
9. Hanaki, A.: Character of association schemes and normal closed subsets. Graphs Combin. **19**(3), 363–369 (2003)
10. Graham, R.L., Grötschel, M., Lovász, L. (eds.): Handbook of Combinatorics, vol. 1. The MIT Press, North-Holland, Cambridge (1995)

11. Horn, R., Johnson, C.R.: Matrix Analysis. Cambridge University Press, Cambridge (1985)
12. Horn, R., Johnson, C.R.: Topics in Matrix Analysis. Cambridge University Press, Cambridge (1991)
13. Scott Jr., L.L.: A condition on Higman's parameters. Notices Am. Math. Soc. **20**, A-97, 721-20-45 (1973)

Reducing Waiting Time for Orthopaedic Consultation Through a Continuous Improvement Approach

José Dinis-Carvalho[1](✉), Elisa Vieira[1], Rui M. Lima[1], and Bruno S. Gonçalves[1,2]

[1] University of Minho, Centro Algoritmi, 4800-058 Guimaraes, Portugal
{dinis,rml,bsg}@dps.uminho.pt

[2] School of Management and Technology of the Polytechnic of Porto, 4610-156 Felgueiras, Portugal

Abstract. The outpatient service of public hospitals has had many difficulties in responding to the high demand that are being subjected to. The waiting times for a medical appointment are well above the clinically recommended waiting time, compromising the patients' health status. To combat this trend, health authorities have set the maximum response time for consultations, financially penalizing hospitals for not meeting these times. However, these penalties only act as an additional problem for hospitals. Given the relevance of improving the care performance provided to patients, in terms of quality and celerity, this article presents the study carried out at the outpatient service of a Portuguese hospital. This study presents a set of actions implemented as the result of a simple Continuous Improvement (CI) approach. That CI approach was inspired on Toyota Kata and the results obtained are translate into a reduction of the waiting time for triage by 90% and the waiting time for the first orthopaedic consultation by 27%, for average values in days. The implemented actions were also planned to promote in the long term an increase in the availability of the service, allowing an additional reduction in the waiting time for consultation.

1 Introduction

Prolonged patient waiting, long waiting lists for elective services, among other constraints have triggered a set of efforts to provide faster access to health services [1–3]. In Portugal, like other countries, Maximum Guaranteed Response Times (MGRT) [4] are established, and penalties are imposed to hospitals for non-compliance with these times. However, it is clear that hospitals are dealing with great difficulties in achieving these goals [5]. Waiting time is also one of the main causes of patient dissatisfaction in health services [3].

The waiting time for first medical appointment is particularly long in the orthopaedic service in certain public hospital in the north of Portugal. That hospital will be referred as Hospital H throughout this article to preserve the anonymity of the involved professionals. This study aims to find ways of reducing patient waiting time for the first medical

J. Machado et al. (Eds.): icieng 2021, LNME, pp. 461–471, 2022.
https://doi.org/10.1007/978-3-030-78170-5_40

appointment in the orthopaedic service as a pilot test of the implemented continuous improvement (CI) approach. If the approach brings interesting results, then the same approach will be applied in other services.

The CI approach will follow for guidance several principles of the Lean Thinking [6], Shingo Model [7], and Toyota Way [8]. Some of the principles are a bit more to the technical side of the models such as "Create Value for the Customer"; "Create flow", "Standardized tasks and processes", "Use visual control so no problems are hidden", and "Go and see for yourself to thoroughly understand the situation". Some other are a bit more to the human side of the models such as "Respect Every Individual", "Embrace Scientific Thinking", and "Develop exceptional people and teams" just to mention some of them. These principles to be followed play a key role in the generations of the appropriate atmosphere for the CI effort become effective. The CI routines, functions and artefacts applied were based on several existing CI approaches such as Toyota Kata [9], Kaizen approach [10] as well as other success practical implementation in local companies.

2 Background

The long waiting lists for health care are transversal to the different hospital health services in several countries [11]), with a significant number of studies demonstrating this reality [2]. In the specific case of outpatient for orthopaedic service, demand substantially exceeds supply, and the prolongation of this imbalance generated long waiting lists [5]. The increase in average life expectancy can contribute to the promotion of this imbalance [12], being more common the emergence of degenerative pathologies of the musculoskeletal system that greatly condition people's quality of life.

The hospital admission process is a distinct experience for patients, as this is where they begin their journey to medical care, usually with a mixture of anxiety and expectation about what will come [12]. In this sense, the waiting time for a first medical appointment from an expert should be taken seriously, considering that the complication rate increases with the prolonged preoperative waiting time in some orthopaedic pathologies [13–15]. Although response time limits are established, established according to clinical criteria, these deadlines are not always met [5].

Given the importance of health care performance, several initiatives have been started as attempts to find a solution to this and other problems, which in the specific case of public health services are conditioned by several restrictions mainly due to existing organizational culture, heavy Bureaucracy, distance between operation and decision, etc. The literature presents a large number of studies with positive results adopting methods of production organization, namely the Lean Thinking [16]. Nevertheless, it is not clear that those CI initiatives resulted in robust and long-term CI sustainability [16, 17].

Continuous improvement (CI) is one of the five principles of Lean Thinking and one of the 10 principles of the Shingo Model (referred as "Seek Perfection"). In the Toyota Way CI is more than just one principle, it is one of its two pillars, being "Respect for people" the other one. If you visit the Toyota website in Europe [18] you will find the following description of the Toyota Way: "Continuous Improvement and Respect for People in everything we do". Continuous Improvement is in the Toyota Way also one of the 4 classes (sections) of principles. In that section called "Continuously Solving Root Problems Drives Organizational Learning" includes the following principles: "Go and see for yourself to thoroughly understand the situation (Genchi Genbutsu)", "Make decisions slowly by consensus, thoroughly considering all options; implement decisions rapidly (Nemawashi)", and "Become a learning organization through relentless reflection (Hansei) and continuous improvement (Kaizen)".

3 Methodology

3.1 Current Situation

This study was carried out in the orthopaedic service of a public hospital Portuguese, and the study was focused on the waiting time for a first medical appointment. The waiting time for a first medical appointment of the orthopaedic service comprises the moment when the hospital receives a request for medical appointment and the time when it is effective. During this waiting period, the request is subject to a medical triage process and the first medical appointment is then scheduled to the patient. There are two types of requests, those from primary health units, called CTH[1] requests, and those from other hospital services, called internal requests. The waiting time for CTH requests is monitored by national health regulators, and a target of 5 days for triage is established and 30 to 120 days for the first medical appointment according to the urgency assigned to the application after medical triage. Internal requests do not have any external monitoring, however in the case of CTH requests the hospital is financially penalized for non-compliance with the target time of 5 days.

At the starting date of this project the waiting list for orthopedics consultations, for CTH requests, was 4,943 patients with an average waiting time of 270 days. Of these patients, 1,023 were still waiting for the medical triage process constituting the waiting list for triage, with an average waiting time of 55 days.

3.2 Methodology

In the development of this study, a methodology based on the development of an IC culture was adopted, based on lean philosophy, conducted through the implementation of improvement cycles inspired by Toyota Kata and its PDCA cycles. Thus, this methodology comprises the following phases:

Phase 1. Creation of an Operational Team, Coaching Routines and Support Team for CI

The operational team was selected by the head of the orthopaedic service and was composed by 6 doctors representing the service. That would be the team responsible to identify problems to be solved, identify improvement opportunities, design the improvement actions, implementing and monitoring key performance indicators. The coaching systems was provided by a coach, expert in CI, meeting the team every week to help

[1] CTH is the Portuguese acronym for "*Consulta a Tempo e Horas*". That designation is used for consultation requests coming from family doctors.

the team in the CI methodology. The hospital administration assigned two top managers and two more people from the management structure to provide support the operational team to improvement implementations that eventually needed external intervention from other hospital support services such as informatics and maintenance.

Phase 2. Selection of KPIs to be Monitored and Improved
The KPIs were selected by consensus with the participation of both operational team and support team having in mind the objective of the study, the interests of the patient, health professionals and the administration of the hospital. The selected KPIs were: Waiting time to medical triage; Waiting time to medical appointment; number of appointment requests screened, number of performed medical appointments.

Phase 3. Goals to be Achieved in the Project Period
According to the discussions and the problems identified in the continuous improvement cycles, the first goals were defined by the teams. So, as a starting point the main goal were a reduction to 30 days for the waiting time to medical triage, and 216 days for the waiting time to medical appointment.

Phase 4. Executing PDCA Cycles
Although in the Toyota Kata methodology the desired is that the operational teams meet with the coach every day, in this study because that was impossible because of several reasons, the only possibility was a meeting every week. In each meeting the KPIs were updated on the team board and the coach would require from the team on the improvements achieved during the week. If the ongoing actions were finished new actions would be defined to solve identified problems.

4 Results

This study presents the work developed between October 2019 and February 2020 by an operational IC team, consisting of 6 orthopaedic physicians, together with researchers and with the hospital administration. The starting point for this project was marked by the definition of KPIs of the external orthopaedic consultation service, by the subsequent analysis of the current state and by the exposure of the main difficulties of the service.

4.1 Continuous Improvement Cycles

Given the high waiting times, the members of the CI operational team admit that requests for medical consultation are accepted for ethical reasons even without real patients' clinical status justification. Thus, the physicians of the specialty state that the orthopaedic service is very conditioned by the presence of patients who do not really need a hospital specialty consultation, something already identified by [5]. The medical triage is a possible means to avoid this situation. This process consists of the analysis, according to clinical criteria, of the clinical status of the patient according to the information contained in the request for consultation. From this may result in the assignment of a degree of priority to the patient (normal, urgent or very urgent) so that the consultation can be scheduled, a refusal of the request, or a return to the entities of provenance to clarify the information about the clinical status of the patient.

In this way, most of the problems identified are related to the medical triage. After some discussion with the team of doctors we came to the conclusion that the waiting time until the medical triage cause impact on the reduction of waiting times for medical consultation, with an impact on financial penalties for non-compliance with the maximum waiting times defined by the Portuguese minister of health. Actions such as the standardization of replies sent to the entities that issue requests for consultation, relocation of resources, change of procedures and updating of computers, are some of the counter measures implemented in the face of the identified problems, presented in, adapted from Lot et al. [19]. The countermeasures presented in Table 1 were implemented in a quite short period of time excepting the amendment of the call (see Table 2). The call is the letter sent to the patient with the information related to its medical consultation. A proposal for the new document was submitted to the administration and it was approved, however, it is not yet implemented because the project had to be stopped due to the covid-19 pandemic situation. It is important to note as well that the consultation service has since that time reduced the services to a minimum, by government imposition, in order to reduce the circulation of people in the hospital.

4.2 Improvements on Some KPIs

These actions had as main objective to reach the goals defined by the team, to be achieved in three months. Initially, the goal for the average waiting time for medical triage was 30 days and for the average waiting time for consultation of 260 days. The results presented in Fig. 1 show that, essentially regarding medical triage, the results were better than the team initially thought possible. Improvements on some KPIs.

The benefits of this project appeared gradually, mainly reflecting on the indicator of the number of triages performed and consequently in the medical triage waiting time. The medical triage as already described had special attention as a means of demand control, when considering cases that should not be seen by the specialty doctor. The benefits of the actions implemented were very positive and resulted in significant changes in the medical triage process performance, since in eight weeks the waiting time for medical triage was drastically reduced and remained below 5 days, as shown in Fig. 1.

As the countermeasures were implemented de number of triages performed per week increased significantly up to the highest number (851) achieved on week 7 as shown in the top graph of Fig. 1. This increase in the number of triages resulted in a drastic reduction of the number of medical triage requests waiting to be performed as well as the maximum and average waiting time until medical triage (see bottom graph of Fig. 1). On week 9 the number of medical triage requests waiting in queue were reduced in a way that the established target for the indicator "average waiting time for triage" as achieved (5 days of average waiting time for triage). The countermeasures implemented succeeded in a great deal since the target was reached before the defined 3 months period. Since the average waiting time for medical triage was 55 days at the beginning of this project, the result is around a 90% improvement in this performance indicator.

Table 1. Main identified problems and countermeasures implemented.

Problem	Effect	Countermeasures	Benefits
(1) Consultation requests arrive with unjustified patient's clinical status	Unnecessary use of medical time in ineffective consultations	Definition of standard text to refuse requests, with information on what should be included in a request for consultation and reason for refusal	Easier medical triage process; Fewer requests without a clinical study previously performed by the general practitioner
(2) No doctor time is assigned to medical triage, doctors are not truly committed to doing so	Few triages per day; Long waiting times	Allocate specific doctor time for triage; At an early stage allocate one doctor just for triage	Shorter waiting time Reduction in the number of ineffective consultations
(3) Returned consultation requests unanswered from the health centre	Negative effect on the patient waiting time fort medical triage and for doctor consultation	Instead of returning requests, choose to refuse them and ask for new ones	Reduction of waiting times for medical triage and consultation; Improvement on the quality of information in the requests
(4) Outdated computer systems	Slow medical triage process (around 6 min); Doctors avoiding triage	Update computer systems	Medical triage taking less than 2 min
(5) Existence of requests for consultation that are not administratively closed	Negative effect on the information regarding waiting times	Analyse and eliminate illegitimate entities in the waiting lists	More reliable information regarding waiting lists and performance indicators

On the other hand, the indicator "maximum waiting time for triage" had a reduction from 612 days to 12 days in 9 weeks, representing a reduction of around 98%.

As expected, the increased number of medical triages performed reduced the waiting time for triage, however the same did not happen in the waiting time for the medical consultation shows the relationship between the waiting lists for the triage and consultation processes and the respective waiting times, before the actions implemented. Through the analysis of this scheme (Fig. 2) it becomes noticeable that by increasing the number of triages, it is possible to achieve a reduction in waiting time for triage, however without necessarily reducing the waiting time for the consultation. This action, implemented in isolation, allows only the transfer of patients from one queue to another.

Table 2. The countermeasure waiting to be implemented.

Problem	Effect	Countermeasures	Benefits
(5) Some patients do not bring with them the required complementary means of diagnosis	Ineffective consultations; Cancelation of consultations	Amendment of the call, reinforcing the importance of bringing the required complementary means of diagnosis	Reduction of ineffective consultations; Reduction of cancelations

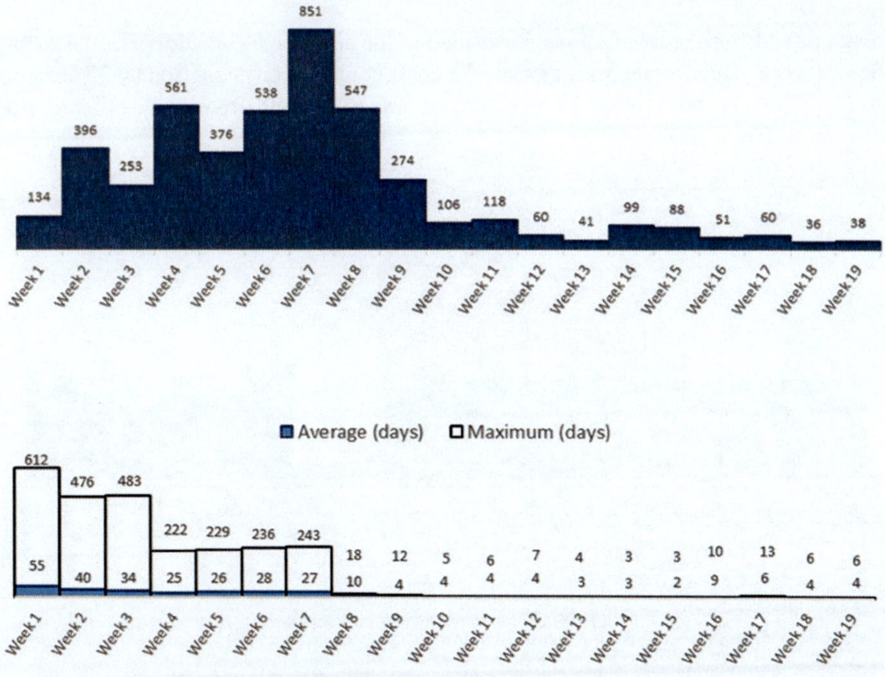

Fig. 1. Number of medical triages (top) and waiting time for medical triage (bottom).

In this sense, and since the objectives of this project are aligned with the objectives of hospital management, it was possible to promote an additional capacity period in terms of doctor*hours for the consultation process. The results of this initiative are shown in Fig. 3.

The both performance indicators shown (Fig. 3) represents two of the indicators monitored in the continuous improvement cycles.

The period of additional capacity took place between weeks 6 and 11 according to what was presented. In weeks 12 and 13 there was a sharp drop in the number of consultations carried out as it coincided with weeks when the service was partially closed. It is possible to observe a reduction in the maximum and average waiting times as

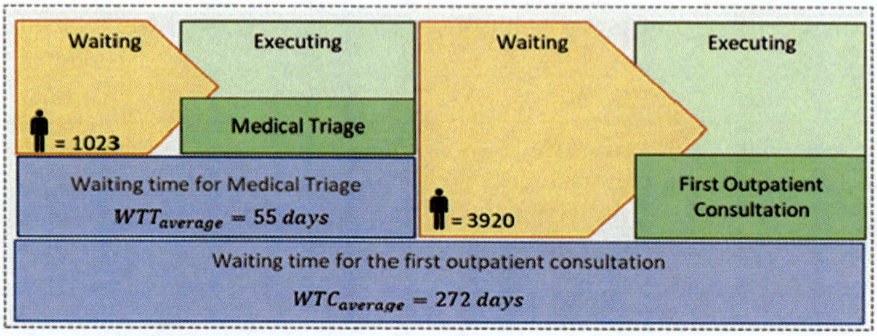

Fig. 2. Patient flow in the first medical appointment.

a result of additional consultations performed in the additional capacity period, starting from 7th week. The average waiting time for CTH consultation decreased by 27%, being initially 270 days and in week 19 of 196 days, and 16% at maximum times (decreasing from 762 to 640 days).

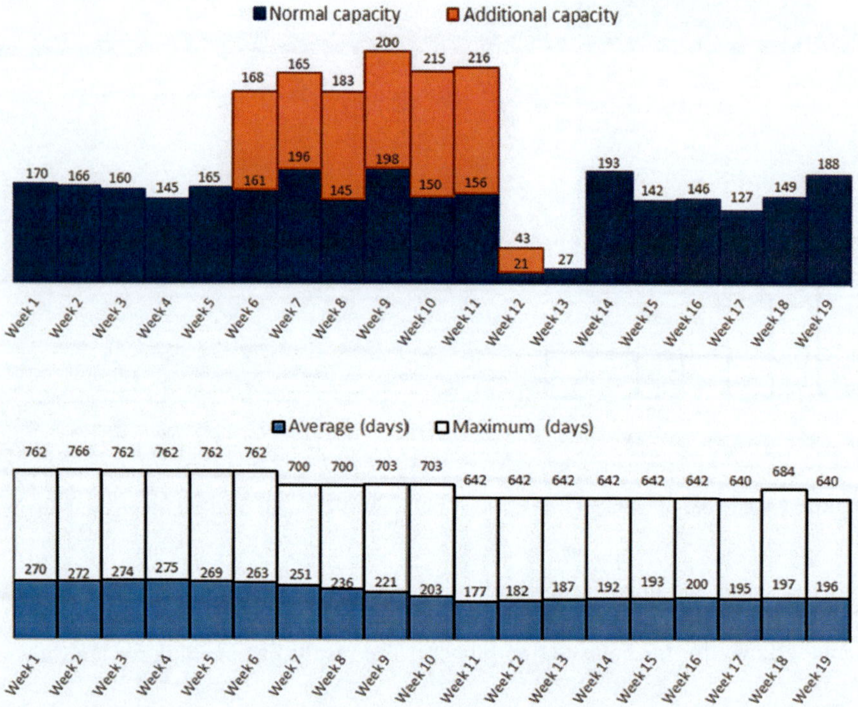

Fig. 3. Number of CTH medical consultations (top) and waiting time for first medical consultation (bottom).

The reduction in the waiting time for the CTH medical consultation may be only reflex from the additional consultations, but it is expected that it could reduce, in the long term, as a result of other actions implemented. Since these have already been reflected in the waiting list for consultation of internal requests (Fig. 4), however only in the last week.

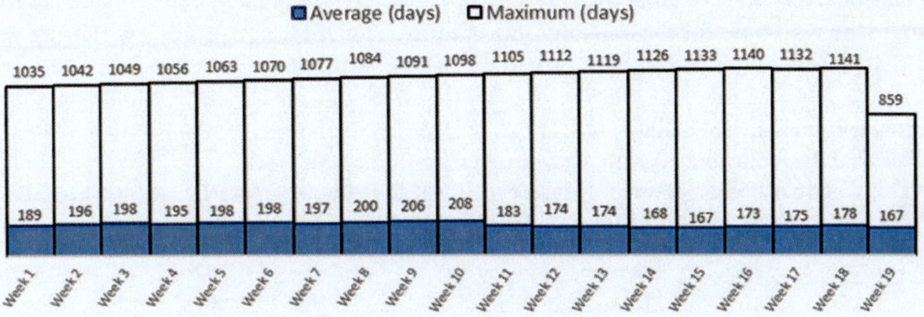

Fig. 4. Waiting time for internal medical consultation.

In the case of internal consultations, the average waiting time at the beginning of the study was 189 days and in the end 167 days, representing a reduction of 12%. Regarding the maximum times there was a reduction of 25%, in the two last weeks.

5 Conclusions

This document presents the work developed in the outpatient orthopaedic specialty service in a Portuguese hospital. This study aimed to find ways to reduce the patient's waiting time until the first consultation of the orthopaedic specialty, as a pilot test to implement the adopted CI approach. Through a team of professionals and after the presentation of some basic concepts of CI concept and Lean Philosophy, it was possible to identify some constraints of the service and define improvement actions.

Initially was notorious the skepticism of some professionals towards some concepts of Lean Philosophy, a limitation already identified in several studies [16] some difficulties in obtaining performance indicators were faced [20], as well as an absence was identified generalization of a CI culture. Nevertheless, although not unanimous, since the participation of team members was not demonstrated in the same way, these limitations were overcome and the team's progressive involvement in the improvement process became more and more evident. This involvement was reflected in the identification of problems in a constructive way and in the presentation and implementation of improvement actions with effect in improving the values of the monitored KPIs. The collaboration of the professionals involved and the hospital administration were essential for the development of this project.

The results achieved are positive with more relevant evidence in the waiting time for triage. The maximum waiting time for triage had a reduction from 612 days to 12 days in 9 weeks, representing a reduction of around 98%, as well as the average waiting time that presented an improvement in around a 90%. The average waiting time for CTH consultation decreased by 27% and 16% at maximum times. In terms of internal consultations, the reductions were 12% and 25% in the average and maximum times, respectively. This study presents a set of actions that seemingly simple, that could never be implemented outside this context [19]. The methodology used planted a seed in health professionals towards the development of an IC culture.

Besides, it is believed that the actions implemented will present better results, in long term, in the wait time for the consultation. A more efficient triage process will bring very positive results in the subsequent processes of this service.

Acknowledgments. The authors would like to thank to the hospital administration and health professional that collaborated with the research team during this work.

This work has been supported by projects UIDCEC003192019 and POCI-01–0145-FEDER-030299, from Fundação para a Ciência e Tecnologia (FCT), Portugal.

References

1. Almomani, I., AlSarheed, A.: Enhancing outpatient clinics management software by reducing patients' waiting time. J. Infect. Public Health **9**(6), 734–743 (2016)
2. Bachelet, V.C., Goyenechea, M., Carrasco, V.A.: Policy strategies to reduce waiting times for elective surgery: a scoping review and evidence synthesis. Int. J. Health Plann. Manage. **34**, e995–e1015 (2019)
3. Lungu, D.A., Grillo Ruggieri, T., Nuti, S.: Decision making tools for managing waiting times and treatment rates in elective surgery. BMC Health Serv. Res. **19**, 369 (2019)
4. Ministério da Saúde,: Portaria n.° 153/2017 de 4 de maio. Diário da República 1ª série, Portugal (2017)
5. Comans, T.A., Chang, A.T., Standfield, L., Knowles, D., O'Leary, S., Raymer, M.: The development and practical application of a simulation model to inform musculoskeletal service delivery in an Australian public health service. Oper. Res. Health Care **15**, 13–18 (2017)
6. Womack, J.P., Jones, D.T.: Lean thinking: Banish waste and create wealth in your corporation. J. Oper. Res. Soc. **48**, 2600967 (2006). Revised and Updated by James P. Womack, Daniel T. Jones. Simon & Schuster
7. Shingo Institute Homepage. https://shingo.org/shingo-model/. Accessed 18 Oct 2020
8. Liker, J.K.: Toyota Way: 14 Management Principles from the World's Greatest Manufacturer. McGraw-Hill Education, New York (2004)
9. Rother, M.: Toyota KATA. McGraw-Hill Education – Europe, New York (2010)
10. Imai, M.: Gemba Kaizen: a commonsense approach to a continuous improvement strategy. Libr. J. **122** (1997)
11. Siciliani, L., Moran, V., Borowitz, M.: Measuring and comparing health care waiting times in OECD countries. Health Policy **118**, 292–303 (2014)
12. Jarman, M.P., Weaver, M.J., Haider, A.H., Salim, A., Harris, M.B.: The national burden of orthopedic injury: cross-sectional estimates for trauma system planning and optimization. J. Surg. Res. **249**, 197–204 (2020)
13. Pincus, D., et al.: Outcomes of after-hours hip fracture surgery. J. Bone Joint Surg. **99**(11), 914–922 (2017)
14. Simunovic, N., et al.: Effect of early surgery after hip fracture on mortality and complications: systematic review and meta-analysis. Can. Med. Assoc. J. **182**, 1609–1616 (2010)
15. Svenøy, S., Watne, L.O., Hestnes, I., Westberg, M., Madsen, J.E., Frihagen, F.: Results after introduction of a hip fracture care pathway: comparison with usual care. Acta Orthop. **91**, 139–145 (2020)

16. Lima, R.M., Dinis-Carvalho, J., Souza, T.A., Vieira, E., Gonçalves, B.: Implementation of lean in health care environments: an update of systematic reviews. Int. J. Lean Six Sigma (2020)
17. Gonzalez Aleu, F., Van Aken, E.M.: Continuous improvement projects: an authorship bibliometric analysis. Int. J. Health Care Qual. Assur. **30**, 467–476 (2017)
18. Toyota_Europe Homepage. https://www.toyota-europe.com/world-of-toyota/this-is-toyota/toyota-global-vision. Accessed 18 Oct 2020
19. Lot, L.T., Sarantopoulos, A., Min, L.L., Perales, S.R., Boin, I.F.S.F., Ataide, E.C.: Using lean tools to reduce patient waiting time. Leadersh. Health Serv. **31**(3), 343–351 (2018)
20. Aleu Gonzalez, F.G.: An empirical investigation of critical success factors for continuous improvement projects in hospitals. In: Proceedings of the International Conference on Industrial Engineering and Operations Management (2017)

Assessment of Mixed-Reality Devices for Production Engineering

Pawel Bun(✉), Damian Grajewski, and Filip Gorski

Poznań University of Technology, Poznan, Poland
pawel.bun@put.poznan.pl

Abstract. The article presents the results of works aimed at evaluation of Mixed Reality equipment for various industrial applications. The technical parameters and the existing limitations of the 15 MR devices that appeared on the market in the last two years was analyzed and compared. As a result, an overall equipment ranking was developed, as well as an expert assessment in terms of their use in industry in supporting operation and maintenance processes of machines and devices, interactive trainings and product and process design. Due to the lack of data, certificates and recommendations from equipment manufacturers, the authors recommend planning and conducting experiments aimed at determining the range of possible working conditions for selected devices. The limits of their use by employees related to the frequency and severity of adverse effects should be considered as well.

Keywords: Mixed reality · Maintenance · Industry 4.0

1 Introduction

Most modern manufacturing companies seek to introduce the Industry 4.0 concept - the paradigm of growing digitization and automation of the production environment [1]. In those companies human workforce is integrated into manufacturing systems (as part of Smart Factories [2] concept), therefore it must be flexible and adaptive [3]. In order to reach that goal Virtual Reality (VR) and Augmented Reality (AR) are used for employee training and work on production line [4]. This is also reflected in industry reports on new technologies - for example, the Gartner Hype Cycle for Emerging Technologies report for 2018 [5]. The progress chart of new technologies no longer includes a point representing VR - Gartner analysts have recognized that this technology has already reached a plateau. Meanwhile, Augmented Reality (AR) and Mixed Reality (MR), which can provide innovative solutions supporting user interaction, and thus can increase efficiency in critical industrial processes are currently in the "valley of disappointment" [5]. This is associated with a cooling of initial enthusiasm and emerging publications indicating existing hardware and software restrictions, especially in the use of AR/MR goggles.

Nevertheless, there are many benefits of using AR/MR solutions as tools supporting supervising and controlling of production processes. In literature there are many examples of using AR for simplification of information exchange [6, 7], reduction of

J. Machado et al. (Eds.): icieng 2021, LNME, pp. 472–483, 2022.
https://doi.org/10.1007/978-3-030-78170-5_41

downtime [8], display of KPI's and current data [9, 10], access to resources and knowledge [11, 12], and maintenance [13–17]. This is possible because AR/MR solutions can exchange data with other IT systems and display complex information [18–20].

The observed reluctance of enterprises to use AR/MR technology may result from the fact that its implementation in the company is difficult [21]. One of challenges is the fact that it is difficult to find studies on the basis of which it would be possible to find out which equipment will meet the specific expectations of the company. The previous attempt to systematize AR solutions was made, among others, in 2011 [22], since then many new devices and software solutions appeared on the market. There is no generally available, comparative list of devices and AR equipment ranking that would include company-specific processes.

The aim of the work carried out at the Virtual and Augmented Reality Laboratory of the Poznan University of Technology was to analyze and then evaluate the available AR and MR equipment and its possible application in industrial (mainly production) conditions in terms of physical and technical conditions.

Due to the lack of data, certificates and recommendations from equipment manufacturers, the authors recommend planning and conducting experiments aimed at determining the range of possible working conditions for selected devices, as well as the limits of their use by employees (maximum working time during a single shift, number and frequency of breaks) related to the frequency and severity of adverse effects.

That is why, among other things, the idea of creating an original ranking emerged. Using the own experience and available on-line materials of producers, reviewers and users of AR/MR equipment, a sample of 15 devices was selected and their characteristics and comparison were made. An expert assessment was also made in terms of their use in industry in supporting work:

1) Operation and maintenance processes of machines and devices.
2) Interactive trainings (with particular emphasis on employees and service technicians).
3) Product and process design.

2 Materials and Methods

2.1 AR/MR Devices Selection Criteria

The first stage of work was the selection of equipment that will be evaluated First, information on AR/MR devices already available on the market was collected. Then, devices were searched for in the development and prototype phases, with potential entry into the market over the next 18 months. Discontinued devices that have been in the prototyping phase for more than 2–3 years and no new reports have been discarded. The dedicated devices (e.g. goggles for sports) have been discarded on the grounds of no possibility of making custom applications for them. Projection devices that are not able to serve as augmented reality devices (i.e. glasses with a display but without a camera, which makes marker recognition impossible) were not considered, which also limited the range of available devices. In this way, a sample of 15 devices was obtained, which were characterized and classified in detail. Device characteristics were divided into 3 groups: Projection and Vision parameters (Table 1), Interaction and Communication parameters (Table 2) and Other (Table 3).

Table 1. Projection and vision characteristics of device

Parameter name	Value
Image parameters	• Resolution, • Field of view
Resolution of the built-in camera for photo/video and marker recognition	
The presence and characteristics of other cameras	
Device type - the following types are distinguished	• Intelligent 2D glasses (no stereoscopy), • Intelligent 3D glasses (stereoscopy), • See-through headset (transparent screen), • Video-see-through helmet (opaque screen + camera image)

Table 2. Interaction and communication characteristics of device

Paramter name	Value
Possibility of independent operation	• yes, • no - operation with a PC, • no - operation with a smartphone
Interaction methods distinguished	• gazing, • hand gestures, • external wireless controller, • voice control, • touchpad
Audio equipment	• headphones, • speakers, • microphone
Other sensors	• depth cameras, • terrain mapping, • light sensors, • GPS, • etc.
Wireless communication	• Wifi • Bluetooth
Presence of wires	• (wired/wireless)

Table 3. Other parameters

Paramter name	Value
Operating system	• Windows, • Android, • Linux, • Other
Computing power and storage	• CPU/GPU characteristics, • RAM, • Storage space
Battery	• Capacity • Working time
Weight (g)	
Price (USD)	
Wireless communication	• Wifi • Bluetooth
Presence of wires	• (wired/wireless)
Device name	
Device manufacturer	
Availability	• commercially - it is on the market terrain mapping, • pre-order • demo
Intended use according to the producer	• general, • consumer, • industrial, • sport

The above-mentioned parameters were collected in a data base. The most important information is generally summarized below:

1) Most of the selected devices are classified as see-through headsets (7 devices) and smart glasses (7 devices), the distinction between glasses and headsets is primarily weight and field of view (larger for sets, smaller for glasses).
2) Most devices are self-contained (12), some require a smartphone, and some work as displays requiring a PC.
3) Two devices classified as stand-alone are connected to the calculation unit (smartphone size) using a cable.
4) In the stand-alone devices, RAM is usually 2 GB, the range among the characterized devices is 512 MB–16 GB.
5) The most common processor types among stand-alone devices are Snapdragon or Intel Atom.
6) Most stand-alone devices allow to store applications and data on an external SD card.

7) The field of view and resolution are the largest (above 90°) in devices using a PC as a calculation unit.
8) For smart glasses, the field of view value is sometimes very small (a dozen or so degrees), some models have projection only in one eye.
9) As mentioned earlier, all devices have a photo-video camera, the size of the matrix is variable in the range of 2–22 mpix.
10) Interaction methods vary depending on whether hand tracking is available; the vast majority of devices can be operated by voice, a large number of devices have dedicated controllers (requiring one-handed operation) or a built-in touchpad (interaction of one finger with one hand, touchpads usually built in around the temples); a smaller part of the headsets has built-in depth cameras, enabling the recognition of hand gestures by the user.
11) Most devices have built-in IMU sensors (accelerometer + gyroscope + magnetometer), there is often a GPS sensor, in more expensive and more advanced devices there are light sensors (ALS) or IR cameras for mapping the environment.
(12) Battery life (for wireless devices) provided by the manufacturers is at least 2–3 h, and the most −12 h.
13) The weight of the devices ranges from several dozen grams (the lightest glasses) to nearly 600 g (the heaviest headsets)
14) The cheapest of the characterized sets are those requiring a smartphone to operate (USD 350–400), the most expensive are independent headsets with gesture and environment recognition (USD 2000–3000); The average cost of buying smart glasses is 700–1200 USD.

2.2 Use of AR/MR Goggles in Production Conditions

Equipment manufacturers rarely state the conditions under which they can safely operate the equipment. It is usually assumed that the conditions are consistent with the working conditions of standard smartphones. The exception from the classified devices is Trivisio Loc. 30 that have high operating temperature range (−40 to +70 °C). In other cases, no data - it should be assumed that the devices are not resistant to moisture, dust, impacts and elevated temperature (which often occurs in production conditions). Several of the classified devices are marked as "industrial" (including Vuzix M300, Epson Moverio BT-300), so it can be assumed that they have been tested and are approved for operation in such conditions.

Lighting conditions are a separate issue - due to translucent displays, AR/MR devices will perform better in conditions where lighting is less bright. Apart from sports glasses, none of the devices is adapted to work in full sun, and as the authors experience shows, in such conditions the visibility of any image on the semi-transparent display is negligible. Some manufacturers (e.g. Epson) add tinted lenses to the glasses, which can be removed - it is worth considering such a solution when the working conditions assume bright lighting.

Regarding work ergonomics there are no special recommendations or reservations from manufacturers. However, the authors' experience shows that during prolonged (regular) use of headsets or glasses the following problems may appear [23, 24]:

- eye strain caused by the need to constantly change accommodation (looking at the change on the digital screen and the real world) in some models of glasses, or with poor image calibration,
- headaches and temporary vision problems caused by stereoscopic projection (symptoms of cyber sickness),
- fatigue of one eye with long use of glasses with a single display,
- imprints and irritation of the scalp (forehead, nose area, etc.) caused by the weight of the set and/or its heating during operation,
- cervical spine pains caused by the weight of some sets.

In wired devices, there are also issues of comfort of use due to the need for appropriate handling of wires. This also applies to devices where glasses and a computing unit are wired, which must also be stored on the user's body during work (e.g. in a pocket, but with easy access due to the need for touch operation).

All the above recommendations should be taken into account when designing applications supporting work in industrial conditions. The issues of ergonomics and resistance to conditions are particularly important in the application for process support (constant work with the use of AR equipment, e.g. throughout the entire work shift) or - less so - in supporting the operation and maintenance of machines (shorter work, but in less controlled conditions, e.g. outdoors, in winter, in high humidity, dust, etc.).

2.3 Methodology of Analysis and Evaluation of Devices

As part of the device analysis, their rankings were developed for specific applications. Team of experts was assembled to conduct the assessment. Two of them were VR/AR specialists with more than 10 years of expertise and two others were industrial engineers with huge expertise on use of AR systems according to the Industry 4.0 concept. The assessment proceeded as follows:

1) For each of the devices, the team of experts awarded points from 1 to 5 in categories identical to the criteria of characterization and possible and sensible for evaluation (some criteria were aggregated into one, e.g. cameras and sensors, the aspect of market availability, naming, etc. was omitted). Points were awarded arbitrarily, based on the experts' own experience.
2) For three of the distinguished industrial applications (operation, process, training), specific weights of specific assessment categories were adopted by expert judgment. The weight values ranged from 1 to 20 for specific criteria. The highest weight values were adopted for those assessment categories that may affect the zero-specified utility of devices in a given application (e.g. cabling and dependence on a PC practically eliminates the use in operation and maintenance processes).
3) Points of individual devices in individual categories were multiplied by weights for particular applications, obtaining 3 point tables.
4) Points were added to four sums: overall score and score for 3 potential industrial applications. The maximal scores to obtain were: 80 for general ranking (no weights), 530 for maintenance, 645 for process and 285 for training.
5) The values have been scaled so that the best device gets 100 points.

6) Devices were arranged according to points, obtaining four rankings. In this way, a standardized, summarized score of 15 devices was created (four rankings).

3 Results and Discussion

General ranking of AR/MR devices ranking (from best to worst) is presented in Table 4.

Table 4. General ranking of AR/MR devices

Place	Device	Points	Price (USD)
1	Thirdeye Gen X2	100	2,250.00
2	Microsoft HoloLens 2	98	3,500.00
3	Magic Leap	98	2,295.00
4	Shadow Creator Action One	91	899.00
5	ODG R-9	89	1,800.00
6	Dynaedge AR100	89	1,900.00
7	Meta 2	89	1,400.00
8	Microsoft HoloLens	88	3,000.00
9	Dimension NXG Ajnalens	88	1,500.00
10	Occipital Bridge	86	399.00
11	Vuzix m300	82	1,285.00
12	Moverio BT-300	79	760.00
13	Trivisio LOC.30	73	-
14	Sony Smart Eyeglass	70	840.00
15	Vuzix Blade	70	1,254.00

The graphical presentation of the ranking is shown in the Fig. 1.

Figures 2, 3 and 4 present rankings in terms of maintenance, production process and industrial training, respectively. Only devices that have obtained a standardized sum of approx. 90 points should be considered useful in a given application - the methodology adopted for calculating points and weights does not allow for very visible differences (the worst devices have slightly less than 70 points).

The general ranking does not contain weighed assessment of usability in certain applications, that is why it should be the last one to take into account while making a selection of a particular device. The three applications have very distinct requirements – the highest being for the process, the lowest (relatively) for the training, however the differences are quite significant.

In terms of raw numbers (not shown in Figs. 1, 2, 3 and 4), the following observations should be emphasized:

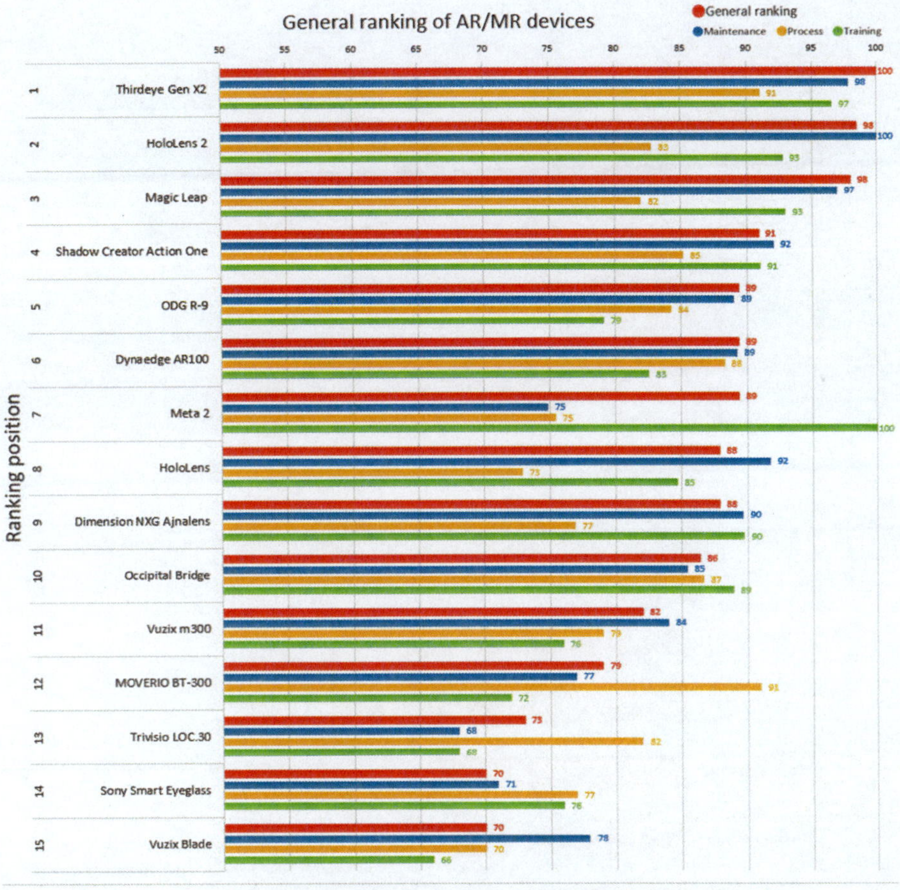

Fig. 1. General ranking of the AR/MR devices

- the highest scoring devices in the general ranking (Thirdeye Gen X2, HoloLens 2, Magic Leap) acquired 65–66 points (80 being the maximum), which is 82,5% of the possible maximum score,
- the highest score for maintenance was 474/530 points (HoloLens 2), which is 89% of possible points to obtain,
- for the process, the highest score was 497/645 (Thirdeye Gen X1), 77% of maximum score,
- for the training, the highest score was 235/285 (Meta 2), which is 82% of the maximum score.

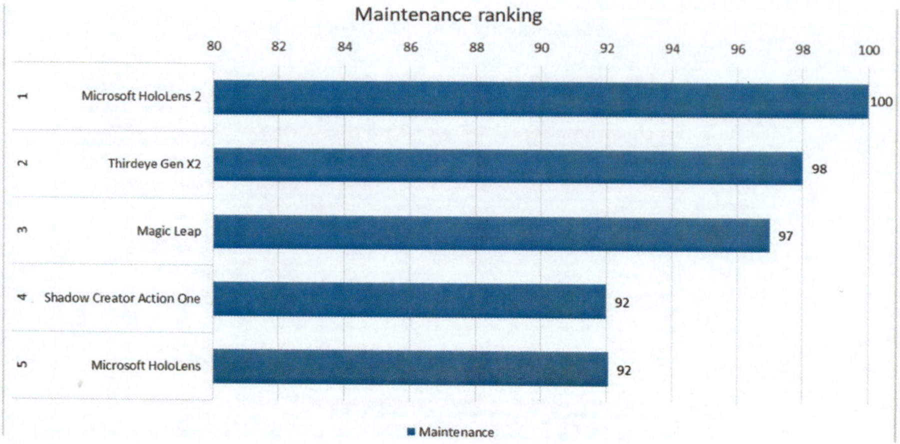

Fig. 2. Ranking of top 5 AR/MR devices for maintenance.

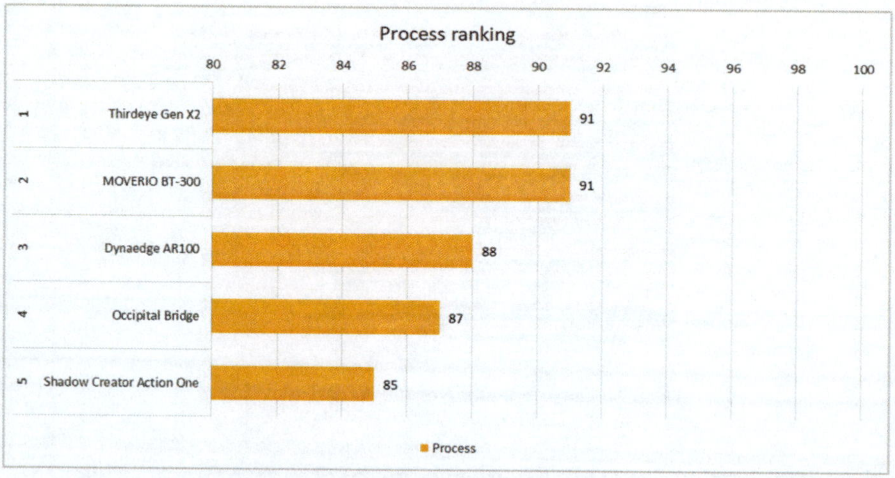

Fig. 3. Ranking of top 5 AR/MR devices for production process.

Observation of ranking tables allows drawing the following conclusions:

1) As the best universal devices (presence in the forefront of all rankings, the first three places in the overall ranking), the following should be mentioned: Thirdeye Gen X2, Microsoft HoloLens 2 and Magic Leap.
2) Microsoft HoloLens 2 is the best solution for maintenance service at industrial level, of the devices taken into consideration. However, apart from this particular ranking, it is not the top device as assessed by the experts. Practical observations and experiments by the authors allow to conclude, that it is not yet ready for full

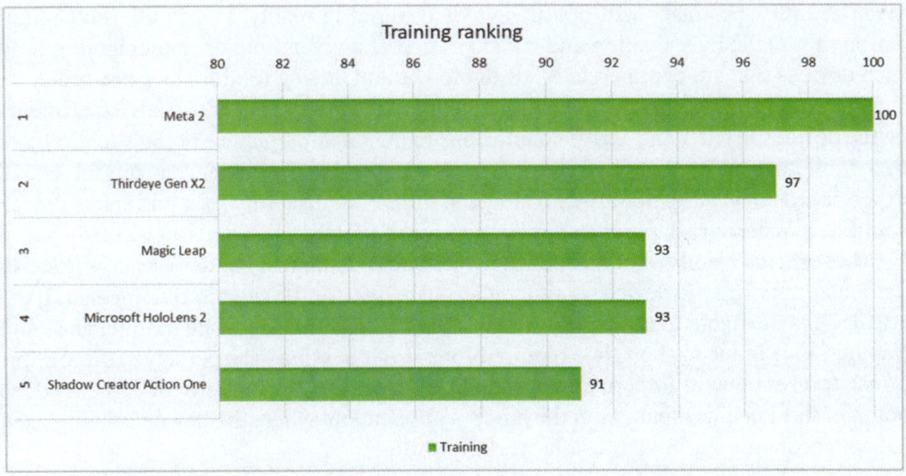

Fig. 4. Ranking of top 5 AR/MR devices for the industrial training.

industrial implementation, mostly due to lack of intuitiveness of use and certain problems with ergonomics.

3) The Thirdeye Gen X2 is a device that on average gets the highest score in all the three rankings. However, it was not commercially available at the time of conducting the studies, so its score must be taken with caution.
4) It is worth paying attention to the low positions in the rankings of intelligent 3D glasses available on the market for some time (manufacturers: Vuzix, Sony) - their weaker parameters are also not correlated with a lower price, as devices topping in rankings sometimes have a similar price.
5) Wired and dependent devices (e.g. Meta 2) have a chance for real industrial application only in the field of training; an alternative solution increasing their potential is the use of a portable computer (laptop) adapted to work in a special backpack worn by the user, recommended by some manufacturers; in this application, work ergonomics is very limited, but full user mobility is gained.
6) The ranking for the process application looks almost entirely different as the other rankings – the top 5 contains mostly smart glasses. That is why the weights are distributed in a different way, with low mass, long battery life and better ergonomics being higher scored than the other qualities.

4 Conclusions

In the paper, the 15 available AR devices for industrial applications were characterized and ranked by experts' judgement. The presented rankings should be generally approached with caution due to the comparison of the scores of devices already available and checked with devices entering the market - the latter generally obtained higher scores. Of course, this is logical due to the constant development of equipment, but the manufacturer's promises will not always be covered in reality. Hence the recommendation of practical verification and consideration at a given time of implementing only such devices that are commercially available without having to order as a pre-order.

The authors recommend conducting practical (laboratory and industrial) experiments on the devices listed in the above conclusions to obtain a full picture of the situation, not based solely on the comparative assessment of measurable parameters and the subjective expert assessment of qualitative parameters. Often, the usability of a particular device could be a matter of adapting an application for it.

However, as results of observations by the authors, none of the devices is perfect. In particular rankings, no device scored higher than 90% for the set criteria. In general, the AR devices available in the market seem to be the best suited for the maintenance and training. This is due to short time required for a single session – the perceived discomfort of use (present due to imperfections such as high mass or low quality of screens) does not manifest itself as much, as in the process application, where the device is being used almost constantly during the whole working shift.

As of 2021, the industrial AR technology must be assessed as being still in development (as opposed to VR technology, which is readily implemented in production conditions). No particular device meets all the requirements of the industrial applications and there are numerous limitations of use. However, the potential is very high and the next generation of devices should be fully suited to requirements of industrial recipients.

Acknowledgments. The research work was done as a part of a project with Apzumi company and was partially supported by Wielkopolska Regional Operational Program for 2014–2020.

References

1. Oesterreich, T.D., Teuteberg, F.: Understanding the implications of digitisation and automation in the context of industry 4.0: a triangulation approach and elements of a research agenda for the construction industry. Comput. Ind. **83**, 121–139 (2016)
2. Żywicki, K., Zawadzki, P., Górski, F.: Virtual reality production training system in the scope of intelligent factory. In: Burduk, A., Mazurkiewicz, D. (eds.) ISPEM 2017. AISC, vol. 637, pp. 450–458. Springer, Cham (2018). https://doi.org/10.1007/978-3-319-64465-3_43
3. Yew, A.W.W., Ong, S.K., Nee, A.Y.: Towards a griddable distributed manufacturing system with augmented reality interfaces. Robot. Comput. Integrat. Manuf. **39**, 43–55 (2016)
4. Azuma, R.T.: A survey of augmented reality. Presence Teleoper. Virtual Environ. **6**(4), 355–385 (1997)
5. Gartner Research: Hype Cycle for Emerging Technologies (2018). https://www.gartner.com/en/documents/3885468/hype-cycle-for-emerging-technologies-2018, Accessed 21 Feb 2021
6. Zhukovskiy, Y., Koteleva, N.: Development of augmented reality system for servicing electromechanical equipment. J. Phys. Conf. Ser. **1015**(4), 042068 (2018)
7. Damiani, L., Demartini, M., Guizzi, G., Revetria, R., Tonelli, F.: Augmented and virtual reality applications in industrial systems: a qualitative review towards the industry 4.0 era. IFAC Papers Online **51**(11), 624–630 (2018)
8. Oliveira, R., Farinha, T., Raposo, H., Pires, N.: Augmented reality and the future of maintenance. In: COIMBRA, pp. 81–88 (2014)

9. Kollatsch, C., Schumann, M., Klimant, P., Wittstock, V., Putz, M.: Mobile augmented reality based monitoring of assembly lines. Procedia CIRP **23**, 246–251 (2014)
10. Jetter, J., Eimecke, J., Rese, A.: Augmented reality tools for industrial applications: what are potential key performance indicators and who benefits? Comput. Hum. Behav. **87**, 18–33 (2018)
11. Gavish, N., et al.: Evaluating virtual reality and augmented reality training for industrial maintenance and assembly tasks. Interact. Learn. Environ. **23**(6), 778–798 (2015)
12. Paelke, V.: Augmented reality in the smart factory. In: Proceedings of the 2014 IEEE Emerging Technology and Factory Automation (ETFA), pp. 1–4 (2014)
13. González, F.C.J., Villegas, O.O.V., Ramírez, D.E.T., Sánchez, V.G.C., Domínguez, H.O.: Smart multi-level tool for remote patient monitoring based on a wireless sensor network and mobile augmented reality. Sensors **14**(9), 17212–17234 (2014)
14. Nazir, S., Totaro, R., Brambilla, S., Colombo, S., Manca, D.: Virtual reality and augmented-virtual reality as tools to train industrial operators. Comput. Aided Chem. Eng. **30**, 1397–1401 (2012)
15. Henderson, S., Feiner, S.: Exploring the benefits of augmented reality documentation for maintenance and repair. IEEE Trans. Visual. Comput. Graph. **17**(10), 1355–1368 (2011)
16. Palmarini, R., Erkoyoncu, J.A., Roy, R.: An innovative process to select augmented reality (AR) technology for maintenance. Procedia CIRP **59**, 23–28 (2017)
17. Erkoyoncu, J.A., Fernandez del Almo, I., Dalle Mura, M., Roy, R., Dini, G.: Improving efficiency of industrial maintenance with context aware adaptive authoring in augmented reality. CIRP Ann. **66**(1), 465–468 (2017)
18. Baumeister, J., Ssin, S.Y., ElSayed, N.A.M., Dorrian, J.: Cognitive cost of using augmented reality displays. IEEE Trans. Visual Comput. Graphics **23**(11), 2378–2388 (2017)
19. Ekren, G., Keskin, N.O.: Existing standards and programs for use in mobile augmented reality, mobile technologies and augmented reality. In: Jared, K. (ed.) Advances in Educational Technologies and Instructional Design Book Series, pp. 118–134. IGI Global, Hershey (2018)
20. Mourtzis, D., Zogopoulos, V., Vlachou, E.: Augmented reality application to support remote maintenance as a service in the robotics industry. Manuf. Syst. Procedia CIRP **63**, 46–51 (2017)
21. Quint, F., Loch, F., Bertram, P.: The challenge of introducing AR in industry-results of a participative process involving maintenance engineers. Procedia Manuf. **11**, 1319–1323 (2017)
22. Carmigniani, J., Furht, B., Anisetti, M., Ceravolo, P., Damiani, E., Ivkovic, M.: Augmented reality technologies, systems and applications. Multimedia Tools Appl. **51**(1), 341–377 (2011)
23. Palmarini, R., Erkoyuncu, J.A., Roy, R., Torabmostaedi, H.: A systematic review of augmented reality applications in maintenance. Robot. Comput. Integr. Manuf. **49**, 215–228 (2018)
24. Uva, A.E., Gattullo, M., Manghisi, V.M., Spagnulo, D., Cascella, G.L., Fiorentino, M.: Evaluating the effectiveness of spatial augmented reality in smart manufacturing: a solution for manual working stations. Int. J. Adv. Manuf. Technol. **94**(1–4), 509–521 (2017). https://doi.org/10.1007/s00170-017-0846-4

Author Index

J. Machado et al. (Eds.): icieng 2021, LNME, pp. 485–487, 2022.
https://doi.org/10.1007/978-3-030-78170-5